# Springer Undergraduate Mathematics Series

The Springer Undergraduate Mathematics Series (SUMS) is a series designed for undergraduates in mathematics and the sciences worldwide. From core foundational material to final year topics, SUMS books take a fresh and modern approach. Textual explanations are supported by a wealth of examples, problems and fully-worked solutions, with particular attention paid to universal areas of difficulty. These practical and concise texts are designed for a one- or two-semester course but the self-study approach makes them ideal for independent use.

More information about this series at http://www.springer.com/series/3423

Christof Eck · Harald Garcke
Peter Knabner

# Mathematical Modeling

 Springer

Christof Eck
Universität Stuttgart
Stuttgart
Germany

Harald Garcke
Fakultät für Mathematik
Universität Regensburg
Regensburg
Germany

Peter Knabner
Department Mathematik
Universität Erlangen-Nürnberg
Erlangen
Germany

ISSN 1615-2085          ISSN 2197-4144   (electronic)
Springer Undergraduate Mathematics Series
ISBN 978-3-319-55160-9          ISBN 978-3-319-55161-6   (eBook)
DOI 10.1007/978-3-319-55161-6

Library of Congress Control Number: 2017935386

Mathematics Subject Classification (2010): 00A71, 34-01, 35-01, 49-01, 74-01, 76-01, 80-01

Printed on acid-free paper

This Springer imprint is published by Springer Nature
The registered company is Springer International Publishing AG
The registered company address is: Gewerbestrasse 11, 6330 Cham, Switzerland

*To the memory of*
*Prof. Dr. Christof Eck*
*(1968–2011)*

# Preface

## Preface to the English Edition

Encouraged by the positive reception to our German textbook on mathematical modeling, we were motivated to create an English version of this book. Although there is a long tradition of English textbooks on mathematical modeling to the best of our knowledge, most of them differ from the one presented here, as our book is ordered according to a hierarchy of mathematical subjects with increasing complexity. We start from simple models given by (linear) equations and end up with complex mathematical models involving nonlinear partial differential equations with free boundaries. In this way, we offer a variety of mathematically more and more inclined subjects from which both elementary undergraduate and sophisticated graduate courses can be composed.[1] In this sense, we hope that our book fills a gap and will be of use also in the curricula for mathematical modeling in English-speaking universities.

We have taken the opportunity of the translation process to correct further misprints and small inconsistencies which have come to our knowledge.

This English version of this book would not have come into existence without support of the following people: Mrs. Eva Rütz, Mrs. Astrid Bigott, and Mrs. Cornelia Weber have created the TEX version on the basis of the first translation provided by the authors, as usual with enormous exactness, speediness, and dedication. This first draft has been considerably improved by Prof. Serge Kräutle and Dr. Kei Fong Lam, who worked through the whole text and removed various inconsistencies and mistakes not only in the usage of the English language but also concerning the mathematical contents. We thank Clemens Heine of Springer-Verlag for his continuous support and encouragement.

---

[1]See the preface to the German edition for more detailed suggestions for the use of this textbook.

It fills us with deep sadness that our scholar, colleague, and friend Christof Eck could not participate in this enterprise. Extremely untimely, just being 43 years of age, he passed away after a long illness. This volume is dedicated to his memory.

Regensburg, Germany                                                      Harald Garcke
Erlangen, Germany                                                         Peter Knabner
July 2016

## Preface to the First German Edition

With the notion *mathematical modeling*, we denote the description of phenomena from nature, technology, or economy by means of mathematical structures. The aim of modeling is the derivation of a meaningful mathematical formulation from which statements and solutions for the original problem can be derived. Here, in principle, every branch of mathematics is "applicable." In technological applications, quite often mathematical problems arise with such complexity that they have to be simplified by neglecting certain influences or they can only be solved with the help of numerical methods.

If one simplifies mathematical models, for example by neglecting certain "small" terms in an equation, then it is the task of mathematicians to check to which extent the behavior of solutions has been changed. Historically, a variety of mathematical concepts that have been developed were driven by the requirements of applications. Having this in mind, it is neither by chance nor astonishing that mathematics and real world "fit together." Mathematical concepts whose development has been triggered from a specific field of application often also proved to be usable in other fields of applications. Furthermore, for a specific problem from applications, there are in general a wide range of mathematical models feasible. The choice of the model certainly depends on the degree of detail in the (temporal or spatial) level of consideration.

Considering this, an education in mathematical modeling appears to be indispensable, if a study in mathematics is also supposed to contain professional qualifications. Often the classical courses of a mathematics curriculum can deliver such an education only insufficiently. Typical courses concentrate on the development of a mathematical theory, in which sample applications—if at all—only play the role of an "optional" motivation. Also, the study in a minor subject from an application field in general cannot close the gap because students in mathematics are often overburdened: On the one hand, they have to learn the results of an application field, and on the other hand, they have to extract mathematical structures from them to make mathematical knowledge useful. Against this background, specific courses in mathematical modeling have entered the curriculum of mathematical studies at various universities. In this context, the present textbook wants to be helpful.

On the one hand, this textbook presents knowledge from an area "between" mathematics and the sciences (e.g., from thermodynamics and from continuum mechanics) which students and lecturers of mathematics need in order to understand models for problems in the sciences and engineering and also to derive them. On the other hand, this book contains a variety of interesting, practically relevant examples for the mathematical theories often only experienced at an abstract level during the study of mathematics and thus answers the question often posed "what do I need this for?". While it cannot substitute any of the textbooks dealing with the underlying mathematical structures such as linear systems of equations/linear algebra or ordinary or partial differential equations, it nevertheless contains essential aspects of the analysis of the models. One aim in particular is to illustrate the interactions between mathematics and applications, which unfortunately are often neglected in mathematics courses.

Furthermore, this textbook also addresses students from the sciences and from engineering and offers them an introduction into the methods of applied mathematics and mechanics.

The content of this book is restricted to deterministic models with continuous scales, as they are in the center of classical natural sciences and engineering. In particular, stochastic models are beyond the scope of this book, and the same applies to processes at very small scales for which particular models or models from quantum mechanics and its approximations are feasible. Also, models from economics are not in the focus of this book, as stochastic approaches play an important role there.

An essential concept of this book consists in using the mathematical structures (and the knowledge about them) as an ordering principle and not the different fields of applications. This reflects the strength of mathematics, lying in the fact that one concept can be used for totally different problem classes and fields of applications. It allows dealing with examples from different fields of applications efficiently without being forced to always repeat the same mathematical basic structures: This line will be followed in Chapter 1 and in Chapters 2, 4, 6, and 7. In this order, one finds embedded Chapters 3 (thermodynamics) and 5 (continuum mechanics). They provide the necessary links to the natural sciences and engineering. Of course, these chapters are also shaped by the application of mathematical tools.

The restriction of the subjects at the level of application corresponds to a restriction at a mathematical level: Throughout this book, we use knowledge from linear algebra and analysis intensively. Chapter 4 relies on the knowledge provided by courses in analysis or in ordinary differential equations, and Chapter 5 makes use in an essential way of the methods of the multidimensional differentiation and integration (integral theorems) and in this way from the more advanced aspects of analysis. In Chapter 7, the foundations of the geometry of curves and surfaces play an important role. It is impossible to define a clear delimitation to the analysis of partial differential equations. Knowledge from this field and also from linear functional analysis certainly is useful for Chapter 6, but not necessary.

A discussion of mathematical results about partial differential equations takes place in this chapter only insofar as there is a tight linkage to the model interpretation. Therefore, the presentation cannot be completely rigorous, but possible gaps and necessary consolidations are pointed out. In this way, this chapter does not necessarily require an intense study of the analysis of partial differential equations, but hopefully, it stimulates to do so. There is no chapter dedicated solely to optimization, but concepts of optimization are used in different chapters. This reflects that optimization problems in modeling often appear as an equivalent formulation of other mathematical problems when expressed in a variational formulation. We have totally excluded the treatment of numerical methods from this book, although they represent a central tool in the practical treatment of technical and scientific problems. We think that this can be justified as there is already a vast amount of excellent textbooks available.

The material of this book can be used in various ways for undergraduate and for graduate students, both at an elementary level and at a more advanced level. The simplest usages are two lecture courses of four hours each week, which cover the whole spectrum of this book. Here, the first part should be placed in the second half of undergraduate studies and the second part in the graduate studies. Alternatively, a course with two hours of lecture per week of introductory character is also possible at an early stage of graduate studies based on parts of Chapters 1, 2, and 4. This can be further complemented by another two-hour-per-week course consisting of parts of Chapters 5 and 6. If there is only room for one course, then it is also possible to build a course of four hours of lecture per week out of parts of Chapters 1, 2, 3, 4, 5, and some aspects of 6. Alternatively, Chapters 5, 6, and 7 can be used for a course about mathematical models and continuum mechanics or from the basic sections of Chapter 5 (derivation of the conservation equations) and Chapters 6 and 7, a course on "applied partial differential equation" can be extracted. Furthermore, this book can be used efficiently for self-study, at undergraduate, graduate, or postgraduate level. Postgraduates working with a mathematical model related to certain fields of application may find the necessary foundations here.

The aim of all the courses described can only be to close a certain gap to the application fields, and they cannot substitute the concrete realization of modeling projects. Intensive treatment of many exercises presented here is helpful, but finally, to our understanding, modeling can only be learned by modeling practice. A possible teaching concept as it has been used by the authors consists of problem seminars in which tasks from the applications are posed without any mathematical material, i.e., the development of feasible mathematical concepts is an essential part of the work. But we hope that courses based on this textbook provide an essential foundation for such a modeling practice.

Finally, this book seems to be also suitable for students of the natural sciences (physics, chemistry) and of engineering. For these students, some of the modeling aspects will be familiar from specific cases, but a rigorous inclusion in a mathematical methodology should lead to further insight.

Of course, we are not going to judge the existing textbooks on mathematical modeling. Certainly, there is a variety of excellent books, but in quantity much less than in other mathematical fields, and many textbooks restrict themselves to an elementary level, in particular also addressing high school students. A textbook which tries to cover the whole spectrum from elementary aspects to recent research is not known to us. Often, textbooks also follow different ordering principles. The present textbook originated from courses which were given by the second author at the University of Regensburg and by the first and third author at the University of Erlangen several times, and therefore, it is the result of a complex developing process. During this process, the authors received important support. The authors express their thanks to Bernd Ammann, Luise Blank, Wolfgang Dreyer, Michael Hinze, and Willi Merz for valuable suggestions. Sincere thanks are given to Barbara Niethammer, who together with the second author has lectured a course on mathematical modeling at the University of Bonn from which much material entered the present book. For careful proofreading, we thank Martin Butz, Daniel Depner, Günther Grün, Robert Haas, Simon Jörres, Fabian Klingbeil, David Kwak, Boris Nowak, Andre Oppitz, Alexander Prechtel, and Björn Stinner. In TEX writing, we were supported by Mrs. Silke Berghof and in particular by Mrs. Eva Rütz who typed a large part of the manuscript and worked on the numerous figures with strong dedication—to both our cordial thanks. We thank Serge Kräutle who provided the figure at the cover—a numerical simulation of the Kármán vortex street. We would also like to thank cordially Ulrich Weikard for providing Figure 6.14 and James D. Murray for providing Figure 6.10.

Bielefeld, Germany                                                          Christof Eck
Regensburg, Germany                                                      Harald Garcke
Erlangen, Germany                                                          Peter Knabner
December 2007

# Contents

# Chapter 1
# Introduction

## 1.1  What Do We Mean by (Mathematical) Modeling?

With *(mathematical) modeling* we denote the translation of a specific problem from
the natural sciences (experimental physics, chemistry, biology, geosciences) or the
social sciences, or from technology, into a well-defined mathematical problem. The
mathematical problem may range in complexity from a single equation to a system
of several equations, to an ordinary or partial differential equation or a system of
such equations, to an optimization problem, where the state is described by one
of the aforementioned equations. In more complicated cases we can also have a
combination of the problems mentioned. A mathematical problem is *well-posed*, if
it has a unique solution and if the solution of the problem depends continuously
on its data, where continuity has to be measured in such a way that the results are
meaningful for the application problem in mind. In general the phenomena to be
described are very complex and it is not possible or sensible to take all its aspects
into account in the process of modeling, because for example

- not all the necessary data are known,
- the model thus achieved cannot be solved anymore, meaning that its (numerical)
  solution is expensive and time consuming, or it is not possible to show the well-
  posedness of the model.

Therefore nearly every model is based on *simplifications* and *modeling assumptions*.
Typically the *influence of unknown data* are neglected, or only taken into account
in an approximative fashion. Usually *complex effects* with only *minor influences* on
the solution are neglected or strongly simplified. For example if the task consists
of the computation of the ballistic trajectory of a soccer ball then it is sensible to
use classical Newtonian mechanics without taking into account relativity theory. In
principle using the latter one would be more precise, but the difference in results for
a typical velocity of a soccer ball is negligible. In particular this holds true if one
takes into account that there are errors in the data, for example slight variations in the
size, the weight, and the kickoff velocity of the soccer ball. Typically available data

© Springer International Publishing AG 2017
C. Eck et al., *Mathematical Modeling*, Springer Undergraduate
Mathematics Series, DOI 10.1007/978-3-319-55161-6_1

are measured and therefore afflicted with measurement errors. Furthermore in this example certainly the gravitational force of the Earth has to be taken into account, but its dependence on the flight altitude can be neglected. In a similar way the influence of the rotation of the Earth can be neglected. On the other hand the influence of air resistance cannot be neglected. The negligible effects are exactly those which make the model equations more complex and require additional data, but do not improve the accuracy of the results significantly.

In deriving a model one should make oneself clear what is the question to be answered and which effects are of importance and have to be taken into account in any case and which effects are possibly negligible. The aim of the modeling therefore plays a decisive role. For example the model assumptions mentioned above are sensible for the flight trajectory of a soccer ball, but certainly not for the flight trajectory of a rocket in an orbit around the Earth. Another aspect shows the following example from weather forecasting: An exact model to compute the future weather for the next seven days from the data of today cannot serve for the purposes of weather forecast if the numerical solution of this model would need nine days of computing time of the strongest available supercomputer. Therefore often a balance between the accuracy required for the predictions of a model and the costs to achieve a solution is necessary. The costs can be measured for example by the time which is necessary to achieve a solution of the model and for numerical solutions also by the necessary computer capacities. Thus at least in industrial applications costs often mean *financial* costs. Because of these reasons there can be no clear separation between *correct* or *false* models, a given model can be sensible for certain applications and aims but not sensible for others.

An important question in the construction of models is: Does the *mathematical structure* of a model change by neglecting certain terms? For example in the initial value problem

$$\varepsilon\, y'(x) + y(x) = 0\,, \quad y(0) = 1$$

with the small parameter $\varepsilon$ one could think about omitting the term $\varepsilon y'$. However, this would lead to an obviously unsolvable algebraic system of equations

$$y(x) = 0\,, \quad y(0) = 1\,.$$

The term neglected is decisive for the mathematical structure of the problem independent of the smallness of parameter $\varepsilon$. Therefore sometimes terms which are identified as small, cannot be neglected. Hence, constructing a good mathematical model also means to take aspects of analysis (well-posedness) and numerics (costs) of the model into account.

The essential ingredients of a *mathematical model* are

- an *application problem* to be described,
- a number of *model assumptions*,

- a mathematical problem formulation, for example in the form of a mathematical *relation*, specifically an equation, an inequality, or differential equation, or several coupled relations, or an optimization problem.

The knowledge of the model assumptions is of importance to estimate the scope of applications and the accuracy of the predictions of a model. The aim of a good model is, starting from known but probably only estimated data and accepted laws of nature to give an answer as good as possible for a given question in an application field. A sensible model should only need data which are known or for which at least plausible approximations can be used. Therefore the task consists in extracting as much as possible information from known data.

## 1.2 Aspects of Mathematical Modeling: Example of Population Dynamics

To illustrate some important aspects of modeling in this section we consider a very simple example: A farmer has a herd of 200 cattle and he wants to increase this herd to 500 cattle, but only by natural growth, i.e., without buying additional animals. After a year the cattle herd has grown to 230 animals. He wants to estimate how long it lasts till he has reached his goal.

A sensible modeling assumption is the statement that the growth of the population depends on the size of the population, as a population of the double size should also have twice as much offspring. The data available are

- the initial number $x(t_0) = 200$ of animals at the initial time $t_0$,
- the increment in time $\Delta t = 1$ year,
- a growth factor of $r = 230/200 = 1.15$ per animal and per time increment $\Delta t$.

If one sets $t_n = t_0 + n \Delta t$ and if $x(t)$ denotes the number of animals at time $t$, then knowing the growth factor leads to the recursion formula

$$x(t_{n+1}) = r x(t_n). \tag{1.1}$$

From this recursion formula one gets

$$x(t_n) = r^n x(t_0).$$

Therefore the question can be formulated as:

Find a number $n$ such that $x(t_n) = 500$.

The solution is

$$n \ln(r) = \ln\left(\frac{x(t_n)}{x(t_0)}\right), \quad \text{or } n = \frac{\ln\left(\frac{500}{200}\right)}{\ln(1.15)} \approx 6.6.$$

Hence, the farmer has to wait for 6.6 years.

This is a simple *population model* which in principle can also be applied to other problems from biology, for example the growth of other animal population, of plants or bacteria. But it can also be used in apparently totally different fields of applications, for example the computation of interests or the cooling of bodies (see Exercises 1.1 and 1.2). Without possibly noticing, in deriving the above model we have used several important modeling assumptions, which are fulfilled sometimes, but which are not fulfilled in a lot of cases. In particular the influence of the following effects has been neglected:

- the *spatial distribution* of the population,
- limited *resources*, for example limited nutrients,
- a loss of population by natural enemies.

Further details which also have been neglected, are for example the age distribution in the population, which has influence on the death rate and the birth rate, and the subdivision in female and male animals. Additionally the model leads to non-integer population quantities, which strictly speaking is not correct. The simplifications and deficiencies do not render the model worthless but they have to be recognized and taken into account to assess the result correctly. In particular the specific result of 6.6 years should not be taken too seriously, and an appropriate interpretation rather is that the farmer presumably will reach his goal in the $7^{th}$ year.

An aspect, which is not optimal for intrinsic mathematical reasons, is the time increment of one year, because it is chosen arbitrarily. For the application under consideration it has a sensible meaning, nevertheless also an increment of three months or of two years could have been chosen. Furthermore we need two data, the increment in time and the growth rate. Both data depend on each other, meaning that the growth characteristics possibly can only be described by one number. As a first approach one can conjecture that the growth rate depends *linearly* on the time increment, i.e.,

$$r = 1 + \Delta t \, p$$

with a factor $p$ still unknown. From $r = 1.15$ for $\Delta t = 1$ year we conclude that $p = 0.15/\text{year}$. Taking this for granted then for $\Delta t = 2$ years one has $r = 1.3$. Therefore after 6 years, ($6 = 3$ times 2) the farmer has

$$200 \cdot 1.3^3 = 439.4$$

cattle. But in the "old" model with $\Delta t = 1$ year he has

$$200 \cdot 1.15^6 \approx 462.61$$

animals. Therefore the assumption of a linear relation between $r$ and $\Delta t$ is wrong.

A better approach can be gained by the *limiting process* $\Delta t \to 0$:

$$x(t + \Delta t) \approx (1 + \Delta t \, p) \, x(t) \text{ for "small" } \Delta t \,,$$

or more precisely

$$\lim_{\Delta t \to 0} \frac{x(t + \Delta t) - x(t)}{\Delta t} = p \, x(t) \,,$$

i.e.,

$$x'(t) = p \, x(t) \,. \tag{1.2}$$

This is a *continuous* model in the form of an *ordinary differential equation*, which does not contain an arbitrarily chosen time increment anymore. It possesses the exact solution

$$x(t) = x(t_0) \, e^{p(t - t_0)} \,.$$

If the data are as above, i.e., a time increment of $\Delta t = 1$ year and a growth rate $r = 1.15$, this means

$$e^{p \cdot 1 \text{ year}} = 1.15$$

and therefore

$$p = \ln(1.15)/\text{year} \approx 0.1398/\text{year} \,.$$

This is a *continuous exponent of growth*.

The discrete model (1.1) can be perceived as a special *numerical discretization* of the continuous model. An application of the explicit Euler method with time step $\Delta t$ to (1.2) leads to

$$x(t_{i+1}) = x(t_i) + \Delta t \, p \, x(t_i) \,, \quad \text{or } x(t_{i+1}) = (1 + \Delta t \, p) \, x(t_i) \,,$$

this is (1.1) with $r = 1 + \Delta t \, p$. In the case $p < 0$ a time increment of the size $\Delta t < (-p)^{-1}$ has to be chosen to achieve a sensible sequence of numbers. On the other hand using the implicit Euler method one gets

$$x(t_{i+1}) = x(t_i) + \Delta t \, p \, x(t_{i+1}) \,, \quad \text{or } x(t_{i+1}) = (1 - \Delta t \, p)^{-1} x(t_i) \,,$$

i.e., (1.1) with $r = (1 - \Delta t \, p)^{-1}$. Here for $p > 0$ the time step has to be chosen such that $\Delta t < p^{-1}$. By Taylor series expansion one can see that the different growth factors coincide for *small* $\Delta t$ "up to an error of the order $O((\Delta t)^2)$":

$$(1 - \Delta t \, p)^{-1} = 1 + \Delta t \, p + O((\Delta t \, p)^2) \,.$$

The connection between the continuous and the discrete model therefore can be established by an analysis of the *convergence properties* of the numerical method. For the (explicit or implicit) Euler method one gets for example

$$|x(t_i) - x_i| \le C(t_e) \, \Delta t \,,$$

where $x(t_i)$ is the exact solution of (1.2) at time $t_i$ and $x_i$ is the approximate solution of the numerical method, assuming that $t_i \leq t_e$, where $t_e$ is the given final time for the model. For details about the analysis of numerical methods for ordinary differential equations we refer to the textbooks of Stoer and Bulirsch [123] and Deuflhard and Bornemann [28].

Both models, the discrete and the continuous, have the seeming disadvantage that they also allow *non-integer* solutions, which obviously are not realistic for the considered example. The model describes — as it is true for every other model — not the total reality but only leads to an idealized picture. For *small* populations the model is not very precise, as in general population growth also depends heavily on stochastic effects and therefore cannot be computed precisely in a deterministic way. In addition for small populations the model assumptions are questionable, in particular one neglects the age and the sex of the animals. In the extreme case of a herd of two animals obviously the growth will depend heavily on the fact whether there is a male and a female animal, or not. For large populations on the other hand one can assume with a certain qualification that it possesses a characteristic uniform distribution in age and in sex, such that the assumption of a growth proportional to population size make sense.

The substitution of integer values by real numbers reflects the inaccuracy of the model. Therefore it is not sensible to change the model such that integer values in the solutions are enforced. This would only lead to an unrealistic perception of high accuracy of the model. For a small population a *stochastic* model, which then "only" provides statements about the *probability distribution* of the population size, makes sense instead of deterministic models.

**Nondimensionalization**

The quantities in a mathematical model generally have a *physical dimension*. In the population model (1.2) we have the units *number* and *time*. We denote the physical dimension of a quantity $f$ with $[f]$ and abbreviate the units number of entities by $A$ and time by $T$. Therefore we have

$$[t] = T \, ,$$
$$[x(t)] = A \, ,$$
$$[x'(t)] = \frac{A}{T} \, ,$$
$$[p] = \frac{1}{T} \, .$$

The specification of a physical dimension is not yet a decision about the physical unit of measurements. As a unit of measurement for time one can use seconds, minutes, hours, days, weeks, or years, for example. If we measure time in years, then $t$ is indicated in years, $x(t)$ by a number, $x'(t)$ in number/years and $p$ in number/years.

To get models as simple as possible and furthermore in order to determine characteristic quantities in a model, one can *nondimensionalize* the model equations.

For this aim one defines a characteristic value for every appearing dimension and correspondingly a unit of measurement. Here it is not necessary to choose one of the common units as for example seconds or hours but it is more appropriate to choose a unit *adapted to the problem*. For the population model there are two dimensions, therefore two characteristic values are needed, the characteristic number $\bar{x}$ and the characteristic time $\bar{t}$. These are chosen in such a way that the *initial data $t_0$* and $x_0 = x(t_0)$ are as simple as possible. Therefore a convenient unit of measurement for time is given by

$$\tau = \frac{t - t_0}{\bar{t}},$$

where $\bar{t}$ denotes a unit of time which still has to be specified, and as a unit for number we choose

$$\bar{x} = x_0.$$

Setting

$$y = \frac{x}{\bar{x}}$$

and expressing $y$ as a function of $\tau$,

$$y(\tau) = \frac{x(\bar{t}\tau + t_0)}{\bar{x}},$$

one obtains

$$y'(\tau) = \frac{\bar{t}}{\bar{x}} x'(t)$$

and therefore the model becomes

$$\frac{\bar{x}}{\bar{t}} y'(\tau) = p \bar{x} y(\tau).$$

This model gets its most simple form for the choice

$$\bar{t} = \frac{1}{p}. \tag{1.3}$$

The model thus derived is the initial value problem

$$y'(\tau) = y(\tau),$$
$$y(0) = 1. \tag{1.4}$$

This model has the solution

$$y(\tau) = e^{\tau}.$$

From this solution all solutions of the original model (1.2) can be achieved by using a transformation:

$$x(t) = \overline{x}\, y(\tau) = x_0\, y(p(t - t_0)) = x_0\, e^{p(t-t_0)} \, .$$

The advantage of the nondimensionalization therefore is the reduction of the solution of all population models of a given type by the choice of units to *one single problem*. Note that this holds true *independent of the sign of p*, although the behavior of the solutions for $p > 0$ and $p < 0$ is different. For $p < 0$ the solution of (1.2) is given by the solution (1.4) for the range $\tau < 0$.

   The scaling condition (1.3) also can be obtained by means of *dimensional analysis*. In this procedure the characteristic time $\overline{t}$ to be determined is expressed as a product of the other characteristic parameters in the model,

$$\overline{t} = p^n x_0^m \quad \text{with } n, m \in \mathbb{Z} \, .$$

By computing the dimension one obtains

$$[\overline{t}] = [p]^n [x_0]^m \quad \text{and therefore} \quad T = \left(\frac{1}{T}\right)^n A^m \, .$$

The only possible solution of this equation is given by $n = -1, m = 0$, if the number of animals is interpreted as a dimension of its own. Thus we get exactly (1.3).

   In more complex models typically the model cannot be reduced to a single problem by nondimensionalization but the *number of relevant parameters* can be strongly reduced and the characteristic parameters can be identified. This also relates to the corresponding experiments: For instance, from the nondimensionalization of the equations for airflows one can conclude how the circulation around an airplane can be experimentally measured by using a (physical) model for the airplane much smaller in scale. We will explain dimensional analysis in one of the following sections using a more meaningful example.

## 1.3   Population Models with Restricted Resources

For large populations in nature a constant growth rate is not realistic anymore. A restriction of the habitat, or the available nutrients, or other mechanisms impose limitations on the growth. To construct a model it is feasible for such situations to assume that there is a certain capacity $x_M > 0$ for which the resources of the habitat are still sufficient. For population quantities $x$ smaller than $x_M$ the population still can grow, but for values larger than $x_M$ the population decreases. This means that the growth rate $p$ now depends on the population $x$, $p = p(x)$, and that

$$p(x) > 0 \text{ for } 0 < p < x_M,$$
$$p(x) < 0 \text{ for } p > x_M$$

have to hold true. The most simple functional form satisfying these conditions is given by a *linear* ansatz for $p$, i.e.,

$$p(x) = q(x_M - x) \text{ for all } x \in \mathbb{R}$$

with a parameter $q > 0$. With this ansatz we obtain the differential equation

$$x'(t) = q\, x_M\, x(t) - q\, x(t)^2 \tag{1.5}$$

as a model. The additional term $-q\, x(t)^2$ is proportional to the probability for the *number of encounters* of two specimens of the population per unit of time. The term represents the more competitive situation if the population size increases, the so-called "social friction". The Eq. (1.5) has been proposed by the Dutch biomathematician Verhulst and is called *logistic differential equation* or *equation of limited growth*.

Equation (1.5) also can be solved in closed form (compare Exercise 1.3). From

$$\frac{x'}{x(x_M - x)} = q$$

we conclude using the partial fraction decomposition

$$\frac{1}{x(x_M - x)} = \frac{1}{x_M}\left(\frac{1}{x} + \frac{1}{x_M - x}\right)$$

and by integration

$$\ln(x(t)) - \ln|x_M - x(t)| = x_M q t + c_1, \quad c_1 \in \mathbb{R}.$$

After the choice of an appropriate constant $c_2 \in \mathbb{R}$ we obtain

$$\frac{x(t)}{x_M - x(t)} = c_2 e^{x_M q t},$$

and

$$x(t) = \frac{c_2 x_M e^{x_M q t}}{1 + c_2 e^{x_M q t}} = \frac{x_M}{1 + c_3 e^{-x_M q t}}.$$

Incorporating the initial condition $x(t_0) = x_0$ we obtain

$$x(t) = \frac{x_M x_0}{x_0 + (x_M - x_0)e^{-x_M q(t - t_0)}}. \tag{1.6}$$

From this exact solution the following properties can be easily derived:

- If $x_0$ is positive, the solution always stays positive.
- If $x_0$ is positive, then for $t \to +\infty$ the solution converges to the equilibrium point $x_\infty = x_M$.

The graph of $x$ can be sketched also without knowing the exact solution. From (1.5) first we conclude

$$x' > 0, \quad \text{if } x < x_M,$$
$$x' < 0, \quad \text{if } x > x_M.$$

Furthermore we have

$$x'' = (x')' = (q\,(x_M - x)\,x)' = q(x_M - x)\,x' - q\,x\,x'$$
$$= q(x_M - 2x)x' = q^2(x_M - 2x)(x_M - x)\,x.$$

From these results we conclude

$$x'' > 0, \quad \text{if } x \in (0, x_M/2) \cup (x_M, \infty),$$
$$x'' < 0, \quad \text{if } x_M/2 < x < x_M.$$

Thus the solution curves have an inflection point at $x_M/2$ and the curves are concave in the interval between $x_M/2$ and $x_M$, and convex otherwise. Solutions of the logistic differential equation are depicted in Fig. 1.1.

**Stationary Solutions**

For more complex time-depending models a closed form solution often cannot be found. Then it is useful to identify *time independent* solutions. Such solutions can be computed using the time dependent model by just setting all time derivatives to zero. For our model with restricted growth one gets

$$0 = qx_M x - qx^2.$$

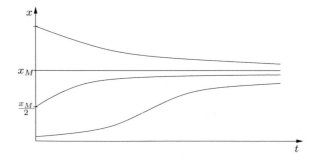

**Fig. 1.1**  Solutions of the logistic differential equation

This equation has the two solutions

$$x_0 = 0 \quad \text{and} \quad x_1 = x_M .$$

These are the solutions of the original model for specific initial data. Often time independent solutions appear as so-called *stationary limits* of arbitrary solutions for large times, meaning that they are solutions constant in time towards which time dependent solutions converge for large times. Typically this only appears if the stationary solution is *stable* in the following sense: If the initial data is only changed slightly then also the solution changes only slightly. Using the exact solution (1.6) the question of stability can be easily answered for the logistic differential equation: The solution for the initial value

$$x(t_0) = \varepsilon$$

with a small $\varepsilon > 0$ is given by

$$x_\varepsilon(t) = \frac{x_M \varepsilon}{\varepsilon + (x_M - \varepsilon)e^{-x_M q(t-t_0)}} ,$$

it converges for $t \to +\infty$ towards $x_M$, therefore the stationary solution $x_0 = 0$ is *not stable*. For

$$x(t_0) = x_M + \varepsilon$$

with a small $\varepsilon \neq 0$ the solution is given by

$$x_\varepsilon(t) = \frac{x_M(x_M + \varepsilon)}{(x_M + \varepsilon) - \varepsilon \, e^{-x_M q(t-t_0)}} ,$$

it converges for $t \to +\infty$ towards $x_M$. From

$$x_\varepsilon'(t) = q \, x_\varepsilon(t)(x_M - x_\varepsilon(t))$$

one can conclude also without knowing the exact solution that the distance to $x_M$ can only decrease for increasing time as from $x_\varepsilon(t) > x_M$ it follows $x_\varepsilon'(t) < 0$ and from $x_\varepsilon(t) < x_M$ it follows $x_\varepsilon'(t) > 0$. Therefore the stationary solution $x_M$ is *stable*. Stability is of importance, as in nature in general no *instable* stationary solution can be observed, therefore they are irrelevant for most practical applications. For more complex models sometimes no closed form solution for the time dependent equation can be derived. However, there are techniques of *stability analysis*, with which often the stability properties of stationary solutions can be deduced. Often this is done by means of a *linearization* of the problem at the stationary solution followed by a computation of the *eigenvalues* of the linearized problem. This will be explained in more detail in Chap. 4.

## 1.4   Dimensional Analysis and Scaling

Now we want to explain dimensional analysis using a slightly more significant example. We consider a body of mass $m$, which is thrown bottom-up in vertical direction with respect to the gravitational field of a planet (for example the Earth). The motion of the body is described by Newton's law

$$a = \frac{F}{m},$$

where $a$ denotes the acceleration of the body and $F$ the force acting on the body. This force is described by Newton's law of gravitation

$$F = -G\frac{m_E\, m}{(x + R)^2},$$

where $G \approx 6.674 \cdot 10^{-11} \mathrm{N} \cdot \mathrm{m}^2/\mathrm{kg}^2$ denotes the gravitational constant, $m_E$ the mass of the planet, $R$ the radius of the planet and $x$ the height of the body, measured from the surface of the planet. We neglect the air resistance in the atmosphere and consider the planet to be a sphere. If one defines the constant $g$ by

$$g = \frac{Gm_E}{R^2},$$

one gets

$$F = -\frac{gR^2m}{(x + R)^2}.$$

For the Earth we have $g = 9.80665 \ \mathrm{m/s}^2$, the gravitational acceleration. The motion of the body then is described by the differential equation

$$x''(t) = -\frac{gR^2}{(x(t) + R)^2}. \tag{1.7}$$

This has to be completed by two initial conditions,

$$x(0) = 0, \quad x'(0) = v_0,$$

where $v_0$ denotes the initial velocity.

   For trajectories expected in our application typically the term $x(t)$ is very small compared to the radius of the Earth and seems to be negligible in the denominator in the right-hand side in (1.7). We want to investigate the validity of this ansatz in a systematic fashion. To do so first we perform a nondimensionalization. As a specific example we use the data

$$g = 10 \, \text{m/s}^2, \quad R = 10^7 \, \text{m}, \quad \text{and} \quad v_0 = 10 \, \text{m/s},$$

which have an order of magnitude corresponding to the application in mind.

The dimensions appearing here are $L$ for the length and $T$ for time. The given data are the initial velocity $v_0$ with dimension $[v_0] = L/T$, the "planet acceleration" $g$ with dimension $[g] = L/T^2$, and the radius $R$ with dimension $[R] = L$. The independent variable is the time $t$ with dimension $[t] = T$, and the quantity to be computed is the height $x$ with dimension $[x] = L$. First we look for all representations of the form

$$\Pi = v_0^a \, g^b \, R^c,$$

which are either dimensionless (case (i)), or have the dimension of a length (case (ii)), or have the dimension of a time (case (iii)). From

$$[\Pi] = \left(\frac{L}{T}\right)^a \left(\frac{L}{T^2}\right)^b L^c = L^{a+b+c} T^{-a-2b}$$

it follows:

Case (i):   We have $a + b + c = 0$, $-a - 2b = 0$, therefore $a = -2b$, $c = b$, and finally

$$\Pi = \left(\frac{gR}{v_0^2}\right)^b.$$

This leads to the identification of

$$\varepsilon = \frac{v_0^2}{gR} \tag{1.8}$$

as a characteristic dimensionless parameter. In fact all other dimensionless parameters are powers of this specific one.

Case (ii):   We have $a + b + c = 1$, $a + 2b = 0$ and therefore $a = -2b$, $c = 1 + b$. As a specific unit for length one obtains

$$\ell = v_0^{-2b} g^b R^{1+b} = R \, \varepsilon^{-b},$$

where $b$ denotes a constant not yet specified.

Case (iii):   We have $a + b + c = 0$, $a + 2b = -1$ and therefore $a = -1 - 2b$, $c = b + 1$. Therefore a characteristic unit for time is given by

$$\tau = v_0^{-1-2b} g^b R^{b+1} = \frac{R}{v_0} \varepsilon^{-b}.$$

We will now try to nondimensionalize Eq. (1.7). To this purpose we consider a unit for length $\bar{x}$ and a unit for time $\bar{t}$ and represent $x(t)$ in the form

$$x(t) = \overline{x}\, y(t/\overline{t})\,.$$

From (1.7) we obtain

$$\frac{\overline{x}}{\overline{t}^2}\, y''(\tau) = -\frac{g R^2}{(\overline{x}\, y(\tau) + R)^2}\,,$$

i.e.,

$$\frac{\overline{x}}{\overline{t}^2 g}\, y''(\tau) = -\frac{1}{((\overline{x}/R)\, y(\tau) + 1)^2}\,. \tag{1.9}$$

This equation has to be provided with the initial conditions

$$y(0) = 0 \ \text{ and } \ y'(0) = \frac{\overline{t}}{\overline{x}} v_0\,.$$

Now we want to choose $\overline{x}$ and $\overline{t}$ in such a way that as many of the appearing parameters as possible equal 1. However, here we have more parameters than scaling units, namely the three parameters

$$\frac{\overline{x}}{\overline{t}^2 g}\,,\quad \frac{\overline{x}}{R}\,,\quad \text{and} \quad \frac{\overline{t}}{\overline{x}} v_0\,.$$

Hence, only two of these parameters can be transformed to one and therefore there are three different possibilities:

(a) $\dfrac{\overline{x}}{\overline{t}^2 g} = 1$ and $\dfrac{\overline{x}}{R} = 1$ are a consequence of $\overline{x} = R, \overline{t} = \sqrt{\dfrac{R}{g}}$, then the parameter

is given by $\dfrac{\overline{t}}{\overline{x}} v_0 = \dfrac{v_0}{\sqrt{Rg}} = \sqrt{\varepsilon}$ using $\varepsilon$ from (1.8). Therefore the model reduces to

$$y''(\tau) = -\frac{1}{(y(\tau) + 1)^2}\,,\quad y(0) = 0\,,\quad y'(0) = \sqrt{\varepsilon}\,. \tag{1.10}$$

(b) $\dfrac{\overline{x}}{R} = 1$ and $\dfrac{\overline{t}}{\overline{x}} v_0 = 1$ can be deduced from $\overline{x} = R$ and $\overline{t} = \dfrac{R}{v_0}$, leading to $\dfrac{\overline{x}}{\overline{t}^2 g} = \dfrac{v_0^2}{Rg} = \varepsilon$ for the third parameter. Then the dimensionless model is given by

$$\varepsilon\, y''(\tau) = -\frac{1}{(y(\tau) + 1)^2}\,,\quad y(0) = 0\,,\quad y'(0) = 1\,.$$

(c) $\dfrac{\overline{x}}{\overline{t}^2 g} = 1$ and $\dfrac{\overline{t}}{\overline{x}} v_0 = 1$ are a consequence of $\overline{t} = \dfrac{v_0}{g}$ and $\overline{x} = \dfrac{v_0^2}{g}$. Then the third parameter is given by $\dfrac{\overline{x}}{R} = \dfrac{v_0^2}{gR} = \varepsilon$. Thus the dimensionless model reads

$$y''(\tau) = -\frac{1}{(\varepsilon\, y(\tau) + 1)^2}\,, \qquad y(0) = 0\,, \quad y'(0) = 1\,. \qquad (1.11)$$

Let us mention that there is a fourth possibility. Besides (1.9) we may also use the equivalent formulation

$$\frac{\bar{x}^3}{\bar{t}^2 g R^2} y''(\tau) = -\frac{1}{(y(\tau) + R/\bar{x})^2}\,.$$

This leads, by setting $\bar{t} = gR^2/v_0^3$ $\bar{x} = gR^2/v_0^2$, to a fourth possibility (d)

$$y''(\tau) = -\frac{1}{(y(\tau) + \varepsilon)^2}\,, \qquad y(0) = 0\,, \quad y'(0) = 1$$

with $\varepsilon = v_0^2/(gR)$.

Now we want to assess and compare the four dimensionless equations for the application example displayed above. For $R = 10^7$ m, $g = 10\,\text{m/s}^2$ and $v_0 = 10\,\text{m/s}$ the parameter $\varepsilon$ is very small,

$$\varepsilon = \frac{v_0^2}{Rg} = 10^{-6}\,.$$

This suggests to neglect terms of the order of $\varepsilon$ in the equations.

The model (a) is then reduced to

$$y''(\tau) = -\frac{1}{(y(\tau) + 1)^2}\,, \qquad y(0) = 0\,, \quad y'(0) = 0\,.$$

Because of $y''(0) < 0$ and $y'(0) = 0$ this model leads to negative solutions and therefore it is extremely inexact and of no use. The reason lies in the scaling within the nondimensionalization: The parameters $\bar{t}$ and $\bar{x}$ here are given by

$$\bar{t} = \sqrt{\frac{R}{g}} = 10^3\,\text{s} \ \ \text{and}\ \bar{x} = 10^7\,\text{m}\,,$$

both scales are much too large for the problem under investigation. The maximal height to be reached and the instance of time for which it is reached are much smaller than the scales $\bar{x}$ for length and $\bar{t}$ for time and therefore are "hardly visible" in the nondimensionalized model.

The model (b) reduces to

$$0 = -\frac{1}{(y(\tau) + 1)^2}\,, \qquad y(0) = 0,\ y'(0) = 1\,.$$

This problem is not well posed, as it has no solution. Also here the chosen scales for time and length are much too large,

$$\bar{t} = \frac{R}{v_0} = 10^6 \, \text{s} \ \text{and} \ \bar{x} = R = 10^7 \, \text{m}.$$

Model (c) reduces to

$$y''(\tau) = -1, \quad y(0) = 0, \quad y'(0) = 1. \tag{1.12}$$

This model has the solution

$$y(\tau) = \tau - \frac{1}{2}\tau^2$$

and thus describes a typical parabola shaped path-time curve for a throw within the gravitational field of the Earth neglecting the air resistance. The back transformation

$$x(t) = \bar{x}\, y(t/\bar{t}) = \frac{v_0^2}{g}\, y(gt/v_0)$$

leads to

$$x(t) = v_0 t - \frac{1}{2}gt^2.$$

This corresponds to the solution of (1.7), if the term $x(t)$ in the denominator of the right-hand side of (1.7) is neglected. The scales in the nondimensionalization here have reasonable values,

$$\bar{t} = \frac{v_0}{g} = 1\,\text{s}, \quad \bar{x} = \frac{v_0^2}{g} = 10\,\text{m}.$$

Model (d) reduces to

$$y''(\tau) = -\frac{1}{y(\tau)^2}, \quad y(0) = 0, \quad y'(0) = 1.$$

This model does not correspond to a constant acceleration force. Also, the initial condition seems to be problematic for this differential equation, and the time and the length scale chosen are much too large.

Hence, for the application considered the nondimensionalization in version (c) is the "correct one". The versions (a), (b), and (d) are equally well mathematically correct, but there the small parameter $\varepsilon$ cannot be neglected anymore because its influence is amplified by the (too) large scaling parameters $\bar{t}$ and $\bar{x}$.

## 1.5  Asymptotic Expansions

Now we will introduce a technique with which the simplified model can be improved. The basic idea is not to neglect the terms of order $\varepsilon$ in the exact model (1.11), but rather to do a series expansion of the solution of (1.11) with respect to $\varepsilon$ to achieve more precise solutions by keeping some of the terms beyond the zeroth order term. The terms of higher order in $\varepsilon$ are determined from equations which we get by substituting a series expansion into (1.11).

We want to discuss this procedure which is called the *method of asymptotic expansion*, first for a simple algebraic example. We consider the equation

$$x^2 + 0.002\,x - 1 = 0. \tag{1.13}$$

The second summand has a small factor in front. Setting $\varepsilon = 0.001 \ll 1$, we obtain

$$x^2 + 2\varepsilon x - 1 = 0. \tag{1.14}$$

Now we want to approximate solutions $x$ of this equation by a series expansion of the form

$$x_0 + \varepsilon^\alpha x_1 + \varepsilon^{2\alpha} x_2 + \cdots \quad \text{with } \alpha > 0. \tag{1.15}$$

Before doing so we first define in general what we mean by an asymptotic expansion. Let $x : (-\varepsilon_0, \varepsilon_0) \to \mathbb{R}, \varepsilon_0 > 0$, be a given function. A sequence $(\phi_n(\varepsilon))_{n\in\mathbb{N}_0}$ is called an asymptotic sequence if and only if

$$\phi_{n+1}(\varepsilon) = o(\phi_n(\varepsilon)) \quad \text{as} \quad \varepsilon \to 0$$

for each $n = 0, 1, 2, 3, \ldots$. An example is the sequence $\phi_n(\varepsilon) = \varepsilon^{n\alpha}$ used above. A series $\sum_{k=0}^{N} \phi_k(\varepsilon)x_k$ is called asymptotic expansion of $x(\varepsilon)$ of the order $N \in \mathbb{N} \cup \{\infty\}$ with respect to the sequence $(\phi_n(\varepsilon))_{n\in\mathbb{N}_0}$, if for $M = 0, 1, 2, 3, \ldots, N$ we have

$$x(\varepsilon) - \sum_{k=0}^{M} \phi_k(\varepsilon)x_k = o(\phi_M(\varepsilon)) \quad \text{as} \quad \varepsilon \to 0.$$

If $\sum_{k=0}^{N} \phi_k(\varepsilon)x_k$ is an asymptotic expansion of $x(\varepsilon)$ we write

$$x \sim \sum_{k=0}^{N} \phi_k(\varepsilon)x_k \quad \text{as} \quad \varepsilon \to 0.$$

If $N = \infty$, we write

$$x(\varepsilon) \sim \sum_{k=0}^{\infty} \phi_k(\varepsilon) x_k \quad \text{as} \quad \varepsilon \to 0.$$

In the special case $\phi_k(\varepsilon) = \varepsilon^k$, we speak of an asymptotic expansion of $x(\varepsilon)$ in powers of $\varepsilon$. Note that asymptotic expansions of arbitrary order can exist, even if the corresponding infinite series are divergent for every $\varepsilon \neq 0$. In particular an asymptotic expansion can exist, although the Taylor expansion for $x(\varepsilon)$ does not converge for any $\varepsilon \neq 0$.

Now we substitute the asymptotic expansion (1.15) into (1.14) and obtain

$$x_0^2 + 2\varepsilon^\alpha x_0 x_1 + \cdots + 2\varepsilon(x_0 + \varepsilon^\alpha x_1 + \cdots) - 1 = 0.$$

If this identity shall hold true, it must hold true in particular for small $\varepsilon$. Therefore all terms which do not contain a factor $\varepsilon$ (or $\varepsilon^\alpha$), must add up to zero. Such terms are of order 1. We write $\mathcal{O}(1)$ or $\mathcal{O}(\varepsilon)$, respectively, and collect only those terms, which are exactly of order 1 or $\varepsilon$, respectively. The equation of order 1 then reads

$$\mathcal{O}(1): \quad x_0^2 - 1 = 0.$$

Its solutions are given by $x_0 = \pm 1$. In particular we see that the equation of order $\mathcal{O}(1)$ has exactly as many solutions as the original problem. This is a condition necessary in order to speak of a *regularly perturbed problem*. Later on we will see when to speak of regular or of *singular* perturbations.

Now we consider the terms of the next higher order in $\varepsilon$. What the next higher order is depends on whether we have $\alpha < 1$, $\alpha > 1$, or $\alpha = 1$. If $\alpha < 1$, then we conclude from the term of order $\varepsilon^\alpha$ that $x_1 = 0$, and from the terms of order $\varepsilon^{j\alpha}$ in a successive fashion $x_j = 0$ for $1 \leq j < 1/\alpha$. If $\alpha = 1/k$ for some $k \in \mathbb{N}$, then it follows from the term of order $k\alpha = 1$ that

$$2x_0 x_k + 2x_0 = 0$$

and therefore $x_k = -1$. Proceeding with the asymptotic expansion one sees that the terms of order $j$ with $\alpha j \notin \mathbb{N}$ always lead to $x_j = 0$. Therefore only the terms $x_{kn}$ for $n \in \mathbb{N}$ remain. For the corresponding powers $\varepsilon^{kn\alpha}$ we have $kn\alpha \in \mathbb{N}$. Therefore the power series ansatz with $\alpha < 1$, $\alpha = 1/k$, for some $k \in \mathbb{N}$, leads to the same result as the ansatz $\alpha = 1$, and therefore it is unnecessarily complicated. If $\alpha \neq 1/k$ for all $k \in \mathbb{N}$, then from the term of order $\varepsilon$ we get

$$2x_0 = 0,$$

but this is in contradiction to the already computed solutions $x_0 = \pm 1$. Therefore the ansatz $\alpha < 1$ is not sensible. In the case $\alpha > 1$ the term of order $\varepsilon$ also leads to $2x_0 = 0$ which is impossible as we have already seen. Therefore $\alpha = 1$ remains as the only sensible choice and we obtain the equation

$$\mathcal{O}(\varepsilon): \quad 2x_0x_1 + 2x_0 = 0$$

for the order $\varepsilon$. Its only solution is given by $x_1 = -1$.

If we also take terms of the next higher order $\varepsilon^2$ into account, we obtain

$$x_0^2 + 2\varepsilon x_0 x_1 + \varepsilon^2 x_1^2 + 2\varepsilon^2 x_2 x_0 + 2\varepsilon(x_0 + \varepsilon x_1 + \varepsilon^2 x_2) - 1 = 0$$

and the terms of order $\varepsilon^2$ lead to the identity

$$\mathcal{O}(\varepsilon^2): \quad x_1^2 + 2x_2x_0 + 2x_1 = 0\,.$$

Therefore we have

$$x_2 = \frac{1}{2}(x_0)^{-1} = \pm\frac{1}{2}\,.$$

The Eq. (1.13) corresponds to (1.14) for $\varepsilon = 10^{-3}$. Therefore we expect that the numbers

$$x_0,\; x_0 + \varepsilon x_1,\; x_0 + \varepsilon x_1 + \varepsilon^2 x_2$$

are good approximations of the solutions of (1.13) if we set $\varepsilon = 10^{-3}$. In fact we have

| $x_0$ | $x_0 + \varepsilon x_1$ | $x_0 + \varepsilon x_1 + \varepsilon^2 x_2$ | exact solutions |
|---|---|---|---|
| 1 | 0.999 | 0.9990005 | $0.9990005\cdots$ |
| $-1$ | $-1.001$ | $-1.0010005$ | $-1.0010005\cdots$ |

Therefore in this simple example the series expansion leads to very good approximations taking only a few terms into account.

This procedure becomes more interesting for complex problems without a closed form solution. Now we want to discuss the method of asymptotic expansion for the example (1.11), a throw in a gravitational field of a planet. We apply Taylor expansion at $z = 0$

$$\frac{1}{(1+z)^2} = 1 - 2z + 3z^2 - 4z^3 \pm \cdots$$

to the right-hand side of the differential equation

$$y_\varepsilon''(\tau) = -\frac{1}{(1 + \varepsilon\, y_\varepsilon(\tau))^2} \tag{1.16}$$

and get

$$y_\varepsilon''(\tau) = -1 + 2\varepsilon\, y_\varepsilon(\tau) - 3\varepsilon^2 y_\varepsilon^2(\tau) \pm \cdots . \tag{1.17}$$

We assume that the solution $y_\varepsilon$ also possesses a series expansion of the form

$$y_\varepsilon(\tau) = y_0(\tau) + \varepsilon^\alpha y_1(\tau) + \varepsilon^{2\alpha} y_2(\tau) + \cdots \tag{1.18}$$

with coefficient functions $y_j(\tau)$ and a parameter $\alpha$ to be specified. This ansatz will be substituted into (1.17) and then the coefficients of the same powers of $\varepsilon$ will be grouped together. The aim of this procedure is to determine a reasonable value for a parameter $\alpha$ and to obtain solvable equations for the coefficient functions $y_j(\tau)$, $j = 0, 1, 2, \ldots$. The substitution of (1.18) into (1.17) leads to

$$\begin{aligned}
y_0''(\tau) &+ \varepsilon^\alpha y_1''(\tau) + \varepsilon^{2\alpha} y_2''(\tau) + \cdots \\
&= - 1 + 2\varepsilon\big(y_0(\tau) + \varepsilon^\alpha y_1(\tau) + \varepsilon^{2\alpha} y_2(\tau) + \cdots\big) \\
&\quad - 3\varepsilon^2\big(y_0(\tau) + \varepsilon^\alpha y_1(\tau) + \varepsilon^{2\alpha} y_2(\tau) + \cdots\big)^2 \pm \cdots .
\end{aligned} \tag{1.19}$$

In the same way the series expansion can be substituted into the initial conditions and one obtains

$$y_0(0) + \varepsilon^\alpha y_1(0) + \varepsilon^{2\alpha} y_2(0) + \cdots = 0 \,,$$
$$y_0'(0) + \varepsilon^\alpha y_1'(0) + \varepsilon^{2\alpha} y_2'(0) + \cdots = 1 \,.$$

By comparing the coefficients of $\varepsilon^{k\alpha}$, $k \in \mathbb{N}$, one immediately gets

$$y_j(0) = 0 \ \text{ for } j \in \mathbb{N} \cup \{0\}, \ \ y_0'(0) = 1 \ \text{ and } \ y_j'(0) = 0 \ \text{ for } j \in \mathbb{N} . \tag{1.20}$$

To compare the coefficients appearing in (1.19) for the same powers of $\varepsilon$ on the left and right-hand side is more complicated. The lowest appearing power of $\varepsilon$ is $\varepsilon^0 = 1$. The comparison of the coefficients of $\varepsilon^0$ leads to

$$y_0''(\tau) = -1 \,.$$

Together with the initial conditions $y_0(0) = 0$ and $y_0'(0) = 1$ we obtain the already known problem (1.12) with its solution

$$y_0(\tau) = \tau - \frac{1}{2}\tau^2 \,.$$

The next exponent to be considered depends on the choice of $\alpha$. For $\alpha < 1$ it is $\varepsilon^\alpha$, comparison of the coefficients leads to

$$y_1''(\tau) = 0 \,.$$

Together with the initial conditions $y_1(0) = y_1'(0) = 0$ we have the unique solution $y_1(\tau) = 0$. The term $2\varepsilon y_0$ in (1.19) can only be compensated by a term of the form $\varepsilon^{k\alpha} y_k''$, $k \in \mathbb{N}, k\alpha = 1$. As in the case of $y_1$ we conclude that $y_j \equiv 0$ for $1 \le j \le k - 1$. Analogously one sees that the terms $y_k$, where $k\alpha \notin \mathbb{N}$, all have to be zero. Hence, one could have started with the ansatz $\alpha = 1$.

For $\alpha > 1$ the next exponent is given by $\varepsilon^1$, and the comparison of coefficients leads to $y_0(\tau) = 0$. This is a contradiction to the solution computed above. Therefore $\alpha > 1$ is the wrong choice.

In summary, the only reasonable exponent is $\alpha = 1$. Then the coefficients of $\varepsilon^1$ are given by

$$y_1''(\tau) = 2\, y_0(\tau) = 2\tau - \tau^2\,.$$

Together with the initial conditions $y_1(0) = y_1'(0) = 0$ one obtains the unique solution

$$y_1(\tau) = \frac{1}{3}\tau^3 - \frac{1}{12}\tau^4.$$

The coefficients of $\varepsilon^2$ lead to the problem

$$y_2''(\tau) = 2\, y_1(\tau) - 3\, y_0^2(\tau) = \frac{2}{3}\tau^3 - \frac{1}{6}\tau^4 - 3\tau^2 + 3\tau^3 - \frac{3}{4}\tau^4$$

together with the initial conditions $y_2(0) = y_2'(0) = 0$. Its solution is given by

$$y_2(\tau) = -\frac{11}{360}\tau^6 + \frac{11}{60}\tau^5 - \frac{1}{4}\tau^4\,.$$

Correspondingly further coefficients $y_3(\tau)$, $y_4(\tau)$, $\cdots$ can be computed, but the effort becomes larger and larger with increasing order. In particular the first three terms of the series expansion are

$$y_\varepsilon(\tau) = \tau - \frac{1}{2}\tau^2 + \varepsilon\left(\frac{1}{3}\tau^3 - \frac{1}{12}\tau^4\right) + \varepsilon^2\left(-\frac{1}{4}\tau^4 + \frac{11}{60}\tau^5 - \frac{11}{360}\tau^6\right) + O\!\left(\varepsilon^3\right).$$

Figure 1.2 shows the graph of the approximations $y_0(\tau)$ of order 0, $y_0(\tau) + \varepsilon\, y_1(\tau)$ of order 1 and of the exact solution for $\varepsilon = 0.2$. One sees that visually the approximation of order 1 can hardly be distinguished from the exact solution, but the approximation of order 0 still contains a clearly visible error.

Now we want to use the series expansion to obtain a better approximation for the height of the throw. For this purpose we first compute an approximation for the instant of time $\tau = \tau_\varepsilon$, at which this maximal height is reached, using the equation

$$y_\varepsilon'(\tau) = 0\,.$$

From $y_\varepsilon(\tau) = y_0(\tau) + \varepsilon\, y_1(\tau) + \varepsilon^2 y_2(\tau) + \cdots$ it follows that

$$y_0'(\tau) + \varepsilon\, y_1'(\tau) + \varepsilon^2 y_2'(\tau) + O\!\left(\varepsilon^3\right) = 0$$

with $y_0, y_1, y_2$ as above. Again we solve this equation in an approximative fashion using the series ansatz

$$\tau_\varepsilon = \tau_0 + \varepsilon\, \tau_1 + \varepsilon^2 \tau_2 + \cdots\,.$$

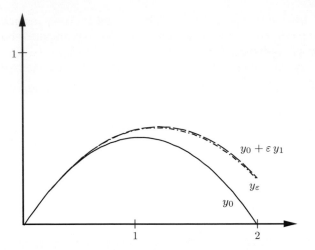

**Fig. 1.2**  Asymptotic expansion for the vertical throw with $\varepsilon = 0.2$

Comparing the coefficients of $\varepsilon^0$ leads to

$$y_0'(\tau_0) = 1 - \tau_0 = 0\,,$$

and therefore $\tau_0 = 1$. From the coefficients of $\varepsilon$ and the expansion $y_i'(\tau_\varepsilon) = y_i'(\tau_0) + \varepsilon\, y_i''(\tau_0)\tau_1 + \cdots, i = 0, 1$, one obtains

$$y_0''(\tau_0)\tau_1 + y_1'(\tau_0) = -\tau_1 + \tau_0^2 - \frac{1}{3}\tau_0^3 = 0$$

and therefore $\tau_1 = 2/3$. Thus, the approximation of first order for $\tau_\varepsilon$ is given by

$$1 + \frac{2}{3}\varepsilon\,.$$

The corresponding height is

$$h_\varepsilon = y_\varepsilon(\tau_\varepsilon) = y_0(\tau_0) + \varepsilon\big(y_0'(\tau_0)\tau_1 + y_1(\tau_0)\big) + O\big(\varepsilon^2\big)$$
$$= y_0(\tau_0) + \varepsilon\, y_1(\tau_0) + O\big(\varepsilon^2\big) = \frac{1}{2} + \frac{1}{4}\varepsilon + O\big(\varepsilon^2\big)\,.$$

If one takes into account that the gravitational force decreases with height, then the maximal height of the throw becomes slightly bigger. For our original example with $\varepsilon = 10^{-6}$ this affects the results in the seventh digit.

A priori it is not clear whether a series expansion of the form (1.18) exists. In order to develop a model in a mathematically rigorous fashion by using such a series expansion makes it necessary to justify the results obtained, for example by the derivation of an error estimate of the form

$$\left| y_\varepsilon(\tau) - \sum_{j=0}^{N} \varepsilon^j y_j(\tau) \right| \le C_N \varepsilon^{N+1} . \tag{1.21}$$

We show such an estimate for $N = 1$, i.e.,

$$|y_\varepsilon(\tau) - y_0(\tau) - \varepsilon\, y_1(\tau)| \le C \varepsilon^2 , \tag{1.22}$$

for $\tau \in (0, T)$ for an appropriate final time $T$, and $\varepsilon$ small enough in the sense that $\varepsilon < \varepsilon_0$ for an appropriate $\varepsilon_0$ to be specified. As a first step we construct a differential equation for the *error*

$$z_\varepsilon(\tau) = y_\varepsilon(\tau) - y_0(\tau) - \varepsilon\, y_1(\tau) .$$

From the differential equations for $y_\varepsilon$, $y_0$, and $y_1$ we obtain

$$z_\varepsilon''(\tau) = y_\varepsilon''(\tau) - y_0''(\tau) - \varepsilon\, y_1''(\tau) = -\frac{1}{(1 + \varepsilon\, y_\varepsilon(\tau))^2} + 1 - 2\varepsilon\, y_0(\tau) .$$

A Taylor expansion with a representation of the remainder leads to

$$\frac{1}{(1 + y)^2} = 1 - 2y + 3\frac{1}{(1 + \vartheta y)^4} y^2,$$

where $\vartheta = \vartheta(y) \in (0, 1)$. Hence, we obtain

$$z_\varepsilon''(\tau) = -1 + 2\varepsilon\, y_\varepsilon(\tau) - 3\varepsilon^2 \frac{1}{(1 + \varepsilon\vartheta\, y_\varepsilon(\tau))^4} y_\varepsilon^2(\tau) + 1 - 2\varepsilon\, y_0(\tau) .$$

Substitution of $y_\varepsilon(\tau) = z_\varepsilon(\tau) + y_0(\tau) + \varepsilon\, y_1(\tau)$ leads to

$$z_\varepsilon''(\tau) = 2\varepsilon\, z_\varepsilon(\tau) + \varepsilon^2 R_\varepsilon(\tau), \tag{1.23}$$

where

$$R_\varepsilon(\tau) = -\frac{3\, y_\varepsilon^2(\tau)}{(1 + \varepsilon\vartheta\, y_\varepsilon(\tau))^4} + 2 y_1(\tau) .$$

Additionally the initial conditions

$$z_\varepsilon(0) = 0 \quad \text{and} \quad z_\varepsilon'(0) = 0$$

are valid. For an estimation of $R_\varepsilon(\tau)$ we need *lower* and *upper* bounds for $y_\varepsilon(\tau)$. These can be derived from the differential equation for $y_\varepsilon$ and the representation

$$y_\varepsilon(\tau) = y_\varepsilon(0) + \int_0^\tau y_\varepsilon'(t)\, dt = \int_0^\tau \left( y_\varepsilon'(0) + \int_0^t y_\varepsilon''(s)\, ds \right) dt$$

$$= \tau + \int_0^\tau \int_0^t y_\varepsilon''(s)\, ds\, dt\,.$$

We set $t_\varepsilon := \inf\{t\,|\,t > 0,\ y_\varepsilon(t) < 0\}$. Obviously it follows from (1.16), that $y_\varepsilon''(\tau) \geq -1$ for $0 < \tau < t_\varepsilon$ and therefore

$$y_\varepsilon(\tau) \geq \tau - \frac{1}{2}\tau^2\,.$$

As $y_\varepsilon$ is continuous, in particular we have $t_\varepsilon \geq 2$. Because of $y_\varepsilon''(\tau) \leq 0$ for $\tau < t_\varepsilon$ we also have $y_\varepsilon(\tau) \leq t_\varepsilon$ for $\tau < t_\varepsilon$. Now let $T \leq t_\varepsilon$ be chosen and fixed, for example $T = 2$. Then we have

$$|R_\varepsilon(\tau)| \leq 3|y_\varepsilon(\tau)|^2 + 2|y_1(\tau)| \leq C_1$$

for $\tau < T$ with a constant $C_1 = C_1(T)$. For a given $C_0 > 0$ we define a further instant of time $\tau_\varepsilon > 0$ by $\tau_\varepsilon := \inf\{t\,|\,t > 0,\ |z_\varepsilon(t)| \geq C_0\varepsilon^2\}$. As $z_\varepsilon$ is continuous and $z_\varepsilon(0) = 0$, we have $\tau_\varepsilon > 0$. For $\tau < \min(T, \tau_\varepsilon)$ we conclude from (1.23) that

$$|z_\varepsilon(\tau)| = \left| \int_0^\tau \int_0^t z_\varepsilon''(s)\, ds\, dt \right| \leq \int_0^\tau \int_0^t |z_\varepsilon''(s)|\, ds\, dt \leq \frac{1}{2}T^2(2C_0\varepsilon + C_1)\varepsilon^2\,.$$

Then for $C_0 > T^2 C_1$ there exists a $\varepsilon_0 > 0$, such that $\frac{1}{2}T^2(2C_0\varepsilon_0 + C_1) = \frac{C_0}{2}$, namely

$$\varepsilon_0 = \frac{1}{2T^2} - \frac{C_1}{2C_0}\,.$$

For all $\varepsilon \leq \varepsilon_0$ and all $t \leq \min\{T, \tau_\varepsilon\}$ it holds true that

$$|z_\varepsilon(t)| \leq \frac{C_0}{2}\varepsilon^2\,.$$

As $z_\varepsilon$ is continuous, in particular we have $\tau_\varepsilon \geq T$. Hence, (1.22) has been shown, for $C = C_0/2$.

The procedure to determine an asymptotic expansion can also be formulated more generally and abstractly in *Banach spaces*, i.e., in complete, normed vector spaces. Let $B_1$, $B_2$ be Banach spaces and

$$F : B_1 \times [0, \varepsilon_0) \to B_2$$

be a smooth mapping, which is sufficiently differentiable for the following considerations. For $\varepsilon \in [0, \varepsilon_0)$ we look for a solution $y_\varepsilon$ of the equation

$$F(y, \varepsilon) = 0 \, .$$

We make the ansatz

$$y_\varepsilon = \sum_{i=0}^{\infty} \varepsilon^i y_i$$

and expand

$$F(y_\varepsilon, \varepsilon) = \sum_{i=0}^{\infty} \varepsilon^i F_i(y_\varepsilon) = \sum_{i=0}^{\infty} \varepsilon^i F_i \left( \sum_{j=0}^{\infty} \varepsilon^j y_j \right)$$

$$= \sum_{i=0}^{\infty} \varepsilon^i \left( F_i(y_0) + DF_i(y_0) \left( \sum_{j=1}^{\infty} \varepsilon^j y_j \right) \right.$$

$$\left. + \tfrac{1}{2} D^2 F_i(y_0) \left( \sum_{j=1}^{\infty} \varepsilon^j y_j, \sum_{j=1}^{\infty} \varepsilon^j y_j \right) + \cdots \right)$$

$$= F_0(y_0) + \varepsilon(F_1(y_0) + DF_0(y_0)(y_1)) +$$

$$+ \varepsilon^2 \left( F_2(y_0) + DF_1(y_0)(y_1) + DF_0(y_0)(y_2) + \tfrac{1}{2} D^2 F_0(y_0)(y_1, y_1) \right)$$

$$+ \cdots .$$

We try to solve these equations successively, with increasing order and obtain

$$F_0(y_0) = 0 \, ,$$

$$DF_0(y_0)(y_1) = -F_1(y_0) \, ,$$

$$DF_0(y_0)(y_2) = -F_2(y_0) - DF_1(y_0)(y_1) - \frac{1}{2} D^2 F_0(y_0)(y_1, y_1) \, ,$$

$$\vdots$$

$$DF_0(y_0)(y_k) = G_k(y_0, \ldots, y_{k-1}) \, .$$

If the linear mapping $DF_0(y_0) : B_1 \to B_2$ possesses an inverse then the values $y_1, y_2, y_3, \ldots$ can be computed successively.

**Definition 1.1** If the values $y_0, \ldots, y_N$ are solutions of the above displayed equations, then the series

$$y_\varepsilon^N := \sum_{i=0}^{N} \varepsilon^i y_i$$

is called asymptotic expansion of order $N$.

An important question now is: Are the solutions for the problems "perturbed" by a small parameter $\varepsilon$ a good approximation for the original problem? A positive answer is encoded in the following definition.

**Definition 1.2** *(Consistency)* The equations

$$F(y, \varepsilon) = 0, \quad \varepsilon > 0,$$

are called *consistent* with

$$F(y, 0) = 0,$$

if for all solutions $y_0$ of $F(y_0, 0) = 0$ it holds true that

$$\lim_{\varepsilon \to 0} F(y_0, \varepsilon) = 0.$$

**Remarks**

1. In general consistency does *not* imply convergence: also in a consistent situation the solutions $y_\varepsilon$ of $F(y, \varepsilon) = 0$ does not need to fulfill

$$y_\varepsilon - y_0 \to 0 \quad \text{in} \quad B_1$$

   (see also Exercise 1.11).
2. An important case in asymptotic analysis is characterized by the fact that the small parameter appears as a factor in a term which is decisive for the mathematical structure of the problem. For differential equations this term in general is the highest order derivative of the unknown function appearing in the equation. In this case one speaks of a *singular perturbation*. At the end of Chap. 6 we will investigate corresponding examples.

**Examples**:

(i) The equation
$$\varepsilon x^2 - 1 = 0$$

changes its order for $\varepsilon \to 0$. In particular for $\varepsilon = 0$ the equation becomes insolvable and the solutions $x_\varepsilon^{\pm}$ of $\varepsilon x^2 - 1 = 0$ converge to infinity for $\varepsilon \to 0$.

(ii) The initial value problem

$$\varepsilon y_\varepsilon'' = \frac{1}{(y_\varepsilon + 1)^2}, \quad y_\varepsilon(0) = 0, \quad y_\varepsilon'(0) = 1$$

changes its character if one sets $\varepsilon = 0$. For $\varepsilon > 0$ one has a differential equation and for $\varepsilon = 0$ one obtains an insolvable algebraic equation.

## 1.6 Applications from Fluid Mechanics

Now we will discuss the introduced notions of dimensional analysis, asymptotic expansion and singular perturbation for a considerably more complex example from fluid mechanics. The models which we use will be derived systematically in the framework of continuum mechanics in Chap. 5.

We consider the following example: a fluid, i.e., a liquid or a gas, flows past a body $K$ (Fig. 1.3). We are interested in the velocity field

$$v = v(t, x) \in \mathbb{R}^3, \ t \in \mathbb{R}, \ x \in \mathbb{R}^3,$$

of the fluid. We assume that for $|x| \to \infty$ the velocity converges to a constant value, i.e.,

$$v(t, x) \to V \in \mathbb{R}^3 \quad \text{as} \quad |x| \to \infty.$$

From conservation principles and using certain constitutive assumptions about the properties of the fluid the *Navier–Stokes equations* can be derived, see Chap. 5. For an incompressible fluid with constant density $\varrho_0$ neglecting exterior forces we obtain

$$\varrho_0(\partial_t v + (v \cdot \nabla)v) = -\nabla p + \mu \Delta v, \tag{1.24}$$

$$\nabla \cdot v = 0, \tag{1.25}$$

where $p$ denotes the pressure and $\mu$ the *dynamic viscosity* of the fluid. The viscosity is caused by *internal friction*. It is high for honey and low for gases. Furthermore, expressed in Cartesian coordinates we have

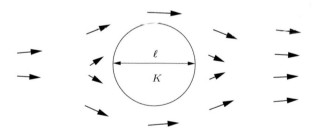

**Fig. 1.3** Flow past an obstacle

$$\nabla \cdot v = \sum_{i=1}^{3} \frac{\partial}{\partial x_i} v_i \in \mathbb{R} \qquad \text{for the \textit{divergence} of a vector field } v,$$

$$\Delta v = \sum_{i=1}^{3} \frac{\partial^2}{\partial x_i^2} v \in \mathbb{R}^3 \qquad \text{the \textit{Laplacian operator} and}$$

$$(v \cdot \nabla)v = \left( \sum_{i=1}^{3} v_i \, \partial_i v_j \right)_{j=1,2,3} \in \mathbb{R}^3 .$$

First we discuss the *dimensions* of the appearing terms.

| variables | dimension |
|---|---|
| $v$ velocity | $L/T$ |
| $\varrho_0$ mass density | $M/L^3$ |
| $p$ pressure = force/area | $(M \cdot L/T^2)/L^2 = M/(LT^2)$ |

Furthermore we have
$$[\mu] = M/(LT) .$$

As a consequence all terms in (1.24) have the dimension $M/(L^2 T^2)$.

As an example for the potential of *dimension analysis* we consider the behavior of a fluid flowing around a large ship. The goal is to perform experiments with a ship whose size is reduced by the factor 100. Under which circumstances the results of such experiments with a ship model can be transferred to the behavior of a large real ship?

For this purpose the equation has to be transferred into a dimensionless form. Here the relevant parameters are the characteristic length $\bar{x}$, for example the length of the ship, the velocity $v$ of the ship, the density $\varrho_0$, and the viscosity $\mu$. We form dimensionless quantities from combinations of these parameters, for example

$$y = \frac{x}{\bar{x}}, \quad \tau = \frac{t}{\bar{t}},$$

where $\bar{t} = \bar{t}(\bar{x}, V, \varrho_0, \mu)$ is a characteristic time still to be determined. Furthermore we set

$$u(\tau, y) = \frac{v}{|V|}$$

and

$$q(\tau, y) = \frac{p}{\bar{p}}, \quad \text{where } \bar{p} \text{ is still to be determined.}$$

Multiplication of the Navier–Stokes equations by $\bar{t}/(\varrho_0 |V|)$ and using the transformation rules $\partial_t = \frac{1}{\bar{t}} \partial_\tau$ and $\nabla_x = \frac{1}{\bar{x}} \nabla_y$ leads to the equation

$$\partial_\tau u + \frac{\bar{t}|V|}{\bar{x}}(u \cdot \nabla)u = -\frac{\bar{p}}{\varrho_0}\frac{\bar{t}}{\bar{x}|V|}\nabla q + \frac{\mu}{\varrho_0}\frac{\bar{t}}{(\bar{x})^2}\Delta u \,.$$

We set $\bar{t} = \bar{x}/|V|$, $\bar{p} = |V|^2\varrho_0$ and $\eta = \mu/\varrho_0$ — this is called the *kinematic viscosity* — and obtain

$$\partial_\tau u + (u \cdot \nabla)u = -\nabla q + \frac{1}{\mathrm{Re}}\Delta u \,,$$
$$u(\tau, y) \to V/|V| \quad \text{as} \quad |y| \to \infty \,.$$

Here $\mathrm{Re} := \bar{x}|V|/\eta$ is called the *Reynolds number*. For large $|y|$ the Euclidean norm of the nondimensionalized velocity converges to 1. Furthermore it still holds that

$$\nabla \cdot u = 0 \,.$$

This means that flow situations with different $\bar{x}$ lead to the *same* dimensionless form, if the Reynolds number is the same for the different situations. If we reduce the size of the ship by the factor 100, then one possibility is to enlarge the approach velocity by the factor 10 and to reduce the kinematic viscosity by the factor 10 to obtain the same Reynolds number.

With the help of the Reynolds number one can estimate which effects are of importance for a flow and which effects are not. We discuss this for the example of two different models for the flow resistance of a body, which we motivate by means of heuristic considerations. For the case of small Reynolds numbers, i.e., for high viscosity or a small approach velocity, the viscous friction dominates the flow resistance. Then the characteristic quantities are the velocity $v$ of the obstacle relative to the flow, a characteristic quantity $\bar{x}$ for the size of the obstacle, and the dynamic viscosity $\mu$ of the fluid. The dimensions are given by $[v] = L/T$, $[\bar{x}] = L$, and $[\mu] = FT/L^2$, where $F$ denotes the dimension of force. Then a combination of these quantities has the dimension

$$\left[v^a\bar{x}^b\mu^c\right] = L^{a+b-2c}T^{-a+c}F^c \,.$$

This is the dimension of force, if

$$a + b - 2c = 0, \quad -a + c = 0, \text{ and } c = 1$$

and therefore $a = b = c = 1$. A law for the friction resistance in a viscous fluid therefore has to have the form

$$F_R = -c_R\,\mu\,\bar{x}\,v \tag{1.26}$$

where $F_R$ is the friction force acting on the body, $v$ the velocity of the body relative to the flow velocity, and $c_R$ is the friction coefficient which is depending on the shape of the body. For a sphere with radius $r$ it can be shown that

$$F_R = -6\pi r \mu v$$

holds true, and this is called *Stokes' law*.

For high Reynolds numbers, however, this part of the flow resistance is dominated by a force which is necessary to accelerate the part of the fluid lying in the direction of the motion of the body. The mass to be accelerated per time interval $\Delta t$ may fulfill

$$\Delta m \approx \varrho \, A \, |v| \, \Delta t \, ,$$

where $\varrho$ is the density of the fluid and $A$ the cross sectional area of the body. Here the term $A \, |v| \, \Delta t$ just describes the volume replaced by the body in the time interval $\Delta t$. This fluid volume is accelerated to velocity $v$. The supplied kinetic energy is

$$\Delta E_{\mathrm{kin}} \approx \tfrac{1}{2} \Delta m \, |v|^2 \approx \tfrac{1}{2} \varrho \, A \, |v|^3 \, \Delta t \, .$$

The friction force is related to the kinetic energy by

$$|F_R| \, |v| \, \Delta t \propto \Delta E_{\mathrm{kin}}$$

where $\propto$ indicates that the two sides are proportional to each other. Hence, we obtain

$$|F_R| \propto \tfrac{1}{2} \varrho \, A \, |v|^2 \, .$$

The corresponding proportionality constant $c_d$ is called drag coefficient and we obtain

$$F_R = -\tfrac{1}{2} c_d \varrho A |v| v \, . \tag{1.27}$$

As this force is proportional to the square of the velocity, for large velocities it dominates the viscous frictional force (1.26), on the other hand for small velocities it can be neglected compared to (1.26). Formula (1.27) can also be justified by a dimensional analysis (see Exercise 1.12). In applications the drag coefficient has to be determined by measurements since for most body shapes a theoretical derivation as for a sphere in the case of Stokes' law does not exist any more. In any case (1.27) is only a relatively coarse approximation to reality, the real dependence of the flow resistance on the velocity is considerably more complex. On the other hand Stokes' law is a relatively good approximation, if only the velocity is sufficiently small.

In order to assess for a given application which of the two laws (1.26) or (1.27) is reasonable, the coefficient of the two frictional forces can be considered:

$$\frac{|F_R^{(1.27)}|}{|F_R^{(1.26)}|} \propto \frac{\varrho \bar{x} |v|}{\mu} = \mathrm{Re} \, .$$

In doing so we choose the scale $\bar{x}$ such that $A = \bar{x}^2$. Therefore Stokes' law (1.26) makes sense for Reynolds numbers $\mathrm{Re} \ll 1$. On the other hand the flow resistance given by (1.27) dominates for Re $\gg$ 1. For Re $\approx$ 1 both effects have the same importance.

The Navier–Stokes equations being a complex model the question arises whether certain terms can be neglected in specific situations. As we have transferred the equation to a nondimensional form it is possible to speak of large or small independent of the choice of units: now the number 1 can be interpreted as a *medium sized* quantity. The only parameter is the Reynolds number and for many problems Re is very large. Then $\varepsilon = 1/\,\mathrm{Re}$ is a small term which suggests to neglect the term $\varepsilon \, \Delta u = \frac{1}{\mathrm{Re}} \, \Delta u$. In this way we obtain the *Euler equations* of fluid mechanics

$$\partial_\tau u + (u \cdot \nabla)u = -\nabla q \,,$$
$$\nabla \cdot u = 0 \,.$$

How good is the description of a real fluid by this reduced model? Later we will see that the Euler equations do not allow for the formation of vortices (see Sect. 6.1.4). Most specifically we have

$$\nabla \times u(t, x) = \begin{pmatrix} \partial_{x_2} u_3 - \partial_{x_3} u_2 \\ \partial_{x_3} u_1 - \partial_{x_1} u_3 \\ \partial_{x_1} u_2 - \partial_{x_2} u_1 \end{pmatrix} = 0$$

for $t > 0$ if $\nabla \times u(0, x) = 0$. Additionally it turns out that

• the term $\varepsilon \, \Delta u$ *cannot* be small in the vicinity of the boundary $\partial K$ of the body,

and if the Euler equations describe the flow,

• a bounded body would exert no resistance to the flow, therefore no forces would act against the flow. This is called *d'Alembert's paradox* and
• no lift force would act on the body (in 3-D).

In reality we see that a flow past an obstacle induces the formation of vortices, which sometimes separate from $K$. Additionally so-called *boundary layers* can be observed. These are "thin" regions of the flow in the vicinity of the body in which the flow field changes drastically.

What is the deficiency of the reduced model?

In the Navier–Stokes equations second derivatives with respect to the spatial variables appear, but in the Euler equations only first derivatives. In the theory of partial differential equations different types of equations are distinguished (see for example Evans, [37]). According to this classification the Navier–Stokes equations are *parabolic*, but the Euler equations are *hyperbolic*. The qualitative behavior of hyperbolic and parabolic differential equations differs considerably. For example in the solutions of the Euler equations even for arbitrary smooth data discontinuities may occur, on the other hand for the Navier–Stokes equations smooth solutions are to be expected in this case, even if the rigorous theoretical proof for three dimensions is still outstanding. The small factor $\varepsilon$ belongs to the term which is decisive for the behavior of the solution. Therefore one speaks of *singular perturbation*. To obtain approximate solutions of the Navier–Stokes equations the method of asymptotic expansion cannot be used in the form we have discussed so far. It breaks down in boundary layers, where the solution changes strongly. In Chap. 6 we will develop the *singular perturbation theory*, to obtain asymptotic expansions also in boundary layers.

## 1.7  Literature

An extensive description and analysis of biological growth models can be found in [105]. For further information on the subjects nondimensionalization, scaling, and asymptotic analysis we recommend [89], Chaps. 6 and 7, for scaling and dimensional analysis also [41], Chap. 1, is a good reference and for various aspects of asymptotic analysis we refer to [68]. A presentation of singular perturbation theory with many examples can be found in [77]. Parts of the presentation in this chapter are based on the lecture notes [116].

## 1.8  Exercises

**Exercise 1.1**  A bank offers four different variants of a savings account:
variant A with monthly payment of interest and an interest rate of 0.3% per month,
variant B with a quarterly payment of interest and an interest rate of 0.9% per quarter,
variant C with a semiannual payment of interest and an interest rate of 1.8% per half-year,
variant D with an annual payment of interest and an interest rate of 3.6% per year.

(a) Compute and compare the *effective* interest rate which is obtained after a year (reinvesting all paid interest).
(b) How must the interest rates be adjusted, such that they lead to the same yearly interest rate of 3.6%?
(c) Develop an interest model that is continuous in time, which does not need a time increment for the payment of interest.

**Exercise 1.2**  A police officer wants to determine the time of death of the victim of a homicide. He measures the temperature of the victim at 12.36 p.m. and obtains 80°F. According to Newton's law of cooling the cooling of a body is proportional to the difference between the body's temperature and the ambient temperature. Unfortunately the proportionality constant is unknown to the officer. Therefore he measures the temperature at 1.06 p.m. once more and now he obtains 77°F. The ambient

temperature is 68°F and it is assumed that the body's temperature at the time of death has been 98°F.
At what time the homicide took place?

**Exercise 1.3** (*Separation of variables, uniqueness, continuation of solutions*) For given functions $f$ and $g$ consider the ordinary differential equation

$$x'(t) = f(t)\,g(x(t)).$$

We look for solutions passing through the point $(t_0, x_0)$, i.e., $x(t_0) = x_0$ holds true.

(a) Show that in the case $g(x_0) \neq 0$ locally a unique solution through the given point exists.
(b) Assume that $g \neq 0$ in the interval $(x_-, x_+)$, where $g(x_-) = g(x_+) = 0$ and let $g$ be differentiable at $x_-$ and $x_+$. Show that the solution of the differential equation through the point $(t_0, x_0)$, where $x_0 \in (x_-, x_+)$, exists globally and is unique. *Hint:* Is the solution through the point $(t_+, x_+)$ unique?

**Exercise 1.4** We consider the model of limited growth of populations

$$x'(t) = q\,x_M\,x(t) - q\,x^2(t), \quad x(0) = x_0.$$

(a) Nondimensionalize the model using appropriate units for $t$ and $x$. Which possibilities exist?
(b) What nondimensionalization is appropriate for $x_0 \ll x_M$ ($x_0$ "much smaller than" $x_M$) in the sense that omitting small terms leads to a reasonable model?

**Exercise 1.5** (*Nondimensionalization, scale analysis*) A body of mass $m$ is thrown upwards in a vertical direction from the Earth's surface with a velocity $v$. The air resistance is supposed to be taken into account by Stokes' law $F_R = -cv$ for the flow resistance in viscous fluids, which is reasonable for small velocities. Here $c$ is a coefficient depending on the shape and the size of the body. The motion is supposed to depend on the mass $m$, the velocity $v$, the gravitational acceleration $g$ and the friction coefficient $c$ with dimension $[c] = M/T$.

(a) Determine the possible dimensionless parameters and reference values for height and time.
(b) The initial value problem for the height is assumed to take the form

$$mx'' + cx' = -mg, \quad x(0) = 0, \quad x'(0) = v.$$

Nondimensionalize the differential equation. Again different possibilities are available.
(c) Discuss the different possibilities of a reduced model if $\beta := cv/(mg)$ is small.

**Exercise 1.6** A model for the vertical throw on the Earth taking into account the air resistance is given by

$$mx''(t) = -mg - c|x'(t)|x'(t), \quad x(t_0) = 0, \quad x'(t_0) = v_0.$$

In this model the gravitational force is approximated by $F = -mg$, the air resistance for a given velocity $v$ is described by $-c|v|v$ with a proportionality constant $c$ depending on the shape and size of the body and the density of the air. This law is reasonable for high velocities.

(a)  Nondimensionalize the model. What possibilities exist?
(b)  Compute the maximal height of the throw for the data $m = 0.1\,\text{kg}$, $g = 10\,\text{m/s}^2$, $v_0 = 10\,\text{m/s}$, $c = 0.01\,\text{kg/m}$ and compare the result with the corresponding result for the model without air resistance.

**Exercise 1.7** (*Nondimensionalization*) We want to compute the power $P$, which is necessary to move a body with known shape (for example a ship) in a liquid (for example water). We assume that the power depends on the length $\ell$ and the velocity $v$ of the ship, the density $\varrho$ and the kinematic viscosity $\eta$ of the liquid, and the gravitational acceleration $g$. The dimensions of the data are $[\ell] = L$, $[\varrho] = M/L^3$, $[v] = L/T$, $[\eta] = L^2/T$, $[P] = ML^2/T^3$, and $[g] = L/T^2$, where $L$ denotes the length, $M$ the mass and $T$ the time. Show that under these assumptions the power $P$ is given by

$$\frac{P}{\varrho\ell^2 v^3} = \Phi(\text{Fr}, \text{Re})$$

with a function $\Phi : \mathbb{R}^2 \to \mathbb{R}$ and the dimensionless quantities

$$\text{Re} = \frac{|v|\ell}{\eta} \text{ (Reynolds number) and Fr} = \frac{|v|}{\sqrt{\ell g}} \text{ (Froude number)}.$$

**Exercise 1.8** (*Formal asymptotic expansion*)

(a)  For the initial value problem

$$x''(t) + \varepsilon x'(t) = -1, \quad x(0) = 0, \quad x'(0) = 1$$

compute the formal asymptotic expansion of the solution $x(t)$ up to the second order in $\varepsilon$.
(b)  Compute the formal asymptotic expansion for the instance of time $t^* > 0$, for which $x(t^*) = 0$ holds true, up to first order in $\varepsilon$, by substituting the series expansion $t^* \sim t_0 + \varepsilon t_1 + O(\varepsilon^2)$ into the approximation obtained for $x$ leading to a determination of $t_0$ and $t_1$.

**Exercise 1.9**  A model already nondimensionalized for the vertical throw with *small* air resistance is given by

$$x''(t) = -1 - \varepsilon(x'(t))^2, \quad x(0) = 0, \ x'(0) = 1.$$

The model describes the throw up to the maximal height.

(a)  Compute the first two coefficients $x_0(t)$ and $x_1(t)$ in the asymptotic expansion

$$x(t) = x_0(t) + \varepsilon\, x_1(t) + \varepsilon^2 x_2(t) + \cdots$$

for small $\varepsilon$.

(b)  Compute the maximal height of the throw up to terms of order $\varepsilon$ using asymptotic expansion.

(c)  Compare the results from (b) for the data of Exercise 1.6(b) with the exact result and the result neglecting the air resistance.

**Exercise 1.10**  (*Multiscale approach*)  The function $y(t)$ is supposed to solve the initial value problem

$$y''(t) + 2\varepsilon\, y'(t) + (1 + \varepsilon^2)y(t) = 0, \quad y(0) = 0, \quad y'(0) = 1,$$

for $t > 0$ and a small parameter $\varepsilon > 0$.

(a)  Compute the approximation of the solution by means of formal asymptotic expansion up to first order in $\varepsilon$.

(b)  Compare the function obtained in (a) with the exact solution

$$y(t) = e^{-\varepsilon t} \sin t.$$

For which times $t$ the approximation from (a) is good?

(c)  To get a better approximation one can try the approach

$$y \sim y_0(t, \tau) + \varepsilon\, y_1(t, \tau) + \varepsilon^2 y_2(t, \tau) + \cdots,$$

here $\tau = \varepsilon t$ is a slow time scale.

Substitute this ansatz in the differential equation and compute $y_0$ such that the approximation becomes better.

*Hint:* The equation of lowest order does not determine $y_0$ uniquely and coefficient functions in $\tau$ appear. Choose them in a clever way such that $y_1$ is easily computable.

**Exercise 1.11**  (*Consistency versus convergence*)  For a parameter $\varepsilon \in [0, \varepsilon_0)$ with $\varepsilon_0 > 0$ we consider the family of operators

$$F(\cdot, \varepsilon) : B_1 := C_b^2([0, \infty)) \rightarrow B_2 := C_b^0([0, \infty)) \times \mathbb{R}^2,$$
$$F(y, \varepsilon) = (y'' + (1 + \varepsilon)y, y(0), y'(0) - 1).$$

Here $C_b^n([0, \infty))$ denotes the vector space of $n$-times differentiable functions with bounded derivatives up to order $n$. The norms of the spaces $B_1$ and $B_2$ are given by

$$\|y\|_{B_1} = \sup_{t \in (0,\infty)} \{|y(t)| + |y'(t)| + |y''(t)|\},$$

$$\|(f, a, b)\|_{B_2} = \sup_{t \in (0,\infty)} \{|f(t)|\} + |a| + |b|.$$

(a)  For the problem $F(y, \varepsilon) = (0, 0, 0)$ compute the exact solution $y_\varepsilon$.
(b)  Show: $F(\cdot, \varepsilon)$ is consistent with $F(\cdot, 0)$, but $y_\varepsilon$ does not converge to $y_0$ in $B_1$ as $\varepsilon \to 0$.

**Exercise 1.12**  Derive the friction law for the flow resistance in the case of high Reynolds numbers,

$$F_R = -\tfrac{1}{2} c_W A \varrho |v| v,$$

by means of a dimensional analysis. Use the assumptions that the frictional force depends on the density $\varrho$ of the liquid, a characteristic quantity $r$ of the body, and the velocity $v$ of the flow. As the drag coefficient depends on the shape of the body choosing $r$ such that $A \approx r^2$ is feasible.

# Chapter 2
# Systems of Linear Equations

Many simple models are based on linear relationships between different quantities. Problem formulations with several variables and linear relations between these variables lead to systems of linear equations. Also more complex processes described by nonlinear relationships between the relevant parameters often can be approximated by linear relationships with a sufficient degree of accuracy for practical applications. In this chapter we will analyze systems of linear equations arising from the description of electrical networks, considering both direct-current circuits and alternating-current circuits, and analyze the structure of the emerging models. A further important application stems from systems of pipelines, for example for the supply of houses or cities with water or gas, this application plays a role in the exercises. As an example from mechanics, space frames are treated. Large systems of linear equations also emerge from numerical discretizations of partial differential equations. Such systems of equations have many features in common with the systems described here.

## 2.1 Electrical Networks

Electrical networks belong to the most elementary building blocks of the modern world. They are essential for the public electric power supply, but also for the mode of operation of many, also small devices and machines. First we discuss the most simple case of an electrical network in the form of a direct-current circuit, which essentially consists of voltage sources or current sources and ohmic resistors. In particular in the beginning we do not consider any electrical components like capacitors, inductors, diodes, or transistors. The network shown in Fig. 2.1 may serve as a specific example.

Essentially a network consists of

- edges in the form of electrical lines,
- nodes, which are connecting points of two or more lines.

© Springer International Publishing AG 2017
C. Eck et al., *Mathematical Modeling*, Springer Undergraduate
Mathematics Series, DOI 10.1007/978-3-319-55161-6_2

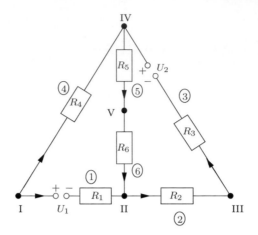

**Fig. 2.1** Electrical network

To describe the network we need the following information about the physics of electric currents:

> The *Kirchhoff current law*, also denoted as Kirchhoff's nodal rule: In every node the sum of all currents is zero. This describes the *conservation of electrical charge*, electrons may move through the network, but they cannot "vanish" in a node and also not be "generated" in a node.
>
> The *Kirchhoff voltage law*, also denoted Kirchhoff's mesh rule: The sum of voltages in every closed conductor loop is zero. This law implies the existence of *potentials* at the nodes, such that voltage applied to a piece of conductor is given by the difference of the potentials at its endpoints.
>
> *Ohm's law*: The voltage drop $U$ at a resistor with resistance $R$ for a given current $I$ equals $U = RI$.

We apply the unit system common in Europe, the so called SI-system (Système International d' Unités), meaning that current is measured in *ampère* (A), voltage is measured in *volt* (V), and the ohmic resistance in *ohm* ($\Omega$), where $1\Omega = (1\text{V})/(1\text{A})$. An ampère corresponds to the flux of one *coulomb* per second, and a coulomb corresponds to $6,24150965 \cdot 10^{18}$ elementary charges. Up to a few exceptions in the exercises, we will omit to note units in this chapter.

For the *modeling of a network* we define:

- A numbering of the nodes—for the example in Fig. 2.1 we have I–V, and equally a numbering of the edges, in Fig. 2.1 of 1–6. These numberings can be chosen arbitrarily. Of course they influence the specific representation of the models constructed from them and their solutions, but not the physical interpretation of these solutions.
- The determination of a *positive direction* for every edge. In Fig. 2.1 this direction is indicated by an arrow: ——▶——
  To call one direction positive of course does not indicate anything about the actual direction of the current which is part of the unknowns.

Furthermore we introduce variables, these are the currents and voltage drops along a conductor and the potentials in the nodes. Hence, there are two different types of variables:

- *Node variables*, namely the *potentials* $x_i$, $i = 1, ..., m$, where $m$ denotes the number of nodes, in the example $m = 5$. They are collected in the *potential vector* $x = (x_1, ..., x_m)^\top$ of dimension $m$.
- *Edge variables*, for example the *currents* $y_j$, $j = 1, ..., n$, where $n$ denotes the number of edges, in the example $n = 6$. They are put together to the *current vector* $y = (y_1, ..., y_n)^\top$ of dimension $n$. Correspondingly the vector $e = (e_1, ..., e_n)^\top$ of voltage drops can be formed. The voltage drops may be computed from the potentials by

$$e_i = x_{u(i)} - x_{o(i)}, \quad i = 1, ..., n, \tag{2.1}$$

where $u(i)$ is the index of the "lower" node and $o(i)$ the index of the "upper" node. "Lower" and "upper" is defined on the basis of the fixed direction of the piece of conductor. For example in Fig. 2.1 the node I is the lower one and II is the upper one of the conductor piece 1.

A necessary important step is to relate the geometry of the network and the physical laws which apply to the variables introduced. This is possible with the help of suitable *matrices*.

The relation (2.1) between potentials and voltage drops can be written in the form

$$e - -Bx,$$

where the matrix $B = \{b_{ij}\}_{i=1\,j=1}^{n\quad m} \in \mathbb{R}^{n,m}$ is defined by

$$b_{ij} = \begin{cases} 1 & \text{if } j = o(i), \\ -1 & \text{if } j = u(i) \text{ and} \\ 0 & \text{otherwise.} \end{cases}$$

In the example of Fig. 2.1 we have

$$B = \begin{pmatrix} -1 & 1 & 0 & 0 & 0 \\ 0 & -1 & 1 & 0 & 0 \\ 0 & 0 & -1 & 1 & 0 \\ -1 & 0 & 0 & 1 & 0 \\ 0 & 0 & 0 & -1 & 1 \\ 0 & 1 & 0 & 0 & -1 \end{pmatrix}.$$

The matrix $B$ is called the *incidence matrix* of the network. It describes solely the *geometry* of the network, more specifically the relations between *nodes* and *edges*. In particular the incidence matrix does not "know" the application for an electrical network. The same incidence matrix is obtained for other applications with the same

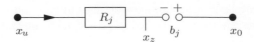

**Fig. 2.2** Illustration of the computation of the current

geometry, for example a system of pipelines. In every row the incidence matrix has exactly once the entry 1 and exactly once the entry $-1$, all other entries are 0. Of course it depends on the numbering of the nodes and of the edges and the directions fixed for the edges.

The relation between the voltage (drop) $e$ and the current $y$ is given by Ohm's law. Here we have to take into account possible voltage sources. Figure 2.2 shows a piece of a conductor with a resistor and a voltage source. For the potentials $x_u$, $x_z$, $x_o$ and the voltage drop $b_j$ at the voltage source, we have

$$x_z = x_u - R_j y_j \text{ and } x_o = x_z + b_j$$

and therefore

$$e_j = x_u - x_o = R_j y_j - b_j .$$

Here $R_j y_j$ denotes the voltage drop at the resistor and $b_j$ denotes the potential difference, generated from the voltage source. In vector notation one gets

$$y = C(e + b),$$

where $C \in \mathbb{R}^{n,n}$ is a *diagonal matrix*, whose entries are the *electric conductance values* $R_j^{-1}$, and $b \in \mathbb{R}^n$ denotes the vector of voltage sources. Concerning the entries of $b$ one has to pay attention to their sign: There has to be a positive sign, if the polarity from $-$ to $+$ corresponds to the positive direction. In our example we have

$$C = \begin{pmatrix} 1/R_1 & 0 & 0 & 0 & 0 & 0 \\ 0 & 1/R_2 & 0 & 0 & 0 & 0 \\ 0 & 0 & 1/R_3 & 0 & 0 & 0 \\ 0 & 0 & 0 & 1/R_4 & 0 & 0 \\ 0 & 0 & 0 & 0 & 1/R_5 & 0 \\ 0 & 0 & 0 & 0 & 0 & 1/R_6 \end{pmatrix} \text{ and } b = \begin{pmatrix} -U_1 \\ 0 \\ U_2 \\ 0 \\ 0 \\ 0 \end{pmatrix} .$$

We have not yet used the *Kirchhoff current law*. In vector notation it takes the form

$$Ay = 0,$$

where $A = \{a_{ij}\}_{i=1\,j=1}^{m\quad n} \in \mathbb{R}^{m,n}$ is defined by

$$a_{ij} = \begin{cases} +1 & \text{if } i = o(j), \\ -1 & \text{if } i = u(j) \text{ and} \\ 0 & \text{otherwise.} \end{cases}$$

In our example we have

$$A = \begin{pmatrix} -1 & 0 & 0 & -1 & 0 & 0 \\ 1 & -1 & 0 & 0 & 0 & 1 \\ 0 & 1 & -1 & 0 & 0 & 0 \\ 0 & 0 & 1 & 1 & -1 & 0 \\ 0 & 0 & 0 & 0 & 1 & -1 \end{pmatrix}.$$

Comparing this matrix with the incidence matrix $B$, one sees that

$$A = B^{\top}.$$

This is *no* coincidence, a comparison of the definitions of the entries $a_{ij}$ and $b_{ij}$ shows that this holds true *for every network*.

The Kirchoff voltage law has already been used in setting up the model by assuming the existence of a potential $x$. Thus we have used all known information about the network. Summing up the above considerations the modeling of an electrical network consisting of $m$ numbered nodes and $n$ numbered conductors with a prescribed direction is given by

- a potential vector $x \in \mathbb{R}^m$,
- a voltage drop vector $e \in \mathbb{R}^n$, which can be derived from the potential by

$$e = -Bx,$$

  using the incidence matrix $B \in \mathbb{R}^{n,m}$,
- a current vector $y \in \mathbb{R}^n$ related to the above quantities by

$$y = C(e + b)$$

  and the conductance matrix $C \in \mathbb{R}^{n,n}$ and the vector $b \in \mathbb{R}^n$ denoting the voltage sources, and finally
- the Kirchhoff current law in the form

$$B^{\top} y = 0.$$

Here the incidence matrix $B$ describes the geometry of the network, the conductance matrix $C$ the material properties and the vector $b$ the exterior "forces".

To compute the currents, voltage drops and the potentials in a network, one chooses a primary variable to be determined, for example $x$, and derives an equation for the primary variable by making use of the above relations. Here one obtains

$$B^\top C(b - Bx) = 0$$

or

$$B^\top CBx = B^\top Cb.  \tag{2.2}$$

The matrix $M = B^\top CB$ is symmetric, if $C$ is symmetric. From

$$\langle x, Mx \rangle = \langle Bx, CBx \rangle$$

with the Euclidean scalar product $\langle \cdot, \cdot \rangle$ one can conclude that $M$ is

- positive semidefinite, if $C$ is positive semidefinite,
- positive definite, if $C$ is positive definite and $B$ only has the trivial kernel Ker $B = \{0\}$.

For most networks in fact $C$ is positive definite. The incidence matrix $B$, however, has a nontrivial kernel. It is easy to be seen that $(1, \ldots, 1)^\top \in \mathbb{R}^m$ belongs to the kernel of $B$. This holds true for every incidence matrix, as always in every row we have exactly one entry $+1$ and one entry $-1$, all the other entries being 0. Consequently the system of linear equations has no unique solution. Physically the reason for this is easy to be understood: The potentials are unique up to a constant, i.e., they are only unique after a "zero point" is fixed for them.

As $B$ has a nontrivial kernel, it is not a priori obvious that the system of equations has a solution at all. However the following theorem holds true.

**Theorem 2.1** *Let $C \in \mathbb{R}^{n,n}$ be symmetric and positive definite and $B \in \mathbb{R}^{n,m}$. Then it holds for $M = B^\top CB$:*

*(i)  Ker $M$ = Ker $B$,*
*(ii)  For every $b \in \mathbb{R}^n$ the system of equations $Mx = B^\top b$ has a solution.*

*Proof* (i): Obviously we have Ker $B \subset$ Ker $M$. On the other hand for $x \in$ Ker $M$ it holds

$$0 = \langle x, B^\top CBx \rangle = \langle Bx, CBx \rangle.$$

As $C$ is positive definite we can conclude $Bx = 0$, meaning $x \in$ Ker $B$.
(ii): The equation is solvable, if we have $B^\top b \perp$ Ker $(M^\top)$. For $x \in$ Ker $(M^\top) =$ Ker $M =$ Ker $B$ we have

$$\langle B^\top b, x \rangle = \langle b, Bx \rangle = 0,$$

and thus the assertion is proved.                                                                     $\square$

In our example we have Ker $B = \text{span}\{(1, 1, 1, 1, 1, 1)^\top\}$. Hence, a system of equations with a positive definite matrix can be achieved by fixing one of the potential values. Setting $x_5 = 0$, one has to cancel the $5^{\text{th}}$ column in the incidence matrix, therefore the definition now reads

$$B = \begin{pmatrix} -1 & 1 & 0 & 0 \\ 0 & -1 & 1 & 0 \\ 0 & 0 & -1 & 1 \\ -1 & 0 & 0 & 1 \\ 0 & 0 & 0 & -1 \\ 0 & 1 & 0 & 0 \end{pmatrix},$$

and instead of $x \in \mathbb{R}^5$ now we have $x = (x_1, ..., x_4)^\top \in \mathbb{R}^4$. For the numerical example $R_1 = R_2 = R_3 = R_4 = R_5 = R_6 = 1$, $U_1 = 2$, $U_2 = 4$ we have $C = I \in \mathbb{R}^{6,6}$, the identity matrix, and $b = (-2, 0, 4, 0, 0, 0)^\top$, therefore

$$B^\top C B = B^\top B = \begin{pmatrix} 2 & -1 & 0 & -1 \\ -1 & 3 & -1 & 0 \\ 0 & -1 & 2 & -1 \\ -1 & 0 & -1 & 3 \end{pmatrix} \quad \text{and} \quad B^\top C b = B^\top b = \begin{pmatrix} 2 \\ -2 \\ -4 \\ 4 \end{pmatrix}.$$

The solution of the system of equations is

$$x = \begin{pmatrix} 1 \\ -1 \\ -2 \\ 1 \end{pmatrix}.$$

From the potentials we can compute the voltage drops and the currents,

$$e = -Bx = \begin{pmatrix} 2 \\ 1 \\ -3 \\ 0 \\ 1 \\ 1 \end{pmatrix} \quad \text{and} \quad y = e + b = \begin{pmatrix} 0 \\ 1 \\ 1 \\ 0 \\ 1 \\ 1 \end{pmatrix}.$$

An obvious question is, whether for the incidence matrix of a network we always have

$$\text{Ker } B = \text{span}\{(1, 1, ..., 1)^\top\}. \tag{2.3}$$

The answer to this question depends on the following geometric property of the network:

**Definition 2.2** A network is called *connected*, if every two nodes can be connected by a path consisting of edges.

It turns out that assertion (2.3) is equivalent to the connectivity of the network, more precisely we have:

**Theorem 2.3** *For a network with incidence matrix B the following assertions are equivalent:*

(i) *The network is connected.*

(ii) *By any reordering of the rows and the columns, B cannot be brought into the form*

$$B = \begin{pmatrix} B_1 & 0 \\ 0 & B_2 \end{pmatrix},$$

*where* $B_1 \in \mathbb{R}^{n_1,m_1}$, $B_2 \in \mathbb{R}^{n_2,m_2}$, $n_1, n_2, m_1, m_2 \geq 1$.

(iii) Ker $B = \text{span} \left\{ (1, 1, \ldots, 1)^\top \right\}$.

Equation (2.2) is not the only possibility to construct a system of linear equations from the physical relations between voltages and currents in a network. For example one can consider also $y$ besides $x$ as a variable to be computed. If one writes Ohm's law in the form

$$Ay = e + b \quad \text{with} \quad A = \text{diag}\,(R_1, \ldots, R_n) = C^{-1},$$

one obtains the system

$$Bx + Ay = b,$$
$$B^\top y = 0,$$

or

$$\begin{pmatrix} A & B \\ B^\top & 0 \end{pmatrix} \begin{pmatrix} y \\ x \end{pmatrix} = \begin{pmatrix} b \\ 0 \end{pmatrix}. \tag{2.4}$$

This formulation in particular make sense if one of the ohmic resistances equals zero, where the matrix $C$ is not well-defined anymore.

### Networks in Alternating Current Circuits

Now we want to extend the described model to alternating current circuits. In an alternating current circuit there are temporarily oscillating currents with a given frequency $\omega$. So we may assume that

$$I(t) = I_0 \cos(\omega t).$$

At an ohmic resistor the voltage drop is given by

$$U(t) = R\, I_0 \cos(\omega t) = U_0 \cos(\omega t) \quad \text{with} \quad U_0 = R\, I_0.$$

An alternating current circuit with ohmic resistors but without further components can be described as an direct current circuit, if one uses the *amplitudes* $I_0$ and $U_0$ for current and voltage instead of the constant currents and voltages in a direct current circuit. But further electrical components add new effects. Here we consider

- Capacitors, denoted with the symbol ⊣⊢. A capacitor can store electrical charges. The amount of the charge stored is proportional to the voltage applied. Therefore during voltage changes a capacitor can absorb or release currents. This is described by the relation

$$I(t) = C\,\dot{U}(t)\,,$$

where $C$ denotes the *capacity* of the capacitor. If we have an alternating current circuit with $I(t) = I_0 \cos(\omega t)$, then it follows

$$U(t) = \frac{I_0}{C\omega} \sin(\omega t) = \frac{I_0}{C\omega} \cos(\omega t - \pi/2)\,.$$

This means that we obtain a *phase shift* of $\pi/2$ between current and voltage.

- Inductors, denoted by the symbol ⎍⎍⎍. An inductor generates a magnetic field, whose magnitude is proportional to the magnitude of the current. A magnetic field stores energy, which must be taken during the setup of the magnetic field from the current of the inductor. This leads to a voltage drop at the inductor, which is proportional to the rate of change of the current,

$$U(t) = L\,\dot{I}(t)$$

where $L$ denotes the *inductivity* of the inductor. Therefore in an alternating current circuit we have

$$U(t) = -L I_0 \omega \sin(\omega t) = L\omega I_0 \cos(\omega t + \pi/2)\,.$$

Here one observes a phase shift of $-\pi/2$.

These possible phase shifts make the computation more complicated than for a direct current circuit. It is helpful to represent the current and the voltage by means of complex numbers. To this end we use *Euler's formula*

$$e^{i\varphi} = \cos\varphi + i\sin\varphi\,.$$

Then we have

$$\cos(\omega t) = \operatorname{Re}(e^{i\omega t}) \quad \text{and} \quad \sin(\omega t) = \operatorname{Re}(-ie^{i\omega t})\,.$$

If we represent the current in the form

$$I(t) = \operatorname{Re}(I_0 e^{i\omega t})\,,$$

then it follows for the voltage drop at an ohmic resistor

$$U(t) = \operatorname{Re}(R I_0 e^{i\omega t})\,,$$

at a capacitor

$$U(t) = \text{Re}\left(-\frac{i}{\omega C} I_0 e^{i\omega t}\right)$$

and at an inductor

$$U(t) = \text{Re}\left(i\omega L I_0 e^{i\omega t}\right).$$

All these relations can be represented in a unified way by introducing the complex *impedances*

$$R, \quad -\frac{i}{\omega C}, \text{ and } i\omega L,$$

which take the role of the real valued resistances. Similarly to the real valued ohmic resistances also the complex valued impedances just have to be added, if there is more than one component at the same piece of conductor. The total impedance of a conductor with an ohmic resistor with resistance $R$, a capacitor with capacity $C$, and an inductor with inductance $L$ therefore is

$$R - \frac{i}{\omega C} + i\omega L.$$

For a direct current circuit we were able to reduce the stationary, i.e., constant in time behavior (which results as a consequence of constant in time voltage or current sources) to the solution of a linear system of equation. Here for an alternating current circuit we obtain a very similar description: assuming now that the voltage and current sources behave as harmonic oscillators, possibly also with phase shifts, currents and voltages also turn out to be harmonic oscillations, possibly with phase shifts, which are computed besides the amplitudes.

**Example**: We consider the network displayed in Fig. 2.3 with $m = 5$ nodes and $n = 6$ edges with the numbering of nodes and edges and also with the positive directions of the edges as indicated there. We assume that voltages $U_1(t)$ and $U_2(t)$ are applied and given by

$$U_1(t) = U_{01} \cos(\omega t) \text{ and}$$
$$U_2(t) = U_{02} \cos(\omega t).$$

Here it is important that the frequencies are the same, because otherwise there would be no alternating current circuit with a fixed frequency and the reduction developed above would be invalid. Here also the phase shifts of $U_1$ and $U_2$ are the same, but it is not difficult to consider also different phases in the model (see below). For the modeling of the network we use

a *potential vector*    $x \in \mathbb{C}^m$,
a *current vector*      $y \in \mathbb{C}^n$, and
a *voltage vector*      $e \in \mathbb{C}^n$.

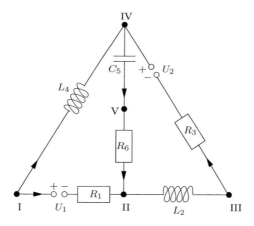

**Fig. 2.3**  Alternating current circuit

We have the relations

$$e = -Bx \quad \text{and}$$
$$y = C(e + b)$$

with the incidence matrix $B$ of a network, the impedance matrix $C^{-1}$ and the vector of the applied voltages $b$. The incidence matrix is the same as for the direct current circuit form Fig. 2.1. The impedance matrix and the vector of applied voltages are given by

$$C^{-1} = \begin{pmatrix} R_1 & 0 & 0 & 0 & 0 & 0 \\ 0 & i\omega L_2 & 0 & 0 & 0 & 0 \\ 0 & 0 & R_3 & 0 & 0 & 0 \\ 0 & 0 & 0 & i\omega L_4 & 0 & 0 \\ 0 & 0 & 0 & 0 & (i\omega C_5)^{-1} & 0 \\ 0 & 0 & 0 & 0 & 0 & R_6 \end{pmatrix} \quad \text{and} \quad b = \begin{pmatrix} -U_{01} \\ 0 \\ U_{02} \\ 0 \\ 0 \\ 0 \end{pmatrix}.$$

As for a direct current circuit again we get the system of equations (2.2), now with the only difference that the coefficients of $C$ and $b$ in general are complex numbers. Of course also for an alternating current circuit the alternative formulation (2.4) can be used.

For the specific example $R_1 = R_3 = R_6 = 1\,\Omega$, $L_2 = L_4 = 0,01\,H$, $C_5 = 0,02\,F$, $\omega = 50/s$ with $U_{01} = 10\,V$, $U_{02} = 5\,V$ we have, omitting the units,

$$C = \text{diag}\,(1, -2i, 1, -2i, i, 1) \quad \text{and} \quad b = (-10, 0, 5, 0, 0, 0)^{\top}.$$

The resulting system of equations $B^{\mathsf{T}}CBx = B^{\mathsf{T}}Cb$ has the form

$$
\begin{pmatrix}
1-2i & -1 & 0 & 2i & 0 \\
-1 & 2-2i & 2i & 0 & -1 \\
0 & 2i & 1-2i & -1 & 0 \\
2i & 0 & -1 & 1-i & -i \\
0 & -1 & 0 & -i & 1+i
\end{pmatrix}
x =
\begin{pmatrix}
10 \\ -10 \\ -5 \\ 5 \\ 0
\end{pmatrix},
$$

and the general solution of this system is

$$
x =
\begin{pmatrix}
2i \\ 2i-6 \\ 2i-5 \\ 0 \\ 4i-2
\end{pmatrix}
+ z
\begin{pmatrix}
1 \\ 1 \\ 1 \\ 1 \\ 1
\end{pmatrix}
\quad \text{where } z \in \mathbb{C}.
$$

From these charges the voltages and currents can be computed:

$$
e = -Bx =
\begin{pmatrix}
6 \\ -1 \\ 2i-5 \\ 2i \\ 2-4i \\ 4+2i
\end{pmatrix}
\quad \text{and} \quad
y = C(e+b) =
\begin{pmatrix}
-4 \\ 2i \\ 2i \\ 4 \\ 4+2i \\ 4+2i
\end{pmatrix}.
$$

From those complex numbers the temporal behavior of the described functions can be decoded now: A quantity given by a complex number $z \in \mathbb{C}$ leads to $z(t) = \mathrm{Re}(z\,e^{i\omega t})$, thus giving both the amplitude and possible phase shifts. As an example, the current in conductor piece number 5 is given by

$$
y_5(t) = \mathrm{Re}\big((4+2i)e^{i\omega t}\big) = 4\cos(\omega t) - 2\sin(\omega t).
$$

If some of the applied voltages also have a phase shift then the corresponding entry in the vector $b$ also is a complex number. For example $U_2(t) = U_{02}\sin(\omega t)$ can be written as $U_2(t) = \mathrm{Re}\big(-iU_{02}\,e^{i\omega t}\big)$ and therefore gives the right-hand side

$$
b =
\begin{pmatrix}
-U_{01} \\ 0 \\ -U_{02}i \\ 0 \\ 0 \\ 0
\end{pmatrix}.
$$

## 2.2 Space Frames

A further technically important example, which leads to systems of linear equations, is given by *elastic space frames*, which are also denoted in other notations as *frameworks* or *trusses*. A space frame is a structure, which consists of (usually "many") connected bars. A typical example is some type of bridge constructions shown schematically in Fig. 2.4, or some tower constructions, for example the Eiffel Tower (here one speaks of *lattice towers*). The construction is deformed under loading, while its material is *elastic*, meaning in particular that the "original" form is regained after removal of the load. The deformation is of importance to determine the "distribution of forces" within the construction, but quite often it is so small that it is hardly visible to the naked eye. In such a case the assumption of a *linear* model often makes sense as it shall be done here also.

For the modeling of a space frame first we consider a *single bar*, as depicted in Fig. 2.5. For simplicity we will specifically consider *two-dimensional* problems. The geometry of the bar is given by its length $L$ and the angle $\theta$ with respect to a fixed direction, for example to a previously fixed $x_1$-axis.

We make the following *modeling assumptions*:

- The bar can only be loaded in *longitudinal direction*, not in transversal direction. This assumption can be justified for example in the following situations:

  - The bar is pivot-mounted in a *frictionless* manner such that every transversal force immediately leads to a rotary motion.
  - The bar is very *thin*, such that every transversal force leads to very large deformation and the structure essentially gains its rigidity by the longitudinal forces.

- Under loading the bar is deformed, i.e., stretched or compressed, the deformation is *proportional to the load*. This is called a *linear material behavior*.

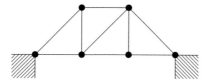

**Fig. 2.4** A simple space frame

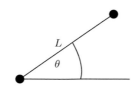

**Fig. 2.5** Geometry of a single bar

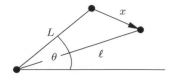

**Fig. 2.6** Deformation of a bar

Now we want to compute the *deformation* of a bar for a given *displacement* of the bar's ends. The displacement is defined by a *displacement vector* $x = (x_1, x_2)^\top$, see Fig. 2.6.

The length $\ell$ of the deformed bar is given by

$$
\begin{aligned}
\ell^2 &= (L \cos \theta + x_1)^2 + (L \sin \theta + x_2)^2 \\
&= L^2 + 2L(\cos \theta\, x_1 + \sin \theta\, x_2) + x_1^2 + x_2^2 .
\end{aligned}
$$

The *force* acting on the bar is supposed to be proportional to the (relative) *strain*

$$
\tilde{e} = \frac{\ell - L}{L} .
$$

However, in general the strain is a *non-linear* function of the displacement vector. But if the displacement $(x_1, x_2)^\top$ is *small* compared with the length, i.e., $\sqrt{x_1^2 + x_2^2} \ll L$, then the strain can be *linearized* according to

$$
\begin{aligned}
\tilde{e} &= (\ell - L)/L = \sqrt{1 + 2L^{-1}(\cos \theta\, x_1 + \sin \theta\, x_2) + L^{-2}(x_1^2 + x_2^2)} - 1 \\
&= 1 + L^{-1}\big(\cos \theta\, x_1 + \sin \theta\, x_2\big) - 1 + \mathcal{O}\big((x_1^2 + x_2^2)/L^2\big) \\
&\approx L^{-1}\big(\cos \theta\, x_1 + \sin \theta\, x_2\big) .
\end{aligned}
$$

Thus one gets a *linear* relation between the displacement vector and the strain. This procedure is called *geometric* linearization. Hence, the linear model does not describe the reality exactly, but only approximately. We have taken into account a (hopefully small) *modeling error*.

For the deformation of a bar a force $y$ is necessary acting at *both* bar ends. For *linear elastic* material this force is proportional to the strain,

$$
y = \tilde{E}\,\tilde{e} ,
$$

where $\tilde{E}$ is a constant, which depends on the *material* and the *thickness* (or more precisely the cross-sectional area) of the bar. The force $y$ is *positive*, if the bar is *stretched*, and *negative*, if the bar is *compressed*. According to Newton's law "actio = reactio" concerning forces one always has to pay attention, "who" exerts the force at "whom" and "in which direction". In Fig. 2.7 this is illustrated for a stretched bar.

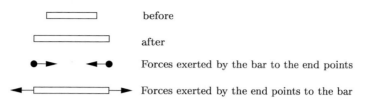

**Fig. 2.7** Forces in a bar under tensile load

The force $y$ defined in this way is also denoted (in the here considered one-dimensional case) as (elastic) *stress*. The stress in a stretched or a compressed bar is positive or negative, respectively.

For a bar in a space frame typically *both* end points will be displaced. If one defines the displacement vector of the "left" end point in Fig. 2.6 to be $(x_1, x_2)^\top$ and the one of the "right" end point to be $(x_3, x_4)^\top$, then the (absolute) deformation $e$ is

$$e = L\widetilde{e} = \cos\theta\,(x_3 - x_1) + \sin\theta\,(x_4 - x_2). \tag{2.5}$$

Then the relation between absolute deformation and stress is

$$y = Ee \quad \text{where} \quad E = \widetilde{E}/L. \tag{2.6}$$

These are the modeling equations for a single bar. The proportionality constant $E$ in the following is also called *modulus of elasticity*.

To model a full space frame one numbers the bars and also the nodes of the space frame and then one defines:

- a (global) *displacement vector* $x \in \mathbb{R}^{2m}$, where $m$ denotes the number of (freely moveable) nodes of the space frame. Here $(x_{2i-1}, x_{2i})$ is the displacement vector of the node with the number $i$,
- a vector of *strains* $e \in \mathbb{R}^n$, where $n$ denotes the number of bars,
- a vector of *stresses* $y \in \mathbb{R}^n$.

Then *global* versions of the model equations (2.5) and (2.6) are defined by means of matrix-vector multiplications, i.e., global relations between displacements and strains

$$e = Bx$$

with a matrix $B \in \mathbb{R}^{n,2m}$, and a global relation between strains and forces,

$$y = Ce$$

with a matrix $C \in \mathbb{R}^{n,n}$ consisting of material parameters. Here $C$ is a diagonal matrix. The entries of $B$ and $C$ are obtained by sorting of the coefficients in (2.5) and (2.6) to the appropriate positions, depending on the numbering of the nodes and of the bars. In order to end up with a system of equations to determine the displacements,

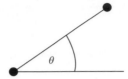

**Fig. 2.8**  Force on a single bar

stresses and stains, a further physical law is necessary, namely the *balance of forces*:

The sum of all forces acting at a specific point is zero.

To express this law in a suitable matrix-vector form, first we consider the force which is exerted from an end point of a single bar having the stress $y$ to this bar. In the situation of Fig. 2.8 this is the force which the upper right point of the bar depicted exerts at this bar.

This force is given by

$$f = \begin{pmatrix} \cos\theta \\ \sin\theta \end{pmatrix} y =: Ay\,, \quad \text{where} \quad A = \begin{pmatrix} \cos\theta \\ \sin\theta \end{pmatrix} \in \mathbb{R}^{2,1}\,.$$

By comparison with the relation between displacement and strain we get

$$e = \cos\theta\, x_1 + \sin\theta\, x_2 =: B \begin{pmatrix} x_1 \\ x_2 \end{pmatrix}\,, \quad \text{where} \quad B = (\cos\theta\ \ \sin\theta) \in \mathbb{R}^{1,2}\,,$$

and we see that $A = B^\top$. This relationship can be transferred from a single bar to an arbitrary space frame:

$$f = B^\top y\,.$$

This equation describes the *internal forces*, which appear as a consequence of the stresses of the bars. In addition there are also *exterior* forces, for example by vehicles on a bridge or people on a tower. Also the gravitational force acting on the construction, i.e., its weight, is an exterior force. These forces will be distributed to *point forces at the nodes* and put together to a force vector $b \in \mathbb{R}^{2m}$. The balance of forces is then described by

$$B^\top y = b\,.$$

In this equation $B^\top y$ describes the forces, which the points exert at the bars, on the other hand $b$ describes the forces exerted at the points from the outside. This explains the different signs respectively the positions left and right of the equality sign. By putting together all three relations one obtains a system of equations for the displacement vector $x$:

$$B^\top C B x = b\,.$$

Now we will explain the theoretical concept by means of several examples and identify possible difficulties showing up.

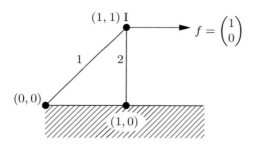

**Fig. 2.9** Example 1

## *Example 1*

We consider the space frame depicted in Fig. 2.9 with the numbering given there, in roman digits for the connecting nodes and in arabic digits for the bars and with the sketched exterior force. The pairs of numbers describe Cartesian coordinates of the points. The space frame consists of two bars and three nodes, two of the nodes, however, are fixed and therefore are not included into the numbering. The variables are the displacement vector $x = (x_1, x_2)^\top$, the strain vector $e = (e_1, e_2)^\top$, and the stress vector $y = (y_1, y_2)^\top$. The displacement-strain relation reads

$$e = Bx \text{ , where } B = \begin{pmatrix} \sqrt{2}/2 & \sqrt{2}/2 \\ 0 & 1 \end{pmatrix}.$$

The strain-stress relation reads

$$y = \begin{pmatrix} E_1 e_1 \\ E_2 e_2 \end{pmatrix} = \begin{pmatrix} E_1 & 0 \\ 0 & E_2 \end{pmatrix} e = Ce \text{ , where } C = \begin{pmatrix} E_1 & 0 \\ 0 & E_2 \end{pmatrix}.$$

The balance of forces at node I is given by

$$\begin{pmatrix} \sqrt{2}/2 \\ \sqrt{2}/2 \end{pmatrix} y_1 + \begin{pmatrix} 0 \\ 1 \end{pmatrix} y_2 = \begin{pmatrix} 1 \\ 0 \end{pmatrix}$$

or

$$B^\top y = f .$$

For the specific choice of material constants

$$E_1 = 100, \ E_2 = 200$$

it follows

$$\begin{aligned} B^\top C B &= \begin{pmatrix} \sqrt{2}/2 & 0 \\ \sqrt{2}/2 & 1 \end{pmatrix} \begin{pmatrix} 100 & 0 \\ 0 & 200 \end{pmatrix} \begin{pmatrix} \sqrt{2}/2 & \sqrt{2}/2 \\ 0 & 1 \end{pmatrix} \\ &= \begin{pmatrix} 50\sqrt{2} & 0 \\ 50\sqrt{2} & 200 \end{pmatrix} \begin{pmatrix} \sqrt{2}/2 & \sqrt{2}/2 \\ 0 & 1 \end{pmatrix} = \begin{pmatrix} 50 & 50 \\ 50 & 250 \end{pmatrix}. \end{aligned}$$

Therefore one gets the system of equations

$$\begin{pmatrix} 50 & 50 \\ 50 & 250 \end{pmatrix} x = \begin{pmatrix} 1 \\ 0 \end{pmatrix}$$

with the unique solution

$$x = \begin{pmatrix} 5/200 \\ -1/200 \end{pmatrix}.$$

From these displacements the strains and stresses can be computed according to

$$e = Bx = \begin{pmatrix} \sqrt{2}/100 \\ -1/200 \end{pmatrix} \quad \text{and} \quad y = Ce = \begin{pmatrix} \sqrt{2} \\ -1 \end{pmatrix}.$$

In this example the distribution of forces to the two bars could also be computed by a much simpler model: To absorb the exterior force vector $f = (1, 0)^\top$ only two directions are available, namely

$$a = \begin{pmatrix} \sqrt{2}/2 \\ \sqrt{2}/2 \end{pmatrix} \quad \text{and} \quad b = \begin{pmatrix} 0 \\ 1 \end{pmatrix}.$$

As $\{a, b\}$ is the basis of $\mathbb{R}^2$, the force $f$ can be uniquely decomposed into

$$f = \alpha a + \beta b,$$

where $\alpha = \sqrt{2}$ and $\beta = -1$. These coefficients are just the stresses.

**Definition 2.4** A space frame in which there is only one possibility of distribution of forces to the bars, is called *statically determinate*.

In our model the distribution of forces to the bars is given by the solution $y$ of the equation

$$B^\top y = f. \tag{2.7}$$

A space frame is statically determinate if, and only if, the set of equations (2.7) is uniquely solvable for an arbitrary right-hand side. This in turn is the case if, and only if, $B^\top$ (or $B$) is a *quadratic, regular* matrix. In particular the number of degrees of freedom for the displacement (here $2m = 2$) has to coincide with the number of degrees of freedom of the stresses (here $n = 2$),

$$2m = n.$$

Denoting with $m$ the number of *all* nodes, with $n$ the number of bars and with $k$ the number of constraints, then the condition reads

$$2m = n + k. \tag{2.8}$$

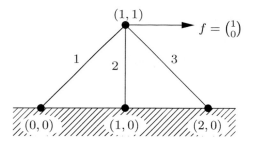

**Fig. 2.10** Example 2

A constraint describes the fixation of a node in one direction. If we also take into account the two fixed nodes in the numbering of Example 1, then we have $m = 3$, $n = 2$, and $k = 4$. An important form of constraint consists in the fixation of a node only in one direction (or in two directions for a three dimensional problem). Technically this corresponds to the sliding of a node in a rail and ignoring possible friction. For spatially three dimensional problems Eq. (2.8) has to be changed to

$$3m = n + k \, ,$$

because every node corresponds to three degrees of freedom now.

### *Example 2*

We consider the space frame from Fig. 2.10 with the material data $E_1 = 200$, $E_2 = 100$, $E_3 = 200$, where $E_i$ denotes the modulus of elasticity for bar $i$.

For the displacement vector $x \in \mathbb{R}^2$, the strain vector $e \in \mathbb{R}^3$, and stress vector $y \in \mathbb{R}^3$ we have the relations

$$e = Bx \, , \text{ where } B = \begin{pmatrix} \sqrt{2}/2 & \sqrt{2}/2 \\ 0 & 1 \\ -\sqrt{2}/2 & \sqrt{2}/2 \end{pmatrix}$$

and

$$y = Ce \, , \text{ where } C = \begin{pmatrix} 200 & 0 & 0 \\ 0 & 100 & 0 \\ 0 & 0 & 200 \end{pmatrix} .$$

Therefore we have

$$B^{\mathsf{T}}CB = \begin{pmatrix} \sqrt{2}/2 & 0 & -\sqrt{2}/2 \\ \sqrt{2}/2 & 1 & \sqrt{2}/2 \end{pmatrix} \begin{pmatrix} 200 & 0 & 0 \\ 0 & 100 & 0 \\ 0 & 0 & 200 \end{pmatrix} \begin{pmatrix} \sqrt{2}/2 & \sqrt{2}/2 \\ 0 & 1 \\ -\sqrt{2}/2 & \sqrt{2}/2 \end{pmatrix}$$

$$= \begin{pmatrix} 100\sqrt{2} & 0 & -100\sqrt{2} \\ 100\sqrt{2} & 100 & 100\sqrt{2} \end{pmatrix} \begin{pmatrix} \sqrt{2}/2 & \sqrt{2}/2 \\ 0 & 1 \\ -\sqrt{2}/2 & \sqrt{2}/2 \end{pmatrix} = \begin{pmatrix} 200 & 0 \\ 0 & 300 \end{pmatrix} .$$

Thus the system of equation reads

$$\begin{pmatrix} 200 & 0 \\ 0 & 300 \end{pmatrix} x = \begin{pmatrix} 1 \\ 0 \end{pmatrix}.$$

It has the unique solution

$$x = \begin{pmatrix} 1/200 \\ 0 \end{pmatrix}.$$

The strains and stresses are obtained from

$$e = Bx = \begin{pmatrix} \sqrt{2}/400 \\ 0 \\ -\sqrt{2}/400 \end{pmatrix} \quad \text{and} \quad y = Ce = \begin{pmatrix} \sqrt{2}/2 \\ 0 \\ -\sqrt{2}/2 \end{pmatrix}.$$

Here the space frame is not statically determinate, because $B$ is not a quadratic matrix. There are more bars than degrees of freedom for displacements. Therefore a simple decomposition of the applied force into the available directions is here not possible and hence it is not possible to achieve a solution directly. Nevertheless the model possesses a unique solution, because $B^{\mathrm{T}}CB$ is regular. This is caused by the triviality of Ker $B$, i.e., the *maximality of the rank* of $B$. For a positive definite matrix $C$ it holds according to Theorem 2.1 that Ker $(B^{\mathrm{T}}CB) = $ Ker $B$, and therefore $B^{\mathrm{T}}CB$ is regular, if $C$ is positive definite and $B$ has maximal column rank.

**Definition 2.5**  A space frame is called *statically indeterminate*, if $B \in \mathbb{R}^{n,2m}$ such that $n > 2m$ and $B$ has maximal rank, i.e., $2m$.

*Example 3*

We investigate the space frame from Fig. 2.11 with the data $E_1 = E_2 = E_3 = 100$. In the displacement vector $x \in \mathbb{R}^4$ the components $(x_1, x_2)^{\mathrm{T}}$ describe the displacement of node I and $(x_3, x_4)^{\mathrm{T}}$ the displacement of node II.

We have

$$e = Bx \text{ with } B = \begin{pmatrix} 0 & 1 & 0 & 0 \\ -1 & 0 & 1 & 0 \\ 0 & 0 & 0 & 1 \end{pmatrix} \quad \text{and} \quad y = Ce \text{ with } C = \begin{pmatrix} 100 & 0 & 0 \\ 0 & 100 & 0 \\ 0 & 0 & 100 \end{pmatrix}.$$

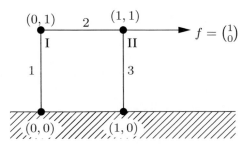

**Fig. 2.11**  Example 3

Therefore it holds

$$B^\top CB = 100 \begin{pmatrix} 0 & -1 & 0 \\ 1 & 0 & 0 \\ 0 & 1 & 0 \\ 0 & 0 & 1 \end{pmatrix} \begin{pmatrix} 0 & 1 & 0 & 0 \\ -1 & 0 & 1 & 0 \\ 0 & 0 & 0 & 1 \end{pmatrix} = 100 \begin{pmatrix} 1 & 0 & -1 & 0 \\ 0 & 1 & 0 & 0 \\ -1 & 0 & 1 & 0 \\ 0 & 0 & 0 & 1 \end{pmatrix}.$$

Thus, the set of equations reads

$$100 \begin{pmatrix} 1 & 0 & -1 & 0 \\ 0 & 1 & 0 & 0 \\ -1 & 0 & 1 & 0 \\ 0 & 0 & 0 & 1 \end{pmatrix} x = \begin{pmatrix} 0 \\ 0 \\ 1 \\ 0 \end{pmatrix}.$$

Considering its first and third row, it is obvious that this system of equations has no solutions. The reason for this can be easily understood in mechanical terms: If the bars are connected in such a way that they are "frictionless pivot-mounted" then the space frame "tilts" to the right, as depicted in Fig. 2.12. Due to its construction it is unable to absorb certain forces. This of course is also a consequence of our modeling assumptions: The same space frame would be able to absorb the force $f$, if the single bars would be able to absorb loads perpendicular to the bar direction, and the links between the bars could transmit torsional moments. But such a situation is not reflected by our model.

A *badly constructed* space frame is here recognized mathematically by a *non-solvable* model equation.

The system matrix $B^\top CB$ is not invertible here, as $B$ has a nontrivial kernel. This motivates the following definition:

**Definition 2.6**  A space frame is called *unstable*, if $B$ has linear dependent columns.

Also for unstable space frames the linear system of equation may have a solution. This is the case, if and only if the vector of exterior forces is orthogonal to the kernel of $B^\top CB$, which for a positive definite $C$ equals the kernel of $B$ according to Theorem 2.1. In our example the condition $b \perp \mathrm{Ker}\ B$ is equivalent to

$$b_1 + b_3 = 0.$$

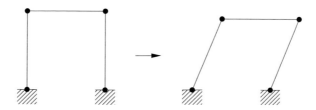

**Fig. 2.12**  Unstable space frame

As $(b_1, b_2)^\top$ and $(b_3, b_4)^\top$ are the forces at the nodes I and II, this condition means that the *resulting force in $x_1$-direction* equals zero.

### Example 4

We consider the space frame in Fig. 2.13, which has no fixed nodes. In particular such a space frame is freely movable in space. As material data we chose $E_1 = E_2 = E_3 = 100$. The global displacement vector $x = (x_1, \ldots, x_6)^\top \in \mathbb{R}^6$ is composed of the local displacement vectors $(x_{2i-1}, x_{2i})^\top$ of the nodes $i = 1, 2, 3$. The strain vector $e \in \mathbb{R}^3$ is given by

$$e = \begin{pmatrix} x_3 - x_1 \\ x_6 - x_2 \\ \sqrt{2}/2 \, (x_3 - x_5 - x_4 + x_6) \end{pmatrix} = Bx \,,$$

$$\text{and} \quad B = \begin{pmatrix} -1 & 0 & 1 & 0 & 0 & 0 \\ 0 & -1 & 0 & 0 & 0 & 1 \\ 0 & 0 & \sqrt{2}/2 & -\sqrt{2}/2 & -\sqrt{2}/2 & \sqrt{2}/2 \end{pmatrix}.$$

Therefore we have

$$B^\top B = \frac{1}{2} \begin{pmatrix} 2 & 0 & -2 & 0 & 0 & 0 \\ 0 & 2 & 0 & 0 & 0 & -2 \\ -2 & 0 & 3 & -1 & -1 & 1 \\ 0 & 0 & -1 & 1 & 1 & -1 \\ 0 & 0 & -1 & 1 & 1 & -1 \\ 0 & -2 & 1 & -1 & -1 & 3 \end{pmatrix}.$$

For a vector $f \in \mathbb{R}^6$ denoting the applied forces, the linear equation therefore is

$$B^\top B x = \frac{1}{100} f \,. \tag{2.9}$$

It is easily to be seen that this matrix is not regular, as for example the fourth and fifth column are identical. Hence, the system of equations does not have a solution

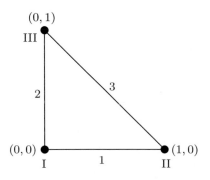

**Fig. 2.13** Example 4

for every right-hand side $f$ and in case of solvability the solution is not unique. As already used in Example 3 the characterization is

$$B^{\mathsf{T}} B x = f \text{ is solvable } \Leftrightarrow f \perp \text{Ker } (B^{\mathsf{T}} B) \Leftrightarrow f \perp \text{Ker } B.$$

The kernel of the matrices $B^{\mathsf{T}} B$ or $B$, respectively, is

$$\text{Ker } (B) = \text{span} \left\{ \begin{pmatrix} 1 \\ 0 \\ 1 \\ 0 \\ 1 \\ 0 \end{pmatrix}, \begin{pmatrix} 0 \\ 1 \\ 0 \\ 1 \\ 0 \\ 1 \end{pmatrix}, \begin{pmatrix} 0 \\ 0 \\ 0 \\ 1 \\ -1 \\ 0 \end{pmatrix} \right\} =: \mathcal{R}.$$

The first two vectors here describe *displacements* of the whole space frame in the $x_1$- and the $x_2$-direction. The third vector is called *rotation*. More precisely a rotation of the space frame around the point zero is given, if every node $p \in \mathbb{R}^2$ of the space frame is mapped to

$$p + x(p) = \begin{pmatrix} \cos \varphi & -\sin \varphi \\ \sin \varphi & \cos \varphi \end{pmatrix} p,$$

where $\varphi$ denotes the rotation angle. Substitution of the node vectors into these local relations and transferring the results to the global displacement vector $x$ leads to

$$x = x(\varphi) = \begin{pmatrix} 0 \\ 0 \\ \cos \varphi \\ \sin \varphi \\ -\sin \varphi \\ \cos \varphi \end{pmatrix} - \begin{pmatrix} 0 \\ 0 \\ 1 \\ 0 \\ 0 \\ 1 \end{pmatrix}.$$

In the linearized theory we only consider *small* displacements, therefore here only small rotation angles $\varphi$. Linearization of $x = x(\varphi)$ at $\varphi = 0$ leads to

$$x(\varphi) \approx x(0) + x'(0)\varphi = \varphi \begin{pmatrix} 0 \\ 0 \\ 0 \\ 1 \\ -1 \\ 0 \end{pmatrix}.$$

Therefore the apparently strange definition of a rotation is a consequence of the modeling simplifications.

The set $\mathcal{R}$ is called the set of *rigid body motions* because it describes those displacement vectors, which do *not* lead to strains of the bars. Of course the specific definition depends on the space frame considered, essentially on its nodes and their numbering. The set of rigid body motions can be constructed from the three local mappings

$$x(p) = \begin{pmatrix} 1 \\ 0 \end{pmatrix}, \quad x(p) = \begin{pmatrix} 0 \\ 1 \end{pmatrix} \text{ and } x(p) = \begin{pmatrix} -p_2 \\ p_1 \end{pmatrix}. \tag{2.10}$$

In summary, the linear system of equations (2.9) has a solution if and only if $f$ is perpendicular to the set of rigid body motions,

$$f \perp \mathcal{R}.$$

Physically this relation means that the *sum of all acting forces* equals zero, and that the acting *torsional moment*, for example with the zero point as center of rotation, also equals zero. Then the solution is not unique, two solutions vary by a rigid body motion. This situation holds true for every space frame without fixed nodes.

These considerations can also be transferred to three space dimensions. Then $\mathcal{R}$ has the dimension 6, as then there are three translations and three rotations. Analogously to (2.10) the rigid body motions can be obtained from the local mappings

$$x(p) = \begin{pmatrix} 1 \\ 0 \\ 0 \end{pmatrix}, \quad x(p) = \begin{pmatrix} 0 \\ 1 \\ 0 \end{pmatrix}, \quad x(p) = \begin{pmatrix} 0 \\ 0 \\ 1 \end{pmatrix},$$

$$x(p) = \begin{pmatrix} -p_2 \\ p_1 \\ 0 \end{pmatrix}, \quad x(p) = \begin{pmatrix} -p_3 \\ 0 \\ p_1 \end{pmatrix}, \text{ and } x(p) = \begin{pmatrix} 0 \\ -p_3 \\ p_2 \end{pmatrix}.$$

## 2.3   Constrained Optimization

The systems of equations derived in the last two sections by modeling electrical networks and elastic frame works all have the same structure.

$$\begin{aligned} Ay + Bx &= b, \\ B^\top y &= f, \end{aligned} \tag{2.11}$$

where $A \in \mathbb{R}^{n,n}$, $B \in \mathbb{R}^{n,m}$ are given matrices, $b \in \mathbb{R}^n$, $f \in \mathbb{R}^m$ are given vectors, and $y \in \mathbb{R}^n$, $x \in \mathbb{R}^m$ denote the unknown vectors. These equations also characterize the solution $y$ of a *quadratic optimization problem with linear constraints*.

There is a close connection between this constrained minimization problem for $y$, the *primal problem*, and a non-constrained maximization problem for $x$, the *dual problem*, and a minimum–maximum-formulation for $(y, x)$, such that (2.11) may also be viewed as a *saddle point problem*. Since the constraints under consideration are linear, the methods of linear algebra are sufficient.

We consider a *quadratic* optimization problem with *linear* constraints in the form

$$\text{Minimize } F(y) := \tfrac{1}{2}\langle y, Ay \rangle - \langle b, y \rangle, \text{ where } y \in \mathbb{R}^n, B^\top y = f. \qquad (2.12)$$

Here $A \in \mathbb{R}^{n,n}$ denotes a symmetric, positive definite matrix, and furthermore we have, $B \in \mathbb{R}^{n,m}, b \in \mathbb{R}^n, f \in \mathbb{R}^m$ as the given data. Here $\langle x, y \rangle := x^\top y$ for $x, y \in \mathbb{R}^n$ denotes the Euclidean scalar product.

We will frequently make use of the following basic results of linear algebra:

**Theorem 2.7** *Let $A \in \mathbb{R}^{n,n}$ be symmetric, positive semidefinite, $b \in \mathbb{R}^n$. Then the following two assertions are equivalent:*

*(i)  $y \in \mathbb{R}^n$ is the solution of $Ay = b$,*
*(ii)  $y \in \mathbb{R}^n$ minimizes the functional*

$$F(y) := \frac{1}{2}\langle y, Ay \rangle - \langle b, y \rangle \text{ in } \mathbb{R}^n.$$

*If $A$ is positive definite, then both problems are uniquely solvable.*

**Theorem 2.8** *Let $( , )$ be a scalar product in $\mathbb{R}^n$ and $\| . \|$ the induced norm. Furthermore let $U \subset \mathbb{R}^n$ be a linear subspace and $W := \widetilde{y} + U$ for some $\widetilde{y} \in \mathbb{R}^n$. Then for $\widehat{y} \in \mathbb{R}^n$ the following assertions are equivalent:*

*(i)  $\overline{y} \in W$ minimizes $y \mapsto \|\widehat{y} - y\|$ in $W$.*
*(ii)  $(\overline{y} - \widehat{y}, u) = 0$ for $u \in U$ (error orthogonality).*

Hence, it follows:

**Theorem 2.9** *Let $A \in \mathbb{R}^{n,n}$ be symmetric, positive definite and $\overline{y} \in \mathbb{R}^n$. Let us assume that the linear system of equations $B^\top y = f$ possesses a solution. Then the following assertions are equivalent:*

*(i)  $\overline{y} \in \mathbb{R}^n$ is a solution of (2.12).*
*(ii)  There exists a so called* Lagrange *multiplier $\overline{x} \in \mathbb{R}^m$, such that $(\overline{y}, \overline{x})$ is a solution of the system of equations (2.11).*

*The solutions $\overline{y}, (\overline{y}, \overline{x})$, respectively, exist, $\overline{y}$ always is unique, and $\overline{x}$ is unique if $B$ has full column rank.*

*Proof* Let $U := \text{Ker } B^\top$ and let $\widetilde{y} \in \mathbb{R}^n$ be a particular solution of $B^\top y = f$. Then the set of constraints $\{ y \in \mathbb{R}^n \mid B^\top y = f \}$ in (2.12) is the affine subspace

$$W := \widetilde{y} + U.$$

Now we use

$$\|y\|_A := (y^\top Ay)^{1/2} \text{ for } y \in \mathbb{R}^n,$$

i.e., the norm induced by the scalar product $\langle x, y \rangle_A := x^\top A y$ for $x, y \in \mathbb{R}^n$, the *energy norm* belonging to $A$.

Because of

$$\frac{1}{2} y^\top A y - b^\top y = \frac{1}{2} \| y - \widehat{y} \|_A^2 - \frac{1}{2} b^\top \widehat{y}$$

for $\widehat{y} := A^{-1} b$, a formulation equivalent to (2.12) is:

$$\text{Minimize } \widetilde{f}(y) = \| y - \widehat{y} \|_A \text{ for } y \in W. \tag{2.13}$$

The minimal value $\overline{y} \in \mathbb{R}^n$ of (2.13) and (2.12), which exists uniquely, can be characterized according to Theorem 2.8 by

$$
\begin{aligned}
& \langle \overline{y} - \widehat{y}, u \rangle_A = 0 \quad \text{for} \quad u \in U \\
\Leftrightarrow \ & \langle A\overline{y} - b, u \rangle = 0 \quad \text{for} \quad u \in U \\
\Leftrightarrow \ & A\overline{y} - b \in U^\perp = (\text{Ker } B^\top)^\perp = \text{Im } B \\
\Leftrightarrow \ & \text{There exists } \overline{x} \in \mathbb{R}^m \text{ such that } A\overline{y} - b = B(-\overline{x}).
\end{aligned}
$$

The preimage $\overline{x}$ is unique if and only if $B$ is injective, i.e., if it has maximal column rank. □

The linear set of equations (2.11) in $\begin{pmatrix} y \\ x \end{pmatrix}$ has block form. Note that we cannot solve the first equation of (2.11) for $x$ and eliminate $x$ thereby. However, we can solve that equation for $y$ and then obtain a linear system only for the Lagrange multiplier $x$ (which is not uniquely solvable).

**Theorem 2.10** *Under the assumptions of Theorem 2.9 the last of the equivalent assertions given there can be extended to*

*(iii)* $\overline{x} \in \mathbb{R}^m$ *is a solution of*

$$B^\top A^{-1} B x = -f + B^\top A^{-1} b \tag{2.14}$$

*and* $\overline{y} \in \mathbb{R}^n$ *is the unique solution of*

$$A y = b - B\overline{x}. \tag{2.15}$$

*(iv)* $\overline{x} \in \mathbb{R}^m$ *is a solution of the maximization problem*

$$
\begin{aligned}
\text{Maximize } F^*(x) := & -\frac{1}{2} \langle B^\top A^{-1} B x, x \rangle + \langle x, B^\top A^{-1} b - f \rangle \\
& -\frac{1}{2} \langle b, A^{-1} b \rangle
\end{aligned}
\tag{2.16}
$$

*and $\overline{y} \in \mathbb{R}^n$ is the unique solution of*

$$Ay = b - B\overline{x}. \tag{2.17}$$

*The maximization problem* (2.16) *is also denoted as the* dual *problem for* (2.12).

*Proof* (ii)⇒(iii): This is an immediate consequence by solving the first equation of (2.11) for $\overline{y}$ and inserting the result into the second equation.

(iii)⇒(ii): If $\overline{x}$ is a solution of (2.14), then we define $\overline{y}$ to be the solution of (2.15). Elimination of $B\overline{x}$ in (2.14) leads to the assertion.

(iii)⇔(iv): As $B^\top A^{-1} B$ is symmetric and positive semidefinite, according to Theorem 2.7 Eq. (2.14) is equivalent to a minimization problem with the functional $-F^*(x) - \frac{1}{2}\langle b, A^{-1}b \rangle$, which in turn is equivalent to the maximization problem (2.16). □

One should note that the dual problem does not contain any constraints. The slightly complicated form of $F^*$ can be rewritten using the *primal* variable $y$, defined by (2.17). To this end let

$$L : \mathbb{R}^n \times \mathbb{R}^m \to \mathbb{R} \quad \text{be defined by } (y, x) \mapsto \frac{1}{2}\langle y, Ay \rangle - \langle y, b \rangle + \langle x, B^\top y - f \rangle,$$

denote the *Lagrangian*.

The functional $L$ is an extension of $F$ by "coupling" the equality constraints with (the multiplier) $x$ to the functional.

If $y$ is the solution of $B^\top y = f$, then obviously we have

$$L(y, x) = F(y). \tag{2.18}$$

Some more elementary rearrangements are necessary to understand the following:

If $y$ and $x$ fulfill $Ay + Bx = b$, then it follows

$$L(y, x) = F^*(x). \tag{2.19}$$

The dual problem also allows for a formulation with constraints. The pair $(\overline{y}, \overline{x})$ may be viewed as the solution of the problem

Maximize $L(y, x)$ for $(y, x) \in \mathbb{R}^n \times \mathbb{R}^m$ such that $Ay + Bx = b$.

This characterization immediately follows from the observation, that pairs $(y, x)$ which fulfill the constraint, also satisfy the identity (2.19).

As $\overline{y} \in \mathbb{R}^n, \overline{x} \in \mathbb{R}^m$ which fulfill the equivalent formulations (i)–(iv) from Theorems 2.9 and 2.10, satisfy both the conditions $B^\top y = f$ and $Ay + Bx = b$, we also have

$$\min \left\{ F(y) : y \in \mathbb{R}^n, \, B^\top y = b \right\} = F(\overline{y}) =$$
$$L(\overline{y}, \overline{x}) = F^*(\overline{x}) = \max \left\{ F^*(x) : x \in \mathbb{R}^m \right\}. \tag{2.20}$$

Furthermore it holds true that:

**Theorem 2.11** *Under the assumptions of Theorem 2.9 the vectors* $\overline{y} \in \mathbb{R}^n$ *and* $\overline{x} \in \mathbb{R}^m$, *characterized there and in Theorem 2.10, fulfill:*

$$\max_{x \in \mathbb{R}^m} \min_{y \in \mathbb{R}^n} L(y, x) = L(\overline{y}, \overline{x}) = \min_{y \in \mathbb{R}^n} \max_{x \in \mathbb{R}^m} L(y, x).$$

*Proof* For an arbitrary, but fixed $x \in \mathbb{R}^m$, define

$$\widetilde{F}(y) = L(y, x).$$

According to Theorem 2.7 $\widetilde{F}$ has a unique minimum $\widehat{y} = \widehat{y}_x$ and this one is characterized by

$$A\widehat{y} = b - Bx.$$

Therefore according to (2.19) we have

$$\min_{y \in \mathbb{R}^n} L(y, x) = L(\widehat{y}, x) = F^*(x)$$

and therefore

$$\max_{x \in \mathbb{R}^m} \min_{y \in \mathbb{R}^n} L(y, x) = F^*(\overline{x}).$$

On the other hand for a fixed $y \in \mathbb{R}^n$ it holds true:

$$\max_{x \in \mathbb{R}^m} L(y, x) = \begin{cases} \infty & \text{if } B^\top y \neq f \\ \frac{1}{2} \langle y, Ay \rangle - \langle y, b \rangle & \text{if } B^\top y = f \end{cases}$$

and therefore

$$\min_{y \in \mathbb{R}^n} \max_{x \in \mathbb{R}^m} L(y, x) = F(\overline{y}).$$

Noting (2.20) leads to the assertion.                                      □

In the following we briefly sketch how this results can be generalized to more general nonlinear optimization problems. We consider an optimization problem of the form

$$\min \left\{ f(y) \mid y \in \mathbb{R}^n, \, g_j(y) = 0 \text{ for } j = 1, \ldots, m \right\}. \tag{2.21}$$

Here $f : \mathbb{R}^n \rightarrow \mathbb{R}$ and $g_j : \mathbb{R}^n \rightarrow \mathbb{R}$, $j = 1, \ldots, m$, denote sufficiently smooth functions. We do not investigate under which condition this optimization problem is solvable.

A necessary condition for the optimal value $y_0$ can be derived with the following considerations: We consider a curve $(-t_0, t_0) \ni t \mapsto y(t)$ through the optimal point $y(0) = y_0$. The curve is supposed to lie completely in the admissible set $\{y \in \mathbb{R}^n \mid g_j(y) = 0 \text{ for } j = 1, \ldots, m\}$, therefore

$$g_j(y(t)) = 0 \text{ for } j = 1, \ldots, m, \; t \in (-t_0, t_0). \qquad (2.22)$$

Considering the limit of the difference quotient

$$\lim_{\substack{t \rightarrow 0 \\ t > 0}} \frac{1}{t} \big(f(y(t)) - f(y_0)\big) \geq 0,$$

one gets

$$\langle \nabla f(y_0), y'(0) \rangle \geq 0.$$

This inequality holds true for all admissible directions $a = y'(0)$. Computing the derivative of the Eq. (2.22) with respect to $t$, it can be verified that a vector $a \in \mathbb{R}^n$ is an admissible direction, if and only if

$$\langle \nabla g_j(y_0), a \rangle = 0 \text{ for all } j = 1, \ldots, m$$

holds true, i.e., if $a$ belongs to the orthogonal complement of

$$U := \text{span} \{\nabla g_1(y_0), \ldots, \nabla g_m(y_0)\}.$$

In total it follows that $\nabla f(y_0)$ has to be orthogonal to the orthogonal complement of $U$ and therefore must be an element of $U$. Thus a necessary condition for an optimum is given by

$$\nabla f(y_0) + \sum_{j=1}^{m} x_j \nabla g_j(y_0) = 0. \qquad (2.23)$$

The coefficients $x_j \in \mathbb{R}$ are called *Lagrange multipliers*. Equation (2.23) has to be amended by the constraints

$$g_j(y_0) = 0 \text{ for } j = 1, \ldots, m. \qquad (2.24)$$

Using the *Lagrangian*

$$L(y, x) = f(y) + \sum_{j=1}^{m} x_j g_j(y),$$

the conditions (2.23) and (2.24) can be written compactly as

$$\nabla_y L(y, x) = 0, \quad \nabla_x L(y, x) = 0.$$

These are optimality criteria for the *saddle point problem*

$$\inf_{y \in \mathbb{R}^n} \sup_{x \in \mathbb{R}^m} L(y, x). \tag{2.25}$$

In many cases the problems (2.25) and (2.21) are equivalent. In problem (2.25) the sequence in the nested optimization can be reversed, and one obtains

$$\sup_{x \in \mathbb{R}^m} \inf_{y \in \mathbb{R}^n} L(y, x). \tag{2.26}$$

From these two equivalent formulations two equivalent optimization problems follow: Problem (2.25) can be rewritten with the help of the function

$$F : \mathbb{R}^n \to \mathbb{R} \cup \{+\infty\},$$

$$F(y) = \sup_{x \in \mathbb{R}^m} L(y, x) = \begin{cases} f(y) & \text{if } g_j(y) = 0 \text{ for } j = 1, \ldots, m, \\ +\infty & \text{otherwise}, \end{cases}$$

in the form

$$\min_{y \in \mathbb{R}^n} F(y). \tag{2.27}$$

This is the *primal* problem, it corresponds exactly to the optimization problem (2.21), where the constraints have been included in the definition of the functional $F$. Alternatively, using

$$F^*(x) = \inf_{y \in \mathbb{R}^n} L(y, x)$$

the problem (2.26) can be rewritten as

$$\max_{x \in \mathbb{R}^m} F^*(x). \tag{2.28}$$

This is the *dual* problem. The variable of the dual problem is the *Lagrange multiplier* of the original problem.

For elastic frames and electric networks one obtains the optimization problem (2.12) with $A := C^{-1}$. Then the problem has the form

$$\min_{y \in \mathbb{R}^n} \left\{ \tfrac{1}{2} y^\top C^{-1} y - b^\top y \mid B^\top y = f \right\}.$$

This optimization problem has the following physical meaning:

- For an elastic frame work $y$ denotes the vector of stresses, $C$ is the diagonal matrix consisting of the module of elasticity $E_i$ of the single bars, and $b = 0$. Therefore it holds

$$\frac{1}{2}y^\top C^{-1}y = \frac{1}{2}\sum_{j=1}^{n} E_j^{-1}y_j^2 = \sum_{j=1}^{n}\frac{1}{2}y_j\,e_j\,,$$

where $e_j = E_j^{-1}y_j$ denotes the strains. The term $\frac{1}{2}y_je_j$ describes the work $W_j$ necessary for the deformation of the bar $j$, because this is given by the integral over the term force times distance:

$$W_j = \int_0^{e_j} E_j e\,de = \frac{1}{2}E_je_j^2 = \frac{1}{2}y_je_j\,.$$

Here $E_j e$ is the stress of a bar elongated by $e$, being equally well the acting force, and $de$ is identical with the path increment. Hence, $\frac{1}{2}y^\top C^{-1}y$ corresponds to the *total* work used for the deformation of the space frame, which is the same as the total *elastic energy* stored in the space frame. The constraint $B^\top y = f$ describes the set of stresses, which can appear under the applied exterior forces $f$. This means that *the elastic energy stored in the space frame is minimized under the constraint of given nodal forces.* The displacement vector $-x$ can be interpreted as the *vector of Lagrange multipliers* of the constraints.
- For an electrical network $y$ is the vector of currents, $C^{-1}$ is the diagonal matrix composed of the resistances $R_i$ and $b$ denotes the vector of voltage sources. Then we have

$$y^\top C^{-1}y = \sum_{j=1}^{n} y_j R_j y_j\,.$$

Here $y_j R_j y_j$ describes the *power* absorbed at the resistor $j$, because $R_j y_j$ is the voltage drop at the resistor $j$ and the electric power is given by voltage drop times current. Taking into account the influence of the voltage sources given by $b$ one obtains the optimization problem

$$\min\left\{\tfrac{1}{2}y^\top C^{-1}y - b^\top y \,\big|\, B^\top y = 0\right\}.$$

The physical interpretation is more clearly seen from the *dual* problem. The dual functional is

$$-\frac{1}{2}x^\top B^\top C B x + x^\top B^\top C b - \frac{1}{2}b^\top C b = -\frac{1}{2}(b - Bx)^\top C(b - Bx)$$

$$= -\frac{1}{2}(b + e)^\top C(b + e) = -\frac{1}{2}y^\top C^{-1}y\,.$$

Hence, in an electrical network the *dissipation of energy* is minimized under the constraint of given voltages.

Here the variables $y$ and $x$, i.e., the electrical currents or elastic stresses on the one hand and the electrical potentials or displacements on the other hand, are dual variables in the sense of constrained optimization.

In these two examples the following principle difference becomes visible:

- The elastic frame work describes a typical *static process*: We are looking for the *minimum of an energy*, in the state given by the solution nothing moves.
- The electrical network describes a typical *stationary process*: There is current, i.e., charges are moved in a continuous fashion, but the current flux is constant in time. Therefore there is an ongoing dissipation of energy which has to be delivered from *the outside*.

Both states can be interpreted as limit states of dynamical, i.e., time-dependent processes for large times, assuming that the exterior influences are constant in time.

## 2.4   Literature

The presentation in this chapter is inspired by Chap. 2 of [124]. More advanced expositions of the applications described one may find in textbooks about technical mechanics, for example [55, 115, 128], and about electrical engineering, for example [64]. The mathematics of linear systems of equations is a standard subject in every textbook about linear algebra, for example in [40, 125]. Solution methods for linear systems of equations are the subject of general textbooks about numerical mathematics, for example [117, 123, 127]. The important subject of iterative solvers is treated in [63, 98, 112, 122].

## 2.5   Exercises

**Exercise 2.1**   Determine the currents and voltages in the following network:

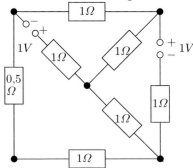

**Exercise 2.2**   Construct a network, without electrical devices for the following incidence matrices:

$$\text{a) } B = \begin{pmatrix} 0 & -1 & 0 & 1 \\ 1 & 0 & -1 & 0 \\ -1 & 1 & 0 & 0 \\ 0 & -1 & 1 & 0 \\ 1 & 0 & 0 & -1 \end{pmatrix} \qquad \text{b) } B = \begin{pmatrix} 1 & 0 & 0 & -1 & 0 & 0 \\ 0 & -1 & 1 & 0 & 0 & 0 \\ -1 & 0 & 0 & 1 & 0 & 0 \\ 0 & 1 & 0 & 0 & -1 & 0 \\ 0 & 0 & 0 & 1 & 0 & -1 \\ 0 & 0 & -1 & 0 & 1 & 0 \end{pmatrix}$$

**Exercise 2.3** Consider the following network with a voltage source and a current source:

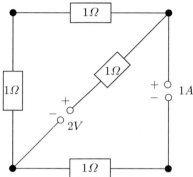

(a) How is it possible to include a current source into the network model? Extend the model from Sect. 2.1 to take also current sources into account.
(b) Compute the voltages and currents in the network.

**Exercise 2.4** Consider a direct current network with incidence matrix $B$, conductance matrix $C$, and the vectors $x$, $y$, $b$ for the potentials, the currents and the voltage sources.

(a) It is well known that the power dissipated at the resistor is $P = UI$, where $U$ is the voltage drop and $I$ the current.
    Find a representation for the total power dissipated in the network.
(b) The power provided by a voltage source is again given by $P = UI$, where $U$ is a supplied voltage and $I$ is the withdrawn current. Find a representation of the power provided by all voltage sources.
(c) Show that the quantities from (a) and (b) are identical.

**Exercise 2.5** Consider the system of equations

$$Mz = f, \text{ where } M = \begin{pmatrix} A & B \\ B^\top & 0 \end{pmatrix}, \quad A \in \mathbb{R}^{n,n}, \ B \in \mathbb{R}^{n,m}, \text{ and } f \in \mathbb{R}^{n+m}.$$

We suppose that $A$ is symmetric and positive semidefinite.

(a) Show: $y \in \text{Ker } A \Leftrightarrow y^\top A y = 0$.
(b) Compute the kernel of $M$ depending on the kernels of $A$, $B$ and $B^\top$.
(c) Characterize the right-hand side vectors $f$, for which the system of equation $Mz = f$ has a solution.

**Exercise 2.6** Let $C \in \mathbb{R}^{n,n}$ be symmetric and regular, but not necessarily positive definite, and $B \in \mathbb{R}^{n,m}$.

For $M = B^{\mathsf{T}} C B$, does the assertion Ker $M$ = Ker $B$ still hold?

If so, give a proof, otherwise give a counterexample.

**Exercise 2.7** Formulate a system of equations for the computation of the voltages and currents in the following alternating current circuit, with angular frequency $\omega = 50/\text{s}$:

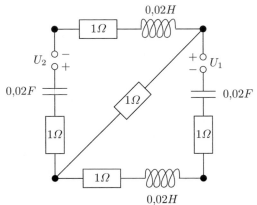

Use the following voltage sources:

(a)  $U_1(t) = 2V \cos(\omega t)$, $U_2(t) = 2V \cos(\omega t)$,
(b)  $U_1(t) = 8V \cos(\omega t)$, $U_2(t) = 8V \sin(\omega t)$,
(c)  $U_1(t) = 8V \cos(\omega t)$, $U_2(t) = 8\sqrt{2}V \cos(\omega t - \pi/4)$.

**Exercise 2.8** Consider an alternating current circuit with angular frequency $\omega$, incidence matrix $B \in \mathbb{R}^{n,m}$, electric conductance matrix $C \in \mathbb{C}^{n,n}$, $x \in \mathbb{C}^m$, current vector $y \in \mathbb{C}^n$, voltage vector $e \in \mathbb{C}^n$, and the vector of voltage sources $b \in \mathbb{C}^n$.

(a)  Determine a representation $P(t)$ for the dissipation of energy at all ohmic resistors at time $t$.
(b)  Determine a representation for the power $Q(t)$ withdrawn from all voltage sources at time $t$.
(c)  Why in general it holds true that $P(t) \neq Q(t)$?
(d)  Show that the power dissipated from all ohmic resistors in one period equals the power withdrawn from the voltage sources in the same interval of time:

$$\int_0^{2\pi/\omega} P(t)\,dt = \int_0^{2\pi/\omega} Q(t)\,dt \,.$$

**Exercise 2.9** Compute, if possible, the displacements, stresses and strains of the following space frames. The moduli of elasticity are indicated at the bars.

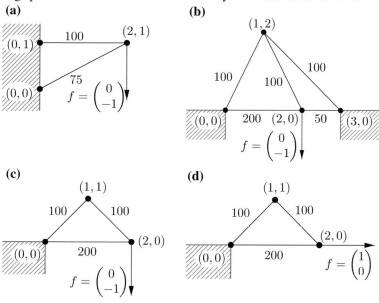

**(a)**

**(b)**

**(c)**

**(d)**

**Exercise 2.10** Decide whether the space frames depicted are statically determinate, statically undeterminate or unstable:

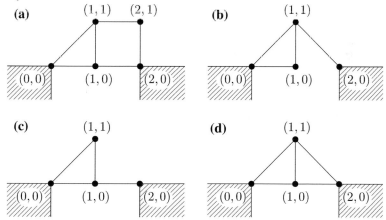

**(a)**

**(b)**

**(c)**

**(d)**

**Exercise 2.11** (a) Determine a representation for the elastic energy, which is stored in a linear elastic space frame, consisting of the sum of the elastic energies of the single bars.

(b) Determine the work done by the nodal forces for the deformation of the space frame.

(c) Show that the quantities from (a) and (b) are identical.

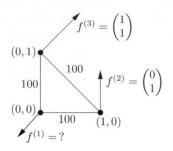

**Exercise 2.12**  Consider the following space frame without any fixed nodes:

Determine the force $f^{(1)}$ such that the linear system of equations for the space frame has a solution and compute the general solution.

**Exercise 2.13**  Suppose that the force $f$ acts at a bar of length $L$ with end points $x^{(1)}$ and $x^{(2)}$ at the position $\alpha x^{(1)} + (1 - \alpha)x^{(2)}$.

(a)  Determine the distribution of force at the two end points in the way

$$f^{(1)} = \beta f, \quad f^{(2)} = (1 - \beta)f,$$

such that the torsional moment of the distributed force (with respect to $x^{(1)}$) equals the torsional moment of the original force. Here $f^{(i)} = f(x^{(i)})$ denotes the force at node $i$.

(b)  How can the weight of a space frame be described by a vector of a nodal forces?

**Exercise 2.14**  In the following three-dimensional space frame all moduli of elasticity are supposed to be 1.

(a)  Compute the distribution of stresses in the following space frame with fixed nodes $(0, 0, 0)$, $(2, 0, 0)$, $(0, 2, 0)$ and a force $f = \left(\sqrt{3}/50, 0, 0\right)^{\mathsf{T}}$, acting at node $(1, 1, 1)$:

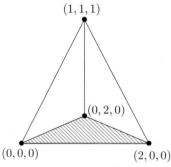

(b)  Compute the distribution of stresses in the following space frame with fixed nodes $(0, 0, 0)$, $(2, 0, 0)$, $(0, 2, 0)$, $(2, 2, 0)$ and a force $f = \left(\sqrt{3}/50, 0, 0\right)^{\mathsf{T}}$, acting at the node $(1, 1, 1)$:

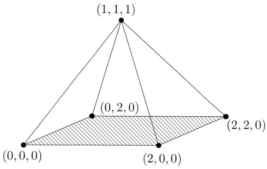

(c) Show that the following space frame with fixed nodes with $(0, 0, 0)$, $(1, 0, 0)$, $(0, 2, 0)$ and $(2, 2, 0)$ is unstable. What conditions must be met by the force vectors acting at the nodes $(0, 1, 1)$ and $(1, 1, 1)$, such that the resulting linear system of equations nevertheless has a solution?

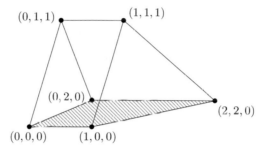

**Exercise 2.15** Determine a basis for the vector space of the rigid body motions of a three-dimensional space frame with the nodes $p^{(1)} = (0, 0, 0)^{\top}$, $p^{(2)} = (1, 0, 0)^{\top}$, $p^{(3)} = (0, 1, 0)^{\top}$, $p^{(4)} = (0, 0, 1)^{\top}$. This includes showing that the vectors selected are linear independent.

**Exercise 2.16** Consider the following system of pipelines:

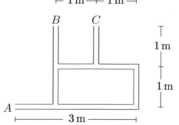

In the opening A a liquid is injected with a rate $q = 0,001$ m³/s, in the openings B and C the pressure is kept fixed to $10^5$ N/m² (this corresponds to 1 bar). According to the *Hagen–Poiseuille* law the flux through a pipe with circular cross-section is given by

$$\Delta V = \frac{\pi R^4 \, \Delta p \, \Delta t}{8 \, \mu \, L},$$

where $\Delta V$ denotes the volume of the fluid flowing through the cross-section in the time $\Delta t$, $L$ the length and $R$ the radius of the pipeline, $\Delta p$ the pressure difference between the end points of the pipeline and $\mu$ the dynamic viscosity of the liquid.

(a) Derive a general model for the computation of flow rates in a pipeline system.
(b) Compute the flow rates at the openings $B$ and $C$ and the pressure at the opening $A$ for a liquid with a viscosity $\mu = 0,001 \, \mathrm{N \, s/m^2}$ (corresponding to water), assuming that the pipes have a radius of $10 \, \mathrm{cm}$.
    *Hint*: To simplify the computation certain scalings are advisable.

**Exercise 2.17** With the help of Lagrange multipliers determine the minima and maxima of the functional

$$f(x, y) = (x + 1)^2 e^y$$

under the constraint

(a) $2(x - 1)^2 + y^2 = 3$,
(b) $2(x - 1)^2 + y^2 \leq 3$.

**Exercise 2.18** Consider the optimization problem

$$\min\left\{ \tfrac{1}{2} y^{\mathsf{T}} A y - b^{\mathsf{T}} y \mid y \in \mathbb{R}^n, \ B^{\mathsf{T}} y = c \right\} \tag{2.29}$$

with a symmetric, positive *semi-definite* matrix $A \in \mathbb{R}^{n,n}$, $B \in \mathbb{R}^{n,m}$, $b \in \mathbb{R}^n$, $c \in \mathbb{R}^m$, and also the following system of linear equations

$$\begin{pmatrix} A & B \\ B^{\mathsf{T}} & 0 \end{pmatrix} \begin{pmatrix} y \\ x \end{pmatrix} = \begin{pmatrix} b \\ c \end{pmatrix}. \tag{2.30}$$

(a) Show that both problems are equivalent in the following sense:

   (i) For every solution $y$ of the optimization problem (2.29) there exists a $x \in \mathbb{R}^m$, such that $(y, x)$ is a solution of the linear system of equations (2.30).
   (ii) If $(y, x)$ is a solution of (2.30), then $y$ is a solution of (2.29).

(b) Under which conditions for $b$, $c$ does the problem (2.29) have a solution?

**Exercise 2.19** Give a proof for Theorems 2.7 and 2.8.

# Chapter 3
# Basic Principles of Thermodynamics

Thermodynamics is concerned with the investigation of certain physically observable properties of matter, for example the temperature, the pressure and the volume, or the density, and their relations. A common characteristic of these quantities is the fact that they describe "macroscopical" measurable "consequences" of motions of atoms or molecules, from which gases, liquids and solids are composed of. The exact motion of a single particle is unknown in this context; the description would be much to complicated because of the large number of particles. The aim of thermodynamics is rather to find a *macroscopic* description of the effects of movement of those particles and their interaction with the help of relatively few macroscopic quantities and to justify this description as rigorously as possible. As the motion of a very large number of particles is unknown in detail, stochastic methods play an important role.

Concepts of thermodynamics are of importance for many processes considered in the natural and engineering sciences, for example phase transitions such as the solidification of liquids or the melting of solids, for diffusion processes, and for chemical reactions and combustion processes. An important requirement for models of such processes is their *thermodynamic consistency*, meaning that they are not in contradiction to the laws of thermodynamics. In particular the compatibility with the second law of thermodynamics has to be verified.

The aim of this chapter is to introduce readers without prerequisites from thermodynamics to the most important notions and laws of thermodynamics, in particular the relations of simple thermodynamic quantities such as temperature, pressure and volume given by the equation of state, the first and second law of thermodynamics, the notion of entropy, and the thermodynamic potentials. Furthermore an outline of the thermodynamics of mixtures shall be given. At the end of this chapter models for the description of chemical reactions will be introduced.

© Springer International Publishing AG 2017
C. Eck et al., *Mathematical Modeling*, Springer Undergraduate
Mathematics Series, DOI 10.1007/978-3-319-55161-6_3

## 3.1   The Model of an Ideal Gas
and the Maxwell–Boltzmann Distribution

An ideal gas consists of material points in motion which only interact by elastic collisions. The material points are enclosed in a possibly variable volume and are reflected elastically from the boundary of this volume. This model is an idealized approximation for gases with relatively low density. Although this model is simple it has proven to be very meaningful for practical questions.

The particles of an ideal gas move with different velocities. It is possible to derive the *probability distribution* of the velocity from the following assumptions:

**(A1)**  The distribution has a *density* $f : \mathbb{R}^3 \to [0, \infty) := \{x \in \mathbb{R} \mid x \geq 0\}$. This means that for every (Lebesgue-measurable) set $B \subset \mathbb{R}^3$ the probability that $v$ belongs to $B$ can be computed as follows

$$P(v \in B) = \int_B f(v)\,dv\,.$$

Hereby we have

$$\int_{\mathbb{R}^3} f(v)\,dv = 1\,.$$

**(A2)**  The probability of every component $v_\ell$ of the velocity also has a density $g_\ell$:

$$P(v_\ell \in B) = \int_B g_\ell(v)\,dv$$

for every Lebesgue-measurable set $B \subset \mathbb{R}$.

**(A3)**  The distributions from (A1) and (A2) are *independent* in the following sense:

    (i)  $g_1 = g_2 = g_3 =: g$,

    (ii)  $P(v \in B_1 \times B_2 \times B_3) = P(v_1 \in B_1)\, P(v_2 \in B_2)\, P(v_3 \in B_3)$, i.e., the distributions of the components $v_1, v_2, v_3$ are *independent* in the sense of probability theory.

    (iii)  There is a function $\Phi : [0, \infty) \to [0, \infty)$, such that $f(v) = \Phi(|v|^2)$.

**(A4)**  The function $\Phi$ is continuously differentiable and $g$ is continuous.

As a side remark, we note that (A1) is already a consequence of (A2) and (A3)-(ii). For a set of the form

$$B = [w_1, w_1 + h] \times [w_2, w_2 + h] \times [w_3, w_3 + h]$$

we then have

$$P(v \in B) = \int_{w_1}^{w_1+h} \int_{w_2}^{w_2+h} \int_{w_3}^{w_3+h} \Phi\left(v_1^2 + v_2^2 + v_3^2\right) dv_3\, dv_2\, dv_1$$

$$= \prod_{j=1}^{3} \int_{w_j}^{w_j+h} g(z)\, dz\,.$$

Division by $h^3$ and considering the limit $h \to 0$ leads to

$$\Phi\left(w_1^2 + w_2^2 + w_3^2\right) = g(w_1)\, g(w_2)\, g(w_3)\,.$$

This is valid for all $w \in \mathbb{R}^3$. In particular the special case $w_1 = z$ and $w_2 = w_3 = 0$ leads to

$$\Phi\left(z^2\right) = g(z)\, g^2(0)\,.$$

For $w_3 = 0$ and taking $y = w_1^2$, $z = w_2^2$ we arrive at

$$\Phi(y + z) = g(w_1)\, g(w_2)\, g(0) = \frac{\Phi(y)}{g^2(0)} \frac{\Phi(z)}{g^2(0)} g(0) = \frac{\Phi(y)\, \Phi(z)}{g^3(0)}\,.$$

This equation can be differentiated with respect to $y$ and we get:

$$\Phi'(y + z) = \frac{\Phi'(y)\, \Phi(z)}{g^3(0)}\,.$$

Substitution of $y = 0$ leads to the following differential equation for $\Phi$:

$$\Phi'(z) = -\hat{K}\, \Phi(z) \quad \text{where} \quad \hat{K} = -\frac{\Phi'(0)}{g^3(0)}\,.$$

The general solution of this equation is

$$\Phi(z) = C^3 e^{-\hat{K}z}$$

with $C \in \mathbb{R}$. The constant $C$ can be computed from the scaling condition

$$1 = \int_{\mathbb{R}^3} \Phi\left(|v|^2\right) dv = C^3 \int_{\mathbb{R}^3} e^{-\hat{K}|v|^2}\, dv = \left(C\, 2 \int_0^{+\infty} e^{-\hat{K}z^2}\, dz\right)^3,$$

and we obtain

$$C = \left(2 \int_0^{+\infty} e^{-\hat{K}z^2}\, dz\right)^{-1} = \sqrt{\frac{\hat{K}}{\pi}}\,.$$

Here and in the following we use the identities

$$\int_0^\infty z^{2n} e^{-\hat{K}z^2}\, dz = \frac{1 \cdot 3 \cdot \cdots \cdot (2n-1)\sqrt{\pi}}{2^{n+1}\hat{K}^{n+1/2}} \quad \text{for } n \in \mathbb{N}_0 = \mathbb{N} \cup \{0\},\ \hat{K} > 0.$$

The probability densities for the velocity therefore reads

$$f(v) = \left(\frac{\hat{K}}{\pi}\right)^{3/2} e^{-\hat{K}|v|^2}$$

and

$$g(z) = \sqrt[3]{\Phi(3z^2)} = \sqrt{\frac{\hat{K}}{\pi}} e^{-\hat{K}z^2}.$$

In the literature one often finds also the distribution of the *absolute value* of the velocity which is achieved from $P(|v| \leq v_0) = \int_{\{|v| \leq v_0\}} \left(\frac{\hat{K}}{\pi}\right)^{3/2} e^{-\hat{K}|v|^2}\, dv$ by applying the transformation theorem as follows:

$$P(|v| \leq v_0) = \int_0^{v_0} F(z)\, dz$$

where

$$F(z) = \frac{4}{\sqrt{\pi}} \hat{K}^{3/2} z^2 e^{-\hat{K}z^2}.$$

All these distributions only depend on a single parameter $\hat{K}$.

The *average kinetic energy* of a particle is

$$u = \frac{1}{2} m_A \int_{\mathbb{R}^3} |v|^2 f(v)\, dv = \frac{1}{2} m_A \left(\frac{\hat{K}}{\pi}\right)^{3/2} 4\pi \int_0^\infty z^4 e^{-\hat{K}z^2}\, dz = \frac{3}{4} m_A \hat{K}^{-1},$$

where $m_A$ denotes the mass of a particle. It turns out that this quantity is proportional to the *absolute temperature* $T$, more specifically we have

$$u = \tfrac{3}{2} k_B T$$

with the *Boltzmann constant* $k_B \approx 1{,}3806504 \cdot 10^{-23}\, J/K$. The absolute temperature is zero exactly in the situation where all particles of the gas are at rest; in the Celsius scale this corresponds to $-273{,}15° \text{C}$. For an ideal gas the absolute temperature is a measure for the average kinetic energy of the particles. The scaling factor $\hat{K}$ therefore has the form

$$\hat{K} = \frac{m_A}{2k_B T}.$$

The total kinetic energy of a gas consisting of $N$ particles with absolute temperature $T$ is

$$U = Nu = \tfrac{3}{2} N k_B T .$$

The probability densities of the velocity distributions have the form

$$f(v) = \left( \frac{m_A}{2\pi k_B T} \right)^{3/2} e^{-m_A |v|^2 / (2 k_B T)}, \tag{3.1}$$

$$F(z) = \sqrt{\frac{2}{\pi}} \left( \frac{m_A}{k_B T} \right)^{3/2} z^2 e^{-m_A z^2 / (2 k_B T)}, \tag{3.2}$$

$$g(z) = \sqrt{\frac{m_A}{2\pi k_B T}} e^{-m_A z^2 / (2 k_B T)} . \tag{3.3}$$

From the considerations above we get

**Theorem 3.1**  (Maxwell–Boltzmann distribution) *Under the assumptions (A1)–(A4) the probability distribution of the velocity of an ideal gas is given by the probability densities (3.1)–(3.3).*

For an *monoatomic* ideal gas the sum of the kinetic energies is identical with the *internal energy* of the gas. For *polyatomic* ideal gases there are further sources of internal energy, for example rotary motions, oscillations of connected atoms with respect to their position of rest, or chemical binding energies; note that not all of these degrees of freedom are active for every temperature. In a model without considering chemical binding energies the internal energy of an ideal gas has the form

$$U = \tfrac{z}{2} N k_B T ,$$

where $z$ denotes the number of active degrees of freedom. In general $z$ can be a real number depending on temperature; however, we will disregard the dependence on the temperature in the following. In the case of a monoatomic gas we have $z = 3$ which corresponds to the three linear independent directions for a motion in space.

From the knowledge of the probability distributions one can compute the *pressure* of an ideal gas. For the sake of convenience we consider a cube of edge length $L$, which contains an ideal gas composed of $N$ material points with mass $m_A$. The side faces of the cube are assumed to be perpendicular to one of the unit vectors. Then we have:

- A particle with velocity component $v_j$ in the direction $j$ performs on average $\frac{|v_j|}{2L}$ fully elastic collisions per time unit with one of the two lateral faces perpendicular to the unit vector $e_j$.
- Every collision results in a momentum transfer of the size $2 m_A v_j$, because the $j^{\text{th}}$ component of the momentum changes from $m_A v_j$ to $-m_A v_j$.
- The force which all "particles" exert on one of the six lateral faces, is then given by

$$F = \frac{Nm_A}{L} \int_{\mathbb{R}} v_j^2 \, g(v_j) \, dv_j = \frac{Nm_A}{L} \sqrt{\frac{\hat{K}}{\pi}} 2 \int_0^{+\infty} s^2 e^{-\hat{K}s^2} \, ds$$

$$= \frac{Nm_A}{L} 2 \sqrt{\frac{\hat{K}}{\pi}} \frac{1}{4} \sqrt{\frac{\pi}{\hat{K}^3}} = \frac{Nm_A}{2L} \hat{K}^{-1} = \frac{Nk_BT}{L} = \frac{2}{3} \frac{U}{L} .$$

Therefore the pressure can be computed to be

$$p = \frac{F}{A} = \frac{2U}{3LA} = \frac{2U}{3V}$$

with the area $A = L^2$ of the lateral face and the volume $V = L^3$. Usually this equation of state is expressed by means of absolute temperature:

$$pV = Nk_BT .$$

The model of an ideal gas is the most simple example for a so-called *pVT system*. Such a system is described by

- three variables: the pressure $p$, the volume $V$, the temperature $T$;
- an *equation of state* $F(p, V, T) = 0$.

Such a system has *two degrees of freedom*, since one of variables is determined from the equation of state by the two other.

One can distinguish two classes for the variables $p$, $V$, $T$: The temperature $T$ and the pressure $p$ do *not* depend on the size of the system, i.e., the number $N$ of particles; the volume however is *proportional* to the size of the system. Variables which are *independent* of the size of the system are denoted *intensive* variables, such variables which are proportional to the size of the system, are denoted as *extensive*.

## 3.2   Thermodynamic Systems and the Thermodynamic Equilibrium

Many results of thermodynamics can be justified by (thought) experiments using so called *thermodynamic systems*. The notation *thermodynamic system* has to be viewed as an abstract superordinate concept for all possible configurations of matter which can be studied in thermodynamics. This notion in particular comprises:

- A given amount of a certain pure substance.
- A given amount of a mixture with known composition.
- *Composite systems*, which consist of several subsystems.

To a thermodynamic system one attributes a *state* which in the most simple case of a *pVT* system consists of the internal energy and the volume of the system. In more complicated cases there can be also several internal energies and several volumes of subsystems. The state of a thermodynamic system can be changed, for example by

- Expansion or compression of subsystems. In this way the energy of the system can be increased or decreased in a "correlated", "mechanically useable" form.
- Supply or leakage of heat, i.e., of energy in "uncorrelated" form.
- Mixing or segregation of subsystems: Using *phase transitions* as it is done for example in the distillation of alcohol or by means of semipermeable membranes.
- Establishing of contacts of different subsystems. The most important possibilities are

  - Contact at a flexible wall or a deformable membrane, across which the exchange of pressure and volume is possible.
  - Contact at a heat permeable wall, which allows for the exchange of energy in the form of heat.

If constant exterior conditions are maintained for a thermodynamic system, for example a given pressure or volume, or a given temperature or heat insulation, then the system approaches a limit state, which is called the *thermodynamic equilibrium*. Nearly all results and concepts of thermodynamics are based on the assumption that the system under consideration is already in equilibrium. This applies for example to the Maxwell–Boltzmann distribution and the equation of state for ideal gases. The assumptions on which the derivation of the Maxwell–Boltzmann distribution is based, can be viewed as a concrete perception how the thermodynamic equilibrium in an ideal gas looks like. Also considerations of changes of state in general are based on the assumption that the system always is "very close" to an equilibrium state and therefore deviations from an equilibrium state can be neglected. This is reasonable if the changes of state are sufficiently *slow*.

## 3.3 The First Law of Thermodynamics

The *first law of thermodynamics* describes the *conservation of energy*. We now consider a change of state in a $pVT$ system from state $(U_1, V_1)$ to state $(U_2, V_2)$ with the equation of state $p = p(U, V)$ for the pressure. This change has form

$$\Delta U = \Delta Q + \Delta W,$$

where $\Delta U = U_2 - U_1$ is the change of internal energy, $\Delta Q$ is the supplied energy in the form of heat and $\Delta W$ is the work performed on the system. If the change of state has the form $[0, 1] \ni t \mapsto (U(t), V(t))$ with continuously differentiable functions $U(t), V(t), p(t) = p(U(t), V(t))$, then the first law gets the form

$$\dot{U} = \dot{Q} + \dot{W},$$

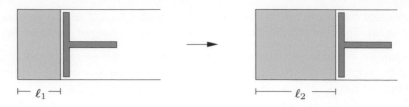

**Fig. 3.1** Example 1

where $U(t)$ is the internal energy for the time $t$, $Q(t)$ the cumulative energy supplied up to time $t$ in form of heat and $W(t)$ denotes the work which has been done on the system up to time $t$. Here the dot denotes the derivative with respect to the time $t$. Often the work performed on the system has the form

$$\dot{W} = -p\,\dot{V}\,, \qquad\qquad (3.4)$$

since work is "force times distance", the pressure is "force by area" and "area times distance" indicates the change of volume. Formula (3.4) is valid, if during the change of state the pressure, which acts from the exterior at the boundary of the thermodynamic system, equals the pressure $p$ of the thermodynamic system itself. This is not the case in all examples and applications. If (3.4) is valid, then the conservation of energy is given by

$$\dot{U} = \dot{Q} - p\dot{V}\,. \qquad\qquad (3.5)$$

As an application of the first law of thermodynamics we consider the following three simple examples:

**Example 1: Adiabatic expansion** We consider an ideal gas enclosed in a piston, consisting of $N$ particles of mass $m_A$, see Fig. 3.1. The piston is supposed to have the cross-sectional area $A$, the length of the gas volume will be increased from $\ell_1$ to $\ell_2$. Then the internal energy changes from $U_1$ to $U_2$ and the pressure changes from $p_1$ to $p_2$. The gas does not exchange heat with the exterior. Such a change of state is called *adiabatic*. We now assume that $U_1$, $p_1$ are known and $U_2$, $p_2$ are to be determined. From the equation of state

$$U = \frac{z}{2}pV$$

and the conservation of energy

$$\dot{U} = -p\dot{V}$$

we derive the following differential equation

$$\dot{U} = \frac{z}{2}(\dot{p}V + p\dot{V}) = -p\dot{V}$$

and therefore

$$\frac{z}{2} \dot{p} V = -\frac{z+2}{2} p \dot{V} .$$

Separation of variables and integration leads to

$$\frac{\dot{p}}{p} = -\frac{z+2}{z} \frac{\dot{V}}{V} \quad \text{and} \quad pV^{\kappa} = \text{constant} \quad \text{with} \quad \kappa = 1 + \frac{2}{z} . \tag{3.6}$$

Here one of the variables can be substituted by the internal energy using the equation of state, for example we also have

$$U V^{\kappa-1} = \text{constant}. \tag{3.7}$$

From the last two equations it follows

$$p_2 = p_1 \left(\frac{V_1}{V_2}\right)^{\kappa} = p_1 \left(\frac{\ell_1}{\ell_2}\right)^{\kappa}$$

and

$$U_2 = U_1 \left(\frac{V_1}{V_2}\right)^{\kappa-1} = U_1 \left(\frac{\ell_1}{\ell_2}\right)^{\kappa-1} .$$

Therefore the work performed by the gas is

$$W = U_1 - U_2 = \left(1 - \left(\frac{\ell_1}{\ell_2}\right)^{\kappa-1}\right) U_1 .$$

**Example 2: Expansion Without Mechanical Energy Gain** We assume that the gas from Example 1 initially is enclosed in the volume $V_1$. By removal of the wall the volume is increased to $V_2$, but the gas does not lose energy. Hence, the internal energy is conserved, but the volume is increased from $V_1$ to $V_2$ and therefore the pressure is decreased from $p_1 = \dfrac{2U_1}{z V_1}$ to

$$p_2 = \frac{2U_1}{z V_2} = p_1 \frac{V_1}{V_2} .$$

This example shows in particular that for adiabatic boundary conditions not every change of volume has to lead to a change of internal energy. In particular the relations (3.4) and (3.5) are not applicable here as during the expansion no external pressure was present against which work had to be performed. Contrary to Example 1 here we do not have a "continuously differentiable" change of state; in reality after the removal of the separating wall it will take some time until thermodynamic equilibrium has been established again (Fig. 3.2).

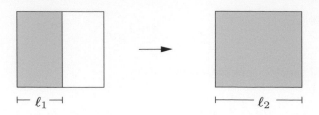

**Fig. 3.2** Example 2

**Example 3: Isothermal Expansion** As in Example 1 we consider the expansion of an ideal gas consisting of $N$ particles with mass $m_A$ from volume $V_1$ to the volume $V_2$, but now the piston is not thermally insulated, but in contact with an ambient medium of temperature $T$. Hence, during the expansion the gas does not cool down, but absorbs heat from the exterior medium. Therefore the pressure is given depending on the volume by

$$p = \frac{Nk_BT}{V} .$$

Consequently the work performed by the system during the expansion is

$$-\Delta W = W_1 - W_2 = \int_{V_1}^{V_2} p(V)\, dV = Nk_BT \ln \frac{V_2}{V_1} .$$

The internal energies before and after the expansions are the same,

$$U_1 = U_2 = \tfrac{3}{2}Nk_BT .$$

Therefore the work performed during the expansion must be gained in terms of a heat quantity from the surrounding medium,

$$\Delta Q = Q_2 - Q_1 = W_1 - W_2 = Nk_BT \ln \frac{V_2}{V_1} . \qquad (3.8)$$

The relations described in Examples 1 and 3 are valid also for corresponding compression processes, where $V_2 < V_1$. The resulting negative sign in the work gained, respectively in the heat gained in Example 3, mean that work has to be performed respectively, heat has to be released. The process from Example 2 cannot be passed through in reverse direction. The processes of Examples 1 and 3 are *reversible*, the process of Example 2 is *irreversible*.

## 3.4 The Second Law of Thermodynamics and the Notion of Entropy

The second law of thermodynamics is a precise formulation of the following two equivalent observations:

(i) Heat does not flow "on its own", i.e., without the expenditure of other forms of energy, from a *cold* to a *warm* body.
(ii) It is impossible to gain mechanical energy or work "just" by cooling a reservoir.

The second law of thermodynamic postulates that there is an essential physical difference between "mechanically useable" energy and "mechanically not useable" energy, a distinction which does not play any role for the first law of thermodynamics. This distinction is also of importance if no energy is transferred as one can see from the two examples above in Sect. 3.3. In Example 2 we had not generated any mechanically useable energy, on the other side the internal energy of the "expanded" gas had stayed the same and as consequence the temperature had been higher than in the outcome of Example 1. According to the second law of thermodynamics it is without further circumstances impossible to reach the final state of Example 1, starting from the final state of Example 2. This is because it would mean to generate mechanically useable energy by cooling a heat reservoir. In this sense the change of state described in Example 2 has led to a loss of mechanically useable energy. Reversely one can start from the final state of Example 1 and using the mechanical energy generated in this example, arrive at the final state of Example 2, for instance by compressing the piston adiabatically and in this way using the gained mechanical energy to achieve the change of state of Example 2.

From this observation we end up with the conclusion that it is possible to "lose" mechanically useable energy by "inappropriate" changes of state, meaning that this energy is transformed to "mechanically not useable" energy. Considering adiabatic changes of state the quantity of $\dot{U} + p\dot{V}$ can be used as a criterium for this loss: If

$$\dot{U} + p\dot{V} = 0 \,, \tag{3.9}$$

is valid then no mechanical energy is lost, because every change of internal energy results from or leads to a mechanical power $p\dot{V}$. On the other hand if

$$\dot{U} + p\dot{V} > 0 \tag{3.10}$$

is valid then we lose mechanically useable energy.

The scientifically exact formulation of this observation leads us to a further important quantity of state, the *entropy*. For the introduction of the notion of entropy we start with the investigation of *heat engines*.

**Heat Engines**

Heat engines generate mechanical energy by extraction of heat from a *warm* reservoir and releasing *a part* of the heat gained to a cold reservoir. They do *not* violate the

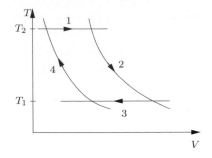

**Fig. 3.3** The Carnot cycle

second law of thermodynamics, because they use a *given difference in temperature* for the gain of mechanical energy.

The heat engine which is most important for theoretical purposes is given by the *Carnot working cycle*. Such a Carnot cycle is characterized by the fact that heat is only released or absorbed at two different temperatures $T_1$ and $T_2$. Therefore the Carnot process consists of two isothermal process steps, at which heat is released or absorbed, and two adiabatic process steps, at which two temperatures $T_1$ and $T_2$ are connected. In Fig. 3.3 it is characterized schematically by a $T$–$V$-diagram and a $p$–$V$-diagram.

The process steps consist of

1. an isothermal expansion,
2. an adiabatic expansion,
3. an isothermal compression,
4. an adiabatic compression.

If the medium is an ideal gas then during the process steps the following happens:

1. The machine expands during the time in contact with the hot reservoir having temperature $T_2$. By this mechanical work $W_1$ is produced, the internal energy stays constant, and the heat quantity $Q_1 = W_1$ is absorbed from the hot reservoir. The temperature stays constant at $T_2$, the volume increases and the pressure decreases.
2. The machine becomes thermally insulated and continues to expand now. At the same time the work $W_2$ is generated. Since there is no heat supply from outside, this leads to a decrease of internal energy by the amount of $W_2$. The temperature falls from $T_2$ to $T_1$, the volume increases, and the pressure decreases.
3. The machine is brought in contact with the cold reservoir of temperature $T_1$ and compresses. By this mechanical work $-W_3$ is fed in. The temperature stays constant at $T_1$ and therefore also the internal energy is not changed. The machine releases the heat quantity $-Q_3 = -W_3$ to the cold reservoir. The volume decreases and the pressure increases.
4. Then the machine again becomes thermally insulated and gets further compressed. This requires to have the work $-W_4$ done. As no heat is released or absorbed this

changes the internal energy by the amount of $W_4$. The temperature increases from $T_1$ to $T_2$, the volume decreases and the pressure increases.

The work gained by one cycle can be easily computed from Example 3 using the result (3.8) and the conservation of energy: In process step 1 the heat quantity

$$Q_1 = Nk_B T_2 \ln \frac{V_2}{V_1}$$

is absorbed, in process step 3 the heat quantity $-Q_2$, where

$$Q_2 = Nk_B T_1 \ln \frac{V_4}{V_3} \, ,$$

is released. From the conservation of energy we can conclude that the mechanical work

$$W = Q_1 + Q_2 = Nk_B \left( T_2 \ln \frac{V_2}{V_1} + T_1 \ln \frac{V_4}{V_3} \right)$$

is gained. The *efficiency factor* of the machine is the ratio of the mechanical energy gained and the heat quantity taken from the hot reservoir:

$$\eta = \frac{W}{Q_1} = 1 + \frac{T_1}{T_2} \frac{\ln(V_4/V_3)}{\ln(V_2/V_1)} \, .$$

The heat which is released to the cold reservoir cannot be used by heat engines in form of mechanical energy.

We will now further simplify the efficiency factor. To this end we need the exact form of the curves in the process diagram. At the isothermal curves we have

$$pV = Nk_B T = \text{constant}.$$

At the adiabatic curves we can conclude using (3.6)

$$pV^{\kappa} = \text{constant},$$

where $\kappa = 1 + 2/z$. In summary we have for the data at the "switching points" of the Carnot cycle

$$Nk_B T_1 = p_3 V_3 = p_4 V_4 , \quad Nk_B T_2 = p_1 V_1 = p_2 V_2$$

and

$$p_1 V_1^{\kappa} = p_4 V_4^{\kappa} , \quad p_2 V_2^{\kappa} = p_3 V_3^{\kappa} .$$

From these identities we arrive at

$$\frac{T_1}{T_2} = \frac{p_3 V_3}{p_2 V_2} = \frac{p_3 V_3^{\kappa} V_3^{1-\kappa}}{p_2 V_2^{\kappa} V_2^{1-\kappa}} = \frac{V_3^{1-\kappa}}{V_2^{1-\kappa}}$$

and analogously

$$\frac{T_1}{T_2} = \frac{p_4 V_4}{p_1 V_1} = \frac{V_4^{1-\kappa}}{V_1^{1-\kappa}} .$$

This implies

$$\frac{V_3}{V_2} = \frac{V_4}{V_1}, \quad \text{or} \quad \frac{V_1}{V_2} = \frac{V_4}{V_3} \quad \text{and thus} \quad \frac{\ln(V_4/V_3)}{\ln(V_2/V_1)} = -1 .$$

Therefore the efficiency factor of the Carnot cycle is

$$\eta = 1 - \frac{T_1}{T_2} .$$

The machine is the more efficient the bigger the temperature difference between the hot and cold reservoir is and the closer $T_1$ is to the absolute zero temperature $T = 0$.

The Carnot working cycle allows us to gain mechanical energy from *existing differences in temperature*. An important requirement is:

If heat flows from a hot to a cold region, then *mechanically useable* energy is transferred to *mechanically not useable* energy and is hence "lost".

An essential property of the Carnot working cycle is the heat exchange at *equal temperature*. In fact this cannot be implemented in reality: if heat is supposed to flow from the hot reservoir to the machine, a difference in temperature is required. The smaller the temperature difference is, the longer is the time period which is necessary for the heat transport. Therefore a cycle in the Carnot process would require an infinite amount of time. On the other hand a heat exchange *without temperature differences* is essential for *reversible* processes: if there is a temperature difference between the machine and the reservoir during the heat exchange, then this process step can *not* be reversed again, as this would require to have heat flowing from colder to hotter regions. In this sense *all reversible processes* during which heat transport occurs, are *idealizations*.

By simple considerations one can conclude from the second law of thermodynamics that *all reversible* heat engines, consisting of two isothermal and two adiabatic process steps with the same temperatures must have the same efficiency factor:

**Theorem 3.2**  *The efficiency factor of every reversible heat engine with two adiabatic and two isothermal process steps and heat exchange at the temperatures $T_1$ and $T_2$ with $T_1 < T_2$ is*

$$\eta = 1 - \frac{T_1}{T_2} . \tag{3.11}$$

**Justification**: We considered two reversible heat engines I and II of the type described. We suppose that machine I absorbes the heat quantity $Q_2^I$ at the temperature $T_2$ and releases the heat quantity $-Q_1^I$ at the temperature $T_1$, the corresponding data for machine II are supposed to be $Q_2^{II}$ and $Q_1^{II}$. The work done during a process step is supposed to be the same for both machines,

$$W = Q_2^I + Q_1^I = Q_2^{II} + Q_1^{II}.$$

Therefore the efficiency factors are $\eta^I = \dfrac{W}{Q_2^I}$ for machine $I$ and $\eta^{II} = \dfrac{W}{Q_2^{II}}$ for machine $II$. If these efficiency factors would be different then also

$$Q_2^{\mathrm{I}} \neq Q_2^{\mathrm{II}}$$

has to be true. If for example $Q_2^{\mathrm{I}} > Q_2^{\mathrm{II}}$ holds, then we can run machine I backward and couple it to machine II. The coupled machine in total does not need energy and transports the heat quantity $Q_2^{\mathrm{I}} - Q_2^{\mathrm{II}} > 0$ from the hot to the cold reservoir which is a contradiction to the second law of thermodynamics.  □

**The Definition of Entropy**

In the following we can derive important consequences from the efficiency factor of a Carnot working cycle which are valid for *all* $pVT$ systems. For this purpose we consider a Carnot cycle which in the $(V, T)$-diagram has the "switching points" $(V_1, T_2)$, $(V_2, T_2)$, $(V_3, T_1)$ and $(V_4, T_1)$ as indicated in Fig. 3.3; the corresponding process curve in the $(V, T)$-diagram will be denoted by $\Gamma$. The heat gained from the hot reservoir can be expressed with the help of the first law (3.5) by means of the line integral

$$Q_1 = \int_{V_1}^{V_2} \left( \frac{\partial U}{\partial V}(T_2, V) + p(T_2, V) \right) dV.$$

The total work done by the system is

$$W = \int_{\Gamma} p(T, V) \, dV.$$

From the efficiency factor (3.11) of the Carnot cycle it follows

$$W = \left( 1 - \frac{T_1}{T_2} \right) Q_1.$$

Therefore we have

$$\int_{V_1}^{V_2} \left( \frac{\partial U}{\partial V}(T_2, V) + p(T_2, V) \right) dV = \frac{T_2}{T_2 - T_1} \int_{\Gamma} p(T, V) \, dV.$$

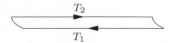

**Fig. 3.4** The limiting process in the Carnot cycle

The limiting process $T_1 \to T_2 =: T$ leads to, as it is sketched in Fig. 3.4,

$$\int_{V_1}^{V_2} \left( \frac{\partial U}{\partial V}(T, V) + p(T, V) \right) dV = T \int_{V_1}^{V_2} \frac{\partial p}{\partial T}(T, V) \, dV .$$

After division by $V_2 - V_1$ the limit process $V_1 \to V_2 =: V$ leads to *Clapeyron's equation*

$$\frac{\partial U}{\partial V}(T, V) + p(T, V) = T \frac{\partial p}{\partial T}(T, V) . \tag{3.12}$$

With the help of this relation we show the following

**Theorem 3.3** *Assuming that the mappings $(T, V) \mapsto U(T, V)$ and $(T, V) \mapsto p(T, V)$ are twice continuously differentiable, it follows that the vector field*

$$\begin{pmatrix} T \\ V \end{pmatrix} \mapsto \begin{pmatrix} \frac{1}{T} \frac{\partial U}{\partial T}(T, V) \\ \frac{1}{T} \left( \frac{\partial U}{\partial V}(T, V) + p(T, V) \right) \end{pmatrix}$$

*possesses a potential $\widetilde{S} = \widetilde{S}(T, V)$.*

*Proof* It is sufficient to show that

$$\frac{\partial}{\partial V} \left( \frac{1}{T} \frac{\partial U}{\partial T}(T, V) \right) = \frac{\partial}{\partial T} \left( \frac{1}{T} \left( \frac{\partial U}{\partial V}(T, V) + p(T, V) \right) \right) .$$

The left-hand side of this equation is given by

$$\frac{1}{T} \frac{\partial^2 U}{\partial T \partial V} ,$$

the right-hand side can be transformed to

$$-\frac{1}{T^2} \left( \frac{\partial U}{\partial V}(T, V) + p(T, V) \right) + \frac{1}{T} \left( \frac{\partial^2 U}{\partial T \partial V}(T, V) + \frac{\partial p}{\partial T}(T, V) \right)$$

$$= \frac{1}{T} \frac{\partial^2 U}{\partial T \partial V}(T, V) .$$

This proves the assertion.                                                   □

Now we want to represent the quantity $\widetilde{S}$ as a function of $U$ and $V$. To be able to distinguish between variables and functions, we will denote the function $(T, V) \mapsto U(T, V)$ now by $\widetilde{U}$ and $(T, V) \mapsto p(T, V)$ by $\widetilde{p}$. For

$$S(U, V) = \tilde{S}\big(\tilde{T}(U, V), V\big),$$

where $(U, V) \mapsto \tilde{T}(U, V)$ is the inverse with respect to $T$ of the function $(T, V) \mapsto \tilde{U}(T, V)$, we have

$$\frac{\partial S}{\partial U}(U, V) = \frac{\partial \tilde{S}}{\partial T}\big(\tilde{T}(U, V), V\big)\frac{\partial \tilde{T}}{\partial U}(U, V) = \frac{1}{T}\frac{\partial \tilde{U}}{\partial T}\big(\tilde{T}(U, V), V\big)\frac{\partial \tilde{T}}{\partial U}(U, V) = \frac{1}{T}$$

and

$$\frac{\partial S}{\partial V}(U, V) = \frac{\partial \tilde{S}}{\partial T}\big(\tilde{T}(U, V), V\big)\frac{\partial \tilde{T}}{\partial V}(U, V) + \frac{\partial \tilde{S}}{\partial V}\big(\tilde{T}(U, V), V\big)$$

$$= \frac{1}{T}\frac{\partial \tilde{U}}{\partial T}\big(\tilde{T}(U, V), V\big)\frac{\partial \tilde{T}}{\partial V}(U, V) + \frac{1}{T}\left(\frac{\partial \tilde{U}}{\partial V}\big(\tilde{T}(U, V), V\big) + \tilde{p}\big(\tilde{T}(U, V), V\big)\right).$$

From

$$0 = \frac{\partial}{\partial V}\big(\tilde{U}\big(\tilde{T}(U, V), V\big)\big) = \frac{\partial \tilde{U}}{\partial T}\big(\tilde{T}(U, V), V\big)\frac{\partial \tilde{T}}{\partial V}(U, V) + \frac{\partial \tilde{U}}{\partial V}\big(\tilde{T}(U, V), V\big)$$

it follows

$$\frac{\partial S}{\partial V}(U, V) = \frac{\tilde{p}\big(\tilde{T}(U, V), V\big)}{T}.$$

Therefore we have

$$\frac{\partial S(U, V)}{\partial U} = \frac{1}{T} \quad \text{and} \quad \frac{\partial S(U, V)}{\partial V} = \frac{p}{T}, \tag{3.13}$$

where $T$ and $p$ here are considered to be functions of $U$ and $V$. These relations are called the *Gibbs formulas*.

**Definition 3.4**  The function $(U, V) \mapsto S(U, V)$ is called *entropy*.

The notation *entropy* had been coined by Rudolf Clausius, it dates back to the Greek word for *transformation*.

With the help of the notion of entropy a more precise formulation of the second law of thermodynamics can be stated. For this purpose we consider a thermodynamic system with an abstract state space $Z$ and an entropy function $S : Z \to \mathbb{R}$. A $pVT$ system with equation of state $f(T, p, V) = 0$ for example has the state space $Z = \{(T, p, V) \mid T, p, V \geq 0, \ f(T, p, V) = 0\}$.

**Definition 3.5**  A change of state $t \mapsto z(t) \in Z$ in a thermodynamic system with state space $Z$ and entropy function $S : Z \to \mathbb{R}$ fulfills the second law of thermodynamics if for $t \mapsto S(t) = S(z(t))$ we have

$$\dot{S} \geq \frac{\dot{Q}}{T}. \tag{3.14}$$

Here $T$ denotes the absolute temperature and $Q(t)$ the heat quantity absorbed by the system up to the time $t$.

For adiabatic changes of state it follows from (3.14)

$$\dot{S} \geq 0.$$

With the help of the notion of entropy one can characterize whether a change of state is *reversible* or *irreversible*: Moreover the entropy can be used to separate reversible and irreversible changes of state.

**Definition 3.6** In a thermodynamic system with a state space $Z$ and entropy $S$ : $Z \to \mathbb{R}$ a continuously differentiable change of state $[0, 1] \to Z$ is called

reversible, if $\dot{S} = \dfrac{\dot{Q}}{T}$,

irreversible, if $S(1) - S(0) > \displaystyle\int_0^1 \dfrac{\dot{Q}(t)}{T(t)}\, dt.$

For an ideal gas the entropy can be computed using the equation of state

$$pV = Nk_B T$$

and the formula

$$U = \frac{z}{2} Nk_B T$$

for the internal energy. We have

$$\frac{\partial S}{\partial U} = \frac{1}{T} = \frac{z}{2} Nk_B \frac{1}{U} \quad \text{and} \tag{3.15}$$

$$\frac{\partial S}{\partial V} = \frac{p}{T} = \frac{Nk_B}{V}. \tag{3.16}$$

From the first equation we conclude $S(U, V) = \frac{z}{2} Nk_B \ln \frac{U}{U_0} + c(V)$ and then by substituting into the second equation $c'(V) = \frac{Nk_B}{V}$, hence, $c(V) = Nk_B \ln \frac{V}{V_0}$. Here the quantities $U_0$ and $V_0$ are reference values for the internal energy and the volume. In particular they are necessary, because $U$ and $V$ are quantities with a physical dimension, and therefore expressions like $\ln U$ and $\ln V$ are not well defined. Furthermore the integration constants can be defined by an appropriate choice of $U_0$ and $V_0$. The values of the integration constants are often unimportant, because in general only *changes in entropy* play a role. In summary we have

$$S(U, V) = \frac{z}{2} Nk_B \ln \frac{U}{U_0} + Nk_B \ln \frac{V}{V_0}.$$

With the help of the equation of state $pV = Nk_B T$ and the formula $U = \frac{z}{2} Nk_B T$ for the internal energy one can express the entropy as a function of other variables, for example as a function of temperature and volume

$$S(T, V) = \frac{z}{2} Nk_B \ln \frac{T}{T_0} + Nk_B \ln \frac{V}{V_0}$$

with the reference temperature $T_0 = \frac{2U_0}{zNk_B}$, or as a function of temperature and pressure

$$S(T, p) = \frac{z+2}{2} Nk_B \ln \frac{T}{T_0} - Nk_B \ln \frac{p}{p_0}$$

with the reference pressure $p_0 = \frac{Nk_B T_0}{V_0}$. The Gibbs formulas (3.13) are valid only for $S$ as a function of $U$ and $V$, however.

As an application we are determining the change in entropy for the Examples 1 and 2 from the end of Sect. 3.3. In Example 1, i.e., the adiabatic expansion, the state $(U_1, V_1)$ with entropy

$$S_1 = \frac{z}{2} Nk_B \ln \frac{U_1}{U_0} + Nk_B \ln \frac{V_1}{V_0}$$

is transformed to the state $(U_2, V_2)$ with $U_2 = U_1 \left(\frac{V_1}{V_2}\right)^{2/z}$ and the entropy

$$S_2 = \frac{z}{2} Nk_B \ln \frac{U_2}{U_0} + Nk_B \ln \frac{V_2}{V_0}.$$

Because of $\ln \frac{U_2}{U_0} = \ln \frac{U_1}{U_0} + \frac{2}{z} \ln \frac{V_1}{V_2}$ it follows that $S_1 = S_2$. Hence, the entropy has not changed, the change of state therefore is *reversible*. In Example 2 we have $U_2 = U_1$ and $V_2 > V_1$ and therefore

$$S_2 = \frac{z}{2} Nk_B \ln \frac{U_1}{U_0} + Nk_B \ln \frac{V_2}{V_0} > S_1.$$

Consequently this change of state is irreversible.

### The Reciprocal of the Temperature as an Integrating Factor

At the beginning of this section we have characterized reversible and irreversible changes of state in a $pVT$ system by means of the relations (3.9) and (3.10). One goal of the consideration has been to find a scalar quantity, which has the same value for two states which are connected by a reversible change of state, but which increases for an irreversible change of state. From (3.9) and (3.10) one can conjecture that a sensible choice of such a quantity is the *potential* $\Sigma$ of the vector field

$$\binom{U}{V} \mapsto \binom{1}{p(U, V)},$$

hence, a function $\Sigma = \Sigma(U, V)$ with

$$\frac{\partial \Sigma}{\partial U} = 1 \text{ and } \frac{\partial \Sigma}{\partial V} = p.$$

With such a function for every differentiable change of state $t \mapsto (U(t), V(t))$ we have

$$\frac{d}{dt}\Sigma(U(t), V(t)) = \dot{U}(t) + p(U(t), V(t))\,\dot{V}(t).$$

Unfortunately not every vector field possesses a potential and in particular $(U, V) \mapsto (1, p(U, V))$ will have no potential in general. But from the definition of entropy it can be concluded that the vector field

$$(U, V) \mapsto \frac{1}{T(U, V)}\begin{pmatrix} 1 \\ p(U, V) \end{pmatrix}$$

has a potential. Here the absolute temperature $T$ has to be considered as a function of $U$ and $V$. In this way the reciprocal temperature $1/T$ plays the role of a so called *integrating factor*.

**Definition 3.7** Let $v : \mathbb{R}^n \mapsto \mathbb{R}^n$ be a vector field. A function $\lambda : \mathbb{R}^n \mapsto \mathbb{R}$ is called *integrating factor* of the vector field $v$, if $\lambda v$ has a potential, i.e., if there is a function $\varphi : \mathbb{R}^n \rightarrow \mathbb{R}$ such that

$$\frac{\partial \varphi}{\partial x_j} = \lambda v_j.$$

Starting from the role of the reciprocal temperature as an integrating factor for $(U, V) \mapsto (1, p(U, V))$ with the entropy as its potential an alternative approach for the definition of entropy is given.

**Entropy and Thermodynamic Equilibrium**

Using the notion of entropy a necessary condition for the thermodynamic equilibrium can be formulated for a system having *adiabatic* boundary conditions, i.e., we consider systems, which are thermally insulated such that no energy in the form of heat is released or absorbed. For this reason we consider a thermodynamic composite system consisting of two subsystems. A typical example is a container, which is separated by a wall or a membrane; there are possibly different substances or one substance but in different phases (solid, liquid, gaseous) present. The separating wall can allow or prevent the exchange of certain quantities, for example it can be thermally insulating or thermally permeable, or also fixed or removable. Our aim is to characterize the thermodynamic equilibrium. An important application of two subsystems being in thermodynamic equilibrium is the determination of the melting point or boiling point of a substance, then the two subsystems should correspond to the two phases (the solid and the liquid or the liquid and the gaseous) of the material.

An important criterion for the determination of the thermodynamic equilibrium is the following theorem:

**Theorem 3.8** (Clausius) *A thermodynamic system with adiabatic boundary conditions is in thermodynamic equilibrium, if the entropy is maximal.*

This assertion is a consequence of the observation that the entropy can at most increase during an adiabatic change of state. Consequently, if the entropy reaches its maximum then no further change of state is possible and this just characterizes the thermodynamic equilibrium.

As an application of the theorem of Clausius we consider a container with a fixed volume $V$, which is separated by a removable membrane in two subvolumes $V_1$ and $V_2 = V - V_1$. We assume that the membrane is permeable for heat such that a heat flow from one subsystem to the other is possible. We denote the temperature, the internal energy, the entropy and pressure in the subsystem $i$ for $i = 1, 2$, by $T_i$, $U_i$, $S_i$ and $p_i$, respectively. The total internal energy is then $U = U_1 + U_2$, and for the entropy we have $S = S_1 + S_2$. Using $S_i = S_i(U_i, V_i)$ we can write the entropy of the total system as follows

$$S(U_1, V_1) = S_1(U_1, V_1) + S_2(U - U_1, V - V_1).$$

A necessary condition for a maximum is

$$\frac{\partial S}{\partial U_1} = \frac{\partial S_1}{\partial U_1} - \frac{\partial S_2}{\partial U_2} = \frac{1}{T_1} - \frac{1}{T_2} = 0 \text{ and}$$

$$\frac{\partial S}{\partial V_1} = \frac{\partial S_1}{\partial V_1} - \frac{\partial S_2}{\partial V_2} = \frac{p_1}{T_1} - \frac{p_2}{T_2} = 0.$$

Hence, the equilibrium is characterized by

$$T_1 = T_2 \text{ and } p_1 = p_2,$$

i.e., the *continuity* of temperature and pressure.

## 3.5 Thermodynamic Potentials

We consider an arbitrary reversible change of state $t \mapsto (U(t), V(t))$ in a $pVT$ system. A rearrangement of the relation

$$\dot{S} = \frac{1}{T}\dot{U} + \frac{p}{T}\dot{V}$$

leads to

$$\dot{U} = T\dot{S} - p\dot{V}.$$

Writing $U$ as a function of $S$ and $V$ we get

$$\dot{U} = \frac{\partial U}{\partial S}\dot{S} + \frac{\partial U}{\partial V}\dot{V}\,.$$

This relation is valid for all changes of state and therefore we can conclude that the equations

$$\frac{\partial U}{\partial S} = T\,, \quad \frac{\partial U}{\partial V} = -p$$

have to be satisfied. We now will use the *Legendre transform*, a simple mathematical transformation which will be shortly explained in the next section. By means of the Legendre transform further energy-like functions can be derived, in which $S$ and $V$ as variables change their roles and so do $T$ and $-p$ as partial derivatives, respectively: For the *free energy*

$$F(T, V) := U - TS$$

as a function of $T$ and $V$ we have:

$$\dot{F} = \dot{U} - \dot{T}S - T\dot{S} = -S\dot{T} - p\dot{V}$$

and hence,

$$\frac{\partial F(T, V)}{\partial T} = -S\,, \quad \frac{\partial F(T, V)}{\partial V} = -p\,.$$

For the *enthalpy*

$$H(S, p) := U + pV$$

as a function of $S$ and $p$ we have:

$$\dot{H} = \dot{U} + \dot{p}V + p\dot{V} = T\dot{S} + V\dot{p}$$

and therefore

$$\frac{\partial H(S, p)}{\partial S} = T\,, \quad \frac{\partial H(S, p)}{\partial p} = V\,.$$

For the *free enthalpy*

$$G(T, p) := U - TS + pV$$

as a function of $T$ and $p$ we have:

$$\dot{G} = \dot{U} - \dot{T}S - T\dot{S} + \dot{p}V + p\dot{V} = -S\dot{T} + V\dot{p}$$

and hence,

$$\frac{\partial G(T, p)}{\partial T} = -S\,, \quad \frac{\partial G(T, p)}{\partial p} = V\,.$$

Another notion for free enthalpy is *Gibbs free energy*. The functions $(T, V) \mapsto F(T, V)$, $(S, p) \mapsto H(S, p)$, $(T, p) \mapsto G(T, p)$ are called *thermodynamic potentials*. For example they are helpful, if *equilibrium conditions* shall be formulated for thermodynamic systems with different *constraints*. For this purpose one combines the first and second law of thermodynamics with time derivatives of $F$, $H$ or $G$, respectively. From the resulting inequalities one can derive the desired equilibrium conditions, similarly as in the justification of the Clausius theorem. The first law is written in the form

$$\dot{U} = \dot{Q} + \dot{W} = \dot{Q} - p_0 \dot{V},$$

where $\dot{W} = -p_0 \dot{V}$ denotes the rate of work fed into the system. This work is implemented by compression or expansion processes and $p_0$ denotes the pressure exerted from the exterior at the boundary of the system. This may or may not coincide with the pressure $p$ of the system, as the two examples of Sect. 3.3 show. In Example 1 we have $p_0 = p$, but in Example 2 there is no release of mechanical energy and therefore $p_0 = 0$ is appropriate. The relation

$$\dot{S} \geq \frac{\dot{Q}}{T}$$

leads to

$$\dot{U} \leq T\dot{S} - p_0 \dot{V}.$$

From this one can conclude further equilibrium conditions:

Isothermal, isochoric case: Assuming $\dot{T} = 0$, $\dot{V} = 0$ it follows from

$$\frac{d}{dt}(U - TS) \leq -S\dot{T} - p_0\dot{V} = 0,$$

that the free energy $F = U - TS$ cannot increase by any change of state. Therefore the equilibrium is characterized by the fact that the *free energy $F = U - TS$ attains its minimum.*

Adiabatic, isobaric case: Assuming $\dot{S} = 0$ and $\dot{p} = 0$ and furthermore $p_0 = p$ it follows that

$$\frac{d}{dt}(U + pV) \leq T\dot{S} + V\dot{p} = 0.$$

Hence, the equilibrium is characterized by the fact that the *enthalpy $H = U + pV$ attains its minimum.*

Isothermal, isobaric case: Assuming $\dot{T} = 0$ and $\dot{p} = 0$ and furthermore $p = p_0$ it follows

$$\frac{d}{dt}(U - TS + pV) \leq -S\dot{T} + V\dot{p} = 0.$$

Hence, the equilibrium is characterized by the fact that the *free enthalpy $G = U - TS + pV$ attains its minimum.*

Therefore for the formulation of the equilibrium conditions one has to choose just the thermodynamic potential for which its "canonical variables" are constant for the constraints considered.

## 3.6 The Legendre Transform

The derivation of thermodynamic potentials from the internal energy in the last section needed the application of a mathematical transformation known as the *Legendre transform*. The Legendre transform is not only important for thermodynamics but for example also in mechanics for the connection between the Lagrangian and the Hamiltonian formalism, which we will explain in Sect. 4.2, or for optimization problems. In the following the definition and the most important properties of the Legendre transform will be shortly sketched.

**Definition 3.9** Let $f : \mathbb{R}^n \to \mathbb{R}$ be a continuously differentiable function. We assume that the derivative $y^\star$ of $f$ with the components

$$y_j^\star(x) := \frac{\partial f}{\partial x_j}(x)$$

is invertible, i.e., that there is a function

$$x^\star : \mathbb{R}^n \to \mathbb{R}^n \text{ with } y^\star(x^\star(y)) = y \text{ for all } y \in \mathbb{R}^n .$$

Then

$$f^\star(y) := x^\star(y) \cdot y - f(x^\star(y))$$

is called the *Legendre transform* of $f$.

For the Legendre transform we have a relation

$$\frac{\partial f^\star}{\partial y_j}(y) = x_j^\star(y) .$$

This can be verified by a simple computation:

$$\frac{\partial f^\star}{\partial y_j}(y) = \sum_{\ell=1}^n \left( \frac{\partial x_\ell^\star}{\partial y_j}(y) \, y_\ell - \frac{\partial f}{\partial x_\ell}(x^\star(y)) \frac{\partial x_\ell^\star}{\partial y_j}(y) \right) + x_j^\star(y) = x_j^\star(y) .$$

The Legendre transform just gives a mean to exchange the role of *variable* and *partial derivative* for sufficiently general functions, a tool which has been of importance for the construction of the thermodynamic potentials in the previous section. The transformation of the internal energy $U$ to the free energy $F$, the enthalpy $H$

and the free enthalpy $G$ corresponds, up to a change of sign, exactly to the Legendre transform with respect to one or two pairs of variables and partial derivatives which are "dual" to each other.

## 3.7   The Calculus of Differential Forms

Many of the previous results can be expressed more concisely using *differential forms of first order*, in the following shortly denoted by *differential forms*.

**Definition 3.10**  A differential form $\omega$ on $U \subset \mathbb{R}^n$ is a continuous mapping

$$\omega : U \to (\mathbb{R}^n)^* ,$$
$$x \mapsto \omega(x) ,$$

where $(\mathbb{R}^n)^*$ is the space of all linear functions from $\mathbb{R}^n$ to $\mathbb{R}$. The mapping

$$(x, t) \mapsto \langle \omega(x), t \rangle := \omega(x)(t)$$

is hence linear in the variable $t \in \mathbb{R}^n$.

An important operation for differential forms of 1st order is given by *line integrals*.

**Definition 3.11**  Let $\omega$ be a differential form and $\Gamma = \{x(s) \mid s \in (0, 1)\}$ a continuously differentiable curve in $\mathbb{R}^n$. Then the line integral of $\omega$ over $\Gamma$ is defined by

$$\int_\Gamma \omega = \int_0^1 \langle \omega(x(s)), x'(s) \rangle \, ds .$$

Due to the linearity of $t \mapsto \langle \omega(x), t \rangle$ this definition is *independent* of the choice of a parametrization, assuming that the orientation is not changed. By means of this definition the different roles of the variables $x$ and $t$ in the expression $\langle \omega(x), t \rangle$ can be perceived: On the one hand $x$ represents a point in space, on the other hand $t$ represents a (possible) *tangential vector* of a curve through $x$.

We will now consider two possibilities to obtain differential forms.

(I)  For a vector field $y : \mathbb{R}^n \to \mathbb{R}^n$ a differential form denoted by

$$\omega = \sum_{j=1}^n y_j \, dx_j$$

is given by

$$\langle \omega(x), t \rangle = \sum_{j=1}^n y_j(x) \, t_j =: y(x) \cdot t .$$

Here $dx_j$ denotes the linear mapping $dx_j(e_i) = \delta_{ij}$, where $e_i$ is the $i$-th unit vector, i.e., $dx_j(t) = t_j$. Obviously *every* differential form can be represented in this way: As for every fixed $x \in \mathbb{R}^n$ the mapping $t \mapsto \langle \omega(x), t \rangle$ is linear, there exists a $y(x)$ such that $\langle \omega(x), t \rangle = y(x) \cdot t$. The line integral of such a differential form over a curve $\Gamma = \{x(s) \mid s \in (0, 1)\}$ is

$$\int_\Gamma \sum_{j=1}^n y_j \, dx_j = \int_0^1 \sum_{j=1}^n y_j(x(s)) \, x_j'(s) \, ds =: \int_\Gamma y(x) \cdot dx .$$

(II)  For every continuously differentiable function $\phi : \mathbb{R}^n \to \mathbb{R}$ let $d\phi$ be defined by

$$d\phi := \sum_{j=1}^n \frac{\partial \phi}{\partial x_j} dx_j .$$

The line integral of $d\phi$ then is

$$\int_\Gamma d\phi = \int_0^1 \sum_{j=1}^n \frac{\partial \phi}{\partial x_j}(x(s)) \, x_j'(s) \, ds = \int_\Gamma \nabla \phi(x) \cdot dx = \phi(x(1)) - \phi(x(0)) .$$

**Definition 3.12**  A differential form $\omega$ on $U \subset \mathbb{R}^n$ is called *complete differential* if there is a differentiable, scalar function $\phi : U \to \mathbb{R}$ such that $\omega = d\phi$.

The notion of a complete differential does not contain anything which we are not yet familiar with: A differential form is a complete differential, if and only if the "corresponding" vector field has a *potential*.

One of the essential advantages of differential forms is the fact that one can "calculate" with them. For this purpose we define a multiplicative composition of a function $\phi : \mathbb{R}^n \to \mathbb{R}$ with a differential form $\omega$ resulting in a new differential form $\phi\omega$ by

$$\langle (\phi\omega)(x), t \rangle = \phi(x) \langle \omega(x), t \rangle .$$

**Theorem 3.13**  *For continuously differentiable functions $\phi, \psi : \mathbb{R}^n \to \mathbb{R}$ and differential forms we have the following relations:*

*(i)*  $d(\phi\psi) = \phi \, d\psi + \psi \, d\phi.$

*(ii)*  *From* $d\phi = \sum_{\ell=1}^n y_\ell \, dx_\ell$ *and* $y_j \neq 0$ *it follows that* $dx_j = \dfrac{1}{y_j} \left( d\phi - \sum_{\substack{\ell=1 \\ \ell \neq j}}^n y_\ell \, dx_\ell \right).$

*Proof*  The relation (i) is a consequence of the product rule:

$$\langle d(\phi\psi), t \rangle = \sum_{j=1}^n \frac{\partial}{\partial x_j}(\phi\psi) \, t_j = \sum_{j=1}^n \left( \phi \frac{\partial \psi}{\partial x_j} + \psi \frac{\partial \phi}{\partial x_j} \right) t_j = \langle \phi \, d\psi, t \rangle + \langle \psi \, d\phi, t \rangle .$$

The implication (ii) can be verified by following computation:

$$\langle dx_j, t \rangle = e_j \cdot t = t_j$$

and, because of $\frac{\partial \phi}{\partial x_\ell} = y_\ell$,

$$\left\langle \frac{1}{y_j} \left( d\phi - \sum_{\substack{\ell=1 \\ \ell \neq j}}^{n} y_\ell \, dx_\ell \right), t \right\rangle = \frac{1}{y_j} \langle d\phi, t \rangle - \sum_{\substack{\ell=1 \\ \ell \neq j}}^{n} \frac{y_\ell}{y_j} t_\ell$$

$$= \sum_{\ell=1}^{n} \frac{1}{y_j} \frac{\partial \phi}{\partial x_\ell} t_\ell - \sum_{\substack{\ell=1 \\ \ell \neq j}}^{n} \frac{y_\ell}{y_j} t_\ell = t_j .$$

$\square$

**Applications in Thermodynamics**

Differential forms are popular in thermodynamics because with their help many relations can be formulated without the need to consider artificial thermodynamic processes together with the change of the corresponding thermodynamic quantities. For example the evolution equation of entropy for reversible changes of state

$$\dot{S} = \frac{1}{T}\dot{U} + \frac{p}{T}\dot{V}$$

shortly can be written as

$$dS = \frac{1}{T}dU + \frac{p}{T}dV .$$

This equation does not need any time derivative of virtual thermodynamic processes and furthermore contains the Gibbs formula (3.13). By applying the calculus rules of Theorem 3.13 one obtains directly

$$dU = T\,dS - p\,dV$$

which replaces the relation $\dot{U} = T\dot{S} - p\dot{V}$. In the same way also using the calculus rules of Theorem 3.13 we derive for the thermodynamic potentials

$$dF = d(U - TS) = dU - T\,dS - S\,dT = -S\,dT - p\,dV ,$$
$$dH = d(U + pV) = dU + p\,dV + V\,dp = T\,dS + V\,dp ,$$
$$dG = d(U - TS + pV) = dU - T\,dS - S\,dT + p\,dV + V\,dp = -S\,dT + V\,dp .$$

Also the Legendre transform can be introduced in a simple fashion by means of the calculus of differential forms: For a function $f$ with a complete differential

$$df = \sum_{j=1}^{n} y_j \, dx_j$$

the Legendre transform is defined by

$$f^\star = \sum_{j=1}^{n} x_j y_j - f \, .$$

The complete differential of the Legendre transform is

$$df^\star = \sum_{j=1}^{n} (y_j \, dx_j + x_j \, dy_j) - df = \sum_{j=1}^{n} x_j \, dy_j \, .$$

## 3.8 Thermodynamics of Mixtures and the Chemical Potential

In reality many substances are in fact a mixture of different components. If the composition of a considered mixture is allowed to change temporally or spatially, then also the influence of this composition to the thermodynamics of the mixture has to be modeled. Among others this is of importance for models for diffusion processes or for chemical reactions.

We consider a mixture consisting of $M$ components with masses $m_1, \ldots, m_M$. Let us assume that the mixture is subdivided into two subsystems, which are separated by a membrane. The membrane is supposed to be permeable for the component 1, permeable for heat and moveable, such that the pressure and the temperature in both subsystems are the same. We now want to determine, how the component 1 is distributed across the two subsystems. Hereby we assume that the temperature $T$ and the pressure $p$ are constant, thus considering the isothermal, isobaric case. Therefore the thermodynamic equilibrium is characterized by the fact that the free enthalpy $G$ assumes its minimum. If, as sketched in Fig. 3.5, the masses of the components in the subsystems 1 and 2 are denoted by $m_j^{(1)}$ and $m_j^{(2)} = m_j - m_j^{(1)}$ and the free enthalpies of the subsystems by $G_1$ and $G_2$, respectively, then the total free enthalpy is given by

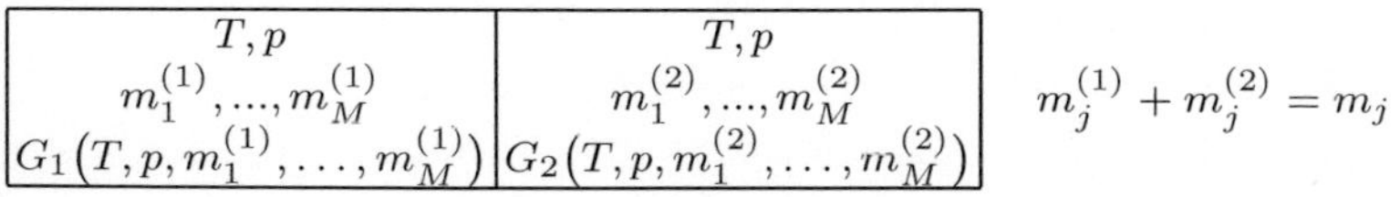

| $T, p$ | $T, p$ | |
|---|---|---|
| $m_1^{(1)}, \ldots, m_M^{(1)}$ | $m_1^{(2)}, \ldots, m_M^{(2)}$ | $m_j^{(1)} + m_j^{(2)} = m_j$ |
| $G_1\big(T, p, m_1^{(1)}, \ldots, m_M^{(1)}\big)$ | $G_2\big(T, p, m_1^{(2)}, \ldots, m_M^{(2)}\big)$ | |

**Fig. 3.5** Notation for the definition of the chemical potentials

$$G\left(T, p, m_1^{(1)}, \ldots, m_M^{(1)}\right) = G_1\left(T, p, m_1^{(1)}, \ldots, m_M^{(1)}\right)$$
$$+ G_2\left(T, p, m_1 - m_1^{(1)}, \ldots, m_M - m_M^{(1)}\right).$$

In equilibrium the component 1 is distributed in such a way that $G$ is minimal. A necessary condition for this is

$$\frac{\partial G}{\partial m_1^{(1)}} = \frac{\partial G_1}{\partial m_1^{(1)}} - \frac{\partial G_2}{\partial m_1^{(2)}} = 0, \ \text{i.e.,} \ \frac{\partial G_1}{\partial m_1^{(1)}} = \frac{\partial G_2}{\partial m_1^{(2)}}.$$

This means that the derivative of the free enthalpy $G$ with respect to the component $m_1$ is *continuous* at thermodynamic equilibrium. For a mixture with free enthalpy $G = G(T, p, m_1, \ldots, m_M)$ the quantity

$$\frac{\partial G}{\partial m_j} =: \mu_j$$

is called the *chemical potential* of the component $j$. If a membrane, which separates the two mixtures, is permeable for the component $j$ of the mixture, then $\mu_j$ attains the same value on both sides of the membrane.

From the definition of the chemical potential we obtain the differential relation

$$dG = -S\, dT + V\, dp + \sum_{j=1}^{M} \mu_j\, dm_j.$$

By using different versions of the Legendre transform one can derive the following relations for mixtures:

$$dU = T\, dS - p\, dV + \sum_{j=1}^{M} \mu_j\, dm_j,$$

$$dF = -S\, dT - p\, dV + \sum_{j=1}^{M} \mu_j\, dm_j,$$

$$dH = T\, dS + V\, dp + \sum_{j=1}^{M} \mu_j\, dm_j,$$

$$dS = \frac{1}{T}\, dU + \frac{p}{T}\, dV - \sum_{j=1}^{M} \frac{\mu_j}{T}\, dm_j.$$

Despite its name the chemical potential is no potential in the mathematical sense. Mathematically it is rather the *derivative* of a potential, namely the free enthalpy, with respect to a variable, namely the mass of the component considered. Derivatives of potentials often are denoted as *driving forces*. This notation reflects the perception that the considered system moves in the direction of the negative gradient of the potential and thus approaches the minimum of the potential over time.

In the following we will compile several important properties of chemical potentials. To simplify the notation the masses of the components are gathered in a mass vector $m = (m_1 \ldots, m_M)^\top$. Assuming that the free enthalpy is a twice continuously differentiable function then the mixed second partial derivatives can be exchanged. Therefore it follows that

$$\frac{\partial \mu_j}{\partial m_k} = \frac{\partial^2 G}{\partial m_j \partial m_k} = \frac{\partial \mu_k}{\partial m_j} \quad \text{for } j, k = 1, \ldots, M.$$

As all thermodynamic potentials, the free enthalpy $G$ is an *extensive* property, i.e., its value is proportional to the size of the system. The size of the system here will be denoted by means of masses. Therefore we have

$$G(T, p, \alpha m) = \alpha\, G(T, p, m) \quad \text{for } \alpha \geq 0.$$

This means that $G$ is 1-*homogeneous* with respect to the variable $m$:

**Definition 3.14**  A function $f : \mathbb{R}^n \to \mathbb{R}$ is called *k-homogeneous*, if

$$f(\alpha x) = \alpha^k f(x) \quad \text{for all } x \in \mathbb{R}^n, \ \alpha > 0.$$

From this simple property several important conclusions can be derived. On the one hand side we have

$$\frac{\partial G(T, p, \alpha m)}{\partial \alpha} = \frac{\partial}{\partial \alpha}\big(\alpha\, G(T, p, m)\big) = G(T, p, m)$$

and on the other hand

$$\frac{\partial G(T, p, \alpha m)}{\partial \alpha} = \sum_{j=1}^{M} \frac{\partial G(T, p, \alpha m)}{\partial m_j}\, m_j.$$

For $\alpha = 1$ using $\mu_j = \mu_j(T, p, m) = \frac{\partial G(T,p,m)}{\partial m_j}$ we get

$$G(T, p, m) = \sum_{j=1}^{M} \mu_j(T, p, m)\, m_j. \qquad (3.17)$$

Differentiation with respect to $m_k$ leads to

$$\mu_k = \frac{\partial G(T, p, m)}{\partial m_k} = \sum_{j=1}^{M} \frac{\partial \mu_j}{\partial m_k} m_j + \mu_k.$$

From this we conclude the *Gibbs–Duhem equation*

$$\sum_{j=1}^{M} \frac{\partial \mu_j}{\partial m_k} m_j = \sum_{j=1}^{M} \frac{\partial \mu_k}{\partial m_j} m_j = 0 \,. \tag{3.18}$$

In addition we have

$$\alpha \mu_j(T, p, \alpha m) = \frac{\partial G(T, p, \alpha m)}{\partial m_j} = \alpha \frac{\partial G(T, p, m)}{\partial m_j} = \alpha \mu_j(T, p, m)$$

and therefore

$$\mu_j(T, p, \alpha m) = \mu_j(T, p, m) \ \text{ for } \alpha > 0 \,.$$

This means that $\mu_j$ is 0-*homogeneous* with respect to $m$. It also means that $\mu_j$ is an intensive quantity as it does not depend on the total mass $m = m_1 + \cdots + m_M$, but only on the mass fractions which are also denoted here as *concentrations* $c_j = m_j/m$,

$$\mu_j(T, p, m) = \widetilde{\mu}_j(T, p, c) \,, \text{ where } c = (c_1, \ldots, c_M)^{\mathsf{T}}, \ c_j = \frac{m_j}{m} \,.$$

**Examples**

The most simple example is a *pure* substance. Then we have $M = 1, m_1 = m$ and

$$\mu_j(T, p, m) = \mu(T, p) = \frac{G(T, p, m)}{m} \,.$$

For an ideal gas with $z$ degrees of freedom the free enthalpy is

$$G(T, p, m) = \tfrac{z+2}{2} N k_B T \left(1 - \ln\left(\tfrac{T}{T_0}\right)\right) + N k_B T \ln\left(\tfrac{p}{p_0}\right) + \alpha N m_0 T$$
$$= \tfrac{z+2}{2} r m T \left(1 - \ln\left(\tfrac{T}{T_0}\right)\right) + r m T \ln\left(\tfrac{p}{p_0}\right) + \alpha m T \,,$$

where $m_0$ is the atomic mass, $N$ is the number of particles, $m = N m_0$ is the total mass of the system, and $r = k_B/m_0$. The term $\alpha m T$ stems from the integration constant in the entropy; here it must be scaled with $m$, because the entropy is an extensive quantity. Therefore the chemical potential reads

$$\mu(T, p) = \tfrac{z+2}{2} r T \left(1 - \ln\left(\tfrac{T}{T_0}\right)\right) + r T \ln\left(\tfrac{p}{p_0}\right) + \alpha T \,.$$

A more significant example is the *mixture of different ideal gases*. The equation of state for ideal gases is

$$pV = N k_B T \,.$$

We consider a mixture of $M$ different ideal gases. We assume that gas $j$ consists of $N_j$ particles, the mass of each particle being $m_j^{(0)}$. Then the total mass of gas $j$ is

$m_j = N_j m_j^{(0)}$. The mixture of these gases will be also considered as an ideal gas and therefore the equation of state of the mixture can be interpreted in two ways: The gas $j$ fills the total volume $V$ and thus exerts a *partial pressure*

$$p_j V = N_j k_B T .$$

Then the total pressure is

$$p = \sum_{j=1}^{M} p_j .$$

On the other hand a *partial volume* $V_j$ can be attributed to every gas such that

$$p V_j = N_j k_B T ,$$

then the total volume is

$$V = \sum_{j=1}^{M} V_j .$$

Comparing both interpretations we arrive at the equation of state

$$pV = N k_B T$$

for the total mixture and obtain the formulas

$$\frac{p_j}{p} = \frac{V_j}{V} = \frac{N_j}{N} .$$

The internal energy is the sum of the internal energy of the single substances, i.e.,

$$U = \sum_{j=1}^{M} \tfrac{z_j}{2} N_j k_B T ,$$

where $z_j$ is the number of degrees of freedom of gas $j$. Similarly the entropy is the sum of the single entropies of the components, but using the partial pressure $p_j$ for component $j$. Thus one obtains

$$S = \sum_{j=1}^{M} \left( \tfrac{z_j+2}{2} N_j k_B \ln \left( \tfrac{T}{T_0} \right) - N_j k_B \ln \left( \tfrac{p_j}{p_0} \right) - N_j m_j^{(0)} \alpha_j \right)$$

with the integration constant $N_j m_j^{(0)} \alpha_j$. This means that we interpret the mixture as a composite system of $M$ different, independent subsystems. Using the formula $\ln(p_j/p_0) = \ln(p_j/p) + \ln(p/p_0)$ and the shorthand notation $r_j = k_B/m_j^{(0)}$ one can write the entropy in the form

$$S = \sum_{j=1}^{M} S_j + S_{\text{mix}}$$

where

$$S_j = \frac{z_j+2}{2} N_j k_B \ln\left(\frac{T}{T_0}\right) - N_j k_B \ln\left(\frac{p}{p_0}\right) - N_j m_j^{(0)} \alpha_j$$

$$= m_j \left(\frac{z_j+2}{2} r_j \ln\left(\frac{T}{T_0}\right) - r_j \ln\left(\frac{p}{p_0}\right) - \alpha_j\right)$$

and the *mixing entropy* is given as

$$S_{\text{mix}} = -\sum_{j=1}^{M} N_j k_B \ln\left(\frac{p_j}{p}\right).$$

Taking into account

$$\frac{p_j}{p} = \frac{N_j}{N} = \frac{\frac{m_j}{m_j^{(0)}}}{\sum_{k=1}^{M} \frac{m_k}{m_k^{(0)}}} =: X_j(m) \tag{3.19}$$

the mixing entropy can be written as a function of $m_1, \ldots, m_M$:

$$S_{\text{mix}}(m) = -\sum_{j=1}^{M} m_j r_j \ln(X_j(m)).$$

The quantity $X_j(m)$ is denoted as the *mole fraction* of component $j$. A mole is defined as $6{,}02214179 \cdot 10^{23}$ particles of a substance. This value is chosen such that a mole of carbon atom nuclei consisting of 6 protons and 6 neutrons has the exact weight of 12 g. The *Avogadro constant* is defined to be $6{,}02214179 \cdot 10^{23}\text{mol}^{-1}$. Then the *mole number* of a substance $j$ is $\nu_j = N_j/N_A$ and the total mole number is $\nu = N/N_A = \sum_{j=1}^{M} \nu_j$. In particular we have $X_j(m) = N_j/N = \nu_j/\nu$. To measure a substance in moles is a commonplace primarily in chemistry. From the mole number and the number of particles in the nucleus (protons and neutrons) of a substance its weight can be computed easily: up to a small deviation, mole number times nuclear particles equal the weight in gram. If particle numbers are measured in mole, then the constant $r_j$ can be represented differently:

$$r_j = \frac{k_B}{m_j^{(0)}} = \frac{R}{M_j},$$

where $M_j = N_A m_j^{(0)}$ is the mass of a mole of particles from substance $j$ and $R = N_A k_B$ denotes the *universal gas constant*.

The free enthalpy of a mixture of ideal gases is given by

$$G = U - TS + pV = \sum_{j=1}^{M} G_j(T, p, m_j) + G_{\text{mix}}(T, p, m)$$

where

$$G_j(T, p, m_j) = m_j \frac{z_j+2}{2} r_j T \left(1 - \ln\left(\frac{T}{T_0}\right)\right) + m_j r_j T \ln\left(\frac{p}{p_0}\right) + \alpha_j m_j T$$

and

$$G_{\text{mix}}(T, p, m) = -T S_{\text{mix}}(m) .$$

To determine the chemical potential first we calculate

$$\sum_{k=1}^{M} m_k r_k \frac{\partial}{\partial m_j} \ln(X_k(m)) = \sum_{k=1}^{M} N k_B \frac{\partial}{\partial m_j} X_k(m) = 0 ,$$

which is true because $m_k r_k / X_k(m) = N k_B$ and $\sum_{k=1}^{M} X_k(m) = 1$. Hence, it follows that

$$\mu_j = \frac{\partial G}{\partial m_j} = \frac{z_j+2}{2} r_j T \left(1 - \ln\left(\frac{T}{T_0}\right)\right) + r_j T \ln\left(\frac{p}{p_0}\right) + r_j T \ln(X_j(m)) + \alpha_j T$$

$$= \mu_j^{(0)}(T, p) + r_j T \ln(X_j(m))$$

with the chemical potentials of the corresponding pure substances

$$\mu_j^{(0)}(T, p) = \frac{z_j+2}{2} r_j T \left(1 - \ln\left(\frac{T}{T_0}\right)\right) + r_j T \ln\left(\frac{p}{p_0}\right) + \alpha_j T . \qquad (3.20)$$

*Ideal mixtures* are a generalization of mixtures of ideal gases. For an ideal mixture we do not assume any more to have ideal gases but nevertheless we postulate the validity of

$$U(T, p, m) = \sum_{j=1}^{M} U_j(T, p, m_j) \quad \text{and}$$

$$S(T, p, m) = \sum_{j=1}^{M} S_j(T, p, m_j) + S_{\text{mix}}(m)$$

for the internal energy and the entropy, respectively, where $U_j$ and $S_j$ are the internal energy and the entropy of the corresponding pure substance and

$$S_{\text{mix}}(m) = -\sum_{j=1}^{M} m_j \, r_j \, \ln(X_j(m))$$

with the mole fraction $X_j(m)$, defined in (3.19). The free enthalpy is

$$G(T, p, m) = \sum_{j=1}^{M} G_j(T, p, m_j) - T S_{\text{mix}}(m)$$

with the free enthalpies $G_j(T, p, m_j)$ of the pure substances. Then the chemical potential reads

$$\mu_j(T, p, m) = \mu_j^{(0)}(T, p) + r_j \, T \, \ln(X_j(m)) \,, \tag{3.21}$$

$$\text{where } \mu_j^{(0)}(T, p) = \frac{\partial G_j}{\partial m_j}(T, p, m_j) \,.$$

The examples discussed so far are still quite simple. For real mixtures sometimes heating or cooling and change of volume is observed during the mixing process. For such mixtures it is not possible to decouple additively the internal energy and the equation of state in the way as it is done above. Often one finds models, where the formula (3.21) for the chemical potential of ideal mixtures is modified. For example this might happen by substituting the mole fraction $X_j(m)$ by a more general function, leading to

$$\mu_j(T, p, m) = \mu_j^{(0)}(T, p) + r_j \, T \, \ln(a_j(T, p, m)) \,. \tag{3.22}$$

Here the functions $a_j(T, p, m)$ are called the *activity*. Often the activity is written in the form

$$a_j(T, p, m) = \gamma_j(T, p, m) \, X_j(m) \,,$$

where the factor $\gamma_j$ is called the *activity coefficient*. It describes the deviation of the mixture from the situation of an ideal mixture, where $\gamma_j(T, p, m) = 1$ is valid.

## 3.9 Chemical Reactions in Multi Species Systems

Chemical substances consist of molecules, which are built from atoms. During a chemical reaction the total mass of an atom has to be conserved. This can be assured by special formulations which will be developed in the following.

A *chemical species* is characterized by

(I)  a *molecular formula*, for example $H_2O$ for water or $C_6H_{12}O_6$ for glucose,
(II)  a certain *molecular structure*, in the case where for the same molecular formula different molecular structure exists. This for example holds true for glucose,

(III)   a *phase*, i.e., being in the solid, liquid or gaseous phase.

One speaks of a *chemical substance*, if it is not necessary to specify the phase. A *chemical element* is a substance, which with respect to the considered problem cannot be further subdivided, for example a specific kind of atoms.

A *chemical system* consists of

- a list of different chemical species,
- a list of elements from which the species are composed.

A chemical system thus is represented by an ordered list of species and of elements. If necessary the molecular formula of a species is amended by the molecular structure or the phase in some short hand notation.

**Example**: The system

$$\{(Na_2O(\ell), \ NaOH(\ell), \ NaCl(\ell), \ H_2O(\ell)), \ (H, \ O, \ Na, \ Cl)\}$$

consists of sodium oxide, sodium hydroxide, sodium chloride (common salt) and water. The attached "$(\ell)$" indicates that the substance is a "liquid" or is a solute in a liquid.

In the following we omit the phase and the molecular structure in the notation. We consider a chemical system with $N_S$ species and $N_E$ elements, where both the species and the elements are numbered. Therefore a species can be described by a *formula vector* $a \in \mathbb{N}_0^{N_E}$ where the $i$-th component denotes the number of atoms of the $i$-th element in the molecular formula. For example the molecular vector for $Na_2O$ in the system given above is $a = (0, 1, 2, 0)^\top$. The molecular vectors can be arranged to form the columns of a matrix leading to the *formula matrix* $A = \left(a^{(1)}, \ldots, a^{(N_S)}\right) \in \mathbb{R}^{N_E, N_S}$. Here $a^{(j)}$ denotes the formula vector of the $j$-th chemical species. For the example above we have

$$A = (a_{ij})_{i=1\,j=1}^{N_E\ N_S} = \begin{pmatrix} 0 & 1 & 0 & 2 \\ 1 & 1 & 0 & 1 \\ 2 & 1 & 1 & 0 \\ 0 & 0 & 1 & 0 \end{pmatrix}.$$

The *quantity* of a chemical species typically is expressed in *mole*. The composition of a chemical species is described by the *element-mole vector* $e = (e_1, \ldots, e_{N_E})^\top$, whose components are the corresponding mole numbers of the elements in the species, and a *species-mole vector* $\nu = (\nu_1, \ldots, \nu_{N_S})^\top$, which contains the mole numbers of the species. The element-mole vector can be computed from the species-mole vector $\nu$: In species $j$ the element $i$ appears exactly $a_{ij}$ times. Hence, we have

$$e_i = \sum_{j=1}^{N_S} a_{ij}\nu_j ,$$

or in short hand notation,

$$e = A\nu.  \tag{3.23}$$

In a *closed* system, for which there is no flow of elements in or out, the element-mole vector is constant, but the species-mole vector can be variable due to chemical reactions. For the *rate of change* $\dot{\nu}$ of a species-mole vector we have

$$A\dot{\nu} = \dot{e} = 0, \quad \text{or } \dot{\nu} \in \text{Ker } A.  \tag{3.24}$$

This relation describes the *conservation of mass* of the single elements despite the chemical reactions. It will be denoted as *element-abundance constraint*.

For every matrix $A \in \mathbb{R}^{N_E, N_S}$ we have the orthogonal decomposition

$$\mathbb{R}^{N_S} = \text{Im}\left(A^\top\right) + \text{Ker } A.$$

Let $\left\{y^{(1)}, \ldots, y^{(N_R)}\right\}$ be a basis of Ker $A$, $\left\{z^{(1)}, \ldots, z^{(N_C)}\right\}$ a basis of Im $\left(A^\top\right)$, where $N_R = \dim \text{Ker } A$, $N_C = \text{rank } A$, and

$$Y = \left(y^{(1)}, \ldots, y^{(N_R)}\right) \in \mathbb{R}^{N_S, N_R}, \quad Z = \left(z^{(1)}, \ldots, z^{(N_C)}\right) \in \mathbb{R}^{N_S, N_C}.$$

In particular, we have $N_R + N_C = N_S$. Then a species-mole vector $\nu \in \mathbb{R}^{N_S}$ can be uniquely expressed as

$$\nu = Y\xi + Z\eta \quad \text{where } \xi \in \mathbb{R}^{N_R}, \ \eta \in \mathbb{R}^{N_C}.$$

From the element-abundance constraint $\dot{\nu} \in \text{Ker } A$ we conclude that $\eta$ cannot be changed by a reaction, therefore we have

$$\nu = Y\xi + Z\eta^{(0)}$$

with a vector $\eta^{(0)} \in \mathbb{R}^{N_C}$. The reaction therefore is described by the vector $\xi \in \mathbb{R}^{N_R}$ in a unique way. The components of $\xi$ are denoted by *reaction coordinates*. The dimension $N_R$ is just the number of *independent* reactions. The components of $\eta$ are called *reaction invariants*.

**Definition 3.15** The matrix $Y$ is called the *stoichiometric* matrix, the entry $y_{ij}$ of $Y$ is the *stoichiometric coefficient* of the $i$-th species in the $j$-th independent chemical reaction. The columns $y^{(1)}, \ldots, y^{(N_R)}$ of $Y$ are called *reaction vectors*.

The space spanned by the columns of $Y$, being Ker $A$, is unique but the matrix $Y$ itself is not unique. In general there is some freedom in the choice of the *independent reactions*. Having the matrix $A$, the matrix $Y$ can be determined by applying the Gaussian elimination method to $A$ to transform it into the form

$$\begin{pmatrix} I_{N_C} & \widehat{A} \\ 0 & 0 \end{pmatrix} \text{ with the identity matrix } I_{N_C} \in \mathbb{R}^{N_C, N_C} \text{ and } \widehat{A} \in \mathbb{R}^{N_C, N_R}.$$

This is possible, if besides the exchange of rows also the exchange of *columns* is allowed. Doing so, one has to take into account that the exchange of columns changes the sequence of chemical species in the list of chemical species, the exchange of rows changes only the sequence of elements. Based on the above form the stoichiometric matrix can be chosen as

$$Y = \begin{pmatrix} -\widehat{A} \\ I_{N_R} \end{pmatrix} \quad \text{with identity matrix } I_{N_R} \in \mathbb{R}^{N_R,N_R} , \tag{3.25}$$

because

$$\begin{pmatrix} I_{N_C} & \widehat{A} \\ 0 & 0 \end{pmatrix} \begin{pmatrix} -\widehat{A} \\ I_{N_R} \end{pmatrix} = 0$$

and $Y$ has maximal rank $N_R$. Equation (3.25) is called the *canonical form* of the stoichiometric matrix.

**Example 1:** We consider the chemical system $\{(H_2O, H, OH), (H, O)\}$. Here we have $N_S = 3$, $N_E = 2$, and the formula matrix has the form

$$A = \begin{pmatrix} 2 & 1 & 1 \\ 1 & 0 & 1 \end{pmatrix} .$$

Gaussian elimination leads to

$$\begin{pmatrix} 1 & 0 & 1 \\ 0 & 1 & -1 \end{pmatrix} = (I \ \widehat{A}) \quad \text{where } \widehat{A} = \begin{pmatrix} 1 \\ -1 \end{pmatrix} .$$

This result can be achieved without exchanging columns. Hence, we have $N_C = 2$, $N_R = 1$, and the stoichiometric matrix consists of the reaction vector

$$y^\top = Y^\top = (-1 \ 1 \ 1) .$$

This describes the only possible reaction

$$H + OH \rightleftharpoons H_2O .$$

**Example 2**: We consider the system

$$\{(Na_2O, CrCl_3, NaOH, NaCl, H_2O, Na_2CrO_4, Cl_2), (H, O, Na, Cr, Cl)\}$$

consisting of sodium oxide $Na_2O$, chromium chloride $CrCl_3$, sodium hydroxide NaOH, sodium chloride NaCl, water, sodium chromate $Na_2CrO_4$ and chlorine. Here we have $N_S = 7$, $N_E = 5$,

$$A = \begin{pmatrix} 0\ 0\ 1\ 0\ 2\ 0\ 0 \\ 1\ 0\ 1\ 0\ 1\ 4\ 0 \\ 2\ 0\ 1\ 1\ 0\ 2\ 0 \\ 0\ 1\ 0\ 0\ 0\ 1\ 0 \\ 0\ 3\ 0\ 1\ 0\ 0\ 2 \end{pmatrix}.$$

Gaussian elimination leads to

$$A \rightarrow \begin{pmatrix} 1\ 0\ \ 1\ 0\ \ 1\ \ \ 4\ \ 0 \\ 0\ 1\ \ 0\ 0\ \ 0\ \ \ 1\ \ 0 \\ 0\ 0\ \ 1\ 0\ \ 2\ \ \ 0\ \ 0 \\ 0\ 0\ -1\ 1\ -2\ -6\ 0 \\ 0\ 0\ \ 0\ 1\ \ 0\ -3\ 2 \end{pmatrix} \begin{matrix} (2) \\ (4) \\ (1) \\ (3) - 2(2) \\ (5) - 3(4) \end{matrix}$$

$$\rightarrow \begin{pmatrix} 1\ 0\ 0\ 0\ -1\ \ 4\ \ 0 \\ 0\ 1\ 0\ 0\ \ 0\ \ \ 1\ \ 0 \\ 0\ 0\ 1\ 0\ \ 2\ \ \ 0\ \ 0 \\ 0\ 0\ 0\ 1\ \ 0\ -6\ 0 \\ 0\ 0\ 0\ 0\ \ 0\ \ \ 3\ \ 2 \end{pmatrix} \begin{matrix} (1) - (3) \\ \ \\ \ \\ (4) + (3) =: (4)^{new} \\ (5) - (4)^{new} \end{matrix}$$

After exchanging the fifth and the seventh column we get as a result

$$\begin{pmatrix} 1\ 0\ 0\ 0\ 0\ \ 4\ \ -1 \\ 0\ 1\ 0\ 0\ 0\ \ 1\ \ \ 0 \\ 0\ 0\ 1\ 0\ 0\ \ 0\ \ \ 2 \\ 0\ 0\ 0\ 1\ 0\ -6\ \ 0 \\ 0\ 0\ 0\ 0\ 1\ 3/2\ \ 0 \end{pmatrix},$$

Hence, the new ordering of the chemical species is

$$(Na_2O, CrCl_3, NaOH, NaCl, Cl_2, Na_2CrO_4, H_2O).$$

Here we have $N_C = 5$, $N_R = 2$, and the canonical form of the stoichiometric matrix reads

$$Y = \begin{pmatrix} -4 & 1 \\ -1 & 0 \\ 0 & -2 \\ 6 & 0 \\ -3/2 & 0 \\ 1 & 0 \\ 0 & 1 \end{pmatrix}.$$

The two column vectors correspond to the two independent reactions

$$12\,NaCl + 2\,Na_2CrO_4 \rightleftharpoons 8\,Na_2O + 2\,CrCl_3 + 3\,Cl_2,$$
$$Na_2O + H_2O \rightleftharpoons 2\,NaOH.$$

Here the first column of the stoichiometric matrix has been scaled by the factor two to transfer the component 3/2 to an integer value.

## 3.10  Equilibria of Chemical Reactions and the Mass Action Law

The point of equilibrium of a chemical system is the distribution of substances, which for its specific chemical reactions is achieved for given temperature and a given pressure for large instances of time. This point of equilibrium is characterized by the fact that the free enthalpy $G = G(T, p, m)$ *attains its minimum*, where $m$ denotes the vector of masses of the chemical substances. These masses can be computed from the mole numbers by means of

$$m_j = m_j(\nu_j) = M_j \nu_j$$

with the mole mass $M_j = N_A m_j^{(0)}$ of a species $j$, where $m_j^{(0)}$ denotes the mass of a single molecule and $N_A \approx 6{,}02214179 \cdot 10^{23}$ mol$^{-1}$ is the *Avogadro constant*. The mole numbers have to fulfill the following constraints:

(I)  $\nu_j \geq 0$ for $j = 1, \dots, N_S$.
(II)  The element-abundance constraint $\nu - \nu^{(0)} \in \mathrm{Ker}\,A$, where $A$ is the formula matrix and $\nu^{(0)}$ a given mole vector. For example this may be the collection of mole numbers of the species present at the beginning of the chemical reaction.

In total one obtains an optimization problem with equality and inequality constraints,

$$\min\left\{ G(T, p, m(\nu)) \,\big|\, \nu_j \geq 0 \text{ for } j = 1, \dots, N_S,\ A(\nu - \nu^{(0)}) = 0 \right\}.$$

If the chemical reaction is written by means of reaction coordinates $\xi$, then we have

$$\nu(\xi) = \nu^{(0)} + Y\xi$$

with the stoichiometric matrix $Y$. Then the optimization problem to be solved reads

$$\min\left\{ G(T, p, m(\xi)) \,\big|\, \nu_j(\xi) \geq 0 \text{ for } j = 1, \dots, N_S \right\}.$$

Now this is an optimization problem only with inequality constraints. The mass vector is given by

$$m(\xi) = M\nu^{(0)} + MY\xi,$$

where $M = \text{diag}(M_1, \ldots, M_{N_S})$ denotes the diagonal matrix composed of the mole masses.

Here we consider the most simple case that at the optimum none of the inequality constraints is *active*, i.e., that the argument of $\xi^*$ of the minimum also satisfies $\nu_j(\xi^*) > 0$ for $j = 1, \ldots, N_S$. Then the necessary optimality criterion takes the form

$$0 = \frac{\partial}{\partial \xi_j} G(T, p, m(\xi)) = \sum_{k=1}^{N_S} \frac{\partial G}{\partial m_k}(T, p, m(\xi)) \frac{\partial m_k(\xi)}{\partial \xi_j}$$

$$= \sum_{k=1}^{N_S} \mu_k(T, p, m(\xi)) M_k y_{kj} \quad \text{for } j = 1, \ldots, N_R,$$

or, written in vector notion,

$$Y^\top M \mu = 0. \tag{3.26}$$

This relation is called the *mass action law*.

Often the chemical potential is given in the form

$$\mu_j(T, p, m) = \mu_j^{(0)}(T, p) + r_j T \ln(a_j(T, p, m)), \tag{3.27}$$

with $\mu_j^{(0)}(T, p)$ being the chemical potential of a pure species, $r_j = k_B / m_j^{(0)}$ being the individual gas constants, and the activities $a_j(T, p, m)$, cf. (3.22). We obtain

$$M_j \mu_j(T, p, m) = M_j \mu_j^{(0)}(T, p) + R T \ln(a_j)$$

with the universal gas constant $R = N_A k_B = M_j r_j$. Application of the exponential function to the mass action law leads to

$$\prod_{k=1}^{N_S} a_k(T, p, m)^{y_{kj}} = \exp\left(-\sum_{k=1}^{N_S} \frac{y_{kj} M_k \mu_k^{(0)}(T, p)}{R T}\right) =: K_j(T, p). \tag{3.28}$$

Here the factor $K_j(T, p)$ is *independent* of $\xi$ and is called the *equilibrium constant*. Using the so called *standard difference of the free enthalpy*

$$\Delta G_j^{(0)}(T, p) = \sum_{k=1}^{N_S} y_{kj} M_k \mu_k^{(0)}(T, p)$$

the equilibrium constant has the form

$$K_j(T, p) = e^{-\Delta G_j^{(0)}(T, p)/(R T)}.$$

In the case of an *ideal mixture* the activities are identical with the *molar fractions*

$$X_k = \frac{\nu_k}{\nu}, \quad \text{where } \nu = \nu_1 + \cdots + \nu_{N_S} .$$

Then the mass action law reads

$$\prod_{k=1}^{N_S} X_k^{y_{kj}} = K_j(T, p) .$$

Here the left-hand side only depends on the composition of the mixture, the right-hand side only on pressure and temperature.

In the case of the *mixture of ideal gases* additionally we have (3.20) and therefore

$$M_k \mu_k^{(0)}(T, p) = M_k \left( \tfrac{z_k+2}{2} r_k T \left( 1 - \ln \left( \tfrac{T}{T_0} \right) \right) + r_k T \ln \left( \tfrac{p}{p_0} \right) + \alpha_k T + \beta_k \right)$$
$$= \tfrac{z_k+2}{2} R T \left( 1 - \ln \left( \tfrac{T}{T_0} \right) \right) + R T \ln \left( \tfrac{p}{p_0} \right) + M_k \left( \alpha_k T + \beta_k \right) ,$$

where $z_k$ denotes the degrees of freedom of a molecule of substance $k$ and $\alpha_k$ and $\beta_k$ are constants. The constant $\beta_k$ has not yet appeared in the previous formulas of the chemical potentials. But considering chemical reactions it is necessary to take chemical binding energies as part of the internal energy into account. This leads to a modified formula for the internal energy

$$U = \frac{z}{2} N k_B T + N \widetilde{\beta}$$

for an ideal gas, where $\widetilde{\beta}$ just describes this chemical binding energy. Therefore an additional constant $\beta = \widetilde{\beta}/m^{(0)}$ appears in the formula of the chemical potential. The equilibrium constant $K_j(T, p)$ then takes the form

$$K_j(T, p) = \exp \left( - \sum_{k=1}^{N_S} y_{kj} \left( \tfrac{z_k+2}{2} \left( 1 - \ln \left( \tfrac{T}{T_0} \right) \right) + \ln \left( \tfrac{p}{p_0} \right) \right) \right.$$
$$\left. - \frac{1}{R} \sum_{k=1}^{N_S} M_k y_{kj} \left( \alpha_k + \tfrac{\beta_k}{T} \right) \right)$$
$$= K_j^{(0)}(T) \, p^{-\overline{Y}_j} ,$$

using the *row sum* $\overline{Y}_j = \sum_{k=1}^{N_S} y_{kj}$ and the temperature dependent constants

$$K_j^{(0)}(T) = p_0^{\overline{Y}_j} \exp \left( - \sum_{k=1}^{N_S} y_{kj} \left( \tfrac{z_k+2}{2} \left( 1 - \ln \left( \tfrac{T}{T_0} \right) \right) - \frac{1}{R} \sum_{k=1}^{N_S} M_k y_{kj} \left( \alpha_k + \tfrac{\beta_k}{T} \right) \right) .$$

Then the mass action law has the form

$$\prod_{k=1}^{N_S} \left(\frac{\nu_k}{\nu}\right)^{y_{kj}} = K_j^{(0)}(T)\, p^{-\overline{Y}_j}.$$

The specific form of $K_j^{(0)}(T)$ usually is determined by measurements.

Alternative to the mole numbers the *molar concentrations* $c_j = \nu_j/V$ can be used, where $V$ denotes the volume of the gas mixture. Then the mole fraction is $X_j = \nu_j/\nu = c_j V/\nu$ and for the mixture of ideal gases it follows

$$\prod_{k=1}^{N_S} \left(\frac{\nu_k}{\nu}\right)^{y_{kj}} = \prod_{k=1}^{N_S} c_k^{y_{kj}} \left(\frac{V}{\nu}\right)^{\overline{Y}_j} = K_j^{(0)}(T)\, p^{-\overline{Y}_j}.$$

Using the equation of state $pV = Nk_BT = \nu R T$ we can conclude

$$\prod_{k=1}^{N_S} c_k^{y_{kj}} = \widetilde{K}_j^{(0)}(T) \ \text{ with } \ \widetilde{K}_j^{(0)}(T) = \frac{K_j^{(0)}(T)}{(RT)^{\overline{Y}_j}}.$$

Here the equilibrium constant $\widetilde{K}_j$ does not depend on the pressure any longer. In the case where the stoichiometric matrix takes its *canonical form*

$$Y = \begin{pmatrix} -\widehat{\Lambda} \\ I \end{pmatrix},$$

then the mass action law can be rewritten as

$$\mu_{N_C+j} = \sum_{k=1}^{N_C} \widehat{a}_{kj} \frac{M_k}{M_{N_C+j}} \mu_k \ \text{ for } j = 1, \ldots, N_R$$

in the general case, or as

$$a_{N_C+j} = K_j(T, p) \prod_{k=1}^{N_C} (a_k(T, p, m))^{\widehat{a}_{kj}}, \quad j = 1, \ldots, N_R$$

for the representation (3.28), or as

$$X_{N_C+j} = K_j(T, p) \prod_{k=1}^{N_C} X_k^{\widehat{a}_{kj}}, \quad j = 1, \ldots, N_R$$

for the case of an ideal mixture, or as

$$c_{N_C+j} = \widetilde{K}_j^{(0)}(T) \prod_{k=1}^{N_C} c_k^{\widehat{a}_{kj}}, \quad j = 1, \ldots, N_R$$

for the case of the mixture of ideal gases.

**Example: Ammonia Synthesis**

We investigate the reaction

$$3\,H_2 + N_2 \rightleftharpoons 2\,NH_3 \,.$$

We assume that at the beginning of the reaction there are 3 mol $H_2$, 1 mol $N_2$ but no $NH_3$, and we assume the gases to be ideal. The chemical system is

$$\{(H_2, N_2, NH_3); (H, N)\} \,.$$

The corresponding formula matrix reads

$$A = \begin{pmatrix} 2 & 0 & 3 \\ 0 & 2 & 1 \end{pmatrix},$$

and using Gaussian elimination one obtains the form

$$\begin{pmatrix} 1 & 0 & 3/2 \\ 0 & 1 & 1/2 \end{pmatrix},$$

therefore the stoichiometric matrix is

$$Y = \begin{pmatrix} -3/2 \\ -1/2 \\ 1 \end{pmatrix}.$$

From the element-abundance constraint

$$\nu - \nu^{(0)} \in \text{Ker}\, A = \text{span}\left\{ \begin{pmatrix} -3/2 \\ -1/2 \\ 1 \end{pmatrix} \right\} \quad \text{where } \nu^{(0)} = \begin{pmatrix} 3 \\ 1 \\ 0 \end{pmatrix},$$

it follows

$$\begin{pmatrix} \nu_1 - 3 \\ \nu_2 - 1 \\ \nu_3 \end{pmatrix} = \lambda \begin{pmatrix} -3/2 \\ -1/2 \\ 1 \end{pmatrix},$$

or, after elimination of $\lambda$

$$\nu_1 + \tfrac{3}{2}\nu_3 = 3\,,$$
$$\nu_2 + \tfrac{1}{2}\nu_3 = 1\,.$$

The mass action law reads

$$\frac{X_3}{X_1^{3/2}X_2^{1/2}} = K(T, p),$$

where $X_j = \nu_j/\nu$, $\nu = \nu_1 + \nu_2 + \nu_3$ and the equilibrium constant has the form

$$K(T, p) = K_0(T)\, p.$$

This equation can be reformulated as

$$\frac{\nu_1^3\, \nu_2}{\nu_3^2\, \nu^2} = (K_0(T))^{-2} p^{-2}.$$

If one expresses all mole numbers in $\nu_3$, then this relation can be rewritten to

$$\frac{3^3}{2^4}\frac{(2 - \nu_3)^4}{\nu_3^2(4 - \nu_3)^2} = (K_0(T))^{-2} p^{-2}.$$

Now assuming a constant temperature $T$ we consider the two limit cases for "large" and "small" pressure $p$.

In the limit case $p \to 0$ the equilibrium condition takes the form

$$\nu_3^2(4 - \nu_3)^2 = 0,$$

and therefore $\nu_3 = 0$ or $\nu_3 = 4$. The solution $\nu_3 - 4$ is not feasible here, because it would lead to negative mole numbers $\nu_1 = 3 - \frac{3}{2}\nu_3$ and $\nu_2 = 1 - \frac{1}{2}\nu_3$. Hence, the correct solution is given by $\nu_3 = 0$, $\nu_1 = 3$ and $\nu_2 = 1$, i.e., no ammonia is produced.

In the other limit case $p \to +\infty$ the equilibrium condition reduces to

$$(2 - \nu_3)^4 = 0,$$

with the solution $\nu_3 = 2$, $\nu_1 = \nu_2 = 0$, i.e., now we have only ammonia.

## 3.11 Kinetic Reactions

The description of reactions using the mass action law only makes sense if the reaction rate is so large such that, during changes in the exterior conditions, in particular of the temperature and the pressure, equilibrium is reached sufficiently faster than these exterior conditions change. Reactions which are slow in this sense will be denoted as *kinetic reactions*. In general the change of the mole numbers will be modeled by *ordinary differential equations*:

$$\dot{\nu}_j = \widetilde{R}_j(T, p, \nu) \text{ for } j = 1, \ldots, N_S. \tag{3.29}$$

The exact form of the function $\widetilde{R}_j$ will be deduced from experimental data in practical applications. But on the basis of the results of Sects. 3.9 and 3.10 we can formulate necessary conditions for $\widetilde{R}$:

(I)  For $\nu_j = 0$ we must have $\widetilde{R}_j(T, p, \nu) \geq 0$ because otherwise for further instances of time the reaction would violate the condition $\nu_j \geq 0$

(II)  Because of $\dot{\nu} \in \text{Ker } A = \text{Im } Y$ we must have $\widetilde{R}(T, p, \nu) = Y R(T, p, \nu)$, where $R(T, p, \nu) \in \mathbb{R}^{N_R}$.

(III)  If $(T, p, \nu)$ is already a point of equilibrium, then $\widetilde{R}_j(T, p, \nu) = 0$ must be satisfied.

(IV)  The evolution of $\nu$ for given constant $T$ and $p$ must not increase the free enthalpy, i.e.,

$$\frac{d}{dt} G(T, p, m(\nu(t))) \leq 0 \text{ for fixed } T, p.$$

The condition (IV) is justified by the perception that also for a kinetic reaction the considered reaction system will move in the direction of a point of equilibrium. The only difference is that this movement is not fast enough to reach the equilibrium instantaneously. Hence, we do not require that the free enthalpy attains its minimum, but nevertheless that it does not increase for given constant exterior conditions.

Instead of the formulation (3.29) one can also use the *stoichiometric formulation*, taking into account the representation

$$\nu = \nu(\xi) = \nu^{(0)} + Y\xi.$$

Then one reaches at the following system of ordinary differential equations

$$\dot{\xi}_j = R_j(T, p, \xi) \text{ for } j = 1, \ldots, N_R,$$

where now the reaction coordinates are the variables. Here the condition (II) is automatically fulfilled.

Now we discuss possible ansatzes for the functions $R_j(T, p, \xi)$ in case that no constraints are active, i.e., that $\nu_j(\xi) > 0$ for $j = 1, \ldots, N_S$ is valid. The ansatzes are constructed in such a way, that the conditions (II), (III) and (IV) are satisfied.

(a)  $R(T, p, \xi) = -K \nabla_\xi G(T, p, m(\xi))$ with a positive definite matrix $K \in \mathbb{R}^{N_R, N_R}$. This matrix may also depend on $T$, $p$ and $\xi$. This form of $R$ satisfies the conditions (III) and (IV), because

$$\frac{d}{dt}G(T, p, m(\xi(t))) = \nabla_\xi G(\cdots) \cdot \dot{\xi} = -K \nabla_\xi G(\cdots) \cdot \nabla_\xi G(\cdots) \leq 0,$$

and in the minimum we have $\nabla_\xi G(T, p, m(\xi)) = 0$. The partial derivative

$$A_j := -\frac{\partial}{\partial \xi_j} G(T, p, m(\xi))$$

is called the *affinity* of the $j$-th reaction. The affinity indicates, how the free enthalpy is changed due to the $j$-th reaction, and therefore measures in which direction and with which velocity the reaction takes place under the given exterior conditions.

(b) In the literature often one finds models which are directly derived from the mass action law for ideal gases

$$\prod_{\ell=1}^{N_S} c_\ell^{y_{\ell j}} = \widetilde{K}_j^{(0)}(T).$$

To this end one subdivides the stoichiometric matrix according to its positive and negative entries, $y_{ij}^+ = \max\{y_{ij}, 0\}$, $y_{ij}^- = -\min\{y_{ij}, 0\}$. Then a possible reaction kinetics compatible with the mass action law is given by

$$R_j(T, p, c) = k_j^f(T, p) \prod_{\ell=1}^{N_S} c_\ell^{y_{\ell j}^-} - k_j^b(T, p) \prod_{\ell=1}^{N_S} c_\ell^{y_{\ell j}^+}. \qquad (3.30)$$

The coefficients $k_j^f(T, p)$ and $k_j^b(T, p)$ describe the rates of the reaction of its two directions. They must satisfy the condition $\frac{k_j^f(T,p)}{k_j^b(T,p)} = \widetilde{K}_j^{(0)}(T)$, as otherwise in the stationary limit $R_j(T, p, c) = 0$ the mass action law would not be valid. For this approach also (III) and (IV) holds true, as it is shown in Exercise 3.16.

If for the considered variable $\xi$ constraints are active, i.e., $\nu_j(\xi) = 0$ for at least one $j \in \{1, \ldots, N_S\}$, then the requirement $\dot{\nu}_j \geq 0$ has to be assured. One possibility to do so may be first to compute the vector $R$ of the reaction rates according to one of the models (a) or (b) and then to project this result to the space of admissible reaction rates

$$\{r \in \mathbb{R}^{N_R} \mid (Yr)_j \geq 0 \text{ for } j \in I_A(\xi)\}.$$

With $I_A(\xi) = \{j \in \{1, \ldots, N_S\} \mid \nu_j(\xi) = 0\}$ the index set of the active constraints is denoted. The implementation of such a projection leads to a quadratic optimization problem which will not be considered here in detail.

**The Coupling of Kinetic and Equilibrium Reactions**

In more complicated chemical systems there will be both kinetic reactions and equilibrium reactions. Hence, the description of equilibrium reactions by means of optimization problems and the description of kinetic reactions by means of ordinary differential equations have to be coupled. To do so one subdivides the vector $\xi$ of reaction coordinates and correspondingly the stoichiometric matrix $Y$ in the part of the kinetic reactions and the part of the equilibrium reactions:

$$\xi = \begin{pmatrix} \xi^K \\ \xi^E \end{pmatrix}, \quad Y = \begin{pmatrix} Y^K & Y^E \end{pmatrix},$$

where $\xi^K \in \mathbb{R}^{N_R^K}$, $\xi^E \in \mathbb{R}^{N_R^E}$, $Y^K \in \mathbb{R}^{N_S, N_R^K}$, $Y^E \in \mathbb{R}^{N_S, N_R^E}$. Here the upper index $K$ denotes the part of the kinetic reactions and the upper index $E$ the part of the equilibrium reactions, $N_R^K$ and $N_R^E$ the numbers of kinetic reactions or equilibrium reactions, respectively. Then we have

$$Y\xi = Y^K \xi^K + Y^E \xi^E.$$

For the kinetic reactions it holds true

$$\dot{\xi}^K = R^K\left(T, p, \xi^K, \xi^E\right), \tag{3.31}$$

where $R^K\left(T, p, \xi^K, \xi^E\right)$ is the vector of the reaction rates. The equilibrium reactions are described by means of the optimization problem

$$\xi^E = \arg\min \left\{ G\left(T, p, m\left(\xi^K, \xi^E\right)\right) \mid \nu_j\left(\xi^K, \xi^E\right) \geq 0 \right.$$
$$\left. \text{for } j = 1, \ldots, N_S \right\}. \tag{3.32}$$

Here we have $m\left(\xi^K, \xi^E\right) = M\left(\nu^{(0)} + Y^K \xi^K + Y^E \xi^E\right)$. If no constraint is active for the minimum, then from (3.32) we conclude

$$Y_E^\top M \, \mu\left(T, p, m\left(\xi^K, \xi^E\right)\right) = 0. \tag{3.33}$$

The coupling of (3.31) and (3.33) then is an differential algebraic equation.

The most simple case appears for a mixture of ideal gases if at the minimum of the equilibrium reactions the constraints are not active. The reaction kinetics (3.30), using $c_\ell = \nu_\ell / V$ for the molar concentrations, leads to

$$\dot{\xi}_j^K = k_j^f \prod_{\ell=1}^{N_S} c_\ell^{y_{\ell j}^-} - k_j^b \prod_{\ell=1}^{N_S} c_\ell^{y_{\ell j}^+}, \quad j = 1, \ldots, N_R^K \tag{3.34}$$

with a reaction coefficient $k_j^f = k_j^f(T, p)$ and $k_j^b = k_j^b(T, p)$. The mass action law (3.33) then has the form

$$\prod_{\ell=1}^{N_S} c_\ell^{y_{\ell j}} = \widetilde{K}_j^{(0)}(T), \quad j = N_R^K + 1, \ldots, N_R^K + N_R^E \qquad (3.35)$$

with the equilibrium constants $\widetilde{K}_j^{(0)}(T) = k_j^f(T, p)/k_j^b(T, p)$.

## 3.12 Literature

The exposition of this chapter is influenced in large parts from selected sections of [101], and Sects. 3.9 and 3.11 are motivated from [113]. For a further in-depth study the books [101, 103, 104] are recommended which contain many interesting historical remarks but also the references [83, 106, 126] and in particular for the thermodynamics of mixtures [1].

## 3.13 Exercises

**Exercise 3.1** An ideal gas consisting of $N_1$ particles of mass $m_A$ with an average kinetic energy $u_1 = \frac{1}{2} m_A \overline{|v_1|^2}$ of the particles and an ideal gas consisting of $N_2$ particles of the same substance with average kinetic energy $u_2 = \frac{1}{2} m_A \overline{|v_2|^2}$ are brought together in a container. Here $\overline{|v_j|^2}$ denotes the value of the square of the modulus of the velocity of the particles, respectively. Before being brought into contact the velocity distribution is given by the Maxwell–Boltzmann law in both cases.

(a) Compute the velocity distribution assuming that all particles keep their velocities.
(b) Derive the velocity distribution assuming that after contact there is a Maxwell–Boltzmann distribution and the total internal energy has not been changed by this contact.
(c) How do you explain the obvious differences between this two distributions? Which of the distributions is correct?

**Exercise 3.2** Assume there is an ideal gas in a cylindrically shaped piston with cross-sectional area $0,1\,\mathrm{m}^2$. For a pressure of $p_1 = 10^5\,\mathrm{N/m}^2$ (i.e., 1 bar) and the temperature of $27°\,\mathrm{C}$ the piston is assumed to have a height of 50 cm. Now in addition to the air pressure $p_1$ slowly a force of 10 000 N is applied to the top of the piston.

To which extent the piston gets compressed, if it is heat insulated, and what is the increase in temperature?

*Hint*: Approximately 273 K corresponds to 0°C.

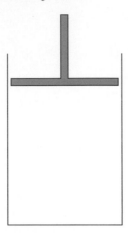

**Exercise 3.3**  We consider the equation

$$\sum_{j=1}^{n} y_j(x(t))\,\dot{x}_j(t) = 0 \tag{3.36}$$

with given functions $y_j : \mathbb{R}^n \to \mathbb{R}$ and unknown functions $x_j : \mathbb{R} \to \mathbb{R}, = 1, \dots, n$.

(a)  Assuming that the vector field $y(x) = (y_j(x))_{j=1}^{n}$ has a potential $\varphi$, construct a representation of the solution (3.36).
(b)  Find a representation of the solution in the case that $y(x)$ has no potential, but an integrating factor $\lambda \neq 0$ with a potential $\varphi$ for $\lambda y$.
(c)  Find the solutions of the equations

   (i)    $2\,x(t)\,y(t)\,\dot{x}(t) + x^2(t)\,\dot{y}(t) = 0,$
   (ii)   $2\,x(t)\,y(t)\,\dot{x}(t) + \dot{y}(t) = 0,$
   (iii)  $y(t)\cos x(t)\,\dot{x}(t) + 2\sin x(t)\,\dot{y}(t) = 0.$

**Exercise 3.4**  The Carnot cycle sketched in the diagram is an approximation for the thermodynamic processes which take place in a spark ignition engine (*Otto engine*). The lines 1 and 3 are adiabatic, the lines 2 and 4 are isochoric. Step 1 corresponds to the compression, step 2 to the combustion at the upper dead point of the piston, step 3 to the expansion and step 4 to the release of the exhaust gas at the lower dead point of the piston.

Assuming an ideal gas, determine the heat quantity transmitted in every process step and the work done and compute the efficiency factor. Show that the efficiency factor only depends on the *ratio of compression* $V_1/V_2$.

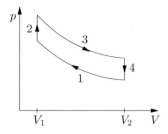

**Exercise 3.5** The following diagram sketches an approximation for the Carnot cycle processes in a compression ignition engine (*Diesel engine*). The essential difference to an Otto engine is the injection of the fuel after the compression, which needs a certain amount of time. During the injection and combustion the piston moves, therefore the combustion phase can be assumed as isobaric, but not as isochoric as for the Otto engine. The circle process consists of an isobaric, an isochoric and two adiabatic steps.

Assuming an ideal gas, determine the heat quantity transmitted in every process step and the work done and give a relation for the efficiency factor depending on the three volumes $V_1$, $V_2$ and $V_3$.

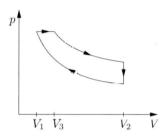

**Exercise 3.6** Compute the Legendre transform of the following functions.

(a) $f : \mathbb{R} \to \mathbb{R}$, $f(x) = \frac{1}{2}ax^2 + bx + c$ with $a, b, c \in \mathbb{R}$, $a > 0$.
(b) $f : \mathbb{R}^n \to \mathbb{R}^n$, $f(x) = \frac{1}{2}x^\top Ax + b^\top x + c$, where $A \in \mathbb{R}^{n,n}$ is symmetric and positive definite and $b \in \mathbb{R}^n$, $c \in \mathbb{R}$.
(c) $f : \mathbb{R}_+ \to \mathbb{R}$, $f(x) = x \ln x$.
(d) $f : \mathbb{R}^2 \to \mathbb{R}$, $f(x, y) = x e^y$.

**Exercise 3.7** A function $f : \mathbb{R}^n \to \mathbb{R}$ is called *convex*, if for all $x, y \in \mathbb{R}^n$ and all $\lambda \in (0, 1)$

$$f(\lambda x + (1 - \lambda)y) \le \lambda f(x) + (1 - \lambda)f(y)$$

holds true. Show:

(a) For a continuously differentiable function $f : \mathbb{R}^n \to \mathbb{R}$ the following three assertions are equivalent:

   (i)   $f$ is convex,
   (ii)   $f(y) - f(x) \ge \nabla f(x) \cdot (y - x)$ for all $x, y \in \mathbb{R}^n$,

(iii)  $(\nabla f(x) - \nabla f(y)) \cdot (x - y) \geq 0$ for all $x, y \in \mathbb{R}^n$.

(b)  If $f : \mathbb{R}^n \to \mathbb{R}$ is continuously differentiable and convex, the Legendre transform $f^*$ of $f$ coincides with the function

$$\tilde{f}(y) = \sup_{x \in \mathbb{R}^n} (x \cdot y - f(x))$$

on its domain of definition.

(c)  The Legendre transform in the sense of (b) of a convex function $f : \mathbb{R}^n \to \mathbb{R}$ is convex.

**Exercise 3.8**  (*Specific Heat*) If a substance of mass $m$ is heated, then the necessary heat quantity $Q$, assuming an ideal gas, is proportional to the temperature difference,

$$Q = c\,m\,(T_2 - T_1)\,.$$

The proportionality factor $c$ is called the *specific heat*.

(a)  For an ideal gas, compute this specific heat $c = c_V$, assuming that during heating the volume stays constant.

(b)  Determine the specific heat $c = c_p$, assuming that the pressure stays constant.

(c)  Show that for constant pressure the necessary input of heat is identical to the change of one of the thermodynamic potentials (free energy, enthalpy, free enthalpy).

**Exercise 3.9**  Derive from the Gibbs formula

$$dS = \frac{1}{T}\,dU + \frac{p}{T}\,dV$$

and the assumption, that the mappings $(U, V) \mapsto S(U, V)$, $(T, V) \mapsto U(T, V)$ and $(T, V) \mapsto p(T, V)$ are twice continuously differentiable, the validity of the Clapeyron formula

$$\frac{\partial U(T, V)}{\partial V} = -p + T\frac{\partial p(T, V)}{\partial T}\,.$$

*Hint*: First determine the second derivatives

$$\frac{\partial^2}{\partial T\,\partial V}S(U(T, V), V)\ \text{and}\ \frac{\partial^2}{\partial V\,\partial T}S(U(T, V), V)\,.$$

**Exercise 3.10**  The equation of state of a *van der Waals gas* is

$$p = \frac{Nk_B T}{V - Nb} - \frac{aN^2}{V^2}\,.$$

Here $b$ denotes the volume occupied by one particle. Hence, $V - Nb$ is the co-volume of the gas, i.e., the volume which is not occupied by particles. The value of $a$ measures the reduction of pressure due to attractive forces of the molecules. We consider a van der Waals gas, whose specific heat at constant volume equals

$$c_V = \frac{3}{2}\frac{k_B}{m_0},$$

where $m_0$ is the mass of a single gas molecule.

(a)  Show that the internal energy is given by

$$U(T, V) = \frac{3}{2}Nk_BT - \frac{aN^2}{V}.$$

(b)  Compute the entropy of the gas.

**Exercise 3.11**  (*Osmosis*) Assume that two thermodynamic subsystems I and II are given, which are separated by a semipermeable membrane. We assume that the membrane is permeable for the substance $A$ and impermeable for the substance $B$. In subsystem I there is a mixture with masses $m_A^I$ of substance $A$ and $m_B = m_B^I$ of substance $B$, and in subsystem II only the substance $A$ with mass $m_A^{II}$ is present. We assume that the following quantities are known: The total mass $m_A = m_A^I + m_A^{II}$, the mass $m_B$, and the volumes $V^I$ and $V^{II}$ of the subsystems, and the temperature $T$.

(a)  With help of the equation of state for ideal gases and the requirement $\mu_A^I = \mu_A^{II}$ for the chemical potentials of the component $A$ determine the distribution of the mass $m_A$ to the two subsystems I and II and also the (different!) pressures $p^I$ and $p^{II}$ in the subsystems.
(b)  Show that the pressure difference $p^I - p^{II}$ equals the partial pressure of the component $B$ in subsystem I on the membrane.

*Hint*: Use the relations of mixtures for ideal gases.

**Exercise 3.12**  For the following chemical reaction systems determine a stoichio-metric matrix and a corresponding set of independent chemical reactions:

(a)  $\{(CO_2, H_2O, H_2CO_3), (C, H, O)\}$
(b)  $\{(CO_2, O_2, H_2O, C_6H_{12}O_6, C_{57}H_{104}O_6), (C, H, O)\}$
(c)  $\{(N_2, O_2, H_2O, NO, NO_2, NH_3, HNO_3), (H, N, O)\}$

Explanation: $CO_2$ — carbon dioxide, $H_2O$ — water, $H_2CO_3$ — carbonic acid, $O_2$ — oxygen, $C_6H_{12}O_6$ — glucose (a carbohydrate), $C_{57}H_{104}O_6$ — triolein (a biological fat), $N_2$ — nitrogen, $NO$ — nitrogen monoxide, $NO_2$ — nitrogen dioxide, $NH_3$ — ammonia, $HNO_3$ — nitric acd.

**Exercise 3.13**  Some chemical substances are electrically charged. In chemical for-mulas this is indicated by a $-$ or $+$-sign used as superscript. For reaction systems with such substances it can make sense to include an electron $e^-$ in the list of elements. Determine a stoichiometric matrix and a set of independent reactions for this system

$$\{(CO_3^{2-}, HCO_3^-, H^+, H_2O), (C, H, O, e^-)\}$$

consisting of carbonate ($CO_3^{2-}$), hydrogen carbonate ($HCO_3^-$), protons ($H^+$) and water.

**Exercise 3.14**  Consider the chemical reaction

$$2\,CO + O_2 \rightleftharpoons 2\,CO_2.$$

For an initial time and given temperature $T_0$ and given pressure $p_0$ there are 2 mol CO, 1 mol $O_2$ and 2 mol $CO_2$.

(a)  With the help of the mass action law and the relations for ideal gases derive an equation for the mole number of $CO_2$, depending on pressure, temperature and the equilibrium constant $K(T)$.
(b)  Determine the limit values of this mole number for very high pressure ($p \to +\infty$) and for very low pressure ($p \to 0$) for a given temperature $T$.

**Exercise 3.15**  We want to characterize points of equilibrium for a chemical reaction in the adiabatic, isobaric case.

(a)  Show that the derivative of the enthalpy with respect of the masses of the components also in this case is identical with the chemical potential:

$$\frac{\partial H(S, p, m)}{\partial m_j} = \mu_j(T(S, p, m), p, m).$$

(b)  Write the equilibrium condition in the form of an optimization problem.
(c)  Assuming that no constraint is active, write the equilibrium condition as a solution of a nonlinear system of equations. Which difference do you see in comparison to the mass action law from Sect. 3.10 for the isothermal, isobaric case?

**Exercise 3.16**  For ideal gases the following kinetic reaction law

$$\dot{\xi}_j = k_j^f(T, p) \prod_{\ell=1}^{N_S} c_\ell^{y_{\ell j}^-} - k_j^b(T, p) \prod_{\ell=1}^{N_S} c_\ell^{y_{\ell j}^+}$$

can be formulated. Here $\xi$ denotes the vector of reaction coordinates, $c_\ell = \dfrac{\nu_\ell}{V}$ the molar concentrations of species $j$, $y_{ij}^+ = \max\{y_{ij}, 0\}$ and $y_{ij}^- = -\min\{y_{ij}, 0\}$ denote the positive and negative parts of the components in the stoichiometric matrix, and the coefficients $k_j^f(T, p)$ and $k_j^b(T, p)$ fulfill $k_j^f(T, p)/k_j^b(T, p) = K_j(T)$ with the equilibrium constant $K_j(T)$, stemming from the formulation

$$\prod_{\ell=1}^{N_S} c_\ell^{y_{\ell j}} = K_j(T)$$

of the mass action law.

Show that for the reaction law defined above the following conditions holds true

(a)  $\dfrac{d}{dt} G(T, p, m(\xi(t))) \leq 0$ and

(b)  $\nabla_\xi G(T, p, m(\xi)) = 0 \Rightarrow \dot{\xi} = 0$

*Hint:* First determine $\exp\left(\frac{1}{RT} \sum_{\ell=1}^{N_S} y_{\ell j} M_\ell \mu_\ell\right)$ with a molar mass $M_\ell$ and the chemical potential $\mu_\ell$ of a species $\ell$ and the universal gas constant $R$ and then reconsider the derivation of the mass action law for ideal gases formulated in molar concentrations.

# Chapter 4
# Ordinary Differential Equations

Many laws of nature express the *change* of a quantity as the consequence of the action
of other quantities. For example the change of velocity of a body is proportional to
the forces acting on the body, the change of an electric field induces a magnetic field,
and the change of a magnetic field induces an electric field. The change of a quantity
is mathematically expressed by means of derivatives and therefore many laws of
nature are formulated as *differential equations*. In the most simple case, in which the
relevant quantities only depend on one variable, this results in an *ordinary differential
equation*. In Chaps. 1 and 3 we have already seen several examples of models with
ordinary differential equations: population models, models for the movement of a
body in a gravitational field of a planet, and kinetic reactions. In this chapter we will
get acquainted with further models in the form of ordinary differential equations and
by means of these models we will study important qualitative properties of ordinary
differential equations.

## 4.1 One-Dimensional Oscillations

An important phenomenon which can be modeled by ordinary differential equations
are *oscillations* of an ensemble of elastically connected mass points. The most simple
case is one mass point with mass $m$ which is fixed to a spring as depicted in Fig. 4.1.
The motion of a mass point can be deduced from Newton's law

$$m\ddot{x} = F \,, \tag{4.1}$$

where $m$ denotes the mass, $x = x(t)$ the position of the mass point at the time $t$ and
$F = F(t)$ the force acting on the mass point. The spring applies a force at the mass
point which depends on the elongation of the spring. In the most simple case this is
a *linear* relationship

© Springer International Publishing AG 2017
C. Eck et al., *Mathematical Modeling*, Springer Undergraduate
Mathematics Series, DOI 10.1007/978-3-319-55161-6_4

**Fig. 4.1**  One-dimensional oscillation

$$F_F = -kx \tag{4.2}$$

with the *spring constant* $k$. The minus sign appears here, as the force of the spring acts in an opposite direction to the displacement. Further forces can be produced by friction. Here we consider frictional forces of the form

$$F_R = -\beta \dot{x}, \tag{4.3}$$

which depend *linearly* on the velocity. For example such a relation is the consequence of *Stokes' law* for the friction of a sphere in a viscous liquid. Then the constant $\beta$ is given by $\beta = 6\pi \mu r$, where $\mu$ is the dynamic viscosity of the fluid and $r$ the radius of the sphere. For bodies of different shape similar relationships can be applied where the factor $6\pi r$ must be substituted by another one depending on the shape and size of the body. Stokes' law is valid for large viscosities and small velocities, where the viscous friction dominates other effects.

For the movement of a body along a rigid surface, friction is not correctly described by (4.3). A reasonable model is given by *Coulomb's law of friction*,

$$F_R = -c_F F_N \frac{\dot{x}}{|\dot{x}|} \text{ for } \dot{x} \neq 0,$$

$$|F_R| \leq c_F F_N \text{ for } \dot{x} = 0.$$

Here $c_F$ denotes the *friction coefficient* and $F_N$ the force by which the body is pressed against the surface. Here the first line describes the *slide friction*, i.e., in this case the friction force is directed opposite to the movement of the body with strength independent of the velocity. The second line describes *static friction*, i.e., the body is not in motion, the force induced by friction can have any direction and its magnitude is limited by $c_F F_N$. Coulomb's friction law is nonlinear and nonsmooth, therefore it is much more difficult to analyze compared to the simple oscillations considered here.

Hence we consider only oscillations with viscous damping according to (4.3). However, we allow for external forces acting on the spring-mass system, for example gravitational forces. In summary we then have

$$F = F_F + F_R + f \tag{4.4}$$

with an external force $f$. Combining the Eqs. (4.1)–(4.4) one gets the following differential equation for the displacement $x$ of the spring:

$$\ddot{x}(t) + 2a\,\dot{x}(t) + b\,x(t) = g(t) \tag{4.5}$$

Here we have $b = k/m$, $a = \beta/(2m)$ and $g(t) = f(t)/m$. In the general case $x(t)$ is a vectorial variable as the position in space is a vector. If the spring, the external force, the initial velocity, and the initial displacement have the same direction then $x(t)$ will perpetually move in this direction, and then $x(t)$ can be considered as a scalar variable.

Another important application for Eq. (4.4) is given by *electromagnetic oscillations*. We consider an *oscillator circuit* which consist of a capacitor with capacitance $C$, an inductor with inductance $L$, and an ohmic resistor with resistance $R$. We have the following relations for the voltage drops $U_R$ at the ohmic resistor, $U_C$ at the capacitor, $U_L$ at the inductor, and the electrical current $I$

$$U_R(t) = R\,I(t), \quad I(t) = C\,\dot{U}_C(t), \text{ and } U_L(t) = L\,\dot{I}(t). \tag{4.6}$$

According to Kirchhoff's voltage law the sum of all voltages in a closed conductor loop is equal to zero. If $U_0$ denotes the external voltage applied, then we have

$$U_L(t) + U_R(t) + U_C(t) = U_0(t).$$

Differentiating this equation with respect to time and insertion of the relations (4.6) leads to

$$L\,\ddot{I}(t) + R\,\dot{I}(t) + \frac{1}{C}I(t) = \dot{U}_0(t).$$

If the external voltage $U_0(t)$ is known then this equation also has the form (4.5), with the correspondences $x(t) = I(t)$, $a = R/(2L)$, $b = 1/(LC)$, and $f(t) = \dot{U}_0(t)/L$.

The Eq. (4.5) belongs to the class of second order linear differential equations with constant coefficients. As the technique for solving such equations in principle is not dependent on the order let us consider an equation of a *general* order $n$

$$x^{(n)}(t) + a_{n-1}\,x^{(n-1)}(t) + \cdots + a_1\,x'(t) + a_0\,x(t) = f(t)$$

with coefficients $a_0, \ldots, a_{n-1}$. We denote this equation more abstractly in the form

$$\mathcal{L}x = f,$$

where $\mathcal{L}x$ is defined by

$$\mathcal{L}x(t) = x^{(n)}(t) + \sum_{\ell=0}^{n-1} a_\ell\,x^{(\ell)}(t). \tag{4.7}$$

We interpret $\mathcal{L}$ as a mapping of an $n$-times continuously differentiable function $x$ to a continuous function $\mathcal{L}x$. Such a mapping which maps functions by using its derivatives to other functions is denoted as a *differential operator*. The *order* of such a differential operator is the highest order of all appearing derivatives. The domain of definition for a differential operator of order $n$ can be chosen as

$$C^n(\mathbb{R}) := \left\{ f : \mathbb{R} \to \mathbb{R} \mid f^{(k)} \text{ exists and is continuous for } k = 1, \ldots, n \right\}.$$

The differential equation (4.5) is called *linear* if the corresponding differential operator is linear, i.e.,

$$\mathcal{L}(\alpha x + \beta y) = \alpha \, \mathcal{L}x + \beta \, \mathcal{L}y$$

for $x, y \in C^n(\mathbb{R})$ and $\alpha, \beta \in \mathbb{R}$. This property leads to two important consequences:

- If $x_1$ and $x_2$ are solutions of $Lx = f$, then the difference $x_1 - x_2$ is a solution of $Lx = 0$.
- If $x_1$ is a solution of $Lx = f$ and $x_0$ is a solution of $Lx = 0$, then $x_0 + x_1$ is also a solution of $Lx = f$.

If $x_p$ is an arbitrary solution of $\mathcal{L}x = f$, called a *particular solution*, then we can represent the set of solutions in the form

$$\mathbb{L}(f) = \{x_p\} + \mathbb{L}(0),$$

where $\mathbb{L}(0)$ is the set of solutions of the *homogeneous equation*

$$\mathcal{L}x = 0.$$

A homogeneous linear differential equation with constant coefficients can be solved using the ansatz $x(t) = e^{\lambda t}$, where $\lambda$ denotes a parameter to be determined. Because of $\frac{d^\ell}{dt^\ell} e^{\lambda t} = \lambda^\ell e^{\lambda t}$ we have

$$\mathcal{L}e^{\lambda t} = p_L(\lambda) e^{\lambda t}$$

with the *characteristic polynomial*

$$p_L(\lambda) = \lambda^n + a_{n-1}\lambda^{n-1} + \cdots + a_1\lambda + a_0.$$

Therefore the homogeneous equation $\mathcal{L}x = 0$ is satisfied, if and only if $\lambda$ is a *zero* of the characteristic polynomial $p_L$. The most simple case is given if $p_L$ has mutually distinct zeros $\lambda_1, \ldots, \lambda_n$: In this case we obtain $n$ linearly independent solutions of the form

$$x_j(t) = e^{\lambda_j t}.$$

Together they form the basis of $\mathbb{L}(0)$. Such a basis is denoted as a *fundamental system*. If $\lambda_j$ is truly complex, meaning $\lambda_j = \mu_j + i\,\omega_j$ with $\mu_j, \omega_j \in \mathbb{R}$, $\omega_j \neq 0$, where $i$ denotes the imaginary unit, then $x_j$ is a complex-valued function. This can be transformed to a real-valued solution in the following way: If the differential operator has real coefficients, meaning that also the characteristic polynomial has real coefficients, then for every complex zero $\lambda = \mu + i\omega$ there is also the complex conjugate zero $\mu - i\omega$. By taking linear combinations of the corresponding two complex solutions $z_1(t) = e^{(\mu+i\omega)t}$ and $z_2(t) = e^{(\mu-i\omega)t}$ two linearly independent real solutions can be determined,

$$x_1(t) = \tfrac{1}{2}(z_1(t) + z_2(t)) = e^{\mu t}\cos(\omega t) \text{ and}$$
$$x_2(t) = \tfrac{1}{2i}(z_1(t) - z_2(t)) = e^{\mu t}\sin(\omega t)\,.$$

Applying this procedure to all pairs of complex conjugate solutions leads to a *real* fundamental system.

If the characteristic polynomial has zeros with multiplicity greater than 1 then the above ansatz with exponential functions does not lead to a complete fundamental system, as for each such zero $\lambda$ only one solution of the form $e^{\lambda t}$ exists. To gain further solutions first we *factorize* the differential operator. If $\lambda_1, \ldots, \lambda_m$ denote all zeros of $p$, with a corresponding multiplicity denoted by $r_j$, then we have

$$\mathcal{L} = \prod_{j=1}^{m} \left(\tfrac{d}{dt} - \lambda_j\right)^{r_j}.$$

In this representation the sequence of the factors can be arbitrarily permuted. To obtain $r$ different linearly independent solutions for an eigenvalue $\lambda$ with multiplicity $r$ we consider the equation

$$\left(\tfrac{d}{dt} - \lambda\right)^r x(t) = 0\,. \tag{4.8}$$

For the ansatz

$$x(t) = c(t)\,e^{\lambda t}$$

obviously we have

$$x'(t) = c'(t)\,e^{\lambda t} + \lambda c(t)\,e^{\lambda t}\,,$$

or equivalently

$$\left(\tfrac{d}{dt} - \lambda\right)\!\left(c(t)e^{\lambda t}\right) = e^{\lambda t}\tfrac{d}{dt}c(t)\,. \tag{4.9}$$

To satisfy (4.8) using this ansatz therefore we need

$$c^{(r)}(t) = 0\,,$$

i.e., $c$ has to be a polynomial of degree $r - 1$. Furthermore every such polynomial leads to a solution of (4.8). In summary, for the eigenvalue $\lambda$ we obtain these $r$ fundamental solutions

$$e^{\lambda t}, t\,e^{\lambda t}, \ldots, t^{r-1}e^{\lambda t}\,.$$

We want to apply these results to the oscillator equation (4.5). Here we assume $a, b \geq 0$ which is reasonable for the application described. The characteristic polynomial

$$p(\lambda) = \lambda^2 + 2a\lambda + b$$

has the two zeros

$$\lambda_{1/2} = -a \pm \sqrt{a^2 - b}.$$

We distinguish three cases (Fig. 4.2):

Case 1:   $b < a^2$. In this case both zeros are real and negative and we have the two solutions

$$x_1(t) = e^{\lambda_1 t} \quad \text{and} \quad x_2(t) = e^{\lambda_2 t}.$$

Every other solution is obtained as a linear combination of these two solutions. As $\lambda_1, \lambda_2$ are negative both solutions decrease for $t \to +\infty$. Here the damping is so strong that no oscillation develops: After a displacement the system slowly returns to its position of rest. This situation is called the *overdamped case*.

Case 2:   $b > a^2$. Then we have two *complex* eigenvalues $\lambda_{1/2} = -a \pm i\omega$ with $\omega = \sqrt{b - a^2}$. The corresponding real solutions are

$$x_1(t) = e^{-at} \cos(\omega t) \quad \text{and} \quad x_2(t) = e^{-at} \sin(\omega t).$$

Here we have *oscillations* and the oscillator is called *underdamped*. In the case $a = 0$ the amplitude of the oscillation stays constant, the oscillation then is un-damped. If $a > 0$ then the amplitude decreases, i.e., the oscillation is damped. The cause of the damping is friction in the spring-mass system, or the ohmic resistance in the oscillator circuit.

Case 3:   $b = a^2$. In this case the characteristic polynomial has a double zero $\lambda = -a$. A corresponding fundamental system is

$$x_1(t) = e^{-at} \quad \text{and} \quad x_2(t) = t e^{-at}.$$

This is a limit case between the overdamped case 1 and the underdamped case 2. Both fundamental solutions decrease for $t \to +\infty$. The solution $x_2(t)$ however first increases for small $t$. This can be interpreted as half a period of an oscillation which afterwards changes to a nonoscillating solution. This solution is called the *critically damped case*.

With the help of the fundamental solutions the unique solution for arbitrary initial data $x(0) = x_0$ and $\dot{x}(0) = x_1$ can be determined. In the case $x_0 = 1$ and $x_1 = 0$, i.e., for the oscillation with an initial displacement 1 and an initial velocity 0, we obtain

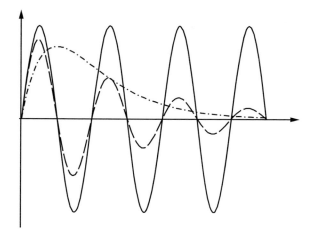

**Fig. 4.2** Undamped Oscillation $(-)$, underdamped oscillation $(---)$, and critically damped oscillation $(-\cdot-\cdot)$

$$x(t) = \frac{\lambda_2}{\lambda_2 - \lambda_1} e^{\lambda_1 t} + \frac{\lambda_1}{\lambda_1 - \lambda_2} e^{\lambda_2 t} \quad \text{for } b < a^2,$$

$$x(t) = e^{-at}\left(\cos(\omega t) + \frac{a}{\omega}\sin(\omega t)\right) \quad \text{for } b > a^2$$

$$x(t) = (1 + at)e^{-at} \quad \text{for } b = a^2.$$

For $x_0 = 0$ and $x_1 = 1$ one has

$$x(t) = \frac{1}{\lambda_1 - \lambda_2}\left(e^{\lambda_1 t} - e^{\lambda_2 t}\right) \quad \text{for } b < a^2,$$

$$x(t) = \frac{1}{\omega}e^{-at}\sin(\omega t) \quad \text{for } b > a^2,$$

$$x(t) = t\,e^{-at} \quad \text{for } b = a^2.$$

These oscillations are also denoted as *free* or *unforced* oscillations as they are solely generated by their initial conditions.

### 4.1.1 Forced Oscillations

Now we consider oscillations which are induced by *periodic* excitations. A typical example for such a situation is an electric oscillating circuit which has a source of alternating current: Then the function $f(t)$ has the form $f(t) = c_1 \cos(\omega t - \varphi)$ with the angular frequency $\omega$ and the *phase* $\varphi$.

First we consider the general case

$$\mathcal{L}x = e^{\mu t}$$

with a differential operator $\mathcal{L}$ of the form (4.7) and a complex parameter $\mu$. Then one can try the *method of undetermined coefficients*, i.e., we make a guess for the solution, using the form of the right-hand side:

$$x(t) = c\, e^{\mu t}\,.$$

Insertion of this ansatz into the differential equation leads to

$$cp(\mu)\, e^{\mu t} = e^{\mu t}$$

with the characteristic polynomial $p$ of $\mathcal{L}$. If $\mu$ is *not* a zero of $p$, then $c = 1/p(\mu)$ is a possibility to satisfy the equation and one gets the *particular solution*

$$x_p(t) = \frac{1}{p(\mu)} e^{\mu t}\,.$$

If $\mu$ is a zero of $p$, for example in the case $p(\lambda) = \prod_{\ell=1}^{m}(\lambda - \lambda_\ell)^{r_\ell}$ where $\mu = \lambda_m$, then the ansatz can be modified to $x(t) = c(t)\, e^{\mu t}$.

Using the relation (4.9), applied to $\lambda = \mu$, we obtain

$$\mathcal{L}x = \prod_{\ell=1}^{m-1}\left(\tfrac{d}{dt} - \lambda_\ell\right)^{r_\ell}\left(\tfrac{d}{dt} - \mu\right)^{r_m}\left(c(t)e^{\mu t}\right) = \prod_{\ell=1}^{m-1}\left(\tfrac{d}{dt} - \lambda_\ell\right)^{r_\ell}\left(c^{(r_m)}(t)e^{\mu t}\right).$$

Now a particular solution is given if $c^{(r_m)}(t) = \tilde{c}$. This holds true for example by using $c(t) = \frac{\tilde{c}}{r_m!}t^{r_m}$. The remaining parameter $\tilde{c}$ can be computed from

$$\prod_{\ell=1}^{m-1}\left(\tfrac{d}{dt} - \lambda_\ell\right)^{r_\ell}\left(\tilde{c}\, e^{\mu t}\right) = e^{\mu t}\,.$$

Setting

$$q(\lambda) := \prod_{\ell=1}^{m-1}(\lambda - \lambda_\ell)^{r_\ell} = \frac{p(\lambda)}{(\lambda - \lambda_m)^{r_m}}$$

this leads to $\tilde{c} = 1/q(\mu)$ and one obtains the particular solution

$$x_p(t) = \frac{1}{r_m!\, q(\mu)} t^{r_m} e^{\mu t}\,.$$

In the following we apply this to the inhomogeneous oscillator equation (4.5) with a right-hand side $g(t) = e^{i\sigma t}$. Here $i\sigma$ is a purely imaginary factor. Let $\lambda_1$ and $\lambda_2$ denote the zeros of the characteristic polynomial. Then in the case $i\sigma \notin \{\lambda_1, \lambda_2\}$ we get the particular solution

$$x_p(t) = \frac{1}{p(i\sigma)} e^{i\sigma t} = \frac{1}{(i\sigma - \lambda_1)(i\sigma - \lambda_2)} e^{i\sigma t}. \tag{4.10}$$

This corresponds to an oscillation with angular frequency $\omega = \sigma$ for the excitation with complex amplitude $1/p(i\sigma) = 1/(i\sigma - \lambda_1)(i\sigma - \lambda_2)$.

The remaining case $i\sigma \in \{\lambda_1, \lambda_2\}$ here only can occur for $a = 0$, i.e., in cases without damping. Then the characteristic polynomial has two complex zeros $\lambda_{1/2} = \pm\sqrt{b}i$. Here one gets the particular solution

$$x_p(t) = \frac{1}{2i\sigma} t\, e^{i\sigma t}.$$

The amplitude of this solution diverges for $t \to +\infty$. This effect is called *resonance*: The external excitation increases the amplitude of the oscillation more and more for increasing times. Resonance in this restricted sense only occurs for *undamped* oscillations. However, for oscillations with only little damping, i.e., for a small $a > 0$ and $\lambda_{1/2} = -a \pm i\omega$ with $\omega = \sqrt{b - a^2}$, the influence of resonance is still perceptible in formula (4.10): for $|\sigma| = \omega$ the factor $1/p(i\sigma)$ is equal to $1/(a(a + 2i\sigma))$, and for small $a$ the modulus of this factor becomes very large. Resonance phenomena are of great importance in many technical applications. The stability of mechanical structures for dynamical loadings, i.e., time-dependent external forces, essentially depends on the question, whether external loads may get in resonance with the so-called eigen-oscillation of the mechanical structure and whether in the case of resonance there is enough damping to balance the energy supply by the external excitation. Resonance phenomena are of paramount importance in acoustics for example in the avoidance of noise.

If one wants to gain a more apparent real representation of the solution for a real right-hand side $g(t)$ then again certain linear combinations of complex solutions can be used. To get the solution of (4.5) for $g(t) = \cos(\sigma t)$ one can represent $g(t)$ as a linear combination of $g_1(t) = e^{i\sigma t}$ and $g_2(t) = e^{-i\sigma t}$,

$$g(t) = \tfrac{1}{2}\big(g_1(t) + g_2(t)\big).$$

If $x_1(t)$ and $x_2(t)$ denote the solutions corresponding to $g_1(t)$ and $g_2(t)$, then

$$x(t) = \tfrac{1}{2}(x_1(t) + x_2(t))$$

is the solution of (4.5) corresponding to $g(t) = \cos(\sigma t)$. Therefore one gets the particular solution

$$x_p(t) = \begin{cases} \dfrac{(b - \sigma^2)\cos(\sigma t) + 2a\sigma \sin(\sigma t)}{(b - \sigma^2)^2 + 4a^2\sigma^2} & \text{for } i\sigma \notin \{\lambda_1, \lambda_2\}, \\ \frac{1}{2\sigma} t\, \sin(\sigma t) & \text{for } a = 0,\ |\sigma| = \sqrt{b}. \end{cases}$$

If one wants to solve an initial value problem then a solution of the homogeneous oscillator equation fulfilling this initial value has to be added to the particular solution satisfying the inhomogeneous right-hand side. In this way *transient oscillations* can be studied, for example the voltage curve in an electric oscillating circuit after the alternating voltage source has been switched on.

### Conservation of Energy

We consider the differential equation (4.5) in its dimensional form

$$m\ddot{x} + \beta\dot{x} + kx = f$$

with an arbitrary time-dependent right-hand side $f$. This equation describes an oscillating mass point having the position $x$, assuming a restoring force $F_F = -kx$, an acting external force $f$ and 'viscous damping' by the force $F_R = -\beta\dot{x}$. Multiplication of the equation by $\dot{x}$ and integration with respect to time from $t_0$ to $t_1$ leads to

$$\frac{1}{2}m\,\dot{x}^2(t_1) + \frac{k}{2}x^2(t_1) = \frac{1}{2}m\,\dot{x}^2(t_0) + \frac{k}{2}x^2(t_0)$$
$$+ \int_{t_0}^{t_1} f\dot{x}\,dt - \int_{t_0}^{t_1} \beta\dot{x}^2(t)\,dt \,. \tag{4.11}$$

This equation describes the *energy balance*. The energy contained in the system consists of the *kinetic energy* $T = \frac{1}{2}m\dot{x}^2$ and the *potential energy* $U = \frac{k}{2}x^2$. The term $\int_{t_0}^{t_1} f\dot{x}\,dt$ describes the energy fed to the system by the external force $f$ and $\int_{t_0}^{t_1} \beta\dot{x}^2(t)\,dt$ describes the energy which is 'dissipated' by friction or viscous damping. In general the latter is transformed to heat and therefore 'vanishes' in purely mechanical balance of energy.

The balance of energy can be easily transferred to general nonlinear constitutive laws for the restoring force $F_F = F_F(x)$ and the viscous damping $F_R = F_R(\dot{x})$. To this end only a *primitive* $V$ of $-F_F$ is needed. This primitive is just the *potential energy*. Furthermore the kinetic energy is denoted by $T = \frac{1}{2}m\dot{x}^2$ and $\Phi = \Phi(\dot{x}) = F_R(\dot{x})\dot{x}$ denotes a *dissipation function*. From the equation

$$m\ddot{x} - F_R(\dot{x}) + V'(x) = f$$

again through multiplication by $\dot{x}$ and integration in time we derive

$$T(\dot{x}(t_1)) + V(x(t_1)) = T(\dot{x}(t_0)) + V(x(t_0)) + \int_{t_0}^{t_1} \left( f\dot{x} + \Phi(\dot{x}) \right) dt \,.$$

## 4.2 The Lagrangian and Hamiltonian Form of Mechanics

In the following we will develop two important more general formulations of the basic equations of mechanics for systems of point masses. As a simple example first we consider an oscillation without damping, with a nonlinear restoring force $F_R(x)$, given by a potential $V$, i.e., $F_R(x) = -V'(x)$. Therefore the corresponding differential equation is

$$m\ddot{x} = -V'(x). \tag{4.12}$$

Here the left-hand side can be interpreted as the time derivative of the derivative of the kinetic energy $T = T(x, \dot{x}) = \frac{1}{2}m\dot{x}^2$ with respect to the velocity $\dot{x}$,

$$m\ddot{x} = \frac{d}{dt}\frac{\partial T}{\partial \dot{x}}.$$

Here we consider $x$ and $\dot{x}$ as *independent* variables and interpret both $T$ and also $V$ as functions of $x$ and $\dot{x}$. Then (4.12) can be expressed in the form

$$\frac{d}{dt}\frac{\partial T}{\partial \dot{x}} = -\frac{\partial V}{\partial x}.$$

In this example $T$ is independent of $x$ and $V$ is independent of $\dot{x}$. Therefore this equation can be reformulated using the *Lagrangian function*

$$L = L(x, \dot{x}) = T(\dot{x}) - V(x)$$

to obtain

$$\frac{\partial L}{\partial x} = \frac{d}{dt}\frac{\partial L}{\partial \dot{x}}.$$

This is the basic equation in the *Lagrangian form* of mechanics for a mass point.

This principle can be easily extended to systems describing a fixed number of mass points. To keep the notation simple also in the case of more than one spatial dimension, we collect the coordinates of the position vectors of all mass points in one single vector $x \in \mathbb{R}^N$. The component $x_j$ is assumed to correspond to the mass $m_j$. If $x_i$ and $x_j$ are different components of the position of the same mass point then of course $m_i = m_j$ has to hold. The kinetic energy of the system is given by

$$T(x, \dot{x}) = \frac{1}{2}\sum_{j=1}^{N} m_j \dot{x}_j^2.$$

For the potential energy we allow for general functions $V = V(x)$. Then the $j$-th component of the force acting on the system is given by

$$-\frac{\partial V}{\partial x_j}.$$

The corresponding Lagrangian is

$$L(x, \dot{x}) = T(x, \dot{x}) - V(x, \dot{x}),$$

and the equation of motion $m_j \ddot{x}_j = -\frac{\partial V}{\partial x_j}$ for the component $j$, is identical to

$$\frac{\partial L}{\partial x_j} = \frac{d}{dt}\frac{\partial L}{\partial \dot{x}_j} \tag{4.13}$$

as can be checked easily. If *external forces* $f_{\text{ext}}$ act on the system then they appear as an additional term in the left-hand side of (4.13),

$$\frac{\partial L}{\partial x_j} + f_{\text{ext},j} = \frac{d}{dt}\frac{\partial L}{\partial \dot{x}_j}. \tag{4.14}$$

These exterior forces may depend on $t, x, \dot{x}$. In principle also the force induced by the potential $V$, i.e., $-\nabla_x V(x)$ can be formulated as an exterior force. Then it does not appear anymore in the Lagrangian function and one gets the equation

$$\frac{\partial T}{\partial x_j} - \frac{\partial V}{\partial x_j} + f_{\text{ext},j} = \frac{d}{dt}\frac{\partial T}{\partial \dot{x}_j}.$$

On the other hand sometimes the external forces can be included in the formulation of the Lagrangian $L$, more precisely if and only if they can be represented as

$$f_{\text{ext},j} = f_{\text{ext},j}(t, x, \dot{x}) = -\frac{\partial U}{\partial x_j}(t, x, \dot{x}) + \frac{d}{dt}\frac{\partial U}{\partial \dot{x}_j}(t, x, \dot{x}) \tag{4.15}$$

with a *generalized potential* $U(t, x, \dot{x})$. In this case the Lagrangian reads as

$$L(t, x, \dot{x}) = T(\dot{x}) - V(x) - U(t, x, \dot{x}).$$

Due to condition (4.15) the additional term $U$ transforms Eq. (4.13) to (4.14) thus introducing the external force $f_{\text{ext},j}$ on the left-hand side. For example the condition (4.15) is satisfied if the force only depends on the time $f_{\text{ext}} = f(t)$. Then the generalized potential is given by $U = U(t, x) = -f(t) \cdot x$. A slightly more general example is given by a force which stems from a time- and space-dependent generalized potential $U = U(t, x)$, thus having a form $f_{\text{ext},j}(t, x) = -\frac{\partial U}{\partial x_j}(t, x)$.

The Eq. (4.13) can be interpreted as a necessary condition for a stationary point of the optimization problem

$$\min\left\{\int_{t_0}^{t_1} L(t, x, \dot{x})\, dt \,\Big|\, x \in C^1\left([t_0, t_1], \mathbb{R}^N\right),\; x(t_0) = x^{(0)},\right.$$

$$\left. x(t_1) = x^{(1)} \right\} \tag{4.16}$$

for fixed initial and end points $x^{(0)}$ and $x^{(1)}$. The functional $x \mapsto \int_{t_0}^{t_1} L(t, x, \dot{x})\, dt$ is also denoted as the *action* of the motion described by $x$. To verify this assertion let $x$ be a solution of (4.16) and let $y \in C^1\left([t_0, t_1], \mathbb{R}^N\right)$, where $y(t_0) = y(t_1) = 0$, be an arbitrary *variation*. Then $x + \varepsilon y$ is an admissible comparison function for the optimization problem. Therefore it holds true that

$$0 = \frac{d}{d\varepsilon} \int_{t_0}^{t_1} L(t, x + \varepsilon y, \dot{x} + \varepsilon \dot{y})\, dt \bigg|_{\varepsilon = 0}$$

$$= \int_{t_0}^{t_1} \sum_{j=1}^{N} \left( \frac{\partial L}{\partial x_j} y_j + \frac{\partial L}{\partial \dot{x}_j} \dot{y}_j \right) dt \tag{4.17}$$

$$= \int_{t_0}^{t_1} \sum_{j=1}^{N} \left( \frac{\partial L}{\partial x_j} - \frac{d}{dt} \frac{\partial L}{\partial \dot{x}_j} \right) y_j\, dt \,,$$

where in the last step integration by parts for the second summand has been performed. If a continuous function $f : [t_0, t_1] \to \mathbb{R}^N$ satisfies

$$\int_{t_0}^{t_1} \sum_{i=1}^{N} f_i(x)\, \varphi_i(x)\, dx = 0 \quad \text{for all} \quad \varphi \in C_0^1\left([t_0, t_1], \mathbb{R}^N\right),$$

then $f \equiv 0$ follows. This assertion is called the *fundamental lemma of calculus of variations* and will be proven in Exercise 4.19. As the functions $y_j$ can be arbitrarily chosen with exception of the fixed boundary data, the Lagrange equations (4.13) follow from (4.17). Please note that solutions of (4.13) in general are only critical points of the optimization problem and do not necessarily represent an extremum. The formulation of the equations of motion of mechanical systems as stationary points (see definition on p. 167) for suitable action functionals is also called *Hamilton's principle of stationary action*.

An important advantage of the Lagrangian formulation is the fact that the equations are independent of the choice of so-called *generalized coordinates*. We consider generalized coordinates

$$q_j = \widehat{q}_j(x), \quad j = 1, \ldots, N$$

and the corresponding Lagrangian

$$\widehat{L}(t, q, \dot{q})\,.$$

Here $\widehat{q} : \mathbb{R}^N \to \mathbb{R}^N$ is supposed to be a bijective mapping whose Jacobian $D_x\widehat{q} = \left(\frac{\partial \widehat{q}_i}{\partial x_j}\right)_{i,j=1}^N$ is invertible. Then because of $\dot{q} = D_x\widehat{q}(x)\dot{x}$ we have a relation between $L$ and $\widehat{L}$ that reads as

$$L(t, x, \dot{x}) = \widehat{L}\bigl(t, \widehat{q}(x), D_x\widehat{q}(x)\dot{x}\bigr).$$

The optimization problem can be written in generalized coordinates in the form

$$\min\left\{ \int_{t_0}^{t_1} \widehat{L}(t, q, \dot{q})\,dt \;\middle|\; q \in C^1([t_0, t_1], \mathbb{R}^N), \; q(t_0) = q^{(0)}, \; q(t_1) = q^{(1)} \right\}$$

with $q^{(0)} = \widehat{q}(x^{(0)})$ and $q^{(1)} = \widehat{q}(x^{(1)})$. This holds true because only another representation of the integrand $L(t, x(t), \dot{x}(t)) = \widehat{L}(t, q(t), \dot{q}(t))$ has been used. The corresponding optimality criterion then is

$$\frac{\partial \widehat{L}}{\partial q_j} = \frac{d}{dt}\frac{\partial \widehat{L}}{\partial \dot{q}_j}.$$

The equivalence of these equations to (4.13) can be shown directly by changing the parametrization provided certain assumptions on the used parametrization are made, see Exercise 4.3.

The generalized Lagrangian formulation is particularly advantageous for problems with *constraints*. Then generalized coordinates can be chosen in such a way that the constraints are fulfilled automatically and solve the corresponding Lagrangian equations of motion. We justify the Lagrangian formalism for problems with constraints by considering a problem with spatial variables $x = x(t) \in \mathbb{R}^N$ and $s$ constraints $g_i(t, x) = 0$, $i = 1, \ldots, s$. The coordinates of the mass points are assumed to form the vector $x$, i.e., having $\ell$ mass points in a $d$-dimensional space this means $N = d\ell$. The constraints define a set of admissible points

$$X(t) = \left\{ x \in \mathbb{R}^N \mid g_i(t, x) = 0 \;\text{ for }\; i = 1, \ldots, s \right\}.$$

We assume that the functions $g_i$ are differentiable, and that the gradients $\{\nabla_x g_i(t, x) \mid i = 1, \ldots, s\}$ are linearly independent for all $x \in X(t)$, and that $X(t)$ is a $(N - s)$-dimensional, sufficiently smooth manifold in $\mathbb{R}^N$.

To fulfill the constraints the equations of motion have to be supplemented by unknown *constraint forces* $Z_1, \ldots, Z_N$. The constraint forces keep the curve of motion in the manifold of admissible points. The vector $Z(t) \in \mathbb{R}^N$ of constraint forces at time $t$ is perpendicular to the set $X(t)$ of admissible points at the point $x(t)$ respectively to the tangential space in the point $x(t)$. This means $Z(t)$ is perpendicular to the tangent space

$$T_{x(t)}X(t) = \{y \in \mathbb{R}^N \mid \nabla_{x(t)}g_i(t, x) \cdot y = 0 \text{ for } i = 1, \ldots, s\}.$$

Written in components we then have

$$m_j \ddot{x}_j = f_j + Z_j .$$

Note that if $x_i$ and $x_j$ denote different components of *the same* mass point, then we have $m_i = m_j$. We write the equations of motion more condensed in a matrix-vector notation

$$M \ddot{x} = f + Z , \qquad (4.18)$$

where $M$ denotes the diagonal matrix of point masses, $f$ the vector of external forces and $Z$ the vector of constraint forces.

For the following considerations it is not necessary that $M$ is a diagonal matrix, it is sufficient to assume that $M$ is symmetric and positive definite. An important example for a mechanical system for which the mass matrix $M$ is not diagonal are dynamical elastic space frames discussed in the next section. We assume that the force $f = f(t, x)$ has a potential $U = U(t, x)$ such that $f = -\nabla_x U$. Furthermore we assume that the set of admissible position vectors $X(t)$ can be parameterized in the form

$$X(t) = \{\widehat{x}(t, q) \,|\, q \in \mathbb{R}^r \} .$$

Here $r = N - s$ denotes the number of *degrees of freedom*. The derivatives of $\widehat{x}(t, \cdot)$ with respect to the components of $q$ then lie in the tangential space,

$$\frac{\partial \widehat{x}}{\partial q_j}(t, q) \in T_{\widehat{x}(t,q)} X(t) \ \text{ for } \ j = 1, \ldots, r .$$

Differentiation of $t \mapsto \widehat{x}(t, q(t))$ for a given curve $t \mapsto q(t)$ leads to

$$\frac{d}{dt} \widehat{x}(t, q(t)) = \partial_t \widehat{x}(t, q(t)) + \nabla_q \widehat{x}(t, q(t)) \cdot \dot{q}(t) .$$

This expression represents the velocity $\dot{x}(t)$ of the curve $x(t) = \widehat{x}(t, q(t))$. Therefore we define

$$\widehat{\dot{x}}(t, q, \dot{q}) = \partial_t \widehat{x}(t, q) + \nabla_q \widehat{x}(t, q) \cdot \dot{q} . \qquad (4.19)$$

Here and in the following $t$, $q$ and $\dot{q}$ will be interpreted as *independent* variables of the Lagrangian, this in particular has to be taken into account in the notation of partial derivatives. From (4.19) we conclude that

$$\frac{\partial \widehat{\dot{x}}}{\partial \dot{q}_j}(t, q, \dot{q}) = \frac{\partial \widehat{x}}{\partial q_j}(t, q) . \qquad (4.20)$$

Multiplying the equation of motion (4.18) by $\frac{\partial \widehat{x}}{\partial q_j}$ leads to an elimination of the unknown constraint force, because $\frac{\partial \widehat{x}}{\partial q_j}(t, q) \in T_{\widehat{x}(t,q)} X(t)$, and $Z$ is perpendicular to $T_{\widehat{x}(t,q)} X(t)$.

Therefore one gets

$$M\ddot{x} \cdot \frac{\partial \widehat{x}}{\partial q_j} = f \cdot \frac{\partial \widehat{x}}{\partial q_j}. \tag{4.21}$$

Here we have $\ddot{x} = \frac{d^2}{dt^2}(\widehat{x}(t, q(t)))$. We write the left-hand side of this equation in the form

$$M\ddot{x} \cdot \frac{\partial \widehat{x}}{\partial q_j} = \frac{d}{dt}\left(M\widehat{x} \cdot \frac{\partial \widehat{x}}{\partial q_j}\right) - M\widehat{x} \cdot \frac{d}{dt}\left(\frac{\partial \widehat{x}}{\partial q_j}\right).$$

We have

$$\frac{d}{dt}\left(\frac{\partial \widehat{x}}{\partial q_j}\right) = \frac{\partial^2 \widehat{x}}{\partial t\, \partial q_j} + \nabla_q \frac{\partial \widehat{x}}{\partial q_j} \cdot \dot{q} = \frac{\partial \dot{\widehat{x}}}{\partial q_j}.$$

Taking into account (4.20), this leads to

$$M\ddot{x} \cdot \frac{\partial \widehat{x}}{\partial q_j} = \frac{d}{dt}\left(M\widehat{x} \cdot \frac{\partial \dot{\widehat{x}}}{\partial \dot{q}_j}\right) - M\widehat{x} \cdot \frac{\partial \dot{\widehat{x}}}{\partial q_j} = \frac{d}{dt}\frac{\partial \widehat{T}}{\partial \dot{q}_j} - \frac{\partial \widehat{T}}{\partial q_j}$$

with the kinetic energy

$$\widehat{T}(t, q, \dot{q}) = \frac{1}{2}\dot{\widehat{x}}(t, q, \dot{q})^\top M\, \dot{\widehat{x}}(t, q, \dot{q}).$$

In addition because of $f(t, x) = -\nabla_x U(t, x)$ it holds true that

$$f \cdot \frac{\partial \widehat{x}}{\partial q_j} = -\frac{\partial \widehat{U}}{\partial q_j}$$

with $\widehat{U}(t, q) = U(t, \widehat{x}(t, q))$. As $\widehat{U}$ is a quantity independent of $\dot{q}$, (4.21) has the form

$$\frac{\partial \widehat{L}}{\partial q_j} = \frac{d}{dt}\frac{\partial \widehat{L}}{\partial \dot{q}_j}$$

with the usual Lagrangian

$$\widehat{L}(t, q, \dot{q}) = \widehat{T}(t, q, \dot{q}) - \widehat{U}(t, q).$$

**Example**: We consider a *pendulum* consisting of a mass point of mass $m$ which is fixed to the origin by means of a massless rigid rod of length $\ell$ and which moves in the gravitational field of the earth, see Fig. 4.3. To describe the position of the pendulum we use the angle $\varphi$ with respect to the vertical axis. In Cartesian coordinates the position is then given by

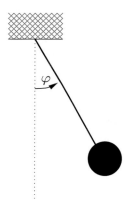

**Fig. 4.3** One-dimensional pendulum swinging

$$x = x(\varphi) = \ell \begin{pmatrix} \sin \varphi \\ -\cos \varphi \end{pmatrix}.$$

Kinetic and potential energy take the form

$$T - \tfrac{1}{2}m|\dot{x}|^2 = \tfrac{1}{2}m\ell^2\dot{\varphi}^2 \ \text{and} \ U = mgx_2 = -mg\ell\cos\varphi.$$

Therefore the Lagrangian is given by

$$L(\varphi, \dot{\varphi}) - T \quad U - \tfrac{1}{2}m\ell^2\dot{\varphi}^2 + mg\ell\cos\varphi.$$

Taking into account

$$\frac{\partial L}{\partial \varphi} = -mg\ell\sin\varphi \ \text{and} \ \frac{\partial L}{\partial \dot{\varphi}} = m\ell^2\dot{\varphi}$$

we get the Lagrangian equation of motion

$$m\ell^2\ddot{\varphi} + mg\ell\sin\varphi = 0.$$

The *Hamiltonian formulation of mechanics* is obtained by means of the *Legendre transform* of the Lagrangian with respect to the generalized velocities. The Legendre transform is explained in Sect. 3.6. Let

$$p_j = \frac{\partial L}{\partial \dot{q}_j}$$

be the *dual variable* with respect to $\dot{q}_j$ and let

$$H(t, q, p) = p \cdot \dot{q}(t, q, p) - L(t, q, \dot{q}(t, q, p))$$

denote the *Hamiltonian*, which is just the Legendre transform of the Lagrangian. Here we assume that $\dot{q}$ can be written as a function of the new variables $t, q, p$, in fact this is one of the prerequisites for the existence of the Legendre transform. Then we have

$$\frac{\partial H}{\partial q_j} = p \cdot \frac{\partial \dot{q}}{\partial q_j} - \frac{\partial L}{\partial q_j} - \sum_{k=1}^{r} \frac{\partial L}{\partial \dot{q}_k}\frac{\partial \dot{q}_k}{\partial q_j} = -\frac{\partial L}{\partial q_j} = -\frac{d}{dt}\frac{\partial L}{\partial \dot{q}_j} = -\dot{p}_j \text{ and}$$

$$\frac{\partial H}{\partial p_j} = \dot{q}_j + \sum_{k=1}^{r} p_k\frac{\partial \dot{q}_k}{\partial p_j} - \sum_{k=1}^{r} \frac{\partial L}{\partial \dot{q}_k}\frac{\partial \dot{q}_k}{\partial p_j} = \dot{q}_j .$$

This leads to the basic equations of the Hamiltonian formulation of mechanics

$$\dot{q}_j = \frac{\partial H}{\partial p_j} \text{ and } \dot{p}_j = -\frac{\partial H}{\partial q_j} \text{ for } j = 1, \ldots, r . \tag{4.22}$$

Similar to the Lagrangian formulation also for the Hamiltonian formulation the choice of arbitrary generalized coordinates is possible. To do so it is only necessary that the kinetic energy and the generalized potential energy can be expressed in the new variables. The Hamiltonian often describes the *energy* of the system, being given as a sum of potential and kinetic energy.

In the example of the one-dimensional, nonlinear oscillation with potential $V$ we have the kinetic energy

$$T(\dot{x}) = \tfrac{1}{2}m\dot{x}^2$$

and the Lagrangian

$$L(t, x, \dot{x}) = \tfrac{1}{2}m\,\dot{x}^2 - V(x) .$$

Therefore we have

$$p = \frac{\partial L}{\partial \dot{x}} = m\dot{x},$$

which is just the *momentum* of a mass point in motion. Setting $q = x, \dot{q} = \dot{x} = \frac{p}{m}$ we obtain

$$H(t, q, p) = p\frac{p}{m} - \frac{m}{2}\left(\frac{p}{m}\right)^2 + V(q) = \frac{p^2}{2m} + V(q) .$$

Here we have $H = T + V$, which in fact is the energy of the system.

The Lagrangian and the Hamiltonian formulation of mechanics not only hold true for systems of mass points, but can be extended to further model classes, for example the so-called *multi-body systems*, which consist of various interconnected rigid bodies. In every case the basis is:

• a description of all possible configurations of the system by suitable generalized coordinates and
• suitable representations of the kinetic and the potential energy of the system.

*Motions of Space Frames*

Now we use the Lagrangian formalism to derive systems of differential equations for the motion of a space frame. To this end only representations of the kinetic and of the potential energy are necessary. We consider a space frame consisting of $n$ numbered bars and $k$ numbered nodes. The coordinates of the nodes are collected to a vector $x \in \mathbb{R}^{dk}$, where $d \in \{2, 3\}$ denotes the considered spatial dimension. Let the mass of the $j^{\text{th}}$ bar be denoted by $m_j$. This mass is assumed to be uniformly distributed over the length of the bar.

The potential energy of a deformed space frame consists of the energy stored in the strained bars. The strains are collected in a vector $e$, which is related to the nodal displacements by the relation

$$e = Bx$$

in which $B \in \mathbb{R}^{n,dk}$ is a matrix depending on the geometry of the bars, see Sect. 2.2. If $e_j$ is the strain and $E_j$ the modulus of elasticity of the $j^{\text{th}}$ bar, then the energy stored in the bar is given by

$$V_j = \tfrac{1}{2} E_j |e_j|^2 .$$

The total potential energy of the space frame is

$$V = \sum_{j=1}^{n} \tfrac{1}{2} E_j |e_j|^2 = \tfrac{1}{2} e^\top C e = \tfrac{1}{2} x^\top K x$$

with a diagonal matrix $C = \text{diag}(E_1, \ldots, E_n)$ consisting of the elasticity constants and with the symmetric so-called *stiffness matrix*

$$K = B^\top C B.$$

To determine the kinetic energy we consider a single bar with mass $m$ and end points $x^{(0)}(t)$ and $x^{(1)}(t)$. This bar is parameterized by

$$x(t, s) = (1 - s)x^{(0)}(t) + s\, x^{(1)}(t) \quad \text{where} \ \ s \in [0, 1] .$$

The kinetic energy of the bar is given by

$$T(t) = \int_0^1 m |\dot{x}(t, s)|^2 \, ds = \tfrac{m}{3} \left( \left| \dot{x}^{(0)}(t) \right|^2 + \dot{x}^{(0)}(t) \cdot \dot{x}^{(1)}(t) + \left| \dot{x}^{(1)}(t) \right|^2 \right) .$$

The kinetic energy of the total space frame corresponds to the sum of the kinetic energies of all bars,

$$T(t) = \sum_{j=1}^{n} \tfrac{m_j}{3} \left( \left| \dot{x}^{(u(j))}(t) \right|^2 + \dot{x}^{(u(j))}(t) \cdot \dot{x}^{(o(j))}(t) + \left| \dot{x}^{(o(j))}(t) \right|^2 \right) ,$$

where $u(j)$ and $o(j)$ denote the indices of the nodes of bar $j$. This term can be expressed as

$$T(t) = \tfrac{1}{2}\dot{x}(t)^\top M\dot{x}(t)$$

with a symmetric *mass matrix* $M$ of dimension $dk \times dk$.

If the external forces acting at the nodes are put together to a vector $f \in \mathbb{R}^{dk}$, then one obtains the Lagrangian

$$L = L(t, x, \dot{x}) = T(\dot{x}) - V(x) + f(t) \cdot x = \tfrac{1}{2}\dot{x}^\top M\dot{x} - \tfrac{1}{2}x^\top Kx + f(t) \cdot x .$$

From (4.13) one obtains the equation of motion

$$M\ddot{x} + Kx = f . \tag{4.23}$$

This is a system of ordinary differential equations. The mass matrix $M$ is positive definite, the stiffness matrix $K$ is positive semidefinite.

We are interested first in the *free oscillations* and therefore set $f = 0$. Using the common exponential ansatz

$$x(t) = e^{\lambda t} v$$

with a complex parameter $\lambda$ and a vector $v \in \mathbb{C}^{dk}$, one obtains the generalized eigenvalue problem

$$Kv = -\lambda^2 Mv . \tag{4.24}$$

As $M$ is positive definite and symmetric, the *square root* $M^{1/2}$ can be defined. Multiplication of (4.24) by $M^{-1/2}$ and using the variable transformation $w = M^{1/2}v$ leads to the eigenvalue problem

$$Aw = -\lambda^2 w \quad \text{with} \quad A = M^{-1/2}KM^{-1/2} .$$

The matrix $A$ here is positive semidefinite and symmetric, therefore it has real and nonnegative eigenvalues. This means that $\lambda$ is purely imaginary

$$\lambda = i\omega \quad \text{with} \quad \omega \in \mathbb{R} .$$

There is a basis of $\mathbb{R}^{dk}$ consisting of eigenvectors $w_1, \ldots, w_{dk}$ of $A$ corresponding to eigenvalues $\omega_1^2, \ldots, \omega_{dk}^2$. The generalized eigenvalue problem (4.24) has the eigenvectors $v_j = M^{-1/2}w_j$ for the same eigenvalues. Assume first that $A$ is positive definite, i.e., $B$ has linearly independent columns. Every positive eigenvalue $-\lambda_j^2$ leads to two admissible values of $\lambda_j$, namely $\lambda_{j,1/2} = \pm i\omega_j$ with $\omega_j > 0$. Therefore the general solution of (4.23) with $f = 0$ has the form

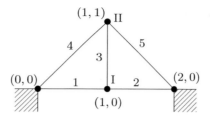

**Fig. 4.4** Example for a space frame

$$x(t) = \sum_{j=1}^{dk} \left( \alpha_j e^{i\omega_j t} + \beta_j e^{-i\omega_j t} \right) v_j$$

with arbitrary complex coefficients $\alpha_j$, $\beta_j$. Real solutions take the form

$$x(t) = \sum_{j=1}^{dk} \left( c_j \cos(\omega_j t) + d_j \sin(\omega_j t) \right) v_j$$

with coefficients $c_j, d_j \in \mathbb{R}$. The values $\omega_j/(2\pi)$ are the *eigenfrequencies* of the space frame, the vectors $v_j$ describe the *vibration modes*.

In the semidefinite case, for every generalized eigenvector $v_j \in \operatorname{Ker} K$ a term of the form $(c_j + t d_j) v_j$ has to be added and the sum has $dk$ summands in total.

**Example**: We consider the space frame depicted in Fig. 4.4 in two spatial dimensions, with a numbering of bars and nodes as indicated there. The moduli of elasticity and the masses of the bars after an appropriate nondimensionalization are assumed to be $E_1 = E_2 = E_3 = 1, E_4 = E_5 = \sqrt{2}/2, m_1 = m_2 = m_3 = 1, m_4 = m_5 = \sqrt{2}$. These values appear if the bars consist of the same material and have the same thickness. Then the elasticity constants scale with the reciprocal length and the masses with the length of the bars. The corresponding matrices then are, compare Chap. 2,

$$B = \begin{pmatrix} 1 & 0 & 0 & 0 \\ -1 & 0 & 0 & 0 \\ 0 & -1 & 0 & 1 \\ 0 & 0 & \sqrt{2}/2 & \sqrt{2}/2 \\ 0 & 0 & -\sqrt{2}/2 & \sqrt{2}/2 \end{pmatrix} \quad \text{and} \quad C = \begin{pmatrix} 1 & 0 & 0 & 0 & 0 \\ 0 & 1 & 0 & 0 & 0 \\ 0 & 0 & 1 & 0 & 0 \\ 0 & 0 & 0 & \sqrt{2}/2 & 0 \\ 0 & 0 & 0 & 0 & \sqrt{2}/2 \end{pmatrix}.$$

Here the columns of $B$ are linearly independent. Therefore we are in the positive definite case. The corresponding stiffness matrix is given by

$$K := B^{\mathsf{T}} C B = \begin{pmatrix} 2 & 0 & 0 & 0 \\ 0 & 1 & 0 & -1 \\ 0 & 0 & \sqrt{2}/2 & 0 \\ 0 & -1 & 0 & 1 + \sqrt{2}/2 \end{pmatrix}.$$

The kinetic energy is given by

$$T = \tfrac{m_1}{3}(\dot{x}_1^2 + \dot{x}_2^2) + \tfrac{m_2}{3}(\dot{x}_1^2 + \dot{x}_2^2) + \tfrac{m_3}{3}(\dot{x}_1^2 + \dot{x}_2^2 + \dot{x}_1\dot{x}_3 + \dot{x}_2\dot{x}_4 + \dot{x}_3^2 + \dot{x}_4^2)$$
$$+ \tfrac{m_4}{3}(\dot{x}_3^2 + \dot{x}_4^2) + \tfrac{m_5}{3}(\dot{x}_3^2 + \dot{x}_4^2)$$
$$= \tfrac{1}{2}\dot{x}^\top M \dot{x} .$$

The mass matrix for the given masses of the bars reads as

$$M = \frac{1}{3}\begin{pmatrix} 6 & 0 & 1 & 0 \\ 0 & 6 & 0 & 1 \\ 1 & 0 & 2+4\sqrt{2} & 0 \\ 0 & 1 & 0 & 2+4\sqrt{2} \end{pmatrix}.$$

The eigenvectors and eigenvalues of the generalized eigenvalue problem

$$Kv = \mu M v$$

are given by

$$\mu_1 \approx 1{,}030682 , \quad \mu_2 \approx 0{,}274782 , \quad \mu_3 \approx 1{,}211484 , \quad \mu_4 \approx 0{,}116887$$

and

$$v_1 \approx \begin{pmatrix} 0{,}9844 \\ 0 \\ -0{,}1758 \\ 0 \end{pmatrix}, \quad v_2 \approx \begin{pmatrix} 0{,}0634 \\ 0 \\ 1{,}0045 \\ 0 \end{pmatrix}, \quad v_3 \approx \begin{pmatrix} 0 \\ 0{,}7096 \\ 0 \\ -0{,}7192 \end{pmatrix}, \quad v_4 \approx \begin{pmatrix} 0 \\ 0{,}8048 \\ 0 \\ 0{,}5935 \end{pmatrix}.$$

The eigenvectors $v_1$ and $v_2$ describe oscillations in horizontal direction, $v_3$ and $v_4$ oscillations in vertical direction. In the motions corresponding to $v_1$ and $v_3$ the points I and II move in opposite directions while for $v_2$ and $v_4$ they move in the same direction.

## 4.3   Examples from Population Dynamics

Now we want to consider further important aspects of modeling with ordinary differential equations, considering selected examples from population dynamics. To this end we first repeat the simple population dynamics models from Chap. 1. The growth of the population without predators and in the presence of unlimited resources is described by

$$x'(t) = p\,x(t) \tag{4.25}$$

with a growth factor $p$. The general solution of this equation is given by

$$x(t) = x(t_0) e^{p(t-t_0)}.$$

Therefore the growth factor can be determined from measurements of the population quantity for two different instances of time, $p\, \Delta t = \ln\left(\frac{x(t+\Delta t)}{x(t)}\right)$. For $p > 0$ and $x(t_0) > 0$ the solution converges for $t \to +\infty$ towards $+\infty$, for $p < 0$ it converges towards 0.

Populations with limited resources can be described by making the growth rate dependent on the population quantity. A simple approach is based on an additional parameter $x_M$, which describes the maximal population which can be sustained by the given resources. Setting $p = p(x) = q(x_M - x)$ we get the *logistic differential equation*

$$x'(t) = q\big(x_M - x(t)\big)x(t).$$

This equation can also be solved in closed form,

$$x(t) = \frac{x_M x_0}{x_0 + (x_M - x_0)e^{-x_M q(t-t_0)}}$$

with $x_0 = x(t_0)$. These solutions are bounded for $x_0 > 0$ and converge towards $x_M$ for $t \to +\infty$.

Both differential equations have the form

$$x'(t) = f(x(t)) \quad \text{with} \quad f : \mathbb{R} \to \mathbb{R}. \tag{4.26}$$

Such a differential equation is called *autonomous* because the right-hand side does not depend on the time $t$, but only on the solution $x(t)$. Autonomous differential equations may have solutions constant in time which are called *stationary* solutions. They can be computed by setting the derivative in the differential equation equal to zero, i.e., one solves the (nonlinear) equation

$$f(x) = 0.$$

For Eq. (4.25) with $f(x) = px$ for $p \neq 0$ there is only one stationary solution: $x(t) = 0$. The logistic differential equation (4.26) (on the other hand) has (for $q \neq 0$) two stationary solutions, $x(t) = 0$ and $x(t) = x_M$.

An important characteristic of a stationary solution is its *stability* or instability. The solution of an ordinary differential equation is stable, if it exhibits only small changes for *small* changes in the data. In general the initial value is taken as the data here. From the closed form solutions of the considered differential equation the stability of the stationary solutions can be investigated easily: The stationary solution $x(t) = 0$ of the Eq. (4.25) is

- stable for $p < 0$,
- unstable for $p > 0$.

For the logistic differential equation with $q \neq 0$ the stationary solution $x(t) = 0$ is always *unstable* whereas the solution $x(t) = x_M$ is *stable*.

Often there exists no closed form for solutions of differential equations. With the help of *linear stability analysis* nevertheless the stability of stationary solutions can be investigated. To this end the differential equation is *linearized* at the stationary solution $x_S$. Because of $f(x_S) = 0$ one obtains the linearized equation

$$y'(t) \approx f'(x_S)\, y(t) = p^* y(t) \ \text{ with } p^* = f'(x_S)$$

for the difference $y(t) = x(t) - x_S$. Here we assume that $f \in C^2(\mathbb{R})$. The equation $y' = p^* y$ has the explicit solution

$$y(t) = c\, e^{p^* t} \ \text{ with } c \in \mathbb{R}.$$

It is to be expected that $y$ is an approximation to the deviation of the perturbed solution from the stationary solution. The modulus of $y$ is increasing or decreasing in time, depending on whether

$$p^* = f'(x_S) > 0$$

or

$$p^* = f'(x_S) < 0$$

holds true. Therefore the stationary solution $x_S$ is called

- linearly stable, if $f'(x^*) < 0$, and
- linearly unstable, if $f'(x^*) > 0$.

It can be shown, see for example [6] or the Sects. 4.5 and 4.6, that for solutions with $x(t_0)$ in a neighborhood of $x^*$

$$x(t) \to x^*, \ \text{ if } \ f'(x^*) < 0\,,$$

and

$$x(t) \nrightarrow x^*, \ \text{ if } \ f'(x^*) > 0\,.$$

For $f'(x^*) = 0$ the linear stability analysis cannot give any conclusions, and in this case terms of higher order have to be incorporated in the considerations. In the Sects. 4.5 and 4.6 we will discuss the linear stability analysis, in particular for *systems* of several ordinary differential equations, in more detail.

The qualitative behavior of solutions of a differential equation also can be deduced from a simple *phase diagram*, see Fig. 4.5. Here to each point $(x, 0)$ the vector $(f(x), 0)$ is attached. The corresponding vector field is called the *direction field*.

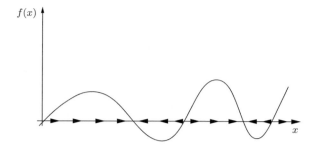

**Fig. 4.5** Phase portrait of a scalar differential equation

It shows in which direction the solution will proceed, and the length of the vector indicates how fast $x$ will be changed. With the help of the direction field it can also be seen nicely, which stationary points (given by the zeros of $f$) are stable and which are unstable. The first case is given if the vector field in the vicinity of the zero points in the direction of the zero and the second case if it points in the opposite direction. These considerations clearly show that a good overview concerning the various solutions and the qualitative behavior of solutions can be obtained without having an explicit solution formula for the differential equation.

## 4.4 Qualitative Analysis, Phase Portraits

Now we want to consider a so called *predator–prey model*, which describes the temporal evolution of two populations. The model is based on the following modeling assumptions:

1. There are two populations, one being a predator and the other a prey population, for example

pikes and carps,
foxes and rabbits,
lions and antelopes.

    Let

$x(t)$ be the quantity of the prey species at time $t$,
$y(t)$ be the quantity of the predator species at time $t$.

2. The predator population exclusively feeds on the prey species. If no prey is present, then the predator has to die of starvation. For the description of this process we assume a constant negative specific growth rate $-\gamma$ with $\gamma > 0$. Then, if $x = 0$, it holds

$$y'(t) = -\gamma\, y(t).$$

In the presence of prey the predators feed on the prey and death by starvation is diminished. The success of the predator is proportional to the size of the predator population and to the size of the prey population, therefore a term of the type

$$\delta x(t) \, y(t) \tag{4.27}$$

has to be added to the equation. This expresses the following facts: if many predators are present then they meet more often species of the prey population. A doubling of their amount leads to a doubling of their success in preying. If many individuals of the prey population are present then the same amount of predators meet more often prey species—a doubling of the prey population doubles the probability that a predator meets prey. Therefore the encounters between predator and prey are multiplied by the factor four if both, the size of the predator and the size of the prey, are doubled. This is expressed by the term (4.27). In summary we obtain

$$y'(t) = -\gamma \, y(t) + \delta x(t) \, y(t) = y(t) \, (-\gamma + \delta x(t)) \, .$$

3. The prey species always has a sufficient amount of food at its disposal and the birth rate is, if no predators are present, higher than the death rate. Therefore, if $y = 0$, we have

$$x'(t) = \alpha \, x(t) \, .$$

In the presence of predators the prey population is diminished through "predator–prey contacts" by the term

$$-\beta \, x(t) \, y(t) \, .$$

As in assumption 2 the number of contacts is proportional to $x(t)$ and to $y(t)$. In summary we obtain the following system of differential equations

$$x'(t) = (\alpha - \beta \, y(t)) x(t), \quad y'(t) = (\delta x(t) - \gamma) y(t) \, . \tag{4.28}$$

Here we have a *system* consisting of two differential equations. Systems of differential equations occur very often in practice, in all the cases where the process is described by more than one variable.

First we compute the stationary points of the system. They are the zeros of the *set of equations*

$$(\alpha - \beta y)x = 0,$$
$$(\delta x - \gamma)y = 0 \, .$$

We obtain the two solutions

$$\begin{pmatrix} x \\ y \end{pmatrix} = \begin{pmatrix} 0 \\ 0 \end{pmatrix} \text{ and } \begin{pmatrix} x \\ y \end{pmatrix} = \begin{pmatrix} \gamma/\delta \\ \alpha/\beta \end{pmatrix} \, .$$

These two points are the constant solutions of (4.28).

In the following we want to determine all possible solutions of the system of differential equations (4.28). To this end it is advisable to reduce the amount of parameters by *nondimensionalization*. The following dimensions appear:

$$[x] = A, \quad [y] = A, \quad [t] = T,$$
$$[\alpha] = [\gamma] = 1/T, \quad [\beta] = [\delta] = 1/(AT).$$

Here $A$ denotes an amount and $T$ a time. If we choose the stationary solution $(\gamma/\delta, \alpha/\beta)$ as an intrinsic reference quantity then the following nondimensionalization is suggested:

$$u = x/(\gamma/\delta), \quad v = y/(\alpha/\beta), \quad \tau = \frac{t - t_0}{\bar{t}}.$$

It turns out that $\bar{t} = 1/\alpha$ is a good choice and then we obtain the following nondimensional system

$$u' = (1 - v)u, \quad v' = a(u - 1)v \text{ with } a = \gamma/\alpha. \tag{4.29}$$

For the analysis of a system of two differential equations

$$u' = f(u, v), \quad v' = g(u, v) \tag{4.30}$$

it is useful to consider the *direction field*. Here at every point in the $(u, v)$ plane the vector $(f(u, v), g(u, v))$ is attached. Then a solution of (4.30) is a parameterized curve in $\mathbb{R}^2$, which has at every point the vector $(f(u, v), g(u, v))$ as the tangential vector. The direction field of the system of differential equations (4.29) is depicted in Fig. 4.6.

In some special cases for a system of two differential equations a so-called *first integral H* exists. This is a function $H$ such that

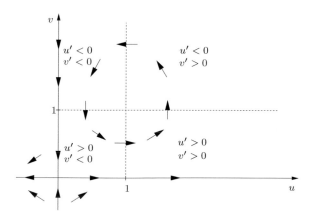

**Fig. 4.6** Direction field for the predator–prey model

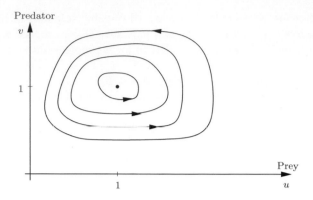

**Fig. 4.7** Phase portrait for the predator–prey model

$$\frac{d}{dt} H(u(t), v(t)) = 0$$

for all solutions $(u, v)$ of the system of differential equations holds true. In our case

$$H(u, v) = -au - v + \ln(u^a) + \ln v$$

is a first integral on the set $(0, \infty)^2$.

The function $H$ has a maximum in the stationary point $(1, 1)$ and approaches $-\infty$ if $u$ or $v$ converge towards 0. The level lines are sketched in Fig. 4.7. They are obtained either by a more precise analysis of the function $H$ or by a computer-aided plot of the function.

Nonconstant solutions of the system of differential equations (4.29) are contained in level lines of $H$. The set of all curves parameterized by solutions of the system of differential equations is called the *phase portrait* of the differential equation. The phase portrait gives a quick overview of the behavior of solutions.

It can be shown that solutions of (4.28) are *periodic*, compare Exercise 4.7. Therefore the populations exhibit periodic oscillations. If many predators and only a few preys are present, then the predator population decreases as not enough food is present. If then only a few predators remain, the prey population again can increase. If the prey population exceeds a certain threshold value, then again enough prey is present for the predators such that their amount can increase. But if then many predators exist, the prey diminishes and the same situation as at the beginning of the period occurs.

### Predator–Prey Models with Restricted Growth

If we include social friction terms into the model (4.28) to describe competition with an increasing amount of species within one population, we obtain

$$x' = (\alpha - \beta y)x - \lambda x^2 = (\alpha - \beta y - \lambda x)x ,$$
$$y' = (\delta x - \gamma)y - \mu y^2 = (\delta x - \gamma - \mu y)y$$

with positive constants $\alpha$, $\beta$, $\gamma$, $\delta$, $\lambda$, $\mu$. To determine the direction field we use that at the straight line

$$\mathcal{G}_x : \alpha - \beta y - \lambda x = 0$$

the right-hand side of the first equation vanishes and at the straight line

$$\mathcal{G}_y : \delta x - \gamma - \mu y = 0$$

the right-hand side of the second equation vanishes. The straight line $\mathcal{G}_x$ has a negative slope as a function of $x$ and the straight line $\mathcal{G}_y$ has a positive slope as a function of $x$.

If these straight lines have no intersection point in the positive quadrant, then a direction field as depicted in Fig. 4.8 is given. In this case there are only two nonnegative stationary solutions, namely $(0, 0)$ and $(\alpha/\lambda, 0)$. It is clear that in terms of the model only nonnegative solutions make sense. It can be shown that all solutions with initial values $(x_0, y_0)$, $x_0 > 0$, $y_0 \geq 0$ converge towards $(\alpha/\lambda, 0)$ for $t \to \infty$. In each case the predator population becomes extinct. This situation is considered in more detail in Exercise 4.9.

If the straight lines have an intersection point in the positive quadrant, then the direction field looks as depicted in Fig. 4.9.

The stationary solutions in $[0, \infty)^2$ are given by $(0, 0)$, $(\alpha/\lambda, 0)$ and the intersection point, namely

$$(\xi, \eta) = \left( \frac{\alpha\mu + \beta\gamma}{\lambda\mu + \beta\delta}, \frac{\alpha\delta - \lambda\gamma}{\lambda\mu + \beta\delta} \right),$$

of the straight lines $\mathcal{G}_x$ und $\mathcal{G}_y$.

Now we want to investigate if the stationary point $(\xi, \eta)$ is *stable*, i.e., if every solution which is close to $(\xi, \eta)$ for a certain instance of time, is also close to $(\xi, \eta)$ for later times. To answer this question we use the *principle of linearized stability*.

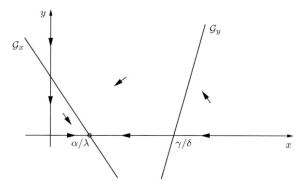

**Fig. 4.8**  Direction field for a predator–prey model with limited growth (1. case)

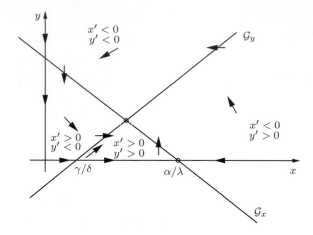

**Fig. 4.9** Direction field for a predator–prey model with limited growth (2. case)

## 4.5   The Principle of Linearized Stability

Let a system of differential equations

$$x' = f(x) \quad \text{with} \ \ f \in C^2(\Omega; \mathbb{R}^n) \ \text{for an open set} \ \Omega \subset \mathbb{R}^n \qquad (4.31)$$

be given. In this section we are interested in the stability of stationary solutions, i.e., time-independent solutions $x^*$ of (4.31). These are exactly the zeros of $f$, $f(x^*) = 0$. The investigation of stability of stationary solutions is an important part of the so-called *qualitative theory of ordinary differential equations*. We want to investigate the stability of stationary solutions without explicitly solving the differential equation. It turns out that it is sufficient to consider the differential equation linearized at $x^*$ to achieve already some results about the stability behavior of the stationary solution. First we have to give a precise formulation of the notion of stability.

**Definition 4.1** A stationary solution $x^*$ of the differential equation $x' = f(x)$ is called

(i)   *stable* if for every neighborhood $U$ of $x^*$ a neighborhood $V$ of $x^*$ exists, such that for every solution of the initial value problems

$$x' = f(x), \quad x(0) = x_0 \in V$$

   it holds true that

$$x(t) \in U \ \text{ for all } t > 0,$$

(ii)  *unstable* if it is not stable.

(iii) A stable stationary solution is called *asymptotically stable* if a neighborhood $W$ of $x^*$ exists such that for every solution of the initial value problems

$$x' = f(x), \quad x(0) = x_0 \in W$$

it holds true that

$$\lim_{t \to \infty} x(t) = x^*.$$

**Examples**:

(i) The stationary solution $x_M$ of the equation of restricted growth

$$x' = q(x_M - x)x$$

is asymptotically stable and the stationary solution $0$ of the same system is unstable. The observation of Sect. 4.3 that the first derivative of the right-hand side is of importance for the stability will be generalized soon.

(ii) The stationary point $(1, 1)$ of the nondimensionalized predator–prey model

$$u' = (1 - v)u,$$
$$v' = a(u - 1)v$$

is stable, but not asymptotically stable, as not every solution which starts in the neighborhood of $(1, 1)$ converges towards $(1, 1)$ for large times.

The principle of linearized stability rests upon the linearization of the differential equation at $x^*$, as in the case of a scalar equation in Sect. 4.3. For

$$y(t) - x(t) - x^*$$

we have

$$y' = x' = f(x) = f(y + x^*) = f(x^*) + Df(x^*)y + \mathcal{O}(|y|^2).$$

Therefore we consider a function $y$ as the solution of the linear approximative problem

$$y' = Df(x^*)\, y.$$

This is a system of *linear differential equations*. The origin $y = 0$ is a stationary solution of this system of differential equations and we can investigate the stability of this solution.

**Definition 4.2** The stationary solution $x^*$ is *linearly stable* (*linearly unstable* or *linearly asymptotically stable*) if $0$ is a stable (unstable or asymptotically stable) solution of the linearized equation.

Now the *principle of linearized stability* reads:

(i) If $x^*$ is linearly asymptotically stable then $x^*$ is asymptotically stable.

(ii) If $Df(x^*)$ possesses an eigenvalue $\lambda$ with $\mathrm{Re}\,\lambda > 0$, and thus $x^*$ is linearly unstable (Theorem 4.4), then $x^*$ is unstable.

Analogous forms of the principle above can be formulated for general evolution equations, for example also for partial differential equations and for integral equations. However, such assertions are often difficult to prove. For ordinary differential equations the following holds true:

**Theorem 4.3** *The principle of linearized stability in the above formulated form is valid.*

The proof of part (i) of the principle of linearized stability will be given in Sect. 4.6.

## 4.6   Stability of Linear Systems

The task to determine the stability of stationary solutions now basically is reduced to investigate the stability of the zero point for linear differential equations.
    Let the system of linear differential equations

$$x' = Ax \quad \text{with} \ \ A \in \mathbb{C}^{n,n} \tag{4.32}$$

be given. In this section we consider differential equations whose solutions have values in $\mathbb{C}^n$. If the matrix is real valued, then the general real-valued solution can be obtained by considering the real and the imaginary part of the solution.

    Let $\widetilde{x}$ be an eigenvector of $A$ for the eigenvalue $\lambda$, i.e.,

$$A\widetilde{x} = \lambda\widetilde{x}.$$

Then

$$x(t) = e^{\lambda t}\widetilde{x}$$

is a solution of (4.32), because

$$x' = e^{\lambda t}\lambda\widetilde{x} = Ae^{\lambda t}\widetilde{x} = Ax.$$

If $|x|$ denotes the Euclidean norm of $x \in \mathbb{C}^n$, then we have

$$|x(t)| = \left|e^{\lambda t}\widetilde{x}\right| = \left|e^{(\mathrm{Re}\,\lambda + i\,\mathrm{Im}\,\lambda)t}\right| \, |\widetilde{x}| = e^{(\mathrm{Re}\,\lambda)t}|\widetilde{x}|.$$

Therefore the sign of $\mathrm{Re}\,\lambda$ determines whether $e^{\lambda t}\widetilde{x}$ converges towards zero or towards infinity for $t \to +\infty$.

**Assumption**: The matrix $A$ is *diagonalizable*, i.e., there exists a basis $\{x_1, \ldots, x_n\}$ of eigenvectors for the eigenvalues $\{\lambda_1, \ldots, \lambda_n\}$, $\lambda_i \in \mathbb{C}$, $x_i \in \mathbb{C}^n$ for $i = 1, \ldots, n$.

Now let $x_0 \in \mathbb{C}^n$ be an arbitrary vector. Then there is a unique representation of $x_0$ in the basis above,

$$x_0 = \sum_{i=1}^{n} \alpha_i x_i, \quad \alpha_1, \ldots, \alpha_n \in \mathbb{C}.$$

Therefore

$$x(t) = \sum_{i=1}^{n} \alpha_i e^{\lambda_i t} x_i$$

is a solution of the initial value problem

$$x' = Ax, \quad x(0) = x_0.$$

If now

$$\text{Re } \lambda_i < 0 \quad \text{for } i = 1, \ldots, n,$$

then we can conclude

$$|x(t)| \le \sum_{i=1}^{n} |\alpha_i| e^{(\text{Re } \lambda_i)t} |x_i| \to 0 \quad \text{for } t \to |\infty. \tag{4.33}$$

This shows that the point 0 is *asymptotically stable* in this case. If otherwise we have

$$\text{Re } \lambda_j > 0 \quad \text{for } a \; j \in \{1, \ldots, n\},$$

then we see for $x_0 = \alpha x_j, \alpha \ne 0$

$$|x(t)| = |\alpha| |x_j| e^{(\text{Re } \lambda_j)t} \to +\infty$$

for $t \to +\infty$. This shows that in this case the point 0 is *unstable*.
The following theorem is valid:

**Theorem 4.4** *Let $A \in \mathbb{C}^{n,n}$ be an arbitrary matrix, in particular not necessarily diagonalizable. Then the following holds true.*

(i) *The stationary point 0 is an asymptotically stable solution of $x' = Ax$ if and only if*
$$\text{Re } \lambda < 0 \quad \text{for all eigenvalues } \lambda \text{ of } A.$$

(ii) *If*
$$\text{Re } \lambda > 0 \quad \text{for some eigenvalue } \lambda \text{ of } A,$$

*then the stationary point 0 is an unstable solution of $x' = Ax$.*

(iii)  *If $A$ is additionally diagonalizable, then it holds true: The stationary point 0 is a stable solution of $x' = Ax$ if and only if*

$$\mathrm{Re}\ \lambda \leq 0 \ \textit{for all eigenvalues}\ \lambda\ \textit{of}\ A\ .$$

**Remarks**

(i)  For diagonalizable matrices the theorem has been proven above. If $A$ is not diagonalizable, the proof of the assertions (i) and (ii) is more involved. We refer for example to the book of Amann [6].

(ii)  Together with the principle of linearized stability the theorem allows the analysis of the stability behavior of stationary points of nonlinear systems. However, remember in the case

$$\begin{cases} \mathrm{Re}\ \lambda \leq 0 & \text{for all eigenvalues } \lambda \text{ of the linearized system with} \\ \mathrm{Re}\ \lambda = 0 & \text{for at least one eigenvalue } \lambda \text{ of the linearized system} \end{cases}$$

in general no statement about the stability of stationary points is possible. For this problem compare also Exercises 4.10 and 4.12.

We now apply the described theory to the stationary points of the predator–prey models. The model with unlimited resources leads to the system of differential equations

$$x' = (\alpha - \beta y)x\ ,$$
$$y' = (\delta x - \gamma)y\ .$$

This has the stationary solutions $(0, 0)$ and $(\gamma/\delta, \alpha/\beta)$. For

$$f(x, y) = \begin{pmatrix} (\alpha - \beta y)x \\ (\delta x - \gamma)y \end{pmatrix}$$

we obtain

$$Df(x, y) = \begin{pmatrix} \alpha - \beta y & -\beta x \\ \delta y & \delta x - \gamma \end{pmatrix}\ .$$

In particular we have:

$$Df(0, 0) = \begin{pmatrix} \alpha & 0 \\ 0 & -\gamma \end{pmatrix}\ .$$

Therefore the point $(0, 0)$ is unstable as $Df(0, 0)$ has a positive eigenvalue. For the second stationary point we get

$$Df(\gamma/\delta, \alpha/\beta) = \begin{pmatrix} 0 & -\beta\gamma/\delta \\ \alpha\delta/\beta & 0 \end{pmatrix}\ .$$

The eigenvalues are

$$\lambda_{1,2} = \pm i\sqrt{\alpha\gamma}.$$

As $\operatorname{Re}\lambda_1 = \operatorname{Re}\lambda_2 = 0$ unfortunately the principle of linearized stabilization does not lead to a statement concerning the stability behavior of the stationary point $(\gamma/\delta, \alpha/\beta)$.

In the case of the predator–prey system with limited growth

$$\left.\begin{array}{l} x' = (\alpha - \beta y - \lambda x)x \\ y' = (\delta x - \gamma - \mu y)y \end{array}\right\} =: f(x, y)$$

we want to investigate the stability of the stationary solution

$$(\xi, \eta) = \left(\frac{\alpha\mu + \beta\gamma}{\lambda\mu + \beta\delta}, \frac{\alpha\delta - \lambda\gamma}{\lambda\mu + \beta\delta}\right).$$

We compute

$$Df(x, y) = \begin{pmatrix} \alpha - \beta y - 2\lambda x & -\beta x \\ \delta y & \delta x - \gamma - 2\mu y \end{pmatrix}$$

and thus get

$$Df(\xi, \eta) = \begin{pmatrix} -\lambda\xi & -\beta\xi \\ \delta\eta & -\mu\eta \end{pmatrix}.$$

The eigenvalues of $Df(\xi, \eta)$ can be easily computed as

$$\lambda_{1,2} = \frac{-(\lambda\xi + \mu\eta) \pm \sqrt{(\lambda\xi + \mu\eta)^2 - 4\xi\eta(\lambda\mu + \delta\beta)}}{2}.$$

For $\xi > 0$ and $\eta > 0$ we have $\xi\eta(\lambda\mu + \delta\beta) > 0$ and thus conclude

$$\operatorname{Re}\lambda_i < 0 \quad \text{for } i = 1, 2.$$

If the straight lines $\mathcal{G}_x$ and $\mathcal{G}_y$ intersect in the positive quadrant, see Fig. 4.9, then the point $(\xi, \eta)$ is stable. The phase portrait for this case is sketched in Fig. 4.10. Qualitatively the solutions in the neighborhood of the stationary point $(\xi, \eta)$ resemble the solutions of the corresponding linearized system.

To prove the principle of linearized stability we need the following important lemma.

**Lemma 4.5** (Gronwall's inequality) *Let* $y : [0, T] \to \mathbb{R}$ *be nonnegative and continuously differentiable such that*

$$y'(t) \le \alpha(t)\, y(t) + \beta(t) \quad \text{for all} \ t \in [0, T], \tag{4.34}$$

*holds with continuous functions* $\alpha, \beta : [0, T] \to \mathbb{R}$. *Then*

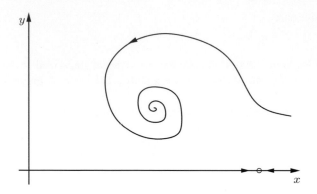

**Fig. 4.10** Phase portrait for the predator–prey model with limited growth (2. case)

$$y(t) \le e^{\int_0^t \alpha(s)\,ds}\left(y(0) + \int_0^t \beta(s)\,ds\right) \quad \textit{for all} \ \ t \in [0, T].$$

*Proof* After multiplying (4.34) with $e^{-\int_0^t \alpha(s)\,ds}$ we obtain

$$\frac{d}{dt}\left(y(t)e^{-\int_0^t \alpha(s)\,ds}\right) \le e^{-\int_0^t \alpha(s)\,ds}\beta(t).$$

Integration yields

$$y(t)\,e^{-\int_0^t \alpha(s)\,ds} \le y(0) + \int_0^t e^{-\int_0^r \alpha(s)ds}\beta(r)\,dr \le y(0) + \int_0^t \beta(s)\,ds$$

and this shows the assertion after multiplication with $e^{\int_0^t \alpha(s)\,ds}$.                   □

*Proof of Theorem* 4.3 *(part (i)).* According to Theorem 4.4 the point $x^*$ is linearly asymptotically stable if and only if

$$\text{Re } \lambda < 0 \quad \text{for all eigenvalues } \lambda \text{ of } A.$$

In this case (see Exercise 4.14) there exists an $\alpha > 0$ and a scalar product $(.,.)$ with corresponding norm $\| . \|$ such that

$$(y, Df(x^*)y) \le -\alpha\|y\|^2 \quad \text{for all} \ \ y \in \mathbb{R}^n.$$

We set $z(t) := x(t) - x^*$. Then

$$\begin{aligned}\tfrac{d}{dt}\tfrac{1}{2}\|z(t)\|^2 &= (z(t), x'(t)) = (z(t), f(x(t)))\\ &= (z(t), [Df(x^*)z(t) + R(z(t))]),\end{aligned} \tag{4.35}$$

where $R(y) = \mathcal{O}(\|y\|^2)$. From this property of $R$ we conclude the existence of a number $\varepsilon > 0$, such that

$$|(y, R(y))| \leq \tfrac{\alpha}{2}\|y\|^2 \quad \text{for all} \quad y \in B_\varepsilon(0) .$$

From (4.35) therefore it follows if $\|z(t)\| < \varepsilon$,

$$\frac{d}{dt}\|z(t)\|^2 \leq -\alpha\|z(t)\|^2 .$$

Gronwall's inequality, see Lemma 4.5, leads to

$$\|z(t)\|^2 \leq e^{-\alpha t}\|z(0)\|^2 .$$

Therefore we can conclude: if $\|z(0)\| < \varepsilon$ then

$$z(t) \to 0 \quad \text{for} \quad t \to +\infty$$

and therefore the asymptotic stability of $x^*$ has been proven. $\qquad\qquad\qquad \square$

## 4.7  Variational Problems for Functions of One Variable

In Sect. 4.2 we have formulated basic equations for mass points in mechanics with the help of a variational problem, more specifically with the help of the necessary conditions of an optimization problem. In fact many mathematical models can be formulated as variational problems. Furthermore in applications one is often interested in maximizing or in minimizing quantities, and also in this case variational problems have to be solved. Calculus of variations is a very extensive area and in this section we only want to discuss a few questions arising from it, which lead to ordinary differential equations. Therefore we will consider variational problems, which can be formulated for functions of one real variable. Aspects of calculus of variations which will involve several spatial variables will be discussed in Chap. 6.

First we want to recall certain facts from the analysis of functions in several variables. If a differentiable function $f : \mathbb{R}^n \to \mathbb{R}$ attains its minimum at a point $\overline{x} \in \mathbb{R}^n$, then we have

$$\nabla f(\overline{x}) = 0 .$$

This condition is necessary, but not sufficient. In general inflection points but also saddle points can appear, as the examples $f(x) = x^3$ with $\overline{x} = 0$ and $f(x_1, x_2) = x_1 x_2$ at $\overline{x} = (0, 0)$ show. Points, in which the derivative vanishes, are called *critical* or *stationary points*.

Often minimization problems have to be solved taking constraints into account, this means, we look for $x_0 \in \mathbb{R}^n$ with $g_1(x_0) = \cdots = g_m(x_0) = 0$, such that

$$f(x_0) \leq f(x) \quad \text{for all} \quad x \in \mathbb{R}^n \quad \text{with} \quad g_1(x) = \cdots = g_m(x) = 0 .$$

Here $g_1, \ldots, g_m : \mathbb{R}^n \to \mathbb{R}$ are differentiable functions. If $\nabla g_1(x_0), \ldots, \nabla g_m(x_0)$ are linearly independent, then there exist Lagrange multipliers $\lambda_1, \ldots, \lambda_m \in \mathbb{R}$ such that

$$\nabla f(x_0) + \lambda_1 \nabla g_1(x_0) + \cdots + \lambda_m \nabla g_m(x_0) = 0$$

holds true. If $\lambda_1, \ldots, \lambda_m$ are known, we can conclude that the point $x_0$ is critical for the function

$$f + \lambda_1 g_1 + \cdots + \lambda_m g_m \,. \tag{4.36}$$

If we consider $x_0$ and $\lambda_1, \ldots, \lambda_m$ as the unknown quantities, then we have that $(x_0, \lambda_1, \ldots, \lambda_m)$ is a critical point of the function

$$F(x_0, \lambda_1, \ldots, \lambda_m) := f(x_0) + \lambda_1 g_1(x_0) + \cdots + \lambda_m g_m(x_0) \,. \tag{4.37}$$

Now we want to solve variational problems, in which the arguments are elements of a function space, and therefore we have to check which of the above facts can be transferred to this case. Here it is important to note that function spaces in general are infinite-dimensional vector spaces, and because of this the considerations above cannot be applied directly.

We start the section with certain classical variational problems.

(i) *Minimize the length of a graph*: We are looking for a function $u : [0, 1] \to \mathbb{R}$ with $u(0) = a$ and $u(1) = b$ such that

$$\mathcal{L}(u) := \int_0^1 \sqrt{1 + (u')^2}\, dx \tag{4.38}$$

is minimal.

(ii) *Minimize the Dirichlet integral*: If we substitute the integrand in (4.38) by the quadratic approximation $\sqrt{1 + (u')^2} \approx 1 + \frac{1}{2}(u')^2$, which seems to be justified if $u'$ is small, then, instead of $\mathcal{L}$, we can minimize the functional

$$\mathcal{D}(u) := \int_0^1 \tfrac{1}{2}(u')^2\, dx \,.$$

Here we neglect the additive constant, which is of no relevance for the minimization problem as it only shifts the minimal value. In addition again the constraints $u(0) = a$ and $u(1) = b$ are required. A variant of the above variational problem in which an additional $x$-dependent term appears, is given by

$$\mathcal{D}_1(u) := \int_0^1 \frac{\mu(x)}{2}(u'(x))^2\, dx \,,$$

where $\mu : [0, 1] \to (0, \infty)$ is a given positive function.

In some applications no boundary conditions for $u$ are required. Additionally cases appear in which the integral term to be minimized contains terms at the boundary and can also involve an external force $f$. This leads for example to the following expression

$$\mathcal{D}_2(u) = \int_0^1 \left( \frac{\mu(x)}{2}(u'(x))^2 - f(x)\,u(x) \right) dx + \frac{\alpha}{2}\,u^2(0) + \frac{\beta}{2}\,u^2(1)\,.$$

(iii) *The brachistochrone curve*: Look for a curve which connects two points $A$ and $B$ in a plane such that a point with mass $m$ moves on this curve under the influence of gravity but neglecting friction in minimal time.
This problem has already been formulated by da Vinci and Galilei and the first who was able to solve it was Johann Bernoulli. Indeed this problem can be called a "classic" of mathematical modeling and it laid the foundations of the calculus of variations. We formulate this problem in the $(x, y)$-plane as follows: Choose $A = (0, 0)$ and $B = (x_B, y_B)$ such that $x_B > 0$ and $y_B < 0$. We are looking for a function $u : [0, x_B] \to \mathbb{R}$ which fulfills in particular

$$u(0) = 0\,, \quad u(x_B) = y_B\,.$$

It is supposed that the gravity acts in the direction of the negative $y$-axis, and the gravitational acceleration is denoted by $g$. Then the potential energy $V$ is given by $mgy$. We assume that the point mass starts at $(0, 0)$, moves along a curve given by the function $u : [0, x_B] \to \mathbb{R}$ and never moves back or stands still such that the motion as a function of time can be described by the mapping

$$t \mapsto (x(t), y(t)) \quad \text{with} \quad \dot{x} > 0 \quad \text{for} \quad t > 0\,.$$

We require

$$(x(0), y(0)) = (0, 0) \quad \text{and} \quad (x(T), y(T)) = (x_B, y_B)$$

and additionally
$$y(t) = u(x(t))\,.$$

We want to compute the final time $T$ in dependence of the given function $u$. To determine the motion $(x(t), y(t))$, we can apply the Hamiltonian principle thereby considering the constraint that $(x(t), y(t))$ has to be at the curve. Therefore the kinetic energy is computed as follows

$$T(x, y, \dot{x}, \dot{y}) = \frac{m}{2}\left(\dot{x}^2 + \dot{y}^2\right) = \frac{m}{2}\left(1 + (u'(x))^2\right)\dot{x}^2\,.$$

For the instance of time $t = 0$ both the kinetic energy $T$ and also the potential energy $V$ vanish.

We obtain from the law of energy conservation, see Exercise 4.18,

$$\frac{m}{2}\left(1 + (u'(x))^2\right)\dot{x}^2 + mg\,u(x) = T(0) + V(0) = 0. \qquad (4.39)$$

From $\dot{x} > 0$ and $u \leq 0$ it follows that

$$\dot{x} = \left(\frac{-2g\,u(x)}{1 + (u'(x))^2}\right)^{1/2}.$$

Because of $\dot{x} > 0$ we can consider $t$ as a function of $x$ and for this function $t(x)$ the following ordinary differential equation holds true

$$\frac{dt}{dx}(x) = \frac{1}{\sqrt{-2g\,u(x)}}\sqrt{1 + (u'(x))^2}.$$

The final time $T = \mathcal{B}(u)$ is now obtained after integration from 0 to $x_B$ as

$$\mathcal{B}(u) = \int_0^{x_B} \frac{1}{\sqrt{-2g\,u(x)}}\sqrt{1 + (u'(x))^2}\,dx.$$

The task is now to minimize $\mathcal{B}$ under the constraints

$$u(0) = 0, \quad u(x_B) = y_B.$$

(iv) We now formulate a task in which a constraint formulated with the help of integrals plays a role. Minimize

$$\int_0^1 u(x)\,dx$$

for all functions such that

$$\int_0^1 \sqrt{1 + (u')^2}\,dx = \ell, \quad u(0) = a, \quad u(1) = b.$$

Considering all graphs with a given length we are looking for the one for which the area under the graph is minimal. This is a classical isoperimetric problem.

(v) As we will see later on, eigenvalue problems for certain differential operators of second order can be written as variational problems of the following kind: We are looking for stationary points of

$$\mathcal{F}(u) = \int_0^1 \left((\mu(x)(u'(x))^2 + q(x)\,u^2(x)\right)dx$$

allowing for all sufficiently smooth functions $u : [0, 1] \to \mathbb{R}$ with $u(0) = u(1) = 0$ and $\int_0^1 u^2(x)\,dx = 1$.

(vi) *Beam bending*:

We consider a beam whose centerline would coincide with the coordinate axis denoted by $x$ in the absence of external forces. It is always assumed that vertical displacements of the beam can be described by a function

$$u : [0, 1] \to \mathbb{R},$$

$$x \mapsto u(x),$$

i.e., the centerline of the beam is described by the points $(x, u(x))$ with $x \in [0, 1]$ after the displacement. It is assumed that there is no dependence on the coordinates perpendicular to the $(x, y)$-plane. In addition we choose an appropriate unit for length such that the beam can be described as a graph over the unit interval, see Fig. 4.11.

The displacement of the beam leads to an elongation or compression of the fibers above and below the centerline. This elongation or compression is also denoted as strain (compare p. 50 and the following and p. 67). It is proportional to the *curvature* $\kappa$ of the beam and to the distance of the fiber of the centerline. In Sect. 5.10 it will be worked out in more detail that the stress in a beam is proportional to the elongation or compression and therefore is given by

$$\alpha \kappa = \alpha \frac{u''}{(1 + (u')^2)^{3/2}} \approx \alpha u'' \text{ for } |u'| \text{ sufficiently small}$$

(compare Fig. 5.16). Here $\alpha$ is a positive parameter which depends on the modulus of elasticity and the thickness of the beam.

As the potential energy density is proportional to the product of strain and stress, see Sect. 5.10, we obtain in the case of a homogeneous material for the potential energy of the beam

$$\frac{\beta}{2} \int_0^1 \left( \frac{u''}{(1 + (u')^2)^{3/2}} \right)^2 dx \approx \frac{\beta}{2} \int_0^1 (u'')^2 dx.$$

If vertical forces act at the beam with a force density $f$ which is supposed to have a positive sign, if it acts upwardly, then the work done by these forces leads to the energy contribution

$$\int_0^1 f(x)\,u(x)\,dx.$$

Analogously a force $g$ acting at the right endpoint leads to the contribution $g\,u(1)$.

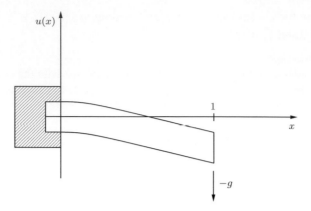

**Fig. 4.11** Beam bending

If we scale the constant $\beta$ to 1 and assume that $|u'|$ is small, then we obtain the energy functional

$$\mathcal{F}(u) = \int_0^1 \left( \frac{1}{2}|u''|^2 - fu \right) dx - g\,u(1) \,.$$

If the beam is rigidly clamped at left endpoint, then the displacement of the beam minimizes the potential energy $\mathcal{F}$ under the constraints

$$u(0) = u'(0) = 0 \,.$$

A fundamental question should be asked for every minimization or maximization problem, namely:

*Is there a solution?*

If one assumes the existence of a solution and then concludes properties of the solution this can lead to serious erroneous conclusions. This is shown by the following *Perron's paradox*:

*Assume that there is a maximal natural number $n$ then we can conclude $n = 1$. This holds true because if otherwise $n > 1$, then $n^2 > n$ would follow in contradiction to the assumption that $n$ is already the maximal number. Therefore $n = 1$ has to hold true.*

In fact many variational problems have no solutions. The task

*Minimize*

$$\mathcal{F}(u) := \int_0^2 \left[ \left( 1 - (u'(x))^2 \right)^2 + u^2(x) \right] dx \tag{4.40}$$

*for all piecewise continuously differential functions* has no solution. For all $u$ the functional $\mathcal{F}(u)$ attains a positive value. If otherwise $\mathcal{F}(u) = 0$ was valid, then we would get $\int_0^2 u^2(x)\, dx = 0$ and therefore $u = 0$ which is in contradiction to $\int_0^2 \big( 1 -$

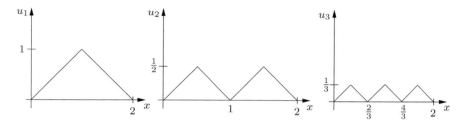

**Fig. 4.12** The first elements of a sequence for which $\mathcal{F}(u_n) \to 0$ for $n \to +\infty$ for $\mathcal{F}$ from (4.40)

$(u'(x))^2)^2 \, dx = 0$. On the other hand $\mathcal{F}$ attains arbitrarily small values, as it is shown by the sawtooth sequence

$$u_n(x) = \frac{1}{n} - \left| x - \frac{2i+1}{n} \right|, \quad \text{if } x \in \left[ \frac{2i}{n}, \frac{2(i+1)}{n} \right], \quad \text{for } i = 0, \dots, n-1,$$

which is depicted in Fig. 4.12. For this sequence we have $u'_n(x) = \pm 1$ up to finitely many points, i.e., almost everywhere and $\int_0^2 u_n^2(x) \, dx \to 0$ for $n \to +\infty$ and therefore we can conclude that $\mathcal{F}(u_n) \to 0$ for $n \to +\infty$.

*Therefore we have* inf $\mathcal{F} = 0$, *but $\mathcal{F}$ does not attain its minimum.*

A sequence $(u_n)_{n \in \mathbb{N}}$ for which $\mathcal{F}(u_n) \to$ inf $\mathcal{F}$ holds true, is called a *minimizing sequence*. Therefore we have just seen an example for which the minimizing sequence converges with respect to the maximum norm but the limit is no minimum of the functional. The reason for this behavior lies in the fact that $\mathcal{F}$ is not continuous with respect to the maximum norm. Continuity or so-called *lower semi-continuity conditions* play an important role in the calculus of variations.

To show the existence of minima there are two possibilities: *the direct* and *the indirect method of the calculus of variations*. We want to shortly explain the direct method of the calculus of variations for a function $f : \mathbb{R}^d \to \mathbb{R}$. We assume that

(i)  $f$ is continuous,
(ii)  $f(x) \to \infty$ for $|x| \to \infty$.

We choose a sequence $(x_n)_{n \in \mathbb{N}}$ in $\mathbb{R}^d$ such that

$$f(x_n) \to \inf_{x \in \mathbb{R}^d} f(x).$$

As we have $\inf_{x \in \mathbb{R}^d} f(x) < \infty$ the sequence $(f(x_n))_{n \in \mathbb{N}}$ is bounded from above, and we conclude from (ii) that the sequence $(x_n)_{n \in \mathbb{N}}$ also has to be bounded. From each bounded sequence $(x_n)_{n \in \mathbb{N}}$ in $\mathbb{R}^d$ a convergent subsequence $(x_{n_j})_{j \in \mathbb{N}}$ can be chosen. This argument is a typical compactness argument, i.e., properties of the sequence are proven which allow for the selection of a convergent subsequence. If we denote the limit of the sequence $(x_{n_j})_{j \in \mathbb{N}}$ by $\bar{x}$, then it follows from the continuity of $f$ that

$$f(\overline{x}) = \lim_{j \to \infty} f(x_{n_j}) = \inf_{x \in \mathbb{R}^d} f(x).$$

From $f(\overline{x}) > -\infty$ it follows $\inf_{x \in \mathbb{R}^d} f(x) > -\infty$ and therefore $f$ attains its minimum at $\overline{x}$. In fact in this proof instead of the assumption of continuity we have only used so-called *lower semi-continuity*

(i') $f(x) \leq \liminf_{n \to \infty} f(x_n)$  for all sequences $(x_n)_{n \in \mathbb{N}}$ such that $\lim_{n \to \infty} x_n = x,$

therefore the assumption (i) can be substituted by (i'). The application of the direct method of the calculus of variations for infinite-dimensional problems leads to many important results. The formulation of necessary continuity, lower semi-continuity, coercivity, and compactness properties requires some functional analysis, however we do not want to go further in this direction and rather refer to the literature [16, 139].

Alternatively we want to point out a method to solve variational problems approximatively. Here the variational problem to be solved is first approximated by a finite-dimensional variational problem which is then solved with standard methods of analysis. For example if we want to minimize

$$\mathcal{D}_1(u) = \int_0^1 \left( \frac{\mu(x)}{2} (u'(x))^2 - f(x)\, u(x) \right) dx$$

under the constraint $u(0) = u(1) = 0$, first we can select the class of admissible functions from a finite-dimensional subspace. Two examples we consider are, for $n \in \mathbb{N}$,

$$V_n = \left\{ u : [0, 1] \to \mathbb{R} \,\middle|\, u \text{ is continuous, } u(0) = u(1) = 0 \text{ and} \right.$$

$$\left. u \text{ is linear on } \left[ \frac{i-1}{n}, \frac{i}{n} \right], \text{ for } i = 1, \ldots, n \right\},$$

or

$$V_n = \left\{ u : [0, 1] \to \mathbb{R} \,\middle|\, u(x) = \sum_{k=1}^n a_k \sin(k\pi x),\ a_k \in \mathbb{R},\ k = 1, \ldots, n \right\}.$$

Now the task to be solved consists of minimizing the functional on the space $V_n$.

Now we want to exemplify the *indirect method of the calculus of variations* for variational problems for functions in one variable. Roughly speaking, the indirect method consists in finding the "zeros of the derivative". To this end we have to first clarify how to extend the notion of a derivative to the infinite-dimensional case. It will turn out that it is sufficient to compute certain directional derivatives. We consider the following variational problem:

Minimize

$$\mathcal{F}(u) = \int_0^1 F(x, u(x), u'(x))\, dx + \alpha(u(0)) + \beta(u(1))$$

for all functions $u \in C^1([0, 1], \mathbb{R}^d)$. Here we assume that $F : \mathbb{R}^{1+2d} \to \mathbb{R}$ and $\alpha, \beta : \mathbb{R}^d \to \mathbb{R}$ are continuously differentiable functions. Similar to Sect. 4.2 we can compute the directional derivatives of $\mathcal{F}$: For $\varphi \in C^1([0, 1], \mathbb{R}^d)$ we consider the following functions for small $\varepsilon$

$$g(\varepsilon) := \mathcal{F}(u + \varepsilon\varphi) \tag{4.41}$$

and differentiate with respect to $\varepsilon$. Now we define

$$\delta\mathcal{F}(u)(\varphi) := g'(0) \tag{4.42}$$

and denote this expression as the *first variation of $\mathcal{F}$ at the point $u$ in direction $\varphi$*. This procedure corresponds to the notion of directional derivatives for functions $f : \mathbb{R}^d \to \mathbb{R}$. If $\mathcal{F}$ takes its minimum at $u$, then we have

$$0 = g'(0) = \delta\mathcal{F}(u)(\varphi)\,.$$

Now we want to compute $\delta\mathcal{F}$. In the following the variables of $F$ are denoted by $(x, z, p) \in \mathbb{R} \times \mathbb{R}^d \times \mathbb{R}^d$, i.e., $F = F(x, z, p)$. Furthermore we use the abbreviations

$$F_{,x} = \partial_x F\,, \quad F_{,z} = (\partial_{z_1} F, \dots, \partial_{z_d} F)\,, \quad F_{,p} = (\partial_{p_1} F, \dots, \partial_{p_d} F)\,.$$

**Theorem 4.6** *Let $F \in C^1(\mathbb{R}^{1+2d}, \mathbb{R})$ and $u \in C^1([0, 1], \mathbb{R}^d)$. Then it holds true that for all $\varphi \in C^1([0, 1], \mathbb{R}^d)$,*

$$\delta\mathcal{F}(u)(\varphi) = \int_0^1 [F_{,z}(x, u(x), u'(x)) \cdot \varphi(x) + F_{,p}(x, u(x), u'(x)) \cdot \varphi'(x)]\, dx$$
$$+ \alpha_{,z}(u(0)) \cdot \varphi(0) + \beta_{,z}(u(1)) \cdot \varphi(1)\,.$$

*If $F_{,p} \in C^1(\mathbb{R}^{1+2d}, \mathbb{R}^d)$ and $u \in C^2([0, 1], \mathbb{R}^d)$, then for all $\varphi \in C^1([0, 1], \mathbb{R}^d)$ it holds true that*

$$\delta\mathcal{F}(u)(\varphi) = \int_0^1 \left[ F_{,z}(x, u(x), u'(x)) - \tfrac{d}{dx} F_{,p}(x, u(x), u'(x)) \right] \cdot \varphi(x)\, dx$$
$$+ \left( F_{,p}(1, u(1), u'(1)) + \beta_{,z}(u(1)) \right) \cdot \varphi(1) \tag{4.43}$$
$$+ \left( - F_{,p}(0, u(0), u'(0)) + \alpha_{,z}(u(0)) \right) \cdot \varphi(0)\,.$$

*Proof* For $g$ from (4.41) we compute

$$
g'(\varepsilon) = \int_0^1 \frac{d}{d\varepsilon} F(x, u(x) + \varepsilon\,\varphi(x), u'(x) + \varepsilon\,\varphi'(x))\,dx + \frac{d}{d\varepsilon}\alpha(u(0) + \varepsilon\,\varphi(0))
$$
$$
+ \frac{d}{d\varepsilon}\beta(u(1) + \varepsilon\,\varphi(1))
$$
$$
= \int_0^1 \big[ F_{,z}(x, u(x) + \varepsilon\,\varphi(x), u'(x) + \varepsilon\,\varphi'(x)) \cdot \varphi(x)
$$
$$
+ F_{,p}(x, u(x) + \varepsilon\,\varphi(x), u'(x) + \varepsilon\,\varphi'(x)) \cdot \varphi'(x) \big]\,dx
$$
$$
+ \alpha_{,z}(u(0) + \varepsilon\,\varphi(0)) \cdot \varphi(0) + \beta_{,z}(u(1) + \varepsilon\,\varphi(1)) \cdot \varphi(1) .
$$

We set $\varepsilon = 0$, and the first assertion is a consequence of the definition in (4.42).

Integration by parts of the integrand $F_{,p}(x, u(x), u'(x)) \cdot \varphi'(x)$ leads to the second assertion in the following way:

$$
\int_0^1 F_{,p}(x, u(x), u'(x)) \cdot \varphi'(x)\,dx = - \int_0^1 \frac{d}{dx} F_{,p}(x, u(x), u'(x)) \cdot \varphi(x)\,dx
$$
$$
+ [F_{,p}(x, u(x), u'(x)) \cdot \varphi(x)]_{x=0}^{x=1} .
$$

$\square$

If $u$ is a minimum of $\mathcal{F}$ then $\delta\mathcal{F}(u)(\varphi)$ has to vanish for all $\varphi \in C^1([0, 1], \mathbb{R}^d)$. From this one concludes that $u$ has to solve a system of ordinary differential equations. This is the content of the following theorem.

**Theorem 4.7**  *Let $F \in C^1(\mathbb{R}^{1+2d}, \mathbb{R})$, $F_{,p} \in C^1(\mathbb{R}^{1+2d}, \mathbb{R}^d)$, $u \in C^2([0, 1], \mathbb{R}^d)$.*

(a)  *If $\delta\mathcal{F}(u)(\varphi) = 0$ for all $\varphi \in C_0^1((0, 1), \mathbb{R}^d)$, then the* Euler–Lagrange *differential equations hold true:*

$$
\frac{d}{dx} F_{,p}(x, u(x), u'(x)) = F_{,z}(x, u(x), u'(x)) \quad \text{for all } x \in (0, 1) . \qquad (4.44)
$$

(b)  *If $\delta\mathcal{F}(u)(\varphi) = 0$ for all $\varphi \in C^1([0, 1], \mathbb{R}^d)$, then in addition the following boundary conditions are valid*

$$
F_{,p}(0, u(0), u'(0)) - \alpha_{,z}(u(0)) = 0 , \qquad (4.45)
$$
$$
F_{,p}(1, u(1), u'(1)) + \beta_{,z}(u(1)) = 0 . \qquad (4.46)
$$

*Proof* The Euler–Lagrange equations are a consequence of (4.43) if we consider functions $\varphi$ which vanish at the boundary and use the fundamental lemma of the calculus of variations (compare Exercise 4.19). Since the Euler–Lagrange equations are valid we conclude that the integral expression in (4.43) vanishes. In (b) the function $\varphi$ can be chosen such that $\varphi(1)$ and $\varphi(0)$ assume arbitrary values. Hence, we conclude that boundary conditions (4.45) and (4.46) must hold. $\qquad\square$

**Remark**. If we minimize

$$\mathcal{F}(u) = \int_0^1 F(x, u(x), u'(x))\, dx$$

under the constraints $u(0) = a$ and $u(1) = b$, then the minimum $u$ has to fulfill

$$\frac{d}{d\varepsilon}\mathcal{F}(u + \varepsilon\varphi)_{|\varepsilon=0} = 0$$

for $\varphi \in C_0^1((0, 1), \mathbb{R}^d)$. The choice of function $\varphi$ implies $\varphi(0) = \varphi(1) = 0$ and therefore $u + \varepsilon\varphi$ satisfies the required constraints. In this case the Euler–Lagrange equations hold true with the boundary conditions $u(0) = a$ and $u(1) = b$. Now we want to analyze the variational problems (i)–(iii) formulated above with the help of the indirect method of the calculus of variations.

(i) Because of $\frac{d}{dp}\sqrt{1 + p^2} = p/\sqrt{1 + p^2}$ the Euler–Lagrange equation for the variational problem corresponding to

$$\mathcal{L}(u) := \int_0^1 \sqrt{1 + (u'(x))^2}\, dx$$

takes the form

$$0 = \frac{d}{dx}\left(\frac{u'(x)}{\sqrt{1 + (u'(x))^2}}\right).$$

Differentiating the term in brackets leads to

$$\frac{u''(x)}{(1 + (u'(x))^2)^{3/2}} = 0,$$

meaning that the curvature has to vanish, compare Appendix B. In particular $u$ is linear. On the other hand every linear function satisfies the Euler–Lagrange equation. This result corresponds to the well-known fact that the shortest connection between two points is given by a straight line.

(ii) For the minimum of

$$\mathcal{D}_2(u) = \int_0^1 \left[\frac{\mu(x)}{2}(u'(x))^2 - f(x)\,u(x)\right] dx + \frac{\alpha}{2}u^2(0) + \frac{\beta}{2}u^2(1)$$

we obtain the Euler–Lagrange equation

$$\frac{d}{dx}(\mu(x)u'(x)) = -f(x).$$

Additionally we conclude from (4.45), (4.46) in Theorem 4.7 the validity of the boundary conditions

$$\mu(0)\, u'(0) - \alpha\, u(0) = 0\,,$$

and

$$\mu(1)\, u'(1) + \beta\, u(1) = 0\,.$$

Before considering the case (iii) we have to deal with integrands of the form $F = F(z, p)$. If we multiply the Euler–Lagrange equation (4.44) by $u'(x)$, we obtain

$$\begin{aligned}0 &= u'(x) \cdot \left[\frac{d}{dx}F_{,p}(u(x), u'(x)) - F_{,z}(u(x), u'(x))\right]\\ &= \frac{d}{dx}\left[u'(x) \cdot F_{,p}(u(x), u'(x)) - F(u(x), u'(x))\right].\end{aligned}$$

Defining

$$G(z, p) := p \cdot F_{,p}(z, p) - F(z, p)\,,$$

we have

$$\frac{d}{dx}G(u(x), u'(x)) = 0\,,$$

i.e., $G$ is a *first integral* of the Euler–Lagrange differential equation.

(iii)  In the case of the Brachistochrone curve we consider

$$F(z, p) = \frac{1}{\sqrt{-z}}\, \sqrt{1 + p^2}\,,$$

where we have neglected the factor $\frac{1}{\sqrt{2g}}$, as it does not lead to other minimizers. As a first integral we obtain with

$$F_{,p} = (-z)^{-1/2}p/(1 + p^2)^{1/2}$$

the function

$$\begin{aligned}G(z, p) &= p \cdot (-z)^{-1/2}p/(1 + p^2)^{1/2} - (-z)^{-1/2}\sqrt{1 + p^2}\\ &= (-z)^{-1/2}\left(-(1 + p^2)^{-1/2}\right).\end{aligned}$$

As $G(u(x), u'(x))$ is constant, there exists a $c \in \mathbb{R}$ such that

$$\sqrt{-u(x)} \, \sqrt{1 + (u'(x))^2} = \frac{1}{c} \,.$$

Squaring of this expression leads to

$$c^2 \big(1 + (u'(x))^2\big) u(x) = -1 \,. \tag{4.47}$$

From this identity we can compute $u$. In Exercise 4.20 the solution of the Brachistochrone problem will be fully determined.

The conditions which we have derived are necessary conditions. In general it is difficult to decide whether a function which fulfills these conditions is a local or even a global minimizer.

There are further necessary conditions which a minimizer has to fulfill. We want to discuss one of these conditions which uses second derivatives. If $u$ is a minimizer of $\mathcal{F}$, then

$$g(\varepsilon) := \mathcal{F}(u + \varepsilon \varphi)$$

satisfies the necessary conditions

$$g'(0) = 0, \ g''(0) \geq 0 \,.$$

The condition for the second derivative can be formulated with the help of $\mathcal{F}$. We want to exemplify it for the case

$$\mathcal{F}(u) = \int_0^1 F(u'(x)) \, dx \,.$$

It holds that

$$0 \leq g''(0) = \int_0^1 \varphi'(x) \cdot F_{,pp}(u'(x)) \, \varphi'(x) \, dx \,.$$

Here $F_{,pp}$ denotes the second derivative of $F$ with respect to the variable $p$. The expression

$$\delta^2 \mathcal{F}(u)(\varphi) := g''(0)$$

is called the *second variation of $\mathcal{F}$ at the point $u$ in direction $\varphi$*. The second variation of $\mathcal{F}$ at the point $u$ therefore has to be non-negative for all possible $\varphi$.

Now we want to consider problems with constraints. In the examples (iv) and (v) the constraints have the form $\int_0^1 \sqrt{1 + (u')^2} \, dx = \ell$ or $\int_0^1 u^2 \, dx = 1$, respectively. Therefore we consider general constraints of the form

$$\mathcal{G}(u) := \int_0^1 G(x, u(x), u'(x)) \, dx = c \,,$$

where $c \in \mathbb{R}$ is given. Similar to the finite-dimensional case we can formulate necessary conditions for the problem

$$\mathcal{F}(u) \to \min \quad \text{with the constraint} \quad \mathcal{G}(u) = c$$

with the help of Lagrange multipliers. This is the content of the following theorem.

**Theorem 4.8** *Let $F, G \in C^1(\mathbb{R}^{1+2d}, \mathbb{R})$ and let $u \in C^1([0, 1], \mathbb{R}^d)$ be a minimizer of the problem: Minimize $\mathcal{F}(v) = \int_0^1 F(x, v(x), v'(x)) \, dx$ for all $v \in C^1([0, 1], \mathbb{R}^d)$ such that $v(0) = a$, $v(1) = b$ and $\mathcal{G}(v) = c$. Furthermore it is assumed that $\delta\mathcal{G}(u)(\psi) \neq 0$ for some $\psi \in C_0^\infty((0, 1), \mathbb{R}^d) \setminus \{0\}$. Then there exists a Lagrange multiplier $\lambda \in \mathbb{R}$ such that*

$$\delta\mathcal{F}(u)(\varphi) + \lambda \, \delta\mathcal{G}(u)(\varphi) = 0 \quad \text{for all} \;\; \varphi \in C_0^\infty((0, 1), \mathbb{R}^d) \,.$$

*If furthermore $F_{,p}, G_{,p} \in C^1(\mathbb{R}^{1+2d}, \mathbb{R}^d)$ and $u \in C^2([0, 1], \mathbb{R}^d)$, then it holds true that*

$$\frac{d}{dx}(F_{,p} + \lambda G_{,p}) = F_{,z} + \lambda G_{,z} \,,$$

*where $F_{,p}, G_{,p}, F_{,z}$ and $G_{,z}$ are dependent on $(x, u(x), u'(x))$.*

*Proof* As assumed there exists a $\psi \in C_0^\infty((0, 1), \mathbb{R}^d)$ such that $\delta\mathcal{G}(u)(\psi) = 1$. For an arbitrary $\varphi \in C_0^\infty((0, 1), \mathbb{R}^d)$ we define

$$\Phi(\varepsilon, \delta) := \mathcal{F}(u + \varepsilon\varphi + \delta\psi), \quad \Psi(\varepsilon, \delta) := \mathcal{G}(u + \varepsilon\varphi + \delta\psi) \,.$$

As $\partial_\delta \Psi(0, 0) = 1$, the implicit function theorem leads to the existence of a continuously differentiable function $\tau : (-\varepsilon_0, \varepsilon_0) \to \mathbb{R}$ with $\tau(0) = 0$ such that

$$\Psi(\varepsilon, \tau(\varepsilon)) = c \quad \text{for all} \;\; \varepsilon \in (-\varepsilon_0, \varepsilon_0)$$

and

$$\tau'(0) = -\partial_\varepsilon \Psi(0, 0) \,. \tag{4.48}$$

This means that $u + \varepsilon\varphi + \tau(\varepsilon)\psi$ satisfies all required constraints and we can conclude that the function

$$\varepsilon \mapsto \Phi(\varepsilon, \tau(\varepsilon))$$

has a minimum at the point $\varepsilon = 0$, leading to

$$\partial_\varepsilon \Phi(0, 0) + \partial_\delta \Phi(0, 0)\tau'(0) = 0 \,. \tag{4.49}$$

If we define

$$\lambda := -\partial_\delta \Phi(0, 0) = -\delta\mathcal{F}(u)(\psi) \,,$$

then it follows from (4.48) and (4.49) that

$$\partial_\varepsilon \Phi(0,0) + \lambda \partial_\varepsilon \Psi(0,0) = 0$$

holds. The last identity leads to

$$\delta \mathcal{F}(u)(\varphi) + \lambda \, \delta \mathcal{G}(u)(\varphi) = 0 \quad \text{for all} \quad \varphi \in C_0^\infty \big((0,1), \mathbb{R}^d\big),$$

which is the first part of the assertion. The second part can be concluded using integration by parts analogously to the proof of Theorem 4.7. $\qquad\square$

The examples (iv) and (v) now can be treated on the basis of Theorem 4.8.

(iv) We consider the case

$$\mathcal{F}(u) := \int_0^1 u(x)\,dx \quad \text{and} \quad \mathcal{G}(u) := \int_0^1 \sqrt{1 + (u')^2}\,dx$$

and use the constraint $\mathcal{G}(u) = \ell$, i.e., the length of the graph is given. Setting $F(x, z, p) = z$ and $G(x, z, p) = \sqrt{1 + p^2}$ Theorem 4.8 leads to the following necessary conditions for a minimizer in form of the following differential equation

$$\lambda \frac{d}{dx}\left( \frac{u'}{\sqrt{1 + (u')^2}} \right) = -1 \,.$$

The expression $\frac{d}{dx}\left( \frac{u'}{\sqrt{1+(u')^2}} \right)$ denotes the curvature of the graph. As $\lambda$ is a constant, we obtain segments of a circle as solutions. If $u$ describes a straight line, then the curvature and thus $\delta \mathcal{G}(u)$ is identically zero, and therefore Theorem 4.8 cannot be applied.

(v) We want to minimize

$$\mathcal{F}(u) := \int_0^1 \big(\mu(x)(u'(x))^2 + q(x)\,u^2(x)\big)\,dx$$

under the constraints $u(0) = u(1) = 0$ and the additional constraint

$$\mathcal{G}(u) := \int_0^1 u^2(x)\,r(x)\,dx = 1 \,.$$

Here we assume that $\mu \in C^1([0,1], \mathbb{R}), q, r \in C^0([0,1], \mathbb{R})$ and $\mu, r > 0$. From Theorem 4.8 we conclude that as a necessary condition for a minimizer $u \in C^2([0,1])$ the boundary value problem

$$-\frac{d}{dx}\left(\mu(x)\,u'(x)\right) + q(x)\,u(x) = -\lambda\,r(x)\,u(x)\,,$$

$$u(0) = u(1) = 0$$

must hold, where $\lambda$ is a constant. The negative Lagrange multiplier $-\lambda$ can be interpreted as a generalized eigenvalue and $u$ is a corresponding eigenfunction which is normalized with respect to the inner product

$$(u,\,v)_r := \int_0^1 u(x)\,v(x)\,r(x)\,dx\,.$$

If integrals have to be minimized in which higher derivatives appear, then the necessary conditions can be formulated with the help of ordinary differential equations of higher order. We will see that in this case more than two boundary conditions appear. We discuss the situation for the example of beam bending. In example (vi) we have to minimize

$$\mathcal{G}(u) = \int_0^1 \left(\tfrac{1}{2}(u'')^2 - fu\right)dx - g\,u(1)$$

under the constraints

$$u(0) = 0\,,\quad u'(0) = 0\,. \tag{4.50}$$

Now let $u : [0,\,1] \to \mathbb{R}$ be a minimizer of $\mathcal{G}$. Then $u + \varepsilon\varphi$ satisfies the required constraints at the boundary if $\varphi(0) = 0$, $\varphi'(0) = 0$. It follows that

$$0 = \frac{\delta\mathcal{G}}{\delta u}(u)(\varphi) = \frac{d}{d\varepsilon}\mathcal{G}(u + \varepsilon\varphi)_{|\varepsilon=0}\,.$$

We compute

$$\frac{\delta\mathcal{G}}{\delta u}(u)(\varphi) = \int_0^1 (u''\varphi'' - f\varphi)\,dx - g\,\varphi(1)\,.$$

If $u \in C^4([0,\,1])$ holds true, then by integration by parts and using the conditions $\varphi(0) = \varphi'(0) = 0$ we obtain:

$$0 = \int_0^1 (-u'''\varphi' - f\varphi)\,dx + u''(1)\,\varphi'(1) - g\,\varphi(1)$$

$$= \int_0^1 (u^{(4)} - f)\varphi\,dx - u'''(1)\,\varphi(1) + u''(1)\,\varphi'(1) - g\,\varphi(1)\,. \tag{4.51}$$

As $\varphi$ is arbitrary we can conclude using the fundamental lemma of the calculus of variations that

$$u^{(4)}(x) = f(x) \quad \text{for} \ \ x \in (0,\,1)\,. \tag{4.52}$$

As the integral expression vanishes in (4.51) and as $\varphi(1)$ and $\varphi'(1)$ can be chosen arbitrarily, it follows:

$$u''(1) = 0 \quad \text{and} \quad u'''(1) + g = 0. \tag{4.53}$$

Summarizing, (4.52) has to be solved with the four boundary conditions (4.50), (4.53).

## 4.8  Optimal Control with Ordinary Differential Equations

Often we would like to *control* processes which can be formulated with differential equations. This means that the process should be steered by the appropriate choice of control quantities such that a certain quantity, for example costs or gain or consumption or the time necessary for the process, is optimized. In this section we want to sketch shortly which questions are of interest in this context and we will formulate important necessary optimality conditions in form of the Pontryagin's maximum principle. First we give three examples.

(i)  *Control of production and consumption.* An enterprise generates revenue, a part of which is distributed as profit and the rest is supposed to be invested again. Here the aim is to reach as much gain as possible up to a given instance of time $T$. It is assumed that the revenue grows proportionally to the reinvested capital. We define

$y(t)$ to be the amount of output produced at time $t \geq 0$

Furthermore let

$u(t)$ be the portion of the gain which is reinvested at the instance of time $t$.

This means that the following inequality constraints have to be posed:

$$0 \leq u(t) \leq 1 \quad \text{for all} \ \ t \geq 0.$$

The function $u$ is the quantity with which the process can be controlled. Now let $k$ be the rate by which the invested capital grows. If a control $u$ is given, then the following initial value problem has to be solved:

$$y'(t) = k\,u(t)\,y(t), \tag{4.54}$$
$$y(0) = y_0. \tag{4.55}$$

For a given instance of time $t$ the amount $(1 - u(t))y(t)$ is the profit and will not be reinvested. Therefore we want to maximize the quantity

$$\mathcal{F}(y, u) := \int_0^T (1 - u(t))\, y(t)\, dt$$

for all $y$ and $u$, which satisfy the constraints (4.54), (4.55).

(ii) *Control of a vehicle.* The aim is to move a vehicle with normalized mass $m = 1$ from one point in the plane along a straight line to another point in the plane in minimal time. Without loss of generality we assume that the vehicle at the time $t = 0$ at the $x$-axis has the position $x = -x_0$ and the aim is to move it as fast as possible to the location $x = x_0$. At the starting time and at the target time the velocity has to be zero. The vehicle has a motor which generates the force $u(t)$ at time $t$. This force can be controlled, but it is limited by a maximal force normalized to 1, i.e., $|u(t)| \leq 1$. This means that we are looking for a minimum of

$$\mathcal{F}(x, u) = \int_0^T 1 \, dt$$

under the constraints

$$\begin{aligned} x''(t) &= u(t) \quad \text{Newton's law}, \\ x(0) &= -x_0, \ x'(0) = 0, \\ x(T) &= x_0, \quad x'(T) = 0, \\ |u(t)| &\leq 1 \qquad \text{for all } t \in [0, T]. \end{aligned}$$

(iii) *Optimal control of fish harvesting.* We consider a fish population which, if there would be no fishing, would grow according to a population model with limited resources. Therefore for the total population $x(t)$ the differential equation

$$x'(t) = q(x_M - x(t)) \, x(t) =: F(x)$$

is valid, one may also compare Sects. 1.3 and 4.3. If we now consider fish harvesting which is modeled by an intensity $u(t)$ this leads to the initial value problem

$$x'(t) = F(x(t)) - u(t), \quad x(0) = x_0.$$

In this case the system can be controlled by the harvesting intensity $u(t)$. The aim is to maximize the total profit. To model this quantity for example the functional

$$\mathcal{F}(x, u) := \int_0^T e^{-\delta t} (p - k(x(t))) u(t) \, dt$$

has been proposed. Hereby we used the following notations:

$p$      for the selling price of fish (per mass unit),
$k(x)$   for the harvesting costs per mass unit for given population quantity $x$,
$\delta$      for a discount factor.

With the discount factor it is taken into account that the fishermen obtain interest for the profit of fish caught at early times, if this profit is invested at a bank. Therefore the discount factor $\delta$ corresponds to the growth rate in a continuous interest model. By the factor $e^{-\delta t}$ all profits are computed back to the instance of time $t = 0$, this is called *discounting*.

There are many more areas in which the optimal control with ordinary differential equations plays an important role. As further examples we mention: control of the motion of airplanes or spacecrafts, control of the motion of robots and in sports, control of chemical processes, control of the spreading of epidemics by vaccination strategies. In this context we refer to the literature for more details. See for example Macki and Strauss [92], Evans [36], Zeidler [139].

Now we want to derive in a formal manner the necessary optimality conditions for the solution of optimization problems in the context of control processes. We formulate necessary conditions for problems with a *fixed final time* for the following general setting. Let $A \subset \mathbb{R}^m$ and

$$f : \mathbb{R}^d \times A \to \mathbb{R}^d, \quad y_0 \in \mathbb{R}^d.$$

The set $A$ is supposed to bound the possible controls, i.e., we consider controls of the form

$$u : [0, T] \to A.$$

For a given $u$ the state $y : [0, T] \to \mathbb{R}^d$ is supposed to be the solution of the following initial value problem

$$y'(t) = f(y(t), u(t)), \quad t \in [0, T], \tag{4.56}$$
$$y(0) = y_0. \tag{4.57}$$

The objective functional to be optimized is given by

$$\mathcal{F}(y, u) := \int_0^T r(y(t), u(t)) \, dt + g(y(T)),$$

where

$$r : \mathbb{R}^d \times A \to \mathbb{R}, \quad g : \mathbb{R}^d \to \mathbb{R}$$

are given smooth functions.

We consider the following task:
(P): *Maximize $\mathcal{F}$ under the constraints (4.56), (4.57) and $u(t) \in A$ for all $t \in [0, T]$.*

In a formal manner we can derive necessary conditions for a solution of (P) with the help of Lagrange multipliers as follows. Analogously to the finite-dimensional case (4.36) we want to describe the solution and the corresponding Lagrangian parameters as critical points of a functional which results from $\mathcal{F}$ by adding a term which is the product of the Lagrangian parameters and a functional characterizing the constraints. We multiply in the Euclidean scalar product the constraint $-y' + f(y, u) = 0$ for all times $t$ by a multiplier

$$p : [0, T] \rightarrow \mathbb{R}^d \, .$$

To get a real quantity instead of a function we integrate with respect to $t$ and obtain

$$\mathcal{L}(y, u, p) := \mathcal{F}(y, u) - \int_0^T p(t) \cdot (y'(t) - f(y(t), u(t))) \, dt \, .$$

First we consider $A = \mathbb{R}^m$. Then all critical points $(\overline{y}, \overline{u}, \overline{p})$ of $\mathcal{L}$, with the constraint $\overline{y}(0) = y_0$, satisfy the following equations

$$\frac{\delta \mathcal{L}}{\delta y}(\overline{y}, \overline{u}, \overline{p})(h) = 0 \, ,$$

$$\frac{\delta \mathcal{L}}{\delta u}(\overline{y}, \overline{u}, \overline{p})(v) = 0 \, ,$$

$$\frac{\delta \mathcal{L}}{\delta p}(\overline{y}, \overline{u}, \overline{p})(q) = 0 \, ,$$

where $v$ and $q$ are arbitrary functions, but for $h$ only functions with $h(0) = 0$ are admissible, as the constraint $y(0) = y_0$ has to be fulfilled.

The first variation of $\mathcal{L}$ can be computed analogously to Sect. 4.6 and we obtain from $\frac{\delta \mathcal{L}}{\delta y} = 0$ the equation

$$\int_0^T r_{,y}(\overline{y}, \overline{u}) \cdot h \, dt + \nabla g(\overline{y}(T)) \cdot h(T) - \int_0^T \overline{p} \cdot (h' - f_{,y}(\overline{y}, \overline{u})h) \, dt = 0$$

for all $h : [0, T] \rightarrow \mathbb{R}^d$ with $h(0) = 0$. Integration by parts leads to

$$\int_0^T h \cdot (\overline{p}' + f_{,y}^\top(\overline{y}, \overline{u}) \, \overline{p} + r_{,y}(\overline{y}, \overline{u})) \, dt + h(T) \cdot (\nabla g(\overline{y}(T)) - \overline{p}(T)) = 0 \, .$$

$$\tag{4.58}$$

Using the fundamental lemma of the calculus of variations we get

$$\overline{p}' = -f_{,y}^\top(\overline{y}, \overline{u}) \, \overline{p} - r_{,y}(\overline{y}, \overline{u}) \quad \text{for} \quad t \in [0, T] \, . \tag{4.59}$$

As the integrand in (4.58) vanishes and $h(T)$ is arbitrary we obtain the "final condition"

$$\overline{p}(T) = \nabla g(\overline{y}(T)) \, . \tag{4.60}$$

The function $\overline{p} : [0, T] \rightarrow \mathbb{R}^d$, for given functions $\overline{y}$ and $\overline{u}$, is the solution of a linear ordinary differential equation, but with a condition at the final time $T$ instead of an initial value condition. The condition $\frac{\delta \mathcal{L}}{\delta p} = 0$ leads to $\int_0^T q \cdot (\overline{y}' - f(\overline{y}, \overline{u})) \, dt = 0$ for all $q$, which is just (4.56).

The identity $\frac{\delta \mathcal{L}}{\delta u} = 0$ together with the fundamental lemma of the calculus of variation leads to

$$r_{,u}(\overline{y}, \overline{u}) + f_{,u}^\top(\overline{y}, \overline{u}) \, \overline{p} = 0 . \tag{4.61}$$

If now $A \neq \mathbb{R}^m$ then $\frac{\delta \mathcal{L}}{\delta u}(\overline{y}, \overline{u}, \overline{p})(v)$ is not well defined for arbitrary $v$. The function $\overline{u} + \varepsilon v$ generally is not pointwise contained in $A$. If $A$ is convex, then at least for all $u : [0, T] \to A$ and $\varepsilon \in [0, 1]$

$$\overline{u} + \varepsilon(u - \overline{u}) = \varepsilon u + (1 - \varepsilon)\overline{u} \in A$$

holds. Then we have

$$\frac{d}{d\varepsilon} \mathcal{L}(\overline{y}, \overline{u} + \varepsilon(u - \overline{u}), \overline{p})_{|\varepsilon=0} \leq 0$$

and therefore

$$\int_0^T (r_{,u}(\overline{y}, \overline{u}) + \overline{p} \cdot f_{,u}(\overline{y}, \overline{u})) \cdot (u - \overline{u}) \, dt \leq 0$$

for all $u : [0, T] \to A$.

The condition (4.61) just means that $\overline{u}$ is a critical point of the functional

$$v \mapsto r(\overline{y}, v) + \overline{p} \cdot f(\overline{y}, v) .$$

The following Pontryagin's maximum principle asserts that $\overline{u}$ not only is a critical point but even the maximizer of the functional above. Before we formulate the maximum principle let us define:

**Definition 4.9** The *Hamiltonian* of optimal control theory is defined for all $y, p \in \mathbb{R}^d$, $u \in A$ as follows:

$$H(y, u, p) := f(y, u) \cdot p + r(y, u) .$$

**Theorem 4.10** (Pontryagin's maximum principle) *Let $(\overline{y}, \overline{u})$ be a solution of problem (P).*

*Then there exists an* adjoint state $\overline{p} : [0, T] \to \mathbb{R}^d$, *such that for all $t \in [0, T]$*

$$\overline{y}'(t) = H_{,p}(\overline{y}(t), \overline{u}(t), \overline{p}(t)), \quad \overline{y}(0) = y_0 , \tag{4.62}$$
$$\overline{p}'(t) = -H_{,y}(\overline{y}(t), \overline{u}(t), \overline{p}(t)), \quad \overline{p}(T) = \nabla g(\overline{y}(T)) , \tag{4.63}$$

*and*

$$H(\overline{y}(t), \overline{u}(t), \overline{p}(t)) = \max_{v \in A} H(\overline{y}(t), v, \overline{p}(t)) \tag{4.64}$$

*hold true. Furthermore the mapping*

$$t \mapsto H(\overline{y}(t), \overline{u}(t), \overline{p}(t)) \tag{4.65}$$

*is constant.*

The proof of this theorem is quite involved and we refer to the literature [36, 92, 139]. We only remark that (4.62) is identical to (4.56), (4.57) because of the definition of $H$. The equations in (4.63) correspond to (4.59) and (4.60). If $A = \mathbb{R}^m$ then (4.61) is a consequence of (4.64).

Continuing with the case $A = \mathbb{R}^m$ and using (4.62)–(4.64) we get

$$\frac{d}{dt} H(\overline{y}, \overline{u}, \overline{p}) = H_{,y}(\overline{y}, \overline{u}, \overline{p}) \cdot \overline{y}' + H_{,u}(\overline{y}, \overline{u}, \overline{p}) \cdot \overline{u}' + H_{,p}(\overline{y}, \overline{u}, \overline{p}) \cdot \overline{p}'$$
$$= -\overline{p}' \cdot \overline{y}' + 0 + \overline{y}' \cdot \overline{p}' = 0 \,.$$

Similar maximum principles can be shown for other problems of optimal control. In particular there is a maximum principle valid for problems in which the final time is variable. Basically it turns out that it is advantageous to work with the adjoint state $p$. Often necessary conditions can be derived easily in a formal manner as above with the help of Lagrange multipliers or adjoint states. However, to make these considerations rigorous often requires a large amount of analytical effort. For details we refer to the literature [36, 92, 139].

For the example of control of production and consumption we want to sketch how Pontryagin's maximum principle can be used to show properties of the solutions of the control problems. We use again the notation from example (i), but for simplicity's sake we set $k = 1$. In this case we have

$$f(y, u) = yu, \quad r(y, u) = (1 - u)y, \quad g = 0, \quad \text{and} \quad A = [0, 1] \,.$$

From (4.59), (4.60) the adjoint problem (omitting the bars over the variables) reads as

$$p' = -up - (1 - u) \quad \text{in } [0, T] \text{ and } p(T) = 0 \,. \tag{4.66}$$

Because of $H(y, u, p) = yup + (1 - u)y$ Pontryagin's maximum principle leads to

$$y(t)\, u(t)\, p(t) + (1 - u(t))\, y(t) = \max_{0 \le v \le 1} [y(t)\, v\, p(t) + (1 - v)\, y(t)] \,.$$

As we have $y(t) > 0$ and $y(t)$ appears as a factor in the line above, we obtain

$$u(t)(p(t) - 1) = \max_{0 \le v \le 1} [v(p(t) - 1)] \,.$$

Therefore it follows that

$$u(t) = \begin{cases} 1 & \text{if } p(t) > 1, \\ 0 & \text{if } p(t) < 1. \end{cases}$$

If we knew the adjoint state $p$ the control $u$ could be computed. As $p(T) = 0$ holds true there exists an instance of time $\bar{t}$ such that

$$p(t) < 1 \quad \text{for } t \in (\bar{t}, T]$$

and therefore we have

$$u(t) = 0 \quad \text{for } t \in (\bar{t}, T].$$

From (4.66) we conclude

$$p'(t) = -1 \quad \text{for } t \in (\bar{t}, T], \quad p(T) = 0$$

and it follows that

$$\bar{t} = T - 1 \quad \text{and} \quad p(t) = T - t \text{ for } t \in (T - 1, T].$$

From (4.66) and the fact that $u$ only can attain values between 0 and 1, we get $p' < 0$ in $[0, T - 1)$. Therefore we have $p(t) > 1$ in $[0, T - 1)$, hence $u(t) = 1$, and from (4.66) one sees that

$$p'(t) = -p(t).$$

As $p(T - 1) = 1$ holds true, it follows that

$$p(t) = e^{T-1-t} > 1 \quad \text{for } t \in [0, T - 1].$$

If $T > 1$ holds true, then we have finally

$$u(t) = \begin{cases} 1 & \text{if } 0 \le t < T - 1, \\ 0 & \text{if } T - 1 < t \le T. \end{cases}$$

Thus the control only switches once. The solution is an example of a *bang–bang* control, meaning that the control always fulfills the upper or lower boundary of the control boundaries.


## 4.9 Literature

The presentation of simple oscillations from Sect. 4.1 can be found in [41] and also in books on electrical engineering or theoretical mechanics, see for example [64]. Lagrangian and Hamiltonian mechanics is part of every textbook about theoretical mechanics, see for example [70], a mathematically quite advanced short exposition is given by Chap. 58 of [140]. For the analysis of ordinary differential equations we refer to [6]. A detailed analysis of convex optimization problems is given in the classical book [16]. A more involved introduction to the theory of steering and control problems can be found in [36, 92, 139]. Numerical methods for the solution of ordinary differential equations are described in [28, 65, 117, 123].

## 4.10   Exercises

**Exercise 4.1**  Solve the initial value problem

$$y''(t) + 2a\, y'(t) + b\, y(t) = \sin(\omega t)\,, \quad y(0) = 1\,, \quad y'(0) = 0$$

for the oscillation equation with the help of an appropriate case distinction for $a$, $b$, and $\omega$. Assume that $a \geq 0$, $b, \omega > 0$.

**Exercise 4.2**  Compute a real fundamental system for the following differential equations:

(a)  $y'''(t) - y''(t) + y'(t) - 1 = 0$,
(b)  $y^{(4)}(t) - 1 = 0$,
(c)  $y^{(4)}(t) + 1 = 0$.

**Exercise 4.3**  Let the Lagrangian $L = L(t, x, \dot{x})$ be given with variables $x = (x_1, \ldots, x_N)$. For new coordinates $q_j = \widehat{q}_j(x), j = 1, \ldots, N$, let the Lagrangian $\widehat{L}(t, q, \dot{q})$ be defined by

$$\widehat{L}\big(t, \widehat{q}(x), D_x\widehat{q}(x)\dot{x}\big) = L(t, x, \dot{x})\,.$$

Transform the equations of motion

$$\frac{\partial L}{\partial x_j} = \frac{d}{dt}\frac{\partial L}{\partial \dot{x}_j} \quad \text{for } j = 1, \ldots, N$$

to the new coordinates $q$. Under which conditions do we obtain the following Lagrangian equations of motion in the new coordinates

$$\frac{\partial \widehat{L}}{\partial q_j} = \frac{d}{dt}\frac{\partial \widehat{L}}{\partial \dot{q}_j} \quad \text{for } j = 1, \ldots, N\,?$$

**Exercise 4.4**  With the help of the Lagrangian formalism develop the equations of motion for the double pendulum with masses $m_1$ and $m_2$ and lengths $\ell_1$ and $\ell_2$ for the (massless) connecting rods, as depicted in the sketch. Use the angles $\varphi_1$ and $\varphi_2$ as defined in the sketch as variables.

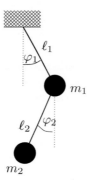

**Exercise 4.5**  Compute the eigen-oscillation of the sketched space frame with elasticity constants $E_1 = E_2 = 100$ and masses $m_1 = m_2 = 1$.

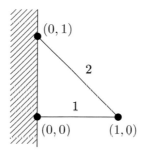

**Exercise 4.6**  (*Stability*) Let us amend the logistic differential equation by a term to model the harvesting or hunt of the population $x$ by another species. For a harvesting rate $e > 0$ we consider the differential equation

$$x' = \alpha x - \beta x^2 - ex, \quad \alpha = x_M \beta.$$

(a)  Determine all stationary solutions. Which are linearly stable?

(b)  Now we only consider harvesting rates $e < \alpha$. Investigate which harvesting rate $e^*$ maximizes the harvesting gain per time interval

$$y(e) = e\, x^*(e)$$

for a stationary solution $x^*(e)$. How large is then the population $x^*(e^*)$?

(c)  A constant harvesting gain $y_0$ per time interval is supposed to be modeled by the differential equation

$$x' = \alpha x - \beta x^2 - y_0.$$

What are the stationary solutions and which of them are stable? Which maximal value is possible for $y_0$?

(d) Discuss the following assertion: The harvesting strategy with constant harvesting rate is potentially less catastrophic for the survival of the population than the strategy with constant harvesting gain.

   *Hint*: In reality small fluctuations of the quantities $e$ and $y_0$ can occur.

**Exercise 4.7** (*Predator–prey model*) Show that the solutions $(x, y)^\top$ of the predator–prey model for positive initial data $x(0) = x_0 > 0$, $y(0) = y_0 > 0$ are periodic. You may argue with the help of the first integral and the uniqueness theorem for solutions of initial value problems.

**Exercise 4.8** (*Predator–prey model*) In this exercise the system

$$x' = x\left(a - by^2\right), \quad y' = y\left(-c + dx^2\right)$$

is discussed. Only solutions $x > 0$, $y > 0$ are of interest.

(a) Determine all stationary solutions and show the existence and uniqueness of a solution for the initial value $(x, y)(0) = (x_0, y_0)$ with $x_0 > 0$, $y_0 > 0$; well-known facts from basic analysis can be used.

(b) Formulate a first integral $H(x, y)$ and sketch the phase portrait.

**Exercise 4.9** (*Predator–prey model with limited growth*) We consider the system

$$x' = x(\alpha - \beta y - \lambda x), \tag{4.67}$$
$$y' = y(-\gamma + \delta x - \mu y), \tag{4.68}$$

where $\alpha$, $\beta$, $\gamma$, $\delta$, $\lambda$, and $\mu$ are positive real constants and $\alpha/\lambda < \gamma/\delta$ is assumed. The functions $x$ and $y$ again describe populations of a prey or predator species, i.e., we are looking for solutions $(x(t), y(t))$ with $x \geq 0$, $y \geq 0$.

Sketch the direction field and the straight lines

$$G_x := \{(x, y) \,|\, \alpha - \beta y - \lambda x = 0\}, \quad G_y := \{(x, y) \,|\, -\gamma + \delta x - \mu y = 0\}.$$

Then show that for a solution $(x(t), y(t))$ for the initial value $(x_0, y_0)$ with $x_0 > 0$ and $y_0 \geq 0$ it holds true that

$$(x(t), y(t)) \to \left(\tfrac{\alpha}{\lambda}, 0\right) \text{ for } t \to +\infty.$$

*Hint*: Consider for example the domain $\{(x, y) \,|\, -\gamma + \delta x - \mu y > 0\}$, which lies "below" the straight line $G_y$. Now give a heuristic argument (and show it also rigorously) why a solution emanating from this domain must leave the domain crossing the straight line $G_y$. Then discuss the other regions.

**Exercise 4.10** (*Parameter-dependent linear differential equation*) Consider for a given real parameter $\alpha$ the linear system

$$z'(t) = \begin{pmatrix} x'(t) \\ y'(t) \end{pmatrix} = \begin{pmatrix} \alpha & \alpha \\ -\alpha & \alpha \end{pmatrix} \begin{pmatrix} x(t) \\ y(t) \end{pmatrix} = Az(t)$$

for $z = (x, y)^T$.

(a) Compute the (complex) eigenvalues of the matrix $A$. Specify two linearly independent solutions of the form $z_j(t) = c_j e^{\lambda_j t}$ where $\lambda_j$ is an eigenvalue of $A$ and the initial value $c_j$ is an eigenvector corresponding to $\lambda_j$. Additionally denote for every solution the real and the imaginary part by setting $\lambda_j = \mu_j + i\nu_j$ and $c_j = a_j + ib_j$, where $i$ denotes the imaginary unit. Give a representation of all real solutions of the differential equation.
(b) Give a sketch of the phase portraits for $\alpha = 0.5$ and $\alpha = -0.5$.
(c) Which assertions can be made about the stability of the point $(0, 0)$ in dependence on $\alpha$?

**Exercise 4.11** (*Competition models*) We consider a situation, in which two species compete for resources, which are necessary for their growth. A simple model for the description of the situation consists of the following two differential equations which are already given in nondimensionalized form:

$$u_1' = u_1(1 - u_1 - auu_2) := f_1(u_1, u_2), \tag{4.69}$$
$$u_2' = \rho u_2(1 - cu_2 - bu_1) := f_2(u_1, u_2). \tag{4.70}$$

The basic model here is the model with limited growth for both populations extended by terms with positive constants $a$ and $b$ to describe the mutual influence by competition.

To simplify the computations, we assume that $\rho = 1$ and $c = 0$. We look for solutions $(u_1, u_2)$ with $u_i \geq 0$, $i = 1, 2$.

(a) Sketch the straight lines

$$G_{u_1} := \{(u_1, u_2) \mid 1 - u_1 - au_2 = 0\}, \ G_{u_2} := \{(u_1, u_2) \mid 1 - cu_2 - bu_1 = 0\}$$

and also the direction field.
(b) Determine the stationary solutions of the differential equation. Are there linearly stable solutions?
(c) Denote the domains for initial values $(u_1^0, u_2^0)$ for which with the help of the directional field definite assertions about the development of the populations for $t \to \infty$ can be made.

**Exercise 4.12** (a)  Solve explicitly the initial value problem

$$x' = Ax, \quad A = \begin{pmatrix} 3 & 8 \\ -0.5 & -1 \end{pmatrix}, \quad x(0) = \begin{pmatrix} 2 \\ -1 \end{pmatrix}.$$

(b)  Give an example of a matrix $A \in \mathbb{R}^{2,2}$ such that for the system $x' = Ax$ the point $(0, 0)$ is not stable although all eigenvalues $\lambda$ of $A$ satisfy $\text{Re}(\lambda) \leq 0$.

**Exercise 4.13**  (a)  For which of the following differential equations zero is a stable stationary solution?

$$\text{(i)} \quad x' = x^2, \quad \text{(ii)} \quad x' = x^3, \quad \text{(iii)} \quad x' = -x^3 \, .$$

(b)  Given a real parameter $c \neq 0$ consider the system

$$x' = -y + c x^5, \quad y' = x + c y^5 \, .$$

Investigate the dependence on $c$: Is $(0, 0)$ (i) stable, (ii) linearly stable?
*Hint*: What is the behavior of the Euclidean distance to the origin?

**Exercise 4.14**  Let $A \in \mathbb{R}^{n,n}$ be given such that $\text{Re} \, \lambda < 0$ for all eigenvalues $\lambda$ of $A$. Show that there exist a constant $\alpha > 0$ and a scalar product $(., .)$ in $\mathbb{R}^n$ such that

$$(y, Ay) \leq -\alpha \|y\|^2 \quad \text{for all} \quad y \in \mathbb{R}^n \, .$$

Here $\| . \|$ denotes the norm induced by the scalar product.

**Exercise 4.15**  For a function $f \in C(\mathbb{R}^n, \mathbb{R}^n)$ we consider the differential equation $x' = f(x)$. Let the smooth function $H$ be a first integral of the differential equation, i.e., if $x(t)$ is a solution then it holds true that

$$0 = \frac{d}{dt} H(x(t)) = DH(x(t))x'(t) = DH(x(t))f(x(t)) \, .$$

Furthermore let $x^*$ be a stationary solution such that

$$H(x^*) = 0, \quad DH(x^*) = 0, \quad D^2 H(x^*) \text{ positive definite}.$$

(a)  Show that $x^*$ is stable.
    *Hint*: Use a Taylor expansion of $H$ at $x^*$ to conclude that for an $\varepsilon > 0$ sufficiently small and a constant $c > 0$ it holds true that

$$\min\{H(x) \,|\, x \in \partial B_\varepsilon(x^*)\} \geq c \, \varepsilon^2.$$

Here $\partial B_\varepsilon(x^*)$ denotes the boundary of the ball with radius $\varepsilon$ around $x^*$. Furthermore there exists (give a justification!) a $\delta > 0$ such that

$$H(x) \leq \frac{1}{2} c \, \varepsilon^2 \quad \forall x \in B_\delta(x^*) \, .$$

What is the consequence for solutions of the differential equation whose initial value is in the ball $B_\delta(x^*)$?

(b)  Apply the facts from (a) to the predator–prey model (4.28).

**Exercise 4.16**  We consider the differential equation

$$x' = y - x^3, \quad y' = -x^3 - y^3,$$

and the function

$$L(x, y) = \frac{1}{2}y^2 + \frac{1}{4}x^4.$$

(a)  Show that $(0, 0)$ is not linearly stable.
(b)  Show that $L$ is a *Lyapunov function*, i.e., if $(x(t), y(t))$ is a solution of the differential equation then we have

$$\frac{d}{dt}L(x(t), y(t)) \le 0.$$

(c)  Show that $(0, 0)$ is stable.
    *Hint*: One can proceed as in Exercise 4.15; $L$ takes the role of $H$. Therefore one first has to investigate how $L$ behaves at the boundary of a circle around $(0, 0)$.
(d)  Is $(0, 0)$ asymptotically stable?

**Exercise 4.17**  Let the function $t \mapsto x(t) \in \mathbb{R}^n$ be the solution of the ordinary differential equation

$$x'(t) = -F(\nabla G(x(t))) \quad \text{for } t > t_0 \tag{4.71}$$

with $F : \mathbb{R}^n \to \mathbb{R}^n$ and $G : \mathbb{R}^n \to \mathbb{R}$. The function $G$ is assumed to be continuously differentiable and $F$ to satisfy $F(0) = 0$ and one of the following conditions

(i)  $F$ is continuously differentiable and $DF(x) = \left(\dfrac{\partial F_i}{\partial x_j}(x)\right)^n_{i,j=1}$ is positive semi-definite for all $x \in \mathbb{R}^n$,
(ii)  $F$ is Lipschitz continuous and *monotone* in the sense

$$(F(x) - F(y)) \cdot (x - y) \ge 0 \quad \text{for all } x, y \in \mathbb{R}^n.$$

Show that $G$ is a Lyapunov function, i.e., that every solution $x(t)$ of (4.71) satisfies the inequality

$$\frac{d}{dt}G(x(t)) \le 0.$$

**Exercise 4.18**  Formulate the Brachistochrone problem in the Lagrangian and in the Hamiltonian formulation. From the Hamiltonian formulation derive the statement (4.39) about energy conservation.

**Exercise 4.19** (*Fundamental lemma of calculus of variations*) Let $f : [0, 1] \to \mathbb{R}^d$ be continuous such that

$$\int_0^1 f(x) \cdot \varphi(x) \, dx = 0 \quad \text{for all} \quad \varphi \in C_0^\infty((0, 1), \mathbb{R}^d).$$

Show that $f(x) = 0$ for all $x \in (0, 1)$.

**Exercise 4.20** Solve the Brachistochrone problem as formulated in Sect. 4.7. To do so, use (4.47) and describe the graph of the solution as a curve $t \mapsto (x(t), u(t))$. Is the ansatz $u(t) = -\kappa(1 - \cos t)$ with $\kappa > 0$ of any help? Interpret the solution geometrically.

**Exercise 4.21** We consider the example (v) from Sect. 4.7.

(a) Show that the Lagrange multiplier $\lambda$ is the smallest eigenvalue of the eigenvalue problem

$$-\frac{d}{dx}\left(p(x)\,u'(x)\right) + q(x)\,u(x) = \lambda r(x)\,u(x), \tag{4.72}$$

$$u(0) = u(1) = 0. \tag{4.73}$$

(b) Consider the minimization problem $\mathcal{F}(u) \to \min$ with the constraints $\int_0^1 r(x)\,u^2(x)\,dx = 1$, $\int_0^1 u(x)\,u_1(x)\,dx = 0$, $u(0) = u(1) = 0$, where $u_1$ is the solution of the minimization problem in (v) from p. 181. Derive necessary optimality conditions and show that a function which satisfies these necessary conditions is just an eigenfunction for the second smallest eigenvalue of problem (4.72), (4.73).

(c) In which way further eigenfunctions can be characterized?

**Exercise 4.22** Consider the minimization problem

$$\min\left\{\int_{-1}^1 \left[\tfrac{1}{2}(u'')^2 - fu\right]dx \mid u(-1) = u(1) = 0, \; u'(-1) = u'(1) = 0\right\}.$$

(a) Determine the boundary value problem which results from the necessary conditions for a minimum.
(b) Determine the solution of the boundary value problem if $f$ is constant. Give a sketch of the solution.
(c) Discuss the necessary conditions if $f$ is concentrated at zero, i.e., if in the minimization problem above the term $\int_{-1}^1 fu \, dx$ is substituted by $\overline{f}\,u(0)\,\overline{f} \in \mathbb{R}$, where $\overline{f} \in \mathbb{R}$ is a constant.

**Exercise 4.23** For the optimal fish harvesting problem from Sect. 4.8 formulate the necessary optimality conditions.

# Chapter 5
# Continuum Mechanics

## 5.1 Introduction

In continuum mechanics one studies processes in a subset of a $d$-dimensional Euclidean space. The relevant quantities, e.g., the mass density, the temperature, the pressure, the velocity, are defined in *each point* of the set. The fact that matter such as water, air or a metal is composed by atoms is neglected.

Continuum mechanics describes many important phenomena in applications such as

- heat conduction,
- flow of gases and liquids,
- deformation of solids, elasticity, plasticity,
- phase transitions,
- processes in which the above effects are coupled.

Hence continuum mechanics is very important in the natural sciences and in engineering.

Mathematical models in continuum mechanics have to be compared with experiments and only those models which can predict experimental results make sense. An important task of mathematics is to provide a detailed analysis of the models. This analysis should result in sufficiently many quantitative and qualitative statements which then have to be compared with experiments.

Models which are based on particles such as atoms or molecules can be related to characteristic quantities in a continuum by a suitable *averaging*. However, it is not necessary for continuum mechanics to be related to a microscopic model which is based on particles. So far it was only possible in very few cases to justify macroscopic models with the help of microscopic models. We can instead take the view point that a continuum model is the "natural" description of macroscopic phenomena. If we consider the flow of a liquid we are not aware of the molecular details. In fact we are often not interested in aspects on the level of particles. A continuum model is often useful because one does not take all the molecular details into account.

© Springer International Publishing AG 2017
C. Eck et al., *Mathematical Modeling*, Springer Undergraduate
Mathematics Series, DOI 10.1007/978-3-319-55161-6_5

Of course typically microscopic details will enter the macroscopic model. For example in a solid the crystal structure (which can be cubic, hexagonal, etc.) will have an influence on the precise form of the equations in continuum mechanics.

Hence, there are good reasons to consider continuum models without a derivation from microscopic models. Nevertheless we will consider at first a simple particle model in order to motivate certain quantities appearing in continuum models. In order to do so we need to introduce some notation.

### *Points and Vectors*

To describe a continuum one introduces in general a three dimensional affine Euclidean point space. An Euclidean space consists of a set $E$ of points, a set $V$ of vectors and a mapping which assigns an ordered pair of points to a vector which connects these points. The set of vectors has the structure of an Euclidean vector space. The vector $u$ which connects points $x$ and $y$ will be denoted by

$$u = y - x.$$

The operation which would sum two points is not possible. For a point $x$ and a vector $u$ we find a point $y$ such that $u$ is the vector connecting $x$ and $y$. This point will formally be denoted as the sum of $x$ and $u$, i.e.,

$$y = x + u.$$

In the following we consider the three dimensional Euclidean space $E^3$ together with the three dimensional $\mathbb{R}$-vector space $V^3$, which we will also denote by $\mathbb{R}^3$. The scalar product of two vectors $u, v \in V^3$ ($u, v \in \mathbb{R}^3$) will be denoted by $u \cdot v$ and the norm is given as $|u| = \sqrt{u \cdot u}$. Choosing a scalar product on $V^3$ also fixes the measurement of *length*. After choosing an *orthonormal basis* $(e_1, e_2, e_3)$ we can always identify $V^3$ with $\mathbb{R}^3$. Each vector $u \in V^3$ can be identified with the help of its coordinates $u_i = u \cdot e_i$. We obtain

$$u = \sum_{i=1}^{3} u_i e_i = \sum_{i=1}^{3} (u \cdot e_i) e_i$$

and

$$u \cdot v = \sum_{i=1}^{3} u_i v_i.$$

The orthonormal basis $(e_1, e_2, e_3)$ of $V^3$ together with the choice of an origin $O \in E^3$ leads to a coordinate system $(O; e_1, e_2, e_3)$ of $E^3$. The coordinates of a point $x \in E^3$ are given as

$$x_i = (x - O) \cdot e_i.$$

In the following we always identify $V^3$ and $\mathbb{R}^3$. There exists a unique vector product $(u, v) \mapsto u \times v$ with the properties

$$u \times v = -v \times u \,,$$
$$u \times u = 0 \,,$$
$$u \cdot (v \times w) = w \cdot (u \times v) = v \cdot (w \times u) \,,$$
$$e_3 = e_1 \times e_2 \,.$$

The number

$$|u \cdot (v \times w)|$$

is the volume of the parallelepiped spanned by $u$, $v$ and $w$. The coordinates of $w = u \times v$ with respect to the basis $\{e_1, e_2, e_3\}$ are given by

$$w_1 = u_2 v_3 - u_3 v_2 \,, \quad w_2 = u_3 v_1 - u_1 v_3 \,, \quad w_3 = u_1 v_2 - u_2 v_1 \,.$$

This can be written in a more compact form with the help of the *Levi–Civita symbol*

$$\varepsilon_{ijk} = \begin{cases} 1 \,, & \text{if } (i, j, k) \text{ is an even permutation,} \\ -1 \,, & \text{if } (i, j, k) \text{ is an odd permutation,} \\ 0 \,, & \text{if } (i, j, k) \text{ is not a permutation.} \end{cases}$$

An even or odd permutation of $(1, 2, 3)$ results if we interchange two numbers in $(1, 2, 3)$ an even or odd number of times. The vector product is given by

$$(u \times v)_i = \sum_{j,k=1}^{3} \varepsilon_{ijk} u_j v_k \,. \tag{5.1}$$

## 5.2   Classical Point Mechanics

To lay the foundations for the understanding of the following chapters we consider the most important physical laws with the help of classical point mechanics. The central object of classical point mechanics is a *mass point*, i.e., a point in $E^3$ to which we assign a given mass. The properties of a mass point at a given time $t$ are characterized by its mass $m$, its position $x(t) \in E^3$ and its velocity $v(t) = x'(t)$. The momentum of a mass point is given as $p(t) = m v(t) = m x'(t)$. Forces, given for example by interaction with other mass points or given through external fields such as the gravitational field or electromagnetic fields, can have an effect on mass points.

### The Newtonian Laws of Point Mechanics

We consider a system of $N$ mass points. We introduce

$x_i(t)$   the position of a mass point $i$ at time $t$,
$m_i$     the mass of the point $i$.

We assume that the following forces have an effect on the mass points:

$f_{ij}(t) \in \mathbb{R}^3$  is the force exerted at time $t$ by particle $j$ on particle $i$
$f_i(t) \in \mathbb{R}^3$  is the external force acting on particle $i$ at time $t$.

*Newton's second law* "force = mass · acceleration" or "force = change of momentum with time" yields the following equality for the momenta $p_i = m_i x_i'$

$$p_i'(t) = \sum_{j \neq i} f_{ij}(t) + f_i(t), \quad i = 1, \ldots, N.$$

*Newton's third law* (the principle of "action and reaction") implies $f_{ij} = -f_{ji}$. The following *central forces* fulfill this requirement:

$$f_{ij}(x_i, x_j) = \frac{x_i - x_j}{|x_i - x_j|} g_{ij}(|x_i - x_j|),$$

where $g_{ij}(x) = g_{ji}(x)$. In the following we assume that the forces $f_{ij}$ are of the above form and remark:

$$g_{ij} > 0, \quad \text{if the forces are } repellent,$$
$$g_{ij} < 0, \quad \text{if the forces are } attractive.$$

**Examples**:

1. Gravitational forces have the form

$$g_{ij}(x) = -G \frac{m_i m_j}{|x|^2}.$$

   Here $G \approx 6{,}67428 \cdot 10^{-11}$ Nm²/kg² is the *gravitational constant*.
2. Forces stemming from electrical fields are given by

$$g_{ij}(x) = K \frac{Q_i Q_j}{|x|^2},$$

   where $Q_i$ is the charge of mass point $i$ and $K$ is a proportionality factor.

*Conservation Laws*

In a many particle system of the form described above there are quantities which only change due to external influences. Without external influences these quantities stay constant during the evolution. Such quantities are called *conserved quantities*. The most important conserved quantities are mass, momentum, angular momentum and energy. In the following we will discuss the most important conservation laws.

**Conservation of Linear Momentum**

For the *total momentum* $p = \sum\limits_{i=1}^{N} p_i$ of a many particle system we have

$$p'(t) = \sum_{i=1}^{N} p_i'(t) = \sum_{i=1}^{N} \left( \sum_{\substack{j=1 \\ j \neq i}}^{N} f_{ij} + f_i \right) = \sum_{i=1}^{N} f_i =: f .$$

Here the property $f_{ij} = -f_{ji}$ for the internal interactions is essential. The term $f$ denotes the *sum of all external forces*. The time derivative of the total momentum is hence equal to the total forces acting on the system and is in particular not depending on the internal interactions. This property is the *balance of linear momentum* for the whole system. In this context one speaks of *conservation of linear momentum*, although the momentum does not necessarily stay constant. If we introduce the total mass of the point masses as

$$m = \sum_{i=1}^{N} m_i$$

and its position as

$$x = \frac{1}{m} \sum_{i=1}^{N} m_i x_i$$

(which is the center of mass) we obtain the total linear momentum as

$$p = \sum_{i=1}^{N} p_i = \sum_{i=1}^{N} m_i x_i' = m x' .$$

For the ensemble of all mass points we then obtain Newton's second law as $p' = f$. If the sum of the external forces is equal to zero, we observe that the velocity of the center of mass does not change and the center moves with a constant velocity on a straight line. Hence Newton's first law is fulfilled.

**Conservation of Angular Momentum**

If the forces $f_{ij}$ are central forces one can derive from Newton's second law a conservation law for the angular momentum. At a fixed resting point $x_0 \in E^3$ we define the angular momentum of the mass points as

$$L_i = (x_i - x_0) \times p_i$$

and the torque as

$$M_i = (x_i - x_0) \times f_i .$$

For the total angular momentum

$$L = \sum_{i=1}^{N} L_i$$

we obtain

$$L' = \sum_{i=1}^{N}(x_i - x_0) \times p_i' = \sum_{i=1}^{N}\sum_{\substack{j=1 \\ j \neq i}}^{N} \left( x_i \times f_{ij} - x_0 \times f_{ij} \right) + \sum_{i=1}^{N}(x_i - x_0) \times f_i$$

$$= -\sum_{i=1}^{N}\sum_{\substack{j=1 \\ j \neq i}}^{N} \frac{x_i \times x_j}{|x_i - x_j|} g_{ij}(|x_i - x_j|) + \sum_{i=1}^{N} M_i = \sum_{i=1}^{N} M_i =: M$$

with the total momentum $M$. Here we used

$$\sum_{\substack{i,j=1 \\ i \neq j}}^{N} f_{ij} = 0$$

which holds since $f_{ij} = -f_{ji}$ and

$$\sum_{\substack{i,j=1 \\ i \neq j}}^{N} \frac{x_i \times x_j}{|x_i - x_j|} g_{ij}(|x_i - x_j|) = 0$$

which holds true since $x_i \times x_j = -x_j \times x_i$. The equation

$$L' = M$$

describes the *conservation of the total angular momentum*. The interaction forces give no contribution to the change of the total angular momentum with respect to an arbitrary point $x_0$ and the total angular momentum is only influenced by the sum of the torques. The conservation of the total angular momentum implies Kepler's second law, see Exercise 5.8.

**Conservation of Energy**

A conservation law also holds for the total energy of the system. The *kinetic energy* of the mass point system is

$$E_{\text{kin}} = \sum_{i=1}^{N} m_i \frac{|x_i'|^2}{2} .$$

The work applied by external forces leads to a change of energy. Defining the power (work per time) as

$$P = \sum_{i=1}^{N} x_i' \cdot f_i$$

we obtain

$$E_{\text{kin}}' = \sum_{i=1}^{N} x_i' \cdot p_i' = \sum_{i=1}^{N} \sum_{\substack{j=1 \\ j \neq i}}^{N} x_i' \cdot f_{ij} + P$$

$$= \sum_{\substack{i,j=1 \\ j<i}}^{N} \frac{(x_i' - x_j') \cdot (x_i - x_j)}{|x_i - x_j|} g_{ij}(|x_i - x_j|) + P$$

$$= \sum_{\substack{i,j=1 \\ j<i}}^{N} g_{ij}(|x_i - x_j|) \frac{d}{dt} |x_i - x_j| + P \, .$$

Let $G_{ij}$ be an antiderivative of $g_{ij}$. We define the *potential energy*

$$E_{\text{pot}} = - \sum_{\substack{i,j=1 \\ j<i}}^{N} G_{ij}(|x_i - x_j|) = -\frac{1}{2} \sum_{\substack{i,j=1 \\ j \neq i}}^{N} G_{ij}(|x_i - x_j|)$$

and the total energy

$$E = E_{\text{kin}} + E_{\text{pot}} \, .$$

We obtain

$$E' = P \, ,$$

which means that the change of the total energy results from the work applied to the system. In particular the interaction forces do not change the total energy.

## 5.3  From Particle Mechanics to a Continuous Medium

We will now apply a suitable averaging process in order to obtain a continuous description. Our presentation will be superficial and only has the purpose to motivate the following continuum models. A more fundamental derivation of models in continuum mechanics, for example in the context of kinetic gas theory, would need a lot of effort.

For the following considerations we always choose an origin $O \in E^3$ and a basis $\{e_1, e_2, e_3\}$ of $\mathbb{R}^3$. In order to identify points in $E^3$ we will always use the coordinates given by the above coordinate system. At a point $x^{(0)} \in \mathbb{R}^3$ and for an edge length $h > 0$ we define the cube

$$W_h\big(x^{(0)}\big) = \left\{ x \in \mathbb{R}^3 \mid \max\left( \big|x_1 - x_1^{(0)}\big|, \big|x_2 - x_2^{(0)}\big|, \big|x_3 - x_3^{(0)}\big| \right) \leq \tfrac{h}{2} \right\}.$$

We now consider a large ensemble of mass points, i.e., $N \gg 1$. In addition we define the *mass density* with the help of a suitable average:

$$\varrho_h\big(t, x^{(0)}\big) = \frac{1}{h^3} \sum_{x_i(t) \in W_h(x^{(0)})} m_i.$$

As a function of $h$ the quantity $\varrho_h$ could look like in Fig. 5.1.

The main assumption of continuum mechanics is:

There is a range of values of $h$ for which $\varrho_h$ is nearly constant.

Values from this range can be used to define a mass density in the continuum model. Such a quantity should have the property that the integral

$$\int_\Omega \varrho_h(t, x)\, dx$$

can be used as an approximation of the mass in the set $\Omega \subset \mathbb{R}^3$. In an analogous way we define an *averaged momentum density*

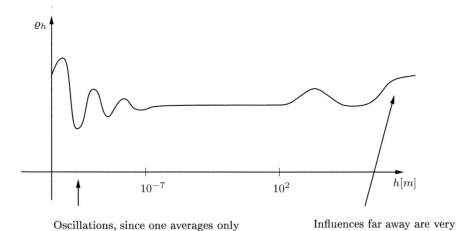

Fig. 5.1  The average mass as a function of the size of the cube

$$p_h(t, x^{(0)}) = \frac{1}{h^3} \sum_{x_i(t) \in W_h(x^{(0)})} m_i \, x_i'(t)$$

and hence an *averaged velocity*

$$v_h = p_h / \varrho_h \, ,$$

a *density of the kinetic energy*

$$E_{\text{kin},h}(t, x^{(0)}) = \frac{1}{h^3} \sum_{x_i(t) \in W_h(x^{(0)})} m_i \, \frac{|x_i'(t)|^2}{2} \, ,$$

a *density of the potential energy*

$$E_{\text{pot},h}(t, x^{(0)}) = -\frac{1}{h^3} \sum_{x_i(t) \in W_h(x^{(0)})} \frac{1}{2} \sum_{\substack{j=1 \\ j \neq i}}^{N} G_{ij}(|x_i(t) - x_j(t)|)$$

and an energy density

$$E_h = E_{\text{kin},h} + E_{\text{pot},h} \, .$$

We will now split the kinetic energy into a *macroscopic* energy contribution and a *microscopic* energy contribution. The macroscopic energy is

$$\varrho_h \frac{|v_h|^2}{2}$$

The term

$$\widetilde{E}_{\text{kin},h}(t, x^{(0)}) = \frac{1}{h^3} \sum_{x_i(t) \in W_h(x^{(0)})} m_i \, \frac{|x_i'(t) - v_h|^2}{2}$$

describes the kinetic energy of *fluctuations* around the macroscopic velocity $v_h$. This contribution can be macroscopically interpreted as *energetic contributions due to heat*. For this part of the kinetic energy we obtain

$$\begin{aligned}
\widetilde{E}_{\text{kin},h}(t, x^{(0)}) &= \frac{1}{h^3} \sum_{x_i(t) \in W_h(x^{(0)})} m_i \, \frac{|x_i'(t) - v_h|^2}{2} \\
&= \frac{1}{h^3} \sum_{x_i(t) \in W_h(x^{(0)})} m_i \, \frac{|x_i'(t)|^2}{2} \\
&\quad - v_h \cdot \frac{1}{h^3} \sum_{x_i(t) \in W_h(x^{(0)})} x_i'(t) \, m_i + \frac{1}{2}|v_h|^2 \frac{1}{h^3} \sum_{x_i(t) \in W_h(x^{(0)})} m_i \\
&= E_{\text{kin},h}(t, x^{(0)}) - \tfrac{1}{2}\varrho_h |v_h|^2 \, .
\end{aligned}$$

With this we obtain the kinetic energy

$$E_{\text{kin},h} = \frac{1}{2}\varrho_h|v_h|^2 + \widetilde{E}_{\text{kin},h} \, ,$$

where the first term $\frac{1}{2}\varrho_h|v_h|^2$ is the macroscopic and the second term $\widetilde{E}_{\text{kin},h}$ is the microscopic contribution to the kinetic energy. Finally we introduce the energy $u_h$ per unit mass through the identity

$$\varrho_h u_h = \widetilde{E}_{\text{kin},h} + E_{\text{pot},h} \, .$$

The energy density now is given as

$$E_h = E_{\text{kin},h} + E_{\text{pot},h} = \varrho_h \left( \frac{|v_h|^2}{2} + u_h \right) .$$

This expression will again appear when we formulate the macroscopic energy conservation law. Finally we want to relate the derivation of the quantities in the continuum case to the considerations on thermodynamics in Chap. 3. If the microscopic fluctuations $x_i'(t) - v_h$ are given by a Maxwell–Boltzmann distribution we observed in Sect. 3.1, that $\widetilde{E}_{\text{kin},h}$ is proportional to the temperature. This simple relation will appear later when we consider constitutive relations between internal energy and temperature.

## 5.4   Kinematics

A main assumption of continuum mechanics is that all relevant quantities are defined on a "continuum", i.e., on an *open subset* of $E^3$. This open set will in general depend on time. *Kinematics* now is the description of bodies or domains which vary in time without taking the *forces* into account which are responsible for the evolution.

In what follows we replace the spatial dimension 3 by a general dimension $d$ and we also replace $E^3$ and $\mathbb{R}^3$ by $\mathbb{R}^d$.

A *material point* in a domain can be described by its position $X$ in a so called *reference configuration* $\Omega \subset \mathbb{R}^d$. Here $\Omega$ is assumed to be open and connected. The reference configuration is arbitrary and depends on the application. Often one chooses the configuration at the initial time as reference configuration. For a deformation of any elastic solid one can also choose a configuration in which no internal forces act (if such a configuration exists at all) or one chooses a configuration which a body would take if no outer forces are applied. The evolution of a point $X \in \Omega$ is now given as a *mapping* (Fig. 5.2)

$$t \mapsto x(t, X) \in \mathbb{R}^d \ \text{ with a time variable } t.$$

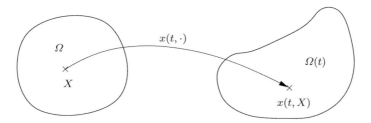

**Fig. 5.2** Deformation of a domain

Here the following assumptions are meaningful:

(A 1) $x(t_0, X) = X$, the point is described by its position at time $t = t_0$.
(A 2) The mapping $(t, X) \mapsto x(t, X)$ is *continuously differentiable*.
(A 3) For all $t \geq t_0$ the mapping $\Omega \ni X \mapsto x(t, X) \in x(t, \Omega)$ is *invertible*.
(A 4) The Jacobi determinant $J(t, X) = \det \left( \frac{\partial x_j}{\partial X_k}(t, X) \right)_{j,k=1}^d$ is positive for all $t \geq t_0$,
$X \in \Omega$.

The terms $X$ and $x$ represent two different types of coordinates:

- *Material* or *Lagrangian* coordinates $X$: One considers one *material point* and follows its evolution.
- *Eulerian* coordinates $x$: One considers one *fixed point in space* and in general one will observe different *material points* at this point in space at different times.

We denote a (physical) variable, e.g., the density or the pressure,

- in Lagrangian coordinates with the help of capital letters $\Phi(t, X)$,
- in Eulerian coordinates with the help of lower-case letters $\varphi(t, x)$.

Both $\Phi(t, X)$ and $\varphi(t, x)$ indicate the value of the same quantity, namely:

- $\Phi(t, X)$ for a material point which was at point $X$ in the reference configuration. This material point can of course be at a different position at time $t$.
- $\varphi(t, x)$ for a material point which at time $t$ is at position $x$.

In particular we have the relation

$$\varphi(t, x(t, X)) = \Phi(t, X).$$

Using the chain rule we obtain for such a quantity

$$\frac{\partial}{\partial t} \Phi(t, X) = \frac{\partial \varphi}{\partial t}(t, x(t, X)) + \nabla_x \varphi(t, x(t, X)) \cdot \frac{\partial x}{\partial t}(t, X).$$

We denote by

$$V(t, X) = \frac{\partial x}{\partial t}(t, X)$$

the *velocity* of a material point $X$ in Lagrangian coordinates and by

$$v(t, x) = V(t, X(t, x))$$

the velocity in Eulerian coordinates. Here, $x \mapsto X(t, x)$ is the inverse of the function $X \mapsto x(t, X)$.

**Definition 5.1**  The quantity

$$D_t \varphi(t, x) = \frac{\partial \varphi}{\partial t}(t, x) + \nabla \varphi(t, x) \cdot v(t, x)$$

is the *material time derivative* of $\varphi$ with respect to $t$.

The material time derivative describes the change in time of the quantity $\varphi$ observed at a material point which at time $t$ has the position $x$ and the velocity $v(t, x)$.

We can now identify two important classes of special curves:

*Pathlines* are solutions of the equation

$$y'(t) = v(t, y(t)),$$

they are the paths a *material* point will follow during the evolution.
*Streamlines* are the solutions of

$$z'(s) = v(t, z(s)),$$

they give a *snapshot* of the velocity field at time $t$.

### Reynolds Transport Theorem

We now consider the deformation of a body or a domain with reference configuration $\Omega \subset \mathbb{R}^d$. Assume that the body occupies the domain

$$\Omega(t) = \{x(t, X) \mid X \in \Omega\}$$

at time $t$. With the help of the determinant $J(t, X) = \det \left( \frac{\partial x_j(t,X)}{\partial X_k} \right)_{j,k=1}^{d}$ we can describe the *change of volume* of the domain as follows

$$|\Omega(t)| = \int_{\Omega(t)} 1 \, dx = \int_{\Omega} J(t, X) \, dX.$$

In order to prove such a result we need the following formula which is due to Euler.

**Theorem 5.2** *Assume* $(t, X) \mapsto x(t, X)$ *fulfills (A1)–(A4) and* $(t, X) \mapsto \dfrac{\partial x}{\partial t}(t, X)$ *is continuously differentiable. Then it holds:*

$$\frac{\partial J}{\partial t}(t, X) = \nabla \cdot v(t, x)_{|x=x(t,X)} J(t, X).$$

In order to prove Theorem 5.2 first we need to prove the following result.

**Lemma 5.3** *Assume that* $t \mapsto A(t) \in \mathbb{R}^{d,d}$ *is differentiable and* $A(t)$ *is invertible. Then it holds:*

$$\frac{d}{dt} \det A(t) = \text{trace}\left(A^{-1}(t)\frac{d}{dt}A(t)\right) \det A(t).$$

*Proof* Let $a_j(t)$ be the $j^{\text{th}}$ column of $A(t)$. Since the determinant is linear with respect to the columns we obtain with the help of the properties of the determinant

$$\frac{d}{dt} \det A(t) = \frac{d}{dt} \det \left(a_1(t), \ldots, a_d(t)\right)$$

$$= \sum_{j=1}^{d} \det \left(a_1(t), \ldots, a_{j-1}(t), \tfrac{d}{dt}a_j(t), a_{j+1}(t), \ldots, a_d(t)\right)$$

$$= \sum_{i,j=1}^{d} (-1)^{i+j} \left(\tfrac{d}{dt}a_{ij}(t)\right) \det A^{(i,j)}(t),$$

where we obtain $A^{(i,j)}(t)$ by eliminating the $i^{\text{th}}$ row and the $j^{\text{th}}$ column of $A(t)$. Cramer's rule now gives

$$\left(A^{-1}(t)\right)_{ij} = \frac{(-1)^{i+j}}{\det A(t)} \det A^{(j,i)}(t).$$

Altogether we obtain

$$\frac{d}{dt} \det A(t) = \sum_{i,j=1}^{d} \det A(t)\frac{d}{dt}a_{ij}(t)\left(A^{-1}(t)\right)_{ji}$$

$$= \text{trace}\left(A^{-1}(t)\tfrac{d}{dt}A(t)\right) \det A(t).$$

□

In order to prove Theorem 5.2 we need to apply Lemma 5.3 for the matrix $A(t) = \dfrac{\partial x}{\partial X}(t, X)$. Using the relation

$$\frac{\partial}{\partial t}\frac{\partial x_i}{\partial X_j}(t,X) = \frac{\partial}{\partial X_j}\frac{\partial x_i}{\partial t}(t,X) = \frac{\partial}{\partial X_j}V_i(t,X) = \frac{\partial}{\partial X_j}v_i(t,x(t,X))$$

$$= \sum_{k=1}^{d}\frac{\partial v_i}{\partial x_k}(t,x(t,X))\frac{\partial x_k}{\partial X_j}(t,X)$$

and the formula

$$\text{trace}\left(A^{-1}(t)\tfrac{\partial}{\partial t}A(t)\right) = \sum_{i,j=1}^{d}(A^{-1}(t))_{ji}\frac{\partial a_{ij}(t)}{\partial t}$$

$$= \sum_{i,j,k=1}^{d}(A^{-1}(t))_{ji}\frac{\partial v_i}{\partial x_k}(t,x(t,X))a_{kj}(t)$$

$$= \sum_{i,k=1}^{d}\delta_{ki}\frac{\partial v_i}{\partial x_k}(t,x(t,X)) = \nabla \cdot v(t,x)\big|_{x=x(t,X)}$$

we obtain the assertion of Theorem 5.2.                                                    □

**Corollary**  *For the volume*

$$|\Omega(t)| = \int_{\Omega(t)} 1\, dx = \int_{\Omega} J(t,X)\, dX$$

*we obtain*

$$\frac{d}{dt}|\Omega(t)| = \int_{\Omega(t)} \nabla \cdot v(t,x)\, dx\,.$$

*With the help of Theorem 5.2 we can prove the following important result.*

**Theorem 5.4**  (Reynolds transport theorem) *We assume that the mapping* $(t,X) \mapsto x(t,X)$ *fulfills* (A1)–(A4) *and that the functions* $(t,X) \mapsto \dfrac{\partial x}{\partial t}(t,X)$ *and* $(t,x) \mapsto \varphi(t,x)$ *are continuously differentiable. Then it holds*

$$\frac{d}{dt}\int_{\Omega(t)}\varphi(t,x)\, dx = \int_{\Omega(t)}\left[\frac{\partial \varphi}{\partial t}(t,x) + \nabla \cdot (\varphi(t,x)v(t,x))\right] dx\,.$$

*Proof*  With the help of integration by substitution, the chain rule and Theorem 5.2 due to Euler we obtain

$$\frac{d}{dt}\int_{\Omega(t)}\varphi(t,x)\, dx = \frac{d}{dt}\int_{\Omega}\varphi(t,x(t,X))J(t,X)\, dX$$

$$= \int_{\Omega}\left(\frac{\partial \varphi}{\partial t}(t,x(t,X)) + \sum_{k=1}^{d}\frac{\partial \varphi}{\partial x_k}(t,x(t,X))V_k(t,X)\right)$$

$$+ \, \varphi(t, x(t, X)) \, \nabla \cdot v(t, x(t, X)) \Big) J(t, X) \, dX$$

$$= \int_{\Omega(t)} \left( \frac{\partial \varphi}{\partial t}(t, x) + \nabla \cdot \big( \varphi(t, x) v(t, x) \big) \right) dx \, .$$

$\square$

In the Reynolds transport theorem the term $\int_{\Omega(t)} \partial_t \varphi(t, x) \, dx$ accounts for the change of $\int_{\Omega(t)} \varphi(t, x) dx$ due to a change of $\varphi$ with respect to time and

$$\int_{\Omega(t)} \nabla \cdot (\varphi(t, x) \, v(t, x)) \, dx = \int_{\partial \Omega(t)} \varphi(t, x) \, v(t, x) \cdot n(t, x) \, ds_x$$

accounts for the change due to the variation of $\Omega(t)$. Here $\partial \Omega(t)$ is the boundary of $\Omega(t)$ and $n(t, x)$ is the outer unit normal to $\partial \Omega(t)$. The total time derivative of $\int_{\Omega} \varphi(t, x)$ is the sum of both contributions.

### Density Variables

In the chapter on thermodynamics we already discussed the notion of an *extensive* variable – its value is "proportional" to the size of the "system". In continuum mechanics we will describe extensive quantities by density variables. Examples are the *mass density*, the *density of the internal energy* and also the *force density*. We distinguish between *mass densities*, *volume densities* and *surface densities*.

We can define a volume density of a quantity $U$ by

$$u_V(x) = \lim_{r \searrow 0} \frac{U(B_r(x))}{V(B_r(x))},$$

where $B_r(x)$ is the ball with radius $r$ and center $x$, $U(B_r(x))$ is the value of the quantity $U$ in the ball $B_r(x)$ and $V(B_r(x))$ is the volume of $B_r(x)$. The simplest example for this case is (cf. Sect. 5.3)

$$\varrho(x) = \lim_{r \to 0} \frac{m(B_r(x))}{V(B_r(x))},$$

where $m(B_r(x))$ is the mass contained in $B_r(x)$. The density of a quantity $U$ per unit mass can be defined via

$$u_m(x) = \lim_{r \to 0} \frac{U(B_r(x))}{m(B_r(x))} \, .$$

If both quantities exist we have

$$u_V(x) = \varrho(x) \, u_m(x) \, .$$

Surface densities are used for quantities which are defined on surfaces and its size is given per unit surface. The most important example for this case are forces on surfaces – in the simplest case this is, e.g., the pressure which a gas exerts on a boundary. In this case we consider a surface $\Gamma$ in space and the surface density of a quantity $U$ in a point $x \in \Gamma$ can be defined as

$$u_\Gamma(x) = \lim_{r \to 0} \frac{U(\Gamma \cap B_r(x))}{A(\Gamma \cap B_r(x))} \,,$$

where $U(\Gamma \cap B_r(x))$ is given as the total value of $U$ on the surface patch $\Gamma \cap B_r(x)$ (for the example of a force density, $U$ is the total force acting on $\Gamma \cap B_r(x)$) and $A(\Gamma \cap B_r(x))$ is the surface area of $\Gamma \cap B_r(x)$.

The importance of density variables comes from the fact that they allow us to describe extensive quantities through an integration with respect to the density variable. The total amount of the quantity $U$ in $\Omega$ is given by

$$U(\Omega) = \int_\Omega u_V(x)\,dx = \int_\Omega \varrho(x)\,u_m(x)\,dx \,.$$

Mostly one uses variables which are defined per unit mass as they often yield a better description of the amount of a quantity than the density per unit volume. If we do not state anything else in the following we will use density variables per unit mass. An exception is of course the mass density $\varrho$ which is the density of mass per unit volume.


## 5.5 Conservation Laws

The fundamental equations of continuum mechanics are based on conservation laws for mass, (linear) momentum, angular momentum and energy. We will formulate these conservation laws in *Eulerian coordinates*. In the following we consider

- $\Omega(t) = \{x(t, X) \mid X \in \Omega\}$ which is a volume transported by the evolution of mass points,
- $v(t, x)$ the corresponding velocity field in Eulerian coordinates, and
- $\varrho(t, x)$ the mass density in Eulerian coordinates.

The sets $\Omega(t)$ contain the same mass points for all times $t$. **Mass conservation** is now given by

$$\frac{d}{dt} \int_{\Omega(t)} \varrho(t, x)\,dx = 0 \,.$$

Using Reynolds transport theorem we obtain

$$\int_{\Omega(t)} \left( \frac{\partial \varrho}{\partial t}(t, x) + \nabla \cdot (\varrho(t, x)\,v(t, x)) \right) dx = 0 \,. \tag{5.2}$$

This equation has to hold for all volumes $\Omega(t)$ and hence in particular for all subsets of a given volume. If we assume that $\varrho$ and $v$ are continuously differentiable we obtain from (5.2) an equivalent formulation given by a differential equation, compare Exercise 5.5,

$$\partial_t \varrho + \nabla \cdot (\varrho\, v) = 0\,. \tag{5.3}$$

This is the *continuity equation*.

An alternative derivation and interpretation of the continuity equation is obtained if we apply the mass conservation on a small cube $Q$ with center $x_0$ and edge length $h$. The mass conservation is now given by the identity

$$\frac{d}{dt} \int_Q \varrho(t, x)\, dx + \int_{\partial Q} \varrho(t, x)\, v(t, x) \cdot n(t, x)\, ds_x = 0\,.$$

The first term describes the change of mass in $Q$ and the second term describes the flux of mass across the surface $\partial Q$. If we approximate all integrals with the help of its value at the midpoint we obtain

$$h^d \left( \partial_t \varrho(t, x_0) + O(h^2) \right)$$
$$+ \sum_{j=1}^{d} h^{d-1} \left( (\varrho v)(t, x_0 + \tfrac{h}{2} e_j) \cdot e_j + (\varrho v)(t, x_0 - \tfrac{h}{2} e_j) \cdot (-e_j) + O(h^2) \right) = 0\,.$$

We now divide by $h^d$ and obtain the continuity equation in the limit $h \to 0$.

In order to formulate the **conservation of (linear) momentum** we need

- a force density $f : \Omega(t) \to \mathbb{R}^d$ per unit mass which describes the outer forces acting on $\Omega(t)$. One example are gravitational forces $f = -g e_3$ (if $d = 3$) with the gravitational acceleration $g$. Other examples are forces due to electromagnetic fields,
- a force density $b : \partial\Omega(t) \to \mathbb{R}^d$ per unit surface area for the forces acting on the boundary $\partial\Omega$ on $\Omega$.

Now the law of conservation of (linear) momentum says that the change of the linear momentum $p = mv$ with mass $m$ and velocity $v$ results from forces $F$ as follows

$$\frac{d}{dt}(mv) = F\,.$$

The corresponding formulation in continuum mechanics is

$$\frac{d}{dt} \int_{\Omega(t)} \varrho(t, x)\, v(t, x)\, dx = \int_{\Omega(t)} \varrho(t, x) f(t, x)\, dx + \int_{\partial\Omega(t)} b(t, x)\, ds_x\,.$$

The first integral describes the change of (linear) momentum for the volume $\Omega(t)$, the second describes the volume forces acting on $\Omega(t)$ and the third accounts for the total surface forces acting on $\partial\Omega(t)$. Reynolds transport theorem now gives

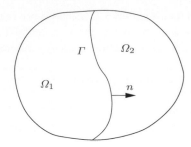

**Fig. 5.3** Definition of the stress tensor

$$\int_{\Omega(t)} \left( \frac{\partial}{\partial t} \big( \varrho(t, x)\, v_j(t, x) \big) + \nabla \cdot \big( \varrho(t, x)\, v_j(t, x)\, v(t, x) \big) \right) dx$$

$$= \int_{\Omega(t)} \varrho(t, x) f_j(t, x)\, dx + \int_{\partial \Omega(t)} b_j(t, x)\, ds_x$$

(5.4)

for all components $j = 1, \dots, d$.

**The Stress Tensor**

We now need a precise knowledge about how forces act in a body. In order to do so we consider a domain $\Omega$, which is divided in two parts $\Omega_1$ and $\Omega_2$, see Fig. 5.3. The surface $\Gamma$ which separates $\Omega_1$ and $\Omega_2$ is assumed to be continuously differentiable. Along the surface $\Gamma$ forces act between the parts $\Omega_1$ and $\Omega_2$.

**Axiom** (Cauchy 1827) *Along $\Gamma$ the set $\Omega_2$ exerts a force on $\Omega_1$ and this force is described by a force density $b(n; x) \in \mathbb{R}^d$:*

$$F_{\Omega_2 \to \Omega_1} = \int_{\Gamma} b(n; x)\, ds_x \, .$$

*Here $n = n(x)$ is the unit normal vector on $\Gamma$ pointing into $\Omega_2$. Note that $b$ does not depend on the specific choice of $\Gamma$ and $\Omega$.*

In the following theorem we obtain that the boundary force density $b$ is a *linear* function of $n$. Here $S_d = \{ x \in \mathbb{R}^d \mid |x| = 1 \}$ is the set of all possible unit vectors.

**Theorem 5.5** (Cauchy, existence of the stress tensor) *We assume that Cauchy's axiom is true and that $\varrho$, $v$, $f$, $b$ are smooth functions which fulfill the momentum balance (5.4) for all $\Omega(t) \subset \Omega$ with piecewise smooth boundary.*
*Then for all $x \in \Omega$ there exists a matrix $\sigma = \big( \sigma_{ij} \big)_{i,j=1}^{d}$ such that*

$$b(n; x) = \sigma(x) n = \left( \sum_{j=1}^{d} \sigma_{ij}(x) n_j \right)_{i=1}^{d} \, .$$

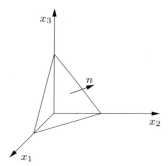

**Fig. 5.4** A tetrahedron which is needed in Cauchy's theorem

Cauchy's theorem is a consequence of the axiom of Cauchy and of the *balance of (linear) momentum.*

*Proof of Cauchy's theorem.* Let $\sigma = \left(b^{(1)}, \ldots, b^{(d)}\right)$ with $b^{(j)} = b(e_j, x) = \sigma(x)e_j$. For $n \in S_d$ with $n \notin \{e_1, \ldots, e_d\}$ we now consider the tetrahedron $V$ which is visualized for $d = 3$ in Fig. 5.4 with faces $S_j = \partial V \cap \{z \in \mathbb{R}^d \mid z_j = 0\}$ and $S = \partial V \setminus \bigcup_{j=1}^{d} S_j$. The face $S$ is assumed to have the outer unit normal $n$.

Without loss of generality we assume $n_j > 0$ for $j = 1, \ldots, d$. For the faces we obtain $|S_j| = |S|n_j, j = 1, \ldots, d$; this relation is shown in Exercise 5.13. Using the balance of (linear) momentum (5.1) and the mean value theorem of integral calculus we obtain

$$|V| \left( \frac{\partial}{\partial t} \left( \varrho(t, y) v_j(t, y) \right) + \nabla \cdot \left( \varrho(t, y)\, v_j(t, y)\, v(t, y) \right) - \varrho(t, y) f_j(t, y) \right)$$

$$= \sum_{k=1}^{d} |S_k|\, b_j\!\left(-e_k, x^{(k)}\right) + |S|\, b_j\!\left(n, x^{(0)}\right)$$

with points $y \in V, x^{(k)} \in S_k, x^{(0)} \in S$. Dividing by $|S|$ in the above identity gives for $|S| \to 0$

$$0 = \sum_{k=1}^{d} n_k\, b(-e_k, x) + b(n, x),$$

where we use $\frac{|V|}{|S|} \to 0$ as $|S| \to 0$ and $\frac{|S_k|}{|S|} = n_k$. Since $b$ was smooth we obtain in the limit $n \to e_j$ for some $j \in \{1, \ldots, d\}$

$$b(-e_j, x) = -b(e_j, x).$$

In particular this implies the force identity $F_{\Omega_2 \to \Omega_1} = -F_{\Omega_1 \to \Omega_2}$. Altogether we obtain for a general $n \in S_d$

$$b(n, x) = \sum_{k=1}^{d} b(e_k, x) n_k = \sigma(x) n.$$

□

We now come back to the formulation of the conservation law for (linear) momentum. With the stress tensor and the divergence theorem we can rewrite the total surface forces in (5.4) as

$$\int_{\partial\Omega(t)} b(t, x) \, ds_x = \int_{\partial\Omega(t)} \sigma(t, x) \, n(t, x) \, ds_x = \int_{\Omega(t)} \nabla \cdot \sigma(t, x) \, dx,$$

where we define the "matrix divergence" $\nabla \cdot \sigma$ as

$$\nabla \cdot \sigma(t, x) = \left( \sum_{k=1}^{d} \partial_{x_k} \sigma_{jk}(t, x) \right)_{j=1}^{d}.$$

We can now restate the conservation of linear momentum as

$$\int_{\Omega(t)} \left( \partial_t(\varrho \, v_j) + \nabla \cdot (\varrho \, v_j \, v) - \varrho f_j - (\nabla \cdot \sigma)_j \right) dx = 0.$$

With the help of the continuity equation it follows

$$\int_{\Omega(t)} \left( \varrho \, \partial_t v_j + \varrho \, v \cdot \nabla v_j - (\nabla \cdot \sigma)_j - \varrho f_j \right) dx = 0;$$

where we used

$$\partial_t(\varrho \, v_j) + \nabla \cdot (\varrho \, v_j \, v) = v_j(\partial_t \varrho + \nabla \cdot (\varrho \, v)) + \varrho(\partial_t v_j + v \cdot \nabla v_j).$$

For continuously differentiable $\varrho$, $v$, $\sigma$ we also obtain a formulation as a differential equation (using the fact that $\Omega(t)$ can be chosen arbitrarily)

$$\varrho(\partial_t v_j + v \cdot \nabla v_j) - (\nabla \cdot \sigma)_j - \varrho f_j = 0.$$

This formulation can be rewritten with the help of the *differential operator* $v \cdot \nabla = \sum_{j=1}^{d} v_j \, \partial_{x_j}$ in the following compact form

$$\varrho(\partial_t v + (v \cdot \nabla)v) - \nabla \cdot \sigma = \varrho f. \tag{5.5}$$

The **law of angular momentum** for a system of mass points can be written as

$$\frac{d}{dt}L(t) = M(t),$$

with the *angular momentum* $L(t)$ and the *torque* $M(t)$. For a mass point with position $x$, velocity $v$ and mass $m$ the angular momentum and the torque with respect to a point $x^{(0)}$ are given as

$$L = \left(x - x^{(0)}\right) \times (mv) \quad \text{and} \quad M = \left(x - x^{(0)}\right) \times F,$$

where $F$ is the force acting on the mass point. For a system of mass points one considers the sum of all mass points and for a solid body we need to consider a corresponding integral representation.

In continuum mechanics one formulates the conservation of angular momentum as

$$\frac{d}{dt} \int_{\Omega(t)} \left(x - x^{(0)}\right) \times (\varrho v)(t, x)\, dx = \int_{\Omega(t)} \left(x - x^{(0)}\right) \times (\varrho f)(t, x)\, dx$$

$$+ \int_{\partial\Omega(t)} \left(x - x^{(0)}\right) \times (\sigma n)(t, x)\, ds_x.$$

The first integral is the angular momentum of the domain $\Omega(t)$, the second and third integral describe the torques given by the volume and surface forces. We now apply Reynolds transport theorem to the left hand side. For computations with the vector product the notation (5.1) is helpful. We obtain

$$\nabla \cdot \left(\left(\left(x - x^{(0)}\right) \times (\varrho\, v)\right)_i v\right) = \sum_{j,k,\ell=1}^{3} \partial_{x_j}\left(\varepsilon_{ik\ell}\left(x_k - x_k^{(0)}\right)\varrho\, v_\ell\, v_j\right)$$

$$= \sum_{j,k,\ell=1}^{3} \varepsilon_{ik\ell}\, \delta_{jk}\, \varrho\, v_\ell\, v_j + \sum_{j,k,\ell=1}^{3} \varepsilon_{ik\ell}\left(x_k - x_k^{(0)}\right)\partial_{x_j}\left(\varrho\, v_\ell\, v_j\right)$$

$$= \left(\left(x - x^{(0)}\right) \times \left(\sum_{j=1}^{3} \partial_{x_j}\left(\varrho\, v_j\, v\right)\right)\right)_i$$

and the Reynolds transport theorem yields

$$\frac{d}{dt} \int_{\Omega(t)} \left(x - x^{(0)}\right) \times (\varrho\, v)\, dx$$

$$= \int_{\Omega(t)} \left(x - x^{(0)}\right) \times \left(\partial_t(\varrho\, v) + \sum_{j=1}^{3} \partial_{x_j}(\varrho\, v_j\, v)\right) dx.$$

After the above manipulations the conservation law for angular momentum now has the form

$$\int_{\Omega(t)} \left(x - x^{(0)}\right) \times \left(\partial_t(\varrho\, v) + \sum_{j=1}^{3} \partial_{x_j}(\varrho\, v_j\, v)\right) dx$$

$$= \int_{\Omega(t)} \left(x - x^{(0)}\right) \times (\varrho f)\, dx + \int_{\partial\Omega(t)} \left(x - x^{(0)}\right) \times (\sigma n)\, ds_x \,.$$

A consequence of the conservation of angular momentum is

**Theorem 5.6** *Let $\varrho, v, f, b$ be smooth and assume that the conservation laws for linear and angular momentum hold. Then the stress tensor necessarily has to be symmetric, i.e., $\sigma_{jk} = \sigma_{kj}$ for $j, k = 1, 2, 3$.*

*Proof* Let $a \in \mathbb{R}^3$. Then the conservation of angular momentum implies

$$a \cdot \int_{\Omega(t)} \left(x - x^{(0)}\right) \times \left(\partial_t(\varrho\, v) + (\nabla \cdot (\varrho\, v\, v_\ell))_{\ell=1}^{3}\right) dx$$

$$= a \cdot \int_{\Omega(t)} \left(x - x^{(0)}\right) \times (\varrho f)\, dx + a \cdot \int_{\partial\Omega(t)} \left(x - x^{(0)}\right) \times (\sigma n)\, ds_x \,.$$

For the last term we obtain, using the formula $a \cdot (b \times c) = (a \times b) \cdot c = \det(a\, b\, c)$ as well as the divergence theorem and the product rule

$$a \cdot \int_{\partial\Omega(t)} \left(x - x^{(0)}\right) \times (\sigma n)\, ds_x = \int_{\partial\Omega(t)} \left(a \times \left(x - x^{(0)}\right)\right) \cdot \sigma n\, ds_x$$

$$= \int_{\Omega(t)} \sum_{i,j=1}^{3} \partial_{x_j}\left(\left(a \times \left(x - x^{(0)}\right)\right)_i \sigma_{ij}\right) dx$$

$$= \int_{\Omega(t)} \left(\left(a \times \left(x - x^{(0)}\right)\right) \cdot (\nabla \cdot \sigma) + \sum_{i,j=1}^{3} \partial_{x_j}\left(a \times \left(x - x^{(0)}\right)\right)_i \sigma_{ij}\right) dx\,.$$

Altogether we obtain

$$\int_{\Omega(t)} a \times \left(x - x^{(0)}\right) \cdot \left(\partial_t(\varrho\, v) + (\nabla \cdot (\varrho\, v\, v_\ell))_{\ell=1}^{3} - \varrho f - \nabla \cdot \sigma\right) dx$$

$$= \int_{\Omega(t)} \sum_{i,j=1}^{3} \partial_{x_j}\left(a \times \left(x - x^{(0)}\right)\right)_i \sigma_{ij}\, dx\,.$$

The conservation of linear momentum implies that the left hand side is equal to zero. Since this is true for *all* $\Omega(t)$ we obtain, see Exercise 5.5,

$$\sum_{i,j=1}^{3} \partial_{x_j}\left(a \times \left(x - x^{(0)}\right)\right)_i \sigma_{ij} = 0\,.$$

For $a = e_1$ it holds $e_1 \times \left(x - x^{(0)}\right) = \begin{pmatrix} 0 \\ -\left(x_3 - x_3^{(0)}\right) \\ x_2 - x_2^{(0)} \end{pmatrix}$ and

$$0 = \sum_{i,j=1}^{3} \partial_{x_j}\left(e_1 \times \left(x - x^{(0)}\right)\right)_i \sigma_{ij} = -\sigma_{23} + \sigma_{32}$$

implies $\sigma_{23} = \sigma_{32}$. In an analogous way we obtain $\sigma_{13} = \sigma_{31}$ (with $a = e_2$) and $\sigma_{12} = \sigma_{21}$ (with $a = e_3$).                                                                            □

**Remark.** When we formulated the conservation of angular momentum we assumed a so-called *non-polar* medium which cannot take a microscopic angular momentum. Media with an internal angular momentum are called *polar* media or *Cosserat continua*. In this case the conservation of angular momentum does not hold as formulated above and the stress tensor is *not* necessarily symmetric.

In order to formulate the **conservation of energy** we introduce a specific internal energy $u = u(t, x)$, a heat flux $q = q(t, x)$ and heat sources $g(t, x)$ which are defined per unit mass. The heat flux $q$ is an energy flux per unit surface and

$$q \cdot n = \lim_{\substack{\Delta t \to 0 \\ \Delta A \to 0}} \frac{\Delta Q}{\Delta t\, \Delta A}$$

describes approximately the amount of heat $\Delta Q$ which flow through a surface with area $\Delta A$ and normal $n$ in a time interval with length $\Delta t$. A specific example for a heat source $g$ is the energy resulting from electromagnetic radiation, e.g., by microwaves. The energy conservation law now states

$$\frac{d}{dt} \int_{\Omega(t)} \varrho\left(\tfrac{1}{2}|v|^2 + u\right) dx = \int_{\Omega(t)} \varrho f \cdot v\, dx + \int_{\partial\Omega(t)} \sigma n \cdot v\, ds_x$$

$$- \int_{\partial\Omega(t)} q \cdot n\, ds_x + \int_{\Omega(t)} \varrho\, g\, dx.$$

The left hand side describes the change of energy which has two contributions, the kinetic energy $\int_{\Omega(t)} \tfrac{1}{2}\varrho|v|^2\, dx$ and the internal energy $\int_{\Omega(t)} \varrho u\, dx$. The first and second term on the right hand side describe the work caused by volume and surface forces, the third term accounts for the heat gain or loss from heat flux across the outer boundary and the fourth term describes changes of the energy due to outer heat sources. With the help of Reynolds transport theorem and the divergence theorem we obtain

$$\int_{\Omega(t)} \Big[\partial_t\big(\varrho\big(\tfrac{1}{2}|v|^2 + u\big)\big) + \nabla \cdot \big(\varrho\big(\tfrac{1}{2}|v|^2 + u\big)v\big)$$

$$- \varrho f \cdot v - \nabla \cdot (\sigma^\top v) + \nabla \cdot q - \varrho g\Big]\, dx = 0.$$

Using the fact that $\Omega(t)$ is arbitrary we now obtain the following formulation of the conservation of energy in terms of a differential equation

$$\partial_t\left(\varrho\left(\tfrac{1}{2}|v|^2+u\right)\right)+\nabla\cdot\left(\varrho\left(\tfrac{1}{2}|v|^2+u\right)v\right)-\varrho f\cdot v-\nabla\cdot(\sigma^\top v)+\nabla\cdot q-\varrho g=0.$$
(5.6)

Using the continuity equation (5.3) we can simplify the above equation as follows

$$\varrho\,\partial_t\left(\tfrac{1}{2}|v|^2+u\right)+\varrho v\cdot\nabla\left(\tfrac{1}{2}|v|^2+u\right)-\varrho f\cdot v-\nabla\cdot(\sigma^\top v)+\nabla\cdot q-\varrho g=0.$$

The kinetic energy can be eliminated by using the conservation of (linear) momentum (5.5)

$$\varrho\,\partial_t\left(\tfrac{1}{2}|v|^2\right)+\varrho v\cdot\nabla\left(\tfrac{1}{2}|v|^2\right)=\sum_{i=1}^{d}\left(\varrho v_i\,\partial_t v_i+\varrho\sum_{j=1}^{d}v_i v_j\,\partial_{x_i}v_j\right)$$

$$=v\cdot\left(\varrho\,\partial_t v+\varrho\,(v\cdot\nabla)v\right)=v\cdot\left(\nabla\cdot\sigma+\varrho f\right).$$

It holds that

$$v\cdot(\nabla\cdot\sigma)-\nabla\cdot(\sigma^\top v)=\sum_{i,j=1}^{d}\left(v_i\,\partial_{x_j}\sigma_{ij}-\partial_{x_j}(\sigma_{ij}v_i)\right)=-\sum_{i,j=1}^{d}\sigma_{ij}\,\partial_{x_j}v_i$$

$$=:-\sigma:Dv,$$

where $Dv=\left(\partial_{x_j}v_i\right)_{i,j=1}^{d}$ is the Jacobi matrix of $v$ and ":" the inner product of two matrices

$$A:B:=\sum_{j,k=1}^{d}a_{jk}b_{jk}.$$

Hence we obtain the following formulation of the energy conservation

$$\varrho\,\partial_t u+\varrho v\cdot\nabla u-\sigma:Dv+\nabla\cdot q-\varrho g=0.$$
(5.7)

**Conservation laws for multi-component systems**: For mixtures consisting of several components we have in addition a conservation law for each component. The composition of the mixture can be described by a *concentration per unit mass* $c_i$ of the component $i\in\{1,\ldots,M\}$. More precisely we have $c_i(x)=\lim\limits_{r\to 0}\frac{m_i(B_r(x))}{m(B_r(x))}$, where $m_i$ is the mass of component $i$ and $m$ is the total mass of all components and $B_r(x)$ is the ball of radius $r$ and center $x$. Obviously

$$\sum_{i=1}^{M}c_i=1.$$

In addition we need the *flux* $j_i$ (typically caused by diffusion) of component $i$ and a rate $r_i$ with which component $i$ is produced or consumed, e.g., by a chemical reaction. The quantity $j_i \cdot n$ describes how much mass of component $i$ flows through a unit surface with normal $n$ per unit time. The value $r_i$ is the change of mass per unit time and per unit total mass. The conservation law for component $i$ is given as

$$\frac{d}{dt} \int_{\Omega(t)} \varrho \, c_i \, dx = \int_{\Omega(t)} \varrho \, r_i \, dx - \int_{\partial \Omega(t)} j_i \cdot n \, ds_x \, .$$

The first integral describes the change of the mass of species $i$ in the domain $\Omega(t)$, the second integral accounts for the production or consumption of species $i$ and the third integral describes the flux of species $i$ across the boundary. Applying Reynolds transport theorem and the divergence theorem now gives

$$\int_{\Omega(t)} \left( \partial_t(\varrho \, c_i) + \nabla \cdot (\varrho \, c_i \, v) - \varrho \, r_i + \nabla \cdot j_i \right) dx = 0.$$

Using the continuity equation we obtain the following local formulation as a differential equation

$$\varrho \, \partial_t c_i + \varrho \, v \cdot \nabla c_i - \varrho \, r_i + \nabla \cdot j_i = 0 \, .$$

The fluxes $j_i$ and the rates $r_i$ need to fulfill side constraints, such that the condition $\sum_{i=1}^{M} c_i - 1$ is fulfilled during the evolution. We require

$$\sum_{i=1}^{M} \nabla \cdot j_i = 0 \quad \text{and} \quad \sum_{i=1}^{M} r_i = 0 \, .$$

For example, the first condition is fulfilled if $\sum_{i=1}^{M} j_i = 0$.

**Summary**

We now derived all fundamental laws of continuum mechanics in *Eulerian* coordinates:

The *continuity equation*, which results from the conservation of mass,

$$\partial_t \varrho + \nabla \cdot (\varrho \, v) = 0 \, ;$$

the *conservation of (linear) momentum*

$$\varrho \, \partial_t v + \varrho \, (v \cdot \nabla) v - \nabla \cdot \sigma = \varrho f \, ;$$

the *conservation of energy*

$$\varrho \, \partial_t u + \varrho \, v \cdot \nabla u - \sigma : Dv + \nabla \cdot q = \varrho \, g$$

and the *conservation of the species*

$$\varrho\, \partial_t c_i + \varrho\, v \cdot \nabla c_i + \nabla \cdot j_i = \varrho\, r_i \ \text{ for } i = 1, \ldots, M\,. \tag{5.8}$$

We need to compute the quantities mass density $\varrho$, velocity $v$, temperature $T$ and concentrations $c := (c_1, \ldots, c_M)$. Further quantities which we need to determine are the stress tensor $\sigma$, the heat flux $q$ and the species fluxes $j_i$. These quantities will be related to each other by *constitutive relations* which describe the specific properties of concrete materials. The constitutive relations have to be determined by experiments.

## 5.6  Constitutive Relations

The conservation laws for mass, (linear) momentum and energy hold for *all* materials and they in particular hold for solids, liquids and gases. The properties of a particular material are expressed by *constitutive relations*. We need to specify so far undetermined quantities in terms of other quantities. For example

- the stress tensor $\sigma$, as a function of the pressure and $Dv$,
- the heat flux $q$, as a function of $\nabla T$,
- the flux $j_i$ of the species $i$, as a function of other quantities,
- or we need to fix thermodynamic relations like $F(T, \varrho, p) = 0$ for the temperature $T$ the mass density $\varrho$ and the pressure $p$ and a constitutive relation for the internal energy which can be, e.g., $u = u(T, \varrho, c_1, \ldots, c_M)$.

Depending on the phenomena under consideration we need some of the above relations and others not.

We now state some examples for constitutive relations and discuss later in the Sects. 5.7 and 5.8 what kind of constraints on constitutive relations we need to take into account.

- Fourier's law of heat conduction

$$q = -K\nabla T \tag{5.9}$$

with the *thermal conductivity $K$*. The thermal conductivity is in general a matrix in $\mathbb{R}^{d \times d}$ but in many cases it suffices to consider $K$ to be scalar. Depending on the application $K$ can depend on the temperature $T$, the density $\varrho$, the temperature gradient $\nabla T$ and the concentration vector $c = (c_i)_{i=1}^{M}$. Fourier's law describes the fact that heat flows from areas with a high temperature to areas with a lower temperature. Here one uses that $-\nabla T$ points into the direction of steepest descent. For *isotropic* materials, for which the properties are the same in all spatial directions, the quantity $K$ is scalar which we will denote by $\lambda$. For anisotropic materials, e.g., fiber materials, $K$ will be a matrix since the thermal conductivity will depend on the direction.

- For a binary system, i.e., $M = 2$ in the multi-component system of Sect. 5.5, we can use the identity $c_1 + c_2 = 1$ in order to describe the two concentrations with the help of a single variable. We set $c = c_1$ and $j = j_1$. For $c$ it holds in the simplest case

$$j = -D\nabla c$$

with a *diffusion constant* $D$. Here $D$ can depend on $T$, $\varrho$, $c$. The reason for postulating this law is similar as in the case of Fourier's law: diffusion leads to a flux from areas with a high concentration to areas with a low concentration. This diffusion law can also be derived using stochastic analysis where a large quantity of "random walkers" are used to describe diffusion. We refer to Sect. 6.2.9 for details.

Another possibility to formulate a constitutive law for the flux $j$ uses the thermodynamics of mixtures. In Sect. 3.8 we already discussed that in isothermal, isobaric situations a mixture tends to minimize its free enthalpy or equivalently its free energy. Hence, the chemical potential which we compute as the derivative of the free energy with respect to the concentration is the driving force for diffusion. If the density of the free energy is given by the relation $f = f(c)$ it is postulated in irreversible thermodynamics that

$$j = -L\nabla\mu \quad \text{with} \quad \mu(c) = f'(c).$$

Here the mobility $L \geq 0$ can depend on $c$. If $c : [0, T] \times \Omega \to \mathbb{R}$ solves the equation

$$\partial_t c - \nabla \cdot (L\nabla\mu) = 0$$

and if $\nabla\mu \cdot n = 0$ holds on $\partial\Omega$ with outer normal $n$, we obtain

$$\frac{d}{dt} \int_\Omega f(c)\, dx \leq 0, \tag{5.10}$$

which tells us that the free energy has to decrease.
- The strain tensor for *inviscid* (also called *non-viscous*) fluids: For inviscid fluids only surface forces due to pressure differences are possible. The force density on a surface with unit normal $n$ is given by $-pn$. The relevant stress tensor is given as

$$\sigma = -pI \quad \text{with the identity matrix } I \text{ and the pressure } p. \tag{5.11}$$

Fluids can be assumed to be inviscid, if atomic or molecular interactions of neighboring particles can be neglected. This is in particular reasonable for a gas under low pressure.
- The stress tensor for viscous fluids. On a microscopic scale there will be diffusion from regions with fast particles to slower ones and vice versa, see Fig. 5.5. If these molecular interactions cannot be neglected one has to take into account that fast particles will accelerate nearby particles with a smaller velocity. Macroscopically

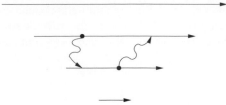

**Fig. 5.5** Explanation for the term viscosity

this leads to additional forces which will enter the stress tensor. The additional term will depend on $Dv$ since $Dv$ describes local velocity differences, i.e.,

$$\sigma = \sigma(p, Dv, \ldots).$$

A detailed analysis which will be discussed in Sect. 5.9 shows:

$\sigma$ only depends on the *symmetric part* of $Dv$, i.e.,

$$\varepsilon(v) = \tfrac{1}{2}\left(Dv + (Dv)^{\top}\right).$$

If $\sigma$ depends affine linearly on $Dv$ we obtain

$$\sigma = 2\,\mu\,\varepsilon(v) + \lambda\ \text{trace}\,(\varepsilon(v))\,I - p\,I\,.$$

We then obtain the stress tensor

$$\sigma = \mu\left(Dv + (Dv)^{\top}\right) + \lambda\,(\nabla \cdot v)\,I - p\,I \qquad (5.12)$$

with the *shear viscosity* $\mu$. The parameter $\mu' = \lambda + \tfrac{2}{3}\mu$ is called bulk or volume viscosity. The viscosities will in general depend on other parameters, e.g., the temperature and the concentration of species.

Based on these constitutive laws we can formulate several important models which lead to partial differential equations.

**The Heat Equation**

A heat diffusion process without fluid flow can be described with the help of the energy conservation law if we set $v = 0$ and $\varrho \equiv const$ in the governing equations. Since $u$ is the density of internal energy per unit mass we obtain from thermodynamics

$$\partial_t u = c_V(T)\,\partial_t T$$

where $c_V(T)$ is the specific heat at a constant volume. This is true since the internal energy of a mass $m$ with temperature $T$ is given as $U(T, m) = m\,u(T)$ and $c_V(T)$ is defined as

$$\frac{\partial U}{\partial T}(T, m) = c_V(T)\, m\,.$$

If the heat flux is now given by Fourier's law, we obtain the heat equation

$$\varrho\, c_V\, \partial_t T - \nabla \cdot (\lambda \nabla T) = \varrho\, g \tag{5.13}$$

with the heat conductivity $\lambda$. This is a typical example for a *parabolic* partial differential equation. The specific heat and the heat conductivity can depend on $T$ – but both are assumed to be constant in many applications. For constant $\varrho$, $c_V$, $\lambda$ the nondimensionalized version of the heat equation is given as

$$\partial_t T - \Delta T = g\,.$$

For boundary conditions which do not depend on time one observes typically that solutions converge for large times to a function which does not depend on $t$. We will discuss this fact in Sect. 6.2.2. The long time limit of (5.13) will be a solution of

$$-\nabla \cdot (\lambda \nabla T) = \varrho\, g\,. \tag{5.14}$$

This is a typical example for an *elliptic partial differential equation*. A nondimensionalized version for a constant heat conductivity is given by the *Poisson equation*

$$-\Delta T = g\,.$$

As a simple example we consider the heat flux through a wall of thickness $a$ and temperature $T_I$ at the interior wall and $T_A$ at the exterior wall. We assume that the temperature varies only in the direction perpendicular to the wall. Hence we can assume that $T = T(x_1)$ where $x_1$ is the coordinate of the direction perpendicular to the wall. Hence we obtain the boundary value problem

$$T''(x_1) = 0 \ \text{ for } x_1 \in (0, a)\,, \quad T(0) = T_I\,, \quad T(a) = T_A$$

with the solution

$$T(x_1) = T_I + \frac{x_1}{a}(T_A - T_I)\,.$$

The heat flux is parallel to the $x_1$-axis and is given by

$$q = -\lambda\, T'(x_1) = \frac{\lambda}{a}(T_I - T_A)\,.$$

Hence the loss of heat is proportional to the temperature difference and also to the *thermal transmittance* $\frac{\lambda}{a}$.

**The Euler Equations of Gas Dynamics**

As first example from fluid mechanics we consider *non-viscous fluids*, i.e., fluids with a stress tensor $\sigma = -pI$, compare (5.11). Then it holds

$$\nabla \cdot \sigma = -\nabla p \text{ and } \sigma : Dv = \sum_{i,j=1}^{d} \sigma_{ij} \, \partial_{x_j} v_i = -p \, \nabla \cdot v \,.$$

Using the continuity equation and the balance laws for (linear) momentum and energy we obtain:

$$\partial_t \varrho + \nabla \cdot (\varrho \, v) = 0 \,,$$
$$\varrho \, \partial_t v + \varrho \, (v \cdot \nabla) v = -\nabla p + \varrho f \,,$$
$$\varrho \, \partial_t u + \varrho \, v \cdot \nabla u + p \, \nabla \cdot v - \nabla \cdot (K \nabla T) = \varrho \, g \,.$$

 Here the heat flux is given by Fourier's law. The density $u$ of internal energy and the pressure $p$ have to be given as functions of $\varrho$ and $T$. These functions describe the thermodynamic properties of the material under consideration. For ideal gases for example it holds $u(\varrho, T) = c_V \, T$ with the specific heat per unit volume $c_V$ which we assume to be independent of $T$. The pressure $p$ is given by $p(\varrho, T) = c_R \, \varrho \, T$ with the gas constant $c_R = k_B / m_0$ where $k_B$ is Boltzmann's constant and $m_0$ the mass of a gas particle. The equation for the pressure follows from the equation of state $pV = Nk_B T$ which holds for a system of $N$ particles where one sets $V = m/\varrho$ with mass $m = Nm_0$. In space dimension $d$ we obtain $d + 2$ equations for $d + 2$ unknowns $\varrho, T, v_1, \ldots, v_d$. A special case of the Euler equations describes *isothermal* compressible non-viscous fluids, where the temperature is assumed to be constant. This is appropriate in the case that density differences are small and that there are no heat sources. The system of Euler equations then simplifies to

$$\partial_t \varrho + \nabla \cdot (\varrho \, v) = 0 \,,$$
$$\varrho \, \partial_t v + \varrho \, (v \cdot \nabla) v = -\nabla p + \varrho f \,.$$

The function $p = p(\varrho)$ follows from the thermodynamic equation of state at the given temperature.

 A further special case are *incompressible* non-viscous flows. A flow is incompressible, if the density at a material point does not change during motion. This is in particular meaningful for fluids for which a *small* change of density will lead to a large change of pressure. For isothermal, incompressible, non-viscous flows we obtain the following system

$$\nabla \cdot v = 0 \,,$$
$$\partial_t v + (v \cdot \nabla) v = -\tfrac{1}{\varrho} \nabla p + f \,.$$

In space dimension $d$ we obtain $d + 1$ equations for $d + 1$ unknowns $v_1, \ldots, v_d, p$. Here, no thermodynamic equation of state for the pressure is needed.

**The Navier–Stokes Equations for Isothermal Compressible Viscous Fluids**

Viscous fluids are fluids for which the internal friction cannot be neglected. The stress tensor is in this case given by (5.12). We obtain

$$(\nabla \cdot \sigma)_i = \sum_{j=1}^{d} \partial_{x_j} \sigma_{ij} = \sum_{j=1}^{d} \left( \mu(\partial_{x_j}\partial_{x_i} v_j + \partial_{x_j}^2 v_i) + \lambda\, \partial_{x_i}\partial_{x_j} v_j \right) - \partial_{x_i} p$$
$$= \mu\, \Delta v_i + (\lambda + \mu)\, \partial_{x_i} \nabla \cdot v - \partial_{x_i} p \,.$$

Combining the continuity equation and the momentum balances yields the Navier–Stokes equations for compressible, isothermal fluids

$$\partial_t \varrho + \nabla \cdot (\varrho\, v) = 0 \,,$$
$$\varrho\, \partial_t v + \varrho\, (v \cdot \nabla)\, v - \mu\, \Delta v - (\lambda + \mu)\nabla \nabla \cdot v = -\nabla p + \varrho f \,.$$

In this case we also need a thermodynamic constitutive law of the form $p = p(\varrho)$. We obtain $d + 1$ equations for $d + 1$ unknowns $\varrho, v_1 \ldots, v_d$.

**The Navier–Stokes Equations for Incompressible Viscous Fluids**

For a constant density $\varrho$ we obtain from the continuity equation and the momentum balance the Navier–Stokes equations for incompressible viscous fluids

$$\nabla \cdot v = 0 \,,$$
$$\partial_t v + (v \cdot \nabla) v - \eta\, \Delta v = -\tfrac{1}{\varrho}\nabla p + f \,. \tag{5.15}$$

Here we multiplied the momentum balance by $\tfrac{1}{\varrho}$ and we introduced $\eta = \mu/\varrho$ which is the *kinematic* viscosity whereas $\mu$ is the *dynamic* viscosity. In the literature one often finds variants of the Navier–Stokes equations with $\nabla p$ instead of $\tfrac{1}{\varrho}\nabla p$. In this case the pressure was rescaled. The term $(\lambda + \mu)\, \nabla \nabla \cdot v$ vanishes since $\nabla \cdot v = 0$. In spatial dimension $d$ we have $d + 1$ equations for the $d + 1$ unknowns $v_1, \ldots, v_d$ and $p$. In particular we do not need a thermodynamic constitutive relation for the pressure $p$. It is still possible to give a meaning to the pressure $p$. This is related to the constraint $\nabla \cdot v = 0$ and will be discussed in Sect. 6.5.

**The Stokes Equations**

For a *stationary* incompressible viscous fluid the time derivative $\partial_t v$ disappears in the Navier–Stokes equations

$$\nabla \cdot v = 0 \,,$$
$$(v \cdot \nabla) v - \eta\, \Delta v = -\tfrac{1}{\varrho}\nabla p + f \,. \tag{5.16}$$

If $|v|$ and $|Dv|$ are small we can neglect the second order term $(v \cdot \nabla)\, v$. Hence we obtain the Stokes equations

$$\nabla \cdot v = 0 \,,$$

$$-\eta \, \Delta v = -\tfrac{1}{\varrho} \nabla p + f \,.$$

As in the Navier–Stokes equations we have $(d + 1)$ equations for $(d + 1)$ unknowns. In order to characterize the applicability of the Stokes equations we non-dimensionalize the stationary Navier–Stokes equations. Let $V$ and $\ell$ be characteristic sizes for the velocity and the length of a flow. Setting $v(x) = V \widetilde{v}(\ell^{-1} x)$ we obtain with $\widetilde{x} = \ell^{-1} x$

$$(v \cdot \nabla)v = \frac{V^2}{\ell} (\widetilde{v} \cdot \widetilde{\nabla}) \widetilde{v}$$

and

$$\eta \, \Delta v = \frac{\eta V}{\ell^2} \widetilde{\Delta} \widetilde{v} \,,$$

where $\widetilde{\nabla}$ and $\widetilde{\Delta}$ are defined with partial derivatives with respect to $\widetilde{x}$. In this case we obtain the condition

$$\frac{V^2}{\ell} \ll \eta \frac{V}{\ell^2} \,, \quad \text{or} \quad \frac{\ell V}{\eta} \ll 1 \,.$$

The dimensionless quantity $\ell V / \eta$ is called the *Reynolds number*. The Stokes equations are applicable for stationary flows with a small Reynolds number. This is, e.g., the case if the viscosity is high, the flow velocities are small and/or in situations were the domain is small.

### Simple Examples

We now compute two simple exact solutions of the Navier–Stokes equations, which are important for applications.

### Couette Flow

In the case of a Couette flow we consider a viscous fluid between two parallel plates with distance $a$, see Fig. 5.6. One of the plates is at rest and the other one moves with velocity $v_0 e_1$. Since the fluid is viscous one has to solve the Navier–Stokes equations in $\Omega = \mathbb{R}^2 \times (0, a)$. The boundary conditions are $v(t, x_1, x_2, 0) = 0$ and $v(t, x_1, x_2, a) = v_0 e_1$. We are interested in the stationary case and assume a flow which is forced only by the moving plate and not by a pressure difference. A natural assumption is that the velocity has only a component in the $e_1$-direction and that the velocity only depends on $x_3$. In addition, we assume that the pressure is constant. This leads to the ansatz

$$v(t, x) = u(x_3) e_1 \quad \text{and} \quad p(t, x) = q$$

with a function $u : (0, a) \to \mathbb{R}$ and a constant $q$. This leads to $\nabla \cdot v = \partial_{x_1} u(x_3) = 0$, $\partial_t v = 0$, $(v \cdot \nabla)v = v_1 \partial_{x_1} v = 0$, $\Delta v = u''(x_3) e_1$ and $\nabla p = 0$. The whole Navier–Stokes equations now reduce to

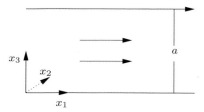

**Fig. 5.6**  Couette flow

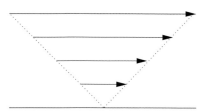

**Fig. 5.7**  Velocity field in the Couette flow

$$-\mu\, u''(x_3) = 0\,.$$

If we combine this with the boundary condition $u(0) = 0$, $u(a) = v_0$ one obtains

$$u(x_3) = \frac{v_0}{a}x_3 \ \text{ and } \ v(x) = \frac{v_0}{a}x_3 e_1\,.$$

The velocity grows linearly with the distance to the plate in rest, see Fig. 5.7

**Poiseuille Flow**

We again consider a flow between two parallel plates with distance $a$. But now both plates do not move. The flow is driven by a pressure difference. Hence, we prescribe two pressure values

$$p(t, x) = p_1 \ \text{if } x_1 = 0\,, \quad p(t, x) = p_2 \ \text{if } x_1 = L \ \text{with } p_1 > p_2\,.$$

We are interested in a stationary solution and we assume that the velocity only has a non-zero component in the $x_1$-direction and that this velocity component only depends on $x_3$, i.e.,

$$v(t, x) = u(x_3)e_1\,.$$

Plugging this ansatz into the Navier–Stokes equations gives

$$-\mu\, u''(x_3)e_1 = -\nabla p(x)$$

and the boundary conditions for $u$ are $u(0) = u(a) = 0$. First we obtain that $p$ only depends on $x_1$. This follows because the second and third component of $\nabla p$ are zero. In the resulting equation the left hand side only depends on $x_3$ and the right hand side only depends on $x_1$. Hence both sides have to equal a constant:

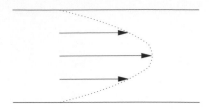

**Fig. 5.8** Velocity profile of the Poiseuille flow

$$\mu\, u''(x_3) = p'(x_1) = c\,.$$

The identities $p'(x_1) = c, p(0) = p_1, p(L) = p_2$ imply

$$p(x_1) = p_1 + \frac{p_2 - p_1}{L} x_1 \ \text{ and } \ c = \frac{p_2 - p_1}{L}\,.$$

In addition $\mu\, u''(x_3) = c$ and $u(0) = u(a) = 0$ imply

$$u(x_3) = \frac{p_1 - p_2}{2\mu L} x_3 (a - x_3)\,.$$

We obtain a parabolic profile of the velocity in $x_3$-direction and the maximal velocity is $v_{\max} = u\left(\frac{a}{2}\right) = \frac{p_1 - p_2}{8\mu L} a^2$ (Fig. 5.8).

We now solve an analogous problem for the *incompressible Euler equations*, namely

$$\nabla \cdot v = 0\,,$$
$$\partial_t v + (v \cdot \nabla)v = -\tfrac{1}{\varrho}\nabla p\,.$$

The Euler equations hold for a medium without internal friction and hence the boundary condition $v(t, x) = 0$ for $x_3 = 0$ and $x_3 = a$ are not meaningful. There is no reason why the velocity should tend to zero towards the boundary. In contrast we search for a solution which is constant in the $x_2$- and $x_3$-directions and which only has a component in $x_1$-direction. This means we make the ansatz

$$v(t, x) = u(t, x_1)e_1\,.$$

The continuity equation

$$\nabla \cdot v(t, x) = \partial_{x_1} u(t, x_1) = 0$$

implies that $u$ does not depend on $x_1$. The momentum equation implies

$$\varrho\, \partial_t u(t) = -\partial_{x_1} p(t, x)\,.$$

Here, the left hand side only depends on $t$ and this implies that $p$ depends on $x_1$ in an affine manner, i.e., $p(t, x_1) = c_1(t) + c_2(t) x_1$. The boundary conditions $p(t, 0) = p_1$ and $p(t, L) = p_2$ imply

$$p(t, x) = p_1 + \frac{p_2 - p_1}{L} x_1$$

and hence

$$u(t) = u(0) + \frac{p_1 - p_2}{\varrho L} t .$$

The velocity grows linearly in time and diverges to $+\infty$ for $t \to +\infty$, provided that $p_1 \neq p_2$.

The difference to the solution for the Navier–Stokes equations results from the fact that viscosity is neglected. Since fluid enters the system under pressure one continuously feeds energy into the system. In the stationary case of the Navier–Stokes equations the energy is used up by internal friction in the fluid (the energy is converted to heat). In the Euler equations there is no internal friction and hence no energy is used up and the kinetic energy in the system grows continuously.

**Potential Flows**

A flow is a *potential flow* if the velocity field $v$ has a potential, i.e., there exists a function $\varphi$ such that

$$v(t, x) = \nabla \varphi(t, x) .$$

Potential flows are *irrotational*. This follows from the fact that second order partial derivatives can be interchanged and hence

$$\nabla \times v = \nabla \times \nabla \varphi = 0 .$$

In a simply connected domain *every* irrotational flow is a potential flow. This is not true for general domains, see the following example: On $\Omega = \mathbb{R}^2 \setminus B_\varepsilon(0)$, $\varepsilon > 0$ we define the velocity field

$$v(x_1, x_2) = \begin{pmatrix} -\dfrac{x_2}{|x|^2} \\ \dfrac{x_1}{|x|^2} \end{pmatrix} .$$

This flow is irrotational which follows from

$$\partial_{x_1} \left( \frac{x_1}{|x|^2} \right) + \partial_{x_2} \left( \frac{x_2}{|x|^2} \right) = \frac{|x|^2 - 2x_1^2}{|x|^4} + \frac{|x|^2 - 2x_2^2}{|x|^4} = 0 .$$

Assume now the existence of a function $\varphi : \Omega \to \mathbb{R}$ with $\nabla \varphi = v$. We parameterize the boundary of the unit circle $C = \partial B_1(0)$ via $\begin{pmatrix} x_1(s) \\ x_2(s) \end{pmatrix} = \begin{pmatrix} \cos s \\ \sin s \end{pmatrix}$. By

$$\int_\gamma f \cdot ds = \int_a^b f(\gamma(s)) \cdot \gamma'(s)\, ds$$

we denote the curve integral of the second kind of a vector field $f : \mathbb{R}^d \to \mathbb{R}^d$ along a curve $\gamma : [a, b] \to \mathbb{R}^d$. We obtain

$$\int_C \nabla\varphi \cdot ds = \int_0^{2\pi} \left( \partial_{x_1}\varphi\, x_1'(s) + \partial_{x_2}\varphi\, x_2'(s) \right) ds$$

$$= \int_0^{2\pi} \frac{d}{ds}\varphi(x_1(s), x_2(s))\, ds = \varphi(1,0) - \varphi(1,0) = 0\,,$$

and

$$\int_C v \cdot ds = \int_0^{2\pi} \left( v_1 x_1' + v_2 x_2' \right) ds$$

$$= \int_0^{2\pi} (\sin^2 s + \cos^2 s)\, ds = 2\pi\,.$$

This shows that $v$ is not a potential flow. For potential flows without internal friction which solve the Euler equations we have the following result.

**Theorem 5.7**  (Bernoulli) *Let $v$ be the velocity field of a potential flow with potential $\varphi$ for which the conservation of momentum with stress tensor $\sigma = -pI$ and external forces $f = 0$ holds. Let $C$ be a curve that connects the points $a$ and $b$ within $\Omega$. Then it holds*

$$\left[ \partial_t \varphi + \tfrac{1}{2}|v|^2 \right]_a^b + \int_C \frac{\nabla p}{\varrho} \cdot ds = 0\,.$$

*Proof* The identity $\nabla \times v = 0$ implies $\partial_{x_i} v_j = \partial_{x_j} v_i$ for all $i, j$ and hence

$$(v \cdot \nabla)v_i = \sum_{j=1}^d v_j\, \partial_{x_j} v_i = \sum_{j=1}^d v_j\, \partial_{x_i} v_j = \frac{1}{2}\partial_{x_i}\left(|v|^2\right).$$

Conservation of momentum now implies

$$0 = \int_C \left( \partial_t v + (v \cdot \nabla)v + \frac{\nabla p}{\varrho} \right) \cdot ds = \int_C \left( \partial_t \nabla\varphi + \nabla\left( \frac{|v|^2}{2} \right) + \frac{\nabla p}{\varrho} \right) \cdot ds$$

$$= \left[ \partial_t \varphi + \frac{|v|^2}{2} \right]_a^b + \int_C \frac{\nabla p}{\varrho} \cdot ds\,.$$

$\square$

The generalization of the Bernoulli theorem for the case $f \neq 0$ is the subject of Exercise 5.25. For such a result one has to assume that $f$ has a potential which is for example true for the gravitational force. The Bernoulli theorem has the following special cases:

(i) If the density $\varrho$ is constant, we obtain

$$\int_C \frac{\nabla p}{\varrho} \cdot ds = \frac{1}{\varrho} \int_C \nabla p \cdot ds = \frac{1}{\varrho}[p]_a^b ,$$

which simplifies the expression in the Bernoulli theorem.

(ii) If $v$ stationary, i.e., $v(t, x) = v(x)$, and $\varrho$ constant we obtain that

$$\frac{1}{2}|v|^2 + \frac{p}{\varrho}$$

is constant. This implies that in areas in which the velocity is higher the pressure is lower and vice versa. One refers to this property as *Bernoulli's principle*

For *incompressible* potential flows the continuity equation reads as

$$\nabla \cdot v = \Delta\varphi = 0 .$$

Hence, the potential $\varphi$ solves *Laplace's equation*. As we will see in Sect. 6.1, Theorem 6.4, such a flow is uniquely determined by the *normal component* $v \cdot n$ of the velocity field on the boundary. To obtain a solution $v \cdot n$ has to fulfill the condition

$$\int_{\partial\Omega} v \cdot n \, ds_x = 0 . \tag{5.17}$$

It is easy to interpret this condition physically. If (5.17) is false then one would obtain that mass would enter or leave the domain. This is not possible for an incompressible flow.

Incompressible, stationary potential flows as special solutions are possible for the (incompressible) Euler equations as well as for the Navier–Stokes equations. The *viscous part* of the stress tensor leads to the term $-\Delta v$ in the Navier–Stokes equations and this term vanishes which follows from

$$-\Delta v_j = -\Delta\partial_j\varphi = -\partial_j\Delta\varphi = 0 .$$

Hence, the momentum balance becomes

$$\varrho(v \cdot \nabla)v + \nabla p = 0 .$$

This implies that the pressure has to be a potential of $\varrho(v \cdot \nabla)v$. For constant density this potential is given as the negative *density of the kinetic energy*

$$p = -\tfrac{\varrho}{2}|v|^2 .$$

## 5.7    The Second Law of Thermodynamics in Continuum Mechanics

The second law of thermodynamics implies that it is impossible to gain mechanical energy by just cooling a heat reservoir. Equivalent to this assertion is that heat cannot flow from cold to hot areas. To understand these statements one has to know about the distinction between energy which can be mechanically used and the energy which cannot be mechanically used. For a precise formulation of these facts one has to use the entropy $S$ as an additional quantity. We now consider a thermodynamic system that can be described by its thermal and mechanical properties. For an evolution the second law requires that

$$\dot{S} \geq \frac{\dot{Q}}{T},$$

where $T$ is the absolute temperature and $Q$ is the heat supplied to the system up to the present time. The thermodynamic equilibrium describes the state which the system attains for given macroscopic data like pressure or temperature for large times. In case that there is no loss of mechanical energy we obtain

$$\dot{S} = \frac{\dot{Q}}{T}, \tag{5.18}$$

and such processes can also take place *backwards* in time and they are called *reversible*. A change of state with

$$\dot{S} > \frac{\dot{Q}}{T}$$

is called *irreversible*. The entropy $S$ of a system is a function of macroscopic quantities such as pressure and temperature. For a gas we have for small densities and small pressures

$$S = S(T, p) = \tfrac{5}{2} N k_B \ln \tfrac{T}{T_0} - N k_B \ln \tfrac{p}{p_0}.$$

Here, $N$ is the number of particles, $k_B$ is the Boltzmann constant, $T_0, p_0$ are reference values for pressure and temperature. The entropy can be interpreted as a measure for the disorder of a system. If in a gas with $N$ particles all particles move with the same velocity and in the same direction we have the most possible degree of order. This corresponds to a pure translational motion of the gas. In particular there is no thermal energy in the system. A good amount of the kinetic energy in this motion can be used mechanically, e.g., in a turbine. In a completely uncorrelated motion of particles there is no macroscopic kinetic energy contained in the gas. In this case all the energy is thermal energy which can be used mechanically only to a limited amount. The translational motion has a low entropy and the uncorrelated motion has a high entropy. The disorder of a system can be interpreted as the number of micro-states which correspond to given macroscopically observed data. For a pure translational motion the micro-states are unique. However, for an uncorrelated motion there are many micro-states which correspond to the same macroscopic data such as temperature, pressure and entropy. One can give the entropy also an interpretation from information theory. In a translational motion with low entropy we know the motion of all particles. For an uncorrelated motion with high entropy we only know something for a few averaged quantities. The second law of thermodynamics in this interpretation says that in the course of time the information about the system cannot increase.

The entropy is a so-called *extensive* quantity and its value is proportional to the number of particles in the system. In particular, the entropy is proportional to the total mass. Therefore, one uses a *mass-based* density of the entropy. This quantity we will denote by s. Models in continuum mechanics that describe thermodynamic processes such as diffusion of heat and chemical substances should be compatible with the second law of thermodynamics. Models for which this can be shown rigorously are called *thermodynamically consistent*.

### Clausius–Duhem Inequality

In order to formulate the second law we need the absolute temperature $T > 0$, the pressure $p$, the specific entropy density $s$ and the specific volume $V = \frac{1}{\varrho}$. In words the second law says

| change of total entropy | $\geq$ | entropy flux across the boundary | $+$ | entropy production by heat sources. |

From (5.18) one can deduce that the entropy flux across the boundary is given as the heat flux divided by temperature. In addition the entropy production by heat sources is given by the heat sources divided by temperature. In conclusion we obtain

$$\frac{d}{dt} \int_{\Omega(t)} \varrho s \, dx \geq - \int_{\partial \Omega(t)} \frac{q \cdot n}{T} \, ds_x + \int_{\Omega(t)} \frac{\varrho g}{T} \, dx \,.$$

We remark that the entropy flux in more general systems, e.g., in systems with several components, might contain additional terms. With the help of the Reynolds transport theorem and by using the fact that $\Omega(t)$ is arbitrary we obtain

$$\frac{\partial}{\partial t} (\varrho s) + \nabla \cdot (\varrho s v) \geq -\nabla \cdot \left( \frac{q}{T} \right) + \varrho \frac{g}{T} \quad \text{for all } t \text{ and } x \,.$$

This is the *Clausius–Duhem inequality*. A fundamental requirement in continuum mechanics is that, in systems which can be described alone by its thermal and mechanical properties, the Clausius–Duhem inequality has to hold for all solutions of the conservation laws. This requirement has fundamental consequences for the way in which quantities can depend on each other.

*Dissipation Inequality*

For the further considerations the following dissipation inequality is important.

**Theorem 5.8** *The conservation laws* (5.3), (5.7) *for mass and energy and the Clausius–Duhem inequality imply the following dissipation inequality*

$$\varrho \left( D_t u - T D_t s \right) - \sigma : Dv + \frac{1}{T} q \cdot \nabla T \leq 0 \,,$$

*where* $D_t = \partial_t + v \cdot \nabla$ *is the material derivative from Definition 5.1.*

*Proof* The Clausius–Duhem inequality and the mass conservation imply:

$$\varrho D_t s \geq \frac{-\nabla \cdot q + \varrho g}{T} + \frac{1}{T^2} \nabla T \cdot q \,. \tag{5.19}$$

We now multiply (5.19) by $(-T)$ and add the result to the energy conservation law

$$\varrho D_t u = -\nabla \cdot q + \sigma : Dv + \varrho g \,.$$

This gives the assertion of the theorem. $\qquad\square$

**Theorem 5.9** *The following inequality for the free energy* $f := u - Ts$ *is true:*

$$\varrho D_t f + \varrho s D_t T - \sigma : Dv + \frac{1}{T} q \cdot \nabla T \leq 0 \,. \tag{5.20}$$

This statement is a simple consequence of Theorem 5.8.

*Consequences for Constitutive Relations*

In our discussion of the second law we will restrict ourselves to thermo-viscoelastic fluids. A *thermo-viscoelastic fluid* is a heat-conductive fluid with viscous friction. Models for thermo-viscoelastic fluids are based on the mass balance (5.3), the momentum balance (5.5), the energy balance (5.7) and constitutive laws that take viscous friction in the stress tensor into account. The constitutive relations have the form

$$u = \widehat{u}(\varrho, T, \nabla T, Dv) \,,$$
$$\sigma = \widehat{\sigma}(\varrho, T, \nabla T, Dv) \,,$$
$$s = \widehat{s}(\varrho, T, \nabla T, Dv) \,,$$
$$q = \widehat{q}(\varrho, T, \nabla T, Dv) \,.$$

Here $\widehat{u}, \widehat{\sigma}, \widehat{s}$ and $\widehat{q}$ are smooth functions. Taking the identity

$$f = u - Ts$$

into account it also holds

$$f = \widehat{f}(\varrho, T, \nabla T, Dv) := \widehat{u}(\varrho, T, \nabla T, Dv) - T\widehat{s}(\varrho, T, \nabla T, Dv).$$

### Coleman–Noll Procedure

The second law in the formulation as Clausius–Duhem inequality implies restrictions on the possible constitutive relations. These restrictions are summarized in the following theorem. In what follows, we denote the variables that correspond to $\nabla T$ and $Dv$ with $X$ and $Y$, i.e., we write $q = \widehat{q}(\varrho, T, X, Y)$. Partial derivatives of constitutive relations will be denoted by lower indices, for example as $\widehat{f},_T = \frac{\partial \widehat{f}}{\partial T}$.

**Theorem 5.10**  *Assume that the Clausius–Duhem inequality holds for all solutions of the conservation laws. In this case the constitutive relations necessarily have the form:*

$$\begin{aligned}
f &= \widehat{f}(\varrho, T), \\
s &= \widehat{s}(\varrho, T) = -\widehat{f},_T (\varrho, T), \\
\sigma &= -\widehat{p}(\varrho, T)I + \widehat{S}(\varrho, T, \nabla T, Dv), \\
\widehat{p}(\varrho, T) &= \varrho^2 \widehat{f},_\varrho (\varrho, T), \\
q &= \widehat{q}(\varrho, T, \nabla T, Dv).
\end{aligned}$$

*In addition it has to hold*

$$-\widehat{S}(\varrho, T, X, Y) : Y + \frac{1}{T}\widehat{q}(\varrho, T, X, Y) \cdot X \le 0$$

*for all $(\varrho, T, X, Y)$ with $\varrho, T > 0$.*

*Proof*  The inequality (5.20) for the free energy implies:

$$\varrho \widehat{f},_\varrho D_t \varrho + \varrho(\widehat{f},_T + \widehat{s})D_t T + \varrho \widehat{f},_X \cdot D_t \nabla T$$
$$+ \varrho \widehat{f},_Y : D_t Dv - \widehat{\sigma} : Dv + \frac{1}{T}\widehat{q} \cdot \nabla T \le 0.$$

Here $\widehat{f},_X$ and $\widehat{f},_Y$ denote the derivatives of $\widehat{f}(\varrho, T, X, Y)$ with respect to the third and fourth variable. The mass balance

$$D_t \varrho + \varrho \nabla \cdot v = 0$$

and the above inequality imply

$$\varrho(\widehat{f},_T + \widehat{s})D_t T + \varrho \widehat{f},_X \cdot D_t \nabla T + \varrho \widehat{f},_Y : D_t Dv$$
$$-(\widehat{\sigma} + \varrho^2 \widehat{f},_\varrho I) : Dv + \frac{1}{T}\widehat{q} \cdot \nabla T \le 0.$$

The above inequality has to hold for all solutions of the conservation laws. Using the fact that outer forces and heat sources and sinks can be chosen arbitrarily, we can choose arbitrary functions for $\varrho$, $T$, $\nabla T$, $Dv$ as solutions of the conservation laws. In a fixed point $(t, x)$ the values for $\varrho$, $T$, $\nabla T$, $Dv$, $D_t T$, $D_t \nabla T$, $D_t Dv$ can be chosen arbitrarily. This can be realized locally through a Taylor series. The fact that $\widehat{f}$ and $\widehat{s}$ are functions of $(\varrho, T, \nabla T, Dv)$ implies that the prefactors of the terms $D_t T$, $D_t \nabla T$, $D_t Dv$ have to vanish. Otherwise we would be able to violate the inequality by choosing $D_t T$, $D_t \nabla T$, $D_t Dv$ appropriately for fixed $(\varrho, T, \nabla T, Dv)$. If, for example $(\widehat{f}_{,T} + \widehat{s}) < 0$ would hold, we could choose $D_t T$ negative and large which would lead to a contradiction. We hence obtain

$$\widehat{f}_{,T} + \widehat{s} = 0\,, \quad \widehat{f}_{,X} = 0 \quad \text{and} \quad \widehat{f}_{,Y} = 0\,.$$

In particular we obtain that $\widehat{f}$ is a function of $T$ and $\varrho$, i.e., $\widehat{f} = \widehat{f}(T, \varrho)$. Setting $\widehat{p}(\varrho, T) = \varrho^2 \widehat{f}_{,\varrho}(\varrho, T)$, $\widehat{S} = \widehat{\sigma} + \widehat{p}I$ and using the fact that $(\varrho, T, \nabla T, Dv)$ can be chosen arbitrarily we obtain the inequality stated in the theorem.                                            □

**Remark.** (i) Theorem 5.10 yields that the stress tensor $\widehat{\sigma}$ can be decomposed into a pressure part and a part $\widehat{S} = \widehat{\sigma} + \widehat{p}I$ for which an inequality has to hold.
(ii) If $\widehat{\sigma}$ is independent of $\nabla T$ and $\widehat{q}$ is independent of $Dv$, we obtain

$$\widehat{S}(\varrho, T, Y) : Y \geq 0\,, \tag{5.21}$$

$$\widehat{q}(\varrho, T, X) \cdot X \leq 0 \tag{5.22}$$

for all $(\varrho, T, X, Y)$ with $\varrho, T > 0$.

### The Gibbs Identities and the Thermodynamic Potentials

The physical quantities which characterize the macroscopic states of a body are called thermodynamic quantities. We now want to introduce several relations between these quantities. These relations will allow it for example to change the variables which were chosen to describe the state of the system. This is often convenient and we refer in this context to the Sects. 3.5–3.7.

In the previous sections $\varrho$ and $T$ have been, besides the velocity $v$, the unknown quantities. With the help of the free energy $f$ we were able to determine the quantities $s$, $u$ and $p$. We have, writing these quantities as functions of $\varrho$ and $T$,

$$\widehat{u} = \widehat{f} + T\widehat{s}\,, \tag{5.23}$$

$$\widehat{s} = -\widehat{f}_{,T}\,, \tag{5.24}$$

$$\widehat{p} = \varrho^2 \widehat{f}_{,\varrho}(\varrho, T)\,. \tag{5.25}$$

In the following we will drop the $\widehat{\phantom{x}}$-notation. With the help of total derivatives (see Sect. 3.7) we can write the above identities (5.24) and (5.25) as follows

$$df = -s\, dT + \frac{p}{\varrho^2}\, d\varrho\,.$$

In some cases one would like to switch from the variables $(\varrho, T)$ to some other variables. We will now discuss two examples:

(a)  Using $d(\frac{1}{\varrho}) = -\frac{1}{\varrho^2}\,d\varrho$ and defining the *specific volume* $V = \frac{1}{\varrho}$, we obtain

$$df = -s\,dT - p\,dV\,.$$

(b)  Using the identity $d(Ts) = T\,ds + s\,dT$, we obtain

$$du = d(f + Ts) = df + d(Ts)$$
$$= -s\,dT - p\,dV + T\,ds + s\,dT = -p\,dV + T\,ds\,.$$

We hence obtain the following:

(a)  For $f = \widetilde{f}(T, V)$ we obtain $\widetilde{f},_T = -s$ and $\widetilde{f},_V = -p$.
(b)  For $u = u^*(V, s)$ we obtain $u^*,_V = -p$ and $u^*,_s = T$.

Another possibility to perform this change of variables uses the Legendre transformation from Sect. 3.6.

### Lagrange Multipliers

At the end of this section we will now discuss an example in which more general entropy fluxes appear. We consider a system at rest, i.e., $v = 0$, with a constant mass density $\varrho$ with two components. As independent variables we choose the temperature $T$ and the concentration $c = c_1$ of one of the components. In this specific system the energy balance

$$\varrho\,\partial_t u + \nabla \cdot q_u = \varrho\,r_u \tag{5.26}$$

holds with the energy flux $q_u$ and a source term $r_u$. In addition the concentration $c$ fulfills the balance equation

$$\varrho\,\partial_t c + \nabla \cdot q_c = \varrho\,r_c \tag{5.27}$$

with the flux $q_c$ and a source term $r_c$. We will now make no specific assumptions on the entropy flux $q_s$ and the entropy source term $r_s$. We require

$$\frac{d}{dt}\int_\Omega \varrho s\,dx \geq -\int_{\partial\Omega} q_s \cdot n\,ds_x + \int_\Omega \varrho r_s\,dx\,.$$

Since $\Omega$ can be chosen arbitrarily, we can deduce the local form

$$\varrho\,\partial_t s + \nabla \cdot q_s \geq \varrho\,r_s\,. \tag{5.28}$$

One possibility to use the entropy inequality without a priori assumptions on the entropy flux uses Lagrange multipliers and goes back to I-Shih Liu [91] and Ingo

Müller [102]. In this approach one subtracts from the inequality (5.28) $\lambda_u$ times the equality (5.26) and $\lambda_c$ times the equality (5.27). We then obtain

$$
\begin{aligned}
\varrho(\partial_t s - \lambda_u\,\partial_t u - \lambda_c\,\partial_t c) + \nabla \cdot q_s - \varrho\, r_s \\
- \lambda_u(\nabla \cdot q_u - \varrho\, r_u) - \lambda_c(\nabla \cdot q_c - \varrho\, r_c) \geq 0 \,.
\end{aligned}
\tag{5.29}
$$

We now assume that $s, u, \lambda_u, \lambda_c, q_u, q_c, q_s$ are arbitrary functions in $(T, c, \nabla T, \nabla c)$ and $r_u, r_c, r_s$ are functions that can depend on $x, t, T$ and $c$. Choosing

$$
r_s = \lambda_u r_u + \lambda_c r_c \,,
$$

we obtain that in the inequality (5.29) all terms which contain $r_u, r_c$ and $r_s$ vanish. The inequality (5.29) now gives

$$
\begin{aligned}
\varrho\big((s_{,T} - \lambda_u u_{,T})\partial_t T + (s_{,c} - \lambda_u u_{,c} - \lambda_c)\partial_t c + (s_{,\nabla T} - \lambda_u u_{,\nabla T})\partial_t \nabla T \\
+ (s_{,\nabla c} - \lambda_u u_{,\nabla c})\partial_t \nabla c\big) + \nabla \cdot q_s - \lambda_u \nabla \cdot q_u - \lambda_c \nabla \cdot q_c \geq 0 \,,
\end{aligned}
\tag{5.30}
$$

here an index $\nabla T$ refers to partial differentiation with respect to the variable related to $\nabla T$ and the same holds for $\nabla c$. As in the Coleman–Noll procedure we can choose the values of $T, c$ and all its derivatives freely. With the help of this observation we can now obtain that the factors of $\partial_t T, \partial_t c, \partial_t \nabla T$ and $\partial_t \nabla c$ have to vanish. We hence obtain

$$
s_{,T} - \lambda_u u_{,T} = 0 \,, \quad s_{,c} - \lambda_u u_{,c} - \lambda_c = 0 \,,
\tag{5.31}
$$
$$
s_{,\nabla T} - \lambda_u u_{,\nabla T} = 0 \,, \quad s_{,\nabla c} - \lambda_u u_{,\nabla c} = 0 \,.
\tag{5.32}
$$

Classically it holds $u_{,T} = T s_{,T}$, cf. Chap. 3, and therefore we choose

$$
\lambda_u = \frac{1}{T} \,.
$$

Defining the free energy $f := u - Ts$ and the chemical potential $\mu := f_{,c}$ we obtain from (5.31), (5.32)

$$
f_{,\nabla T} = f_{,\nabla c} = 0
$$

and

$$
\lambda_c = -\frac{\mu}{T} \,.
$$

The equations $f = u - Ts$ and $u_{,T} = T s_{,T}$ imply $s = -f_{,T}$. This fact and $u = f + Ts$ yield that also $s$ and $u$ do not depend on $\nabla T$ and $\nabla c$. Inequality (5.30) now implies

$$\nabla \cdot \left( q_s - \frac{1}{T} q_u + \frac{\mu}{T} q_c \right) + \nabla \left( \frac{1}{T} \right) \cdot q_u + \nabla \left( -\frac{\mu}{T} \right) \cdot q_c \geq 0 \,.$$

The choice

$$q_s = \frac{1}{T} q_u - \frac{\mu}{T} q_c$$

guarantees the entropy inequality if

$$\nabla \left( \frac{1}{T} \right) \cdot q_u + \nabla \left( -\frac{\mu}{T} \right) \cdot q_c \geq 0 \,. \tag{5.33}$$

The left hand side is the entropy production and (5.33) requires a positive entropy production. We now consider the derivative of the entropy with respect to the quantities which are conserved, i.e., we consider

$$ds = \frac{1}{T} du - \frac{\mu}{T} dc$$

which is equivalent to

$$\partial_u s = \frac{1}{T} \quad \text{and} \quad \partial_c s = -\frac{\mu}{T} \,.$$

The gradients $\nabla(\frac{1}{T})$ and $\nabla(-\frac{\mu}{T})$ are called thermodynamic forces and the entropy production is precisely the sum of products between the thermodynamic forces and the corresponding fluxes.

In the simplest case one postulates in thermodynamics a linear relation between the fluxes and the thermodynamic forces. We hence obtain for the energy flux and the flux of component 1

$$q_u := L_{11} \nabla(\tfrac{1}{T}) + L_{12} \nabla(-\tfrac{\mu}{T}) \,, \tag{5.34}$$

$$q_c := L_{21} \nabla(\tfrac{1}{T}) + L_{22} \nabla(-\tfrac{\mu}{T}) \,. \tag{5.35}$$

The Onsager reciprocity law implies the symmetry of the matrix $L := (L_{ij})_{i,j=1,2}$, see [102], and from (5.33) we obtain that $L$ necessarily has to be positive semidefinite. The fact that the species balance (5.27) has a right hand side is critical, as there is the possibility that components may not be produced or destroyed arbitrarily. In addition, we assume a specific form of the entropy flux and it might be possible that the entropy inequality is also satisfied with other entropy fluxes.

For a more concise discussion we refer, e.g., to the book of Müller [102] or to Alt and Pawlow [5]. The results presented above are a special case of the treatment in these references.

## 5.8   Principle of Frame Indifference

Two observers who observe the same body will record two different movements. For somebody who is moving with the body the body seems to be stationary, whereas somebody who is not moving with the body will notice a movement. An observer is characterized by a coordinate system with an origin at the present position of the observer and coordinate axes which follow from the orientation of the observer. Here we allow that the *coordinate system* changes in time. A change of observers hence corresponds to a change in the coordinate system. Such a change in the observer has certain effects on the basic variables in continuum mechanics. The way in which variables and equations transform will be discussed next.

We now consider two orthonormal, positive oriented coordinate systems in the $d$-dimensional Euclidean space,

$$(O, e_1, \ldots, e_d) \text{ and } (O^*, e_1^*, \ldots, e_d^*),$$

where $O$ and $O^*$ denote the coordinate origins and $e_j$, $e_j^*$ the coordinate vectors. Each point $P$ then has two representations

$$P = O + \sum_{j=1}^{d} x_j \, e_j = O^* + \sum_{j=1}^{d} x_j^* \, e_j^* \tag{5.36}$$

with respect to these two bases with coordinate vectors $x = (x_1, \ldots, x_d)^\top$ and $x^* = (x_1^*, \ldots, x_d^*)^\top$. The relations between these coordinates are

$$x^* = a + Qx$$

with a translational vector $a$ and an orthogonal matrix $Q$. We have

$$O - O^* = \sum_{j=1}^{d} a_j e_j^* \text{ and } e_j = \sum_{i=1}^{d} Q_{ij} e_i^*. \tag{5.37}$$

As both coordinate systems are positively oriented, we have $\det Q = 1$.

In the following we assume that $a$ and $Q$ are as functions of $t$ smooth enough. The coordinates $x$ and $x^*$ are interpreted as Eulerian coordinates over *the same* reference domain, i.e., (Fig. 5.9)

$$(t, X) \mapsto x(t, X),$$
$$(t, X) \mapsto x^*(t, X).$$

We have
$$x^* = F(t, x) \text{ with } F(t, x) = a(t) + Q(t)x.$$

reference configuration (the $X$–coordinates)

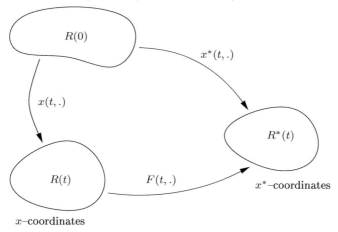

**Fig. 5.9** Observer change

The velocities $v$, $v^*$ in Eulerian coordinates which belong to these coordinates are defined via

$$v(t, x(t, X)) = \partial_t x(t, X) \text{ and}$$
$$v^*(t, x^*(t, X)) = \partial_t x^*(t, X).$$

With this we obtain

$$D_X x^* = D_x F D_X x = Q D_X x,$$
$$v^*(t, x^*) = \partial_t x^* = \partial_t a(t) + \partial_t Q(t) x + Q v(t, x),$$
$$n^* = Q n, \tag{5.38}$$
$$D_{x^*} v^* = \partial_t Q D_{x^*} x + Q D_{x^*} v$$
$$= \partial_t Q Q^\top + Q D_x v Q^\top.$$

The last identity follows with the help of

$$D_{x^*} x = Q^\top \text{ and } D_{x^*} v = D_x v D_{x^*} x = D_x v Q^\top.$$

The principle of frame indifference now makes an assertion on the way how scalar quantities such as the temperature $T$, vector quantities such as the heat flux $q$ and tensors such as the stress tensor $\sigma$ behave under a change of observer. Let $T$, $\varrho$, $q$, $b$, $\sigma$ and $T^*$, $\varrho^*$, $q^*$, $b^*$, $\sigma^*$ be the corresponding descriptions of the temperature, the density, the heat flux, the force density and the stress tensor. The principle of frame indifference requires

$$T(t, x) = T^*(t, x^*) = T^*(t, a(t) + Q(t)x),$$ (5.39)

$$q(t, x) = Q^\top q^*(t, x^*) = Q^\top q^*(t, a(t) + Q(t)x)$$ (5.40)

and

$$\sigma(t, x) = Q^\top \sigma^*(t, x^*)Q = Q^\top \sigma^*(t, a(t) + Q(t)x)Q.$$ (5.41)

Other scalar, vectorial and tensorial quantities change in the same way. Why do we require this? The mass and temperature at a certain material point does not change by a change of observers – just the independent variable changes. This is the reason for (5.39). The condition (5.40) follows from the representations

$$\sum_{j=1}^{d} q_j e_j = \sum_{j=1}^{d} q_j^* e_j^*$$

of the same vector in the Euclidean space through the two coordinate systems with coefficients $q_j$ and $q_j^*$ and the formula $e_j = \sum_{i=1}^{d} Q_{ij} e_i^*$. The condition (5.41) follows from (5.40) applied to the force densities $b = \sigma n$ and $b^* = \sigma^* n^*$, and the normal vectors $n$ and $n^*$:

$$\sigma Q^\top n^* = \sigma n = b = Q^\top b^* = Q^\top \sigma^* n^*.$$

### Isotropy

There are materials, in particular solids, for which properties depend on the "direction" in which the material is rotated, or equivalently, the response on an external influence, e.g., a mechanical loading, depends on the direction of this influence. An example is wood, where the response to stresses in the direction of the fiber or the grains is stronger as in directions orthogonal to the grains. Another example are fiber-reinforced composites such as glass fiber-reinforced plastics whose properties depend on the orientation of the fiber blanket. Such a material is called *anisotropic*. The reason for such anisotropies in continuum mechanics are typically microscopically small structures which one does not consider on the macroscopic scale. However, these microscopic properties enter the macroscopic models through constitutive relations. Materials, which properties *do not* depend on the direction, are called *isotropic*. The word stems from the ancient Greek words "isos" for "equal" and "tropos" for "rotation", "direction". In continuum mechanics an isotropic material is characterized by *constitutive laws* which do not depend on the observer. As an example we consider how for an isotropic material the heat flux depends on the temperature gradient:

$$q = \widehat{q}(\nabla T, \ldots).$$

By a change of observer the quantities $q$ and $\nabla T$ change to $q^* = Qq$ and $\nabla^* T^* = Q \nabla T$. However, the form of the constitutive relation does not change for an isotropic material, i.e.,

$$q^* = \widehat{q}(\nabla^* T^*, \ldots).$$

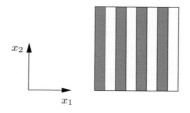

$x_2$

$x_1$

**Fig. 5.10** Layered material

For an anisotropic material the constitutive law in general changes, i.e.,

$$q^* = \widehat{q}^*(\nabla^*T^*, \ldots) \quad \text{with } \widehat{q}^* \neq \widehat{q}.$$

**Example**: We consider a constitutive law for the heat flux in two spatial dimension which is of the form

$$\widehat{q}(\nabla T) = \begin{pmatrix} \varepsilon & 0 \\ 0 & 1 \end{pmatrix} \nabla T$$

with a small parameter $\varepsilon$. Such a relation is for example possible for a layered material as in Fig. 5.10 for which the heat conductivity of the "white" material is very small. The heat flux in $x_2$-direction in this case takes place mainly in the "black" material. In total one obtains an averaged heat conductivity of order 1. On the other hand the heat flux in $x_1$-direction is hindered by the "white" material. This has the consequence that in this direction the heat conductivity is much lower. A change of observer from $x$ to $x^* = Qx$ with $Q = \begin{pmatrix} 0 & 1 \\ -1 & 0 \end{pmatrix}$ rotates the lamellas by 90°. In total we obtain the constitutive law

$$\widehat{q}^*(\nabla^*T^*) = \begin{pmatrix} 1 & 0 \\ 0 & \varepsilon \end{pmatrix} \nabla^*T^*.$$

*Consequences of Isotropy and Invariance Under a Change of Observer*

Isotropy and invariance under a change of observer have consequences for the form of constitutive laws. We will study this in different situations.

(a) We assume that the *stress tensor* is a function of density and temperature, i.e.,

$$\sigma = \widehat{\sigma}(\varrho, T)$$

and we will show that in this case isotropy and invariance under a change of observer implies

$$\sigma = -\widehat{p}(\varrho, T)I.$$

We now fix $(\varrho, T)$. Since $\sigma$ is symmetric we can conclude the existence of an orthogonal matrix $Q$ with $\det Q = 1$ and a diagonal matrix $\Lambda$, such that

$$\widehat{\sigma}(\varrho, T) = Q^\top \Lambda Q.$$

We now choose this $Q$ in the formulation of the invariance under change of observer. Using that under our assumptions the form of the constitutive law does not change under a change of observer we obtain

$$\sigma = Q^{\top} \sigma^* Q$$

and hence

$$Q^{\top} \Lambda Q = \widehat{\sigma}(\varrho, T) = Q^{\top} \widehat{\sigma}(\varrho^*, T^*) Q = Q^{\top} \widehat{\sigma}(\varrho, T) Q.$$

This implies

$$\widehat{\sigma}(\varrho, T) = \Lambda.$$

We now choose the special orthogonal matrices

$$Q = \begin{pmatrix} 0 & 1 & 0 \\ -1 & 0 & 0 \\ 0 & 0 & 1 \end{pmatrix}, \quad Q = \begin{pmatrix} 0 & 0 & 1 \\ 0 & 1 & 0 \\ -1 & 0 & 0 \end{pmatrix} \quad \text{and} \quad Q = \begin{pmatrix} 1 & 0 & 0 \\ 0 & 0 & 1 \\ 0 & -1 & 0 \end{pmatrix}$$

in the formulation of the principle of frame invariance and we can hence conclude that all diagonal arguments of $\Lambda$ have to coincide. We can hence conclude that a function $\widehat{p}(\varrho, T)$ exists, such that

$$\sigma = -\widehat{p}(\varrho, T) I.$$

(b) We now assume that the *heat flux* $q$ in an isotropic material depends on the density $\varrho$, the temperature $T$ and the temperature gradient $\nabla T$, i.e.,

$$q = \widehat{q}(\varrho, T, \nabla T).$$

The principle of frame invariance for an isotropic material requires

$$q^* = Qq$$

and hence

$$\widehat{q}(\varrho^*, T^*, \nabla_{x^*} T^*) = Q \widehat{q}(\varrho, T, \nabla T)$$

has to hold. The identity

$$\nabla_{x^*} T^* = Q \nabla_x T$$

implies

$$\widehat{q}(\varrho, T, Q \nabla T) = Q \widehat{q}(\varrho, T, \nabla T).$$

As $\nabla T$ can be chosen arbitrarily we need to require

$$\widehat{q}(\varrho, T, QX) = Q\widehat{q}(\varrho, T, X) \tag{5.42}$$

for all $\varrho$, $T$, $X$ and all rotations $Q$. With the help of these properties we can prove the following theorem.

**Theorem 5.11** *All constitutive laws which fulfill* (5.42) *can be expressed in the form*

$$\widehat{q}(\varrho, T, X) = -\widehat{k}(\varrho, T, |X|)X$$

**Remark.** This is a generalization of Fourier's law for an isotropic material.

*Proof* In the following we do not state the dependence of $\hat{q}$ on $(\varrho, T)$ explicitly. We need to show that $\hat{q}(a)$ and $a$ lie on a line. For all vectors $a \in \mathbb{R}^3$ we construct an orthogonal matrix $Q$ such that $x \mapsto Qx$ realizes a rotation with rotation axis $a$ by an angle of $\pi$. In this case $a$ is an eigenvector of $Q$ with the eigenvalue 1 and the corresponding eigenspace has the dimension 1. We observe

$$\hat{q}(a) = \hat{q}(Qa) = Q\hat{q}(a),$$

and hence $\hat{q}(a)$ is an eigenvector of $Q$ with eigenvalue 1. Thus

$$\hat{q}(a) = \alpha(a)a \quad \text{with } \alpha(a) \in \mathbb{R}.$$

For an arbitrary orthogonal matrix $Q$ with $\det Q = 1$ we have

$$\alpha(a)Qa = Q\hat{q}(a) = \hat{q}(Qa) = \alpha(Qa)Qa$$

and hence, if $a \neq 0$,

$$\alpha(Qa) = \alpha(a).$$

For two vectors $a$ and $b$ with the same length there always exists an orthogonal matrix $Q$ with $b = Qa$. From this we can conclude that $\alpha(a)$ does only depend on $|a|$.  □

## 5.9  Constitutive Theory for Viscous Fluids

In this section we want to discuss why the stress tensor for viscous fluids has the form (5.12). This is in particular important for the derivation of the Navier–Stokes equations. A fluid is *viscous* in cases where we cannot neglect *internal friction*. Such fluids are modeled by stress tensors with constitutive laws which allow for a dependence on $Dv$:

$$\sigma = \widehat{\sigma}(\varrho, T, Dv).$$

This reflects the experience that variations of the velocity in space leads to friction, see p. 224. We assume that the stress tensor is isotropic. This makes sense as liquids and gases typically have no microstructures which lead to anisotropies. The constitutive law

$$\sigma = \widehat{\sigma}(\varrho, T, Dv)$$

needs to be frame invariant and the constitutive function does not change from observer to observer. This is the case if and only if

$$\widehat{\sigma}(\varrho^*, T^*, D_{x^*}v^*) = Q\widehat{\sigma}(\varrho, T, D_x v)Q^\top$$

and taking (5.38) into account it holds

$$\widehat{\sigma}(\varrho, T, \partial_t Q\, Q^\top + Q D_x v\, Q^\top) = Q\widehat{\sigma}(\varrho, T, D_x v)Q^\top \,.$$

In the following we do not state the dependence on $(\varrho, T)$ explicitly. To make sure that the constitutive law for $\sigma$ does not depend on the observer we need to require

$$\widehat{\sigma}(\partial_t Q\, Q^\top + QAQ^\top) = Q\widehat{\sigma}(A)Q^\top \tag{5.43}$$

for all matrices $A \in \mathbb{R}^{d,d}$.

**Lemma 5.12**  *It holds*

$$\widehat{\sigma}(A) = \widehat{\sigma}\big(\tfrac{1}{2}(A + A^\top)\big)\,,$$

*i.e., the stress tensor only depends on the symmetric part of Dv.*

*Proof*  We choose an *anti-symmetric* matrix $W$, i.e., $W + W^\top = 0$ and define

$$Q(t) := e^{-tW} \quad \text{for } t \in \mathbb{R}\,.$$

The linear mappings $Q(t)$ are rotations. This follows since $WW^\top = W^\top W$, and hence $e^W e^{W^\top} = e^{W+W^\top}$ and we can compute

$$QQ^\top = e^{-tW} e^{-tW^\top} = e^{-t(W+W^\top)} = e^{-t0} = I\,,$$
$$(\det Q)^2 = \det\big(QQ^\top\big) = 1 \quad \text{and} \quad \det Q(0) = \det I = 1\,.$$

We hence have $\det Q(t) = 1$ for all $t \in \mathbb{R}$. Furthermore we have

$$\partial_t Q(t) = -W e^{-tW}$$

and

$$Q(0) = I\,, \quad \partial_t Q(0) = -W\,.$$

The formula (5.43) implies for $t = 0$

$$\widehat{\sigma}(-W + A) = \widehat{\sigma}(A)\,.$$

Setting $W = \frac{1}{2}(A - A^{\top})$, we conclude

$$\widehat{\sigma}\left(\tfrac{1}{2}(A + A^{\top})\right) = \widehat{\sigma}(A)$$

and hence the assertion is verified. □

It is possible to show that in three space dimension the constitutive function $\widehat{\sigma}$ needs to have a very specific structure. This is stated in the following famous theorem.

**Theorem 5.13** (Rivlin–Ericksen theorem) *A function*

$$\widehat{\sigma} : \{A \in \mathbb{R}^{3,3} \mid A \text{ is symmetric, } \det A > 0\}$$
$$\rightarrow \{B \in \mathbb{R}^{3,3} \mid B \text{ is symmetric}\}$$

*has the property*

$$\widehat{\sigma}(QAQ^{\top}) = Q\widehat{\sigma}(A)Q^{\top} \quad \text{for all rotations } Q, \tag{5.44}$$

*if and only if*

$$\widehat{\sigma}(A) = a_0(i_A)I + a_1(i_A)A + a_2(i_A)A^2. \tag{5.45}$$

*Here $a_0$, $a_1$ and $a_2$ are functions of the* principal invariants $i_A$ *of A*.

The *principal invariants* are the *coefficients of the characteristic polynomial* $\det(A - \lambda I)$. A matrix $A \in \mathbb{R}^{3,3}$ has the principal invariants, i.e., $i_A = (i_1(A), i_2(A), i_3(A))$ such that

$$\det(\lambda I - A) = \lambda^3 - i_1(A)\lambda^2 + i_2(A)\lambda - i_3(A).$$

If $\lambda_1, \lambda_2, \lambda_3$ are the eigenvalues of $A$ we compute

$$i_1(A) = \text{trace}(A) = \lambda_1 + \lambda_2 + \lambda_3,$$
$$i_2(A) = \tfrac{1}{2}\left((\text{trace } A)^2 - \text{trace } A^2\right) = \lambda_1\lambda_2 + \lambda_1\lambda_3 + \lambda_2\lambda_3,$$
$$i_3(A) = \det(A) = \lambda_1\lambda_2\lambda_3.$$

If all principal invariants of a matrix are the same we also know that all eigenvalues are the same. We now consider $\sigma = -\widehat{p}I + \widehat{S}$. If a constitutive law for $\widehat{S}$ is not only invariant under a change of observer and isotropic but also *linear*, we know from the Rivlin–Ericksen theorem or (5.45) that $\widehat{S}$ can only dependent on *two* parameters. We necessarily obtain the form

$$\widehat{S}(A) = 2\mu A + \lambda \, \text{trace}(A)I, \tag{5.46}$$

see Exercise 5.32. Here $\mu$ is called *shear viscosity* and the quantity $\mu + \frac{2}{3}\lambda$ is called volume viscosity. In elasticity theory $\mu$ and $\lambda$ are called Lamé coefficients. A fluid with such a stress tensor is called a *Newtonian fluid*.

*Proof of the Theorem of Rivlin–Ericksen.* We first show that (5.45) implies (5.44). This follows directly from the following identities

$$
\begin{aligned}
Q(a_0 I + a_1 A + a_2 A^2)Q^\top &= a_0 I + a_1 Q A Q^\top + a_2 Q A A Q^\top \\
&= a_0 I + a_1 Q A Q^\top + a_2 Q A Q^\top Q A Q^\top \,,
\end{aligned}
$$

noting that $A$ and $QAQ^\top$ have the same eigenvalues. It remains to show that (5.44) implies (5.45). Let $A$ be a symmetric matrix and let

$$
A = Q \Lambda Q^\top
$$

be the transformation of A to a diagonal form with a diagonal matrix $\Lambda$ and an orthogonal matrix $Q$ with $\det Q = 1$. Under the assumption that (5.45) is true for diagonal matrices we obtain

$$
\begin{aligned}
\widehat{\sigma}(A) = \widehat{\sigma}(Q \Lambda Q^\top) = Q \widehat{\sigma}(\Lambda) Q^\top &= Q\big(i_0(\Lambda)I + i_1(\Lambda)\Lambda + i_2(\Lambda)\Lambda^2\big)Q^\top \\
&= i_0(A)I + i_1(A)A + i_2(A)A^2 \,.
\end{aligned}
$$

This follows since the $i_j(A), j = 1, 2, 3$ only depends on the eigenvalues of $A$ and hence we have $i_j(A) = i_j(\Lambda)$. It is hence enough to show the assertion for *diagonal matrices*. Let

$$
A = \begin{pmatrix} \lambda_1 & 0 & 0 \\ 0 & \lambda_2 & 0 \\ 0 & 0 & \lambda_3 \end{pmatrix}
$$

with $\lambda_j \in \mathbb{R}$ and $\lambda = (\lambda_1, \lambda_2, \lambda_3)$. We show that $\widehat{\sigma}(A)$ has to be a diagonal matrix. With

$$
Q = \begin{pmatrix} 1 & 0 & 0 \\ 0 & -1 & 0 \\ 0 & 0 & -1 \end{pmatrix}
$$

it holds $Q^\top = Q$, $Q^\top Q = I$, $\det Q = 1$ and

$$
\widehat{\sigma}(A)e_1 = Q Q^\top \widehat{\sigma}(A) Q e_1 = Q \widehat{\sigma}(Q^\top A Q)e_1 = Q \widehat{\sigma}(A)e_1 \,.
$$

Hence $\widehat{\sigma}(A)e_1$ is an eigenvector of $Q$ with eigenvalue 1. The eigenspace to this eigenvalue is spanned by $e_1$ and this implies

$$
\widehat{\sigma}(A)e_1 = t_1(\lambda)\,e_1
$$

with a suitable function $t_1$ which depends on $\lambda$. Analogously one can show

$$\widehat{\sigma}(A)e_j = t_j(\lambda)\,e_j \quad \text{for } j = 2, 3\,.$$

Altogether we obtain

$$\widehat{\sigma}(A) = \begin{pmatrix} t_1(\lambda) & 0 & 0 \\ 0 & t_2(\lambda) & 0 \\ 0 & 0 & t_3(\lambda) \end{pmatrix}\,.$$

In a next step we show that the permutation of the $\lambda_j$ implies an equivalent permutation of the $t_j$. This means precisely that for each permutation $\pi$ of $\{1, 2, 3\}$ it holds

$$t_{\pi(j)}(\lambda_{\pi(1)}, \lambda_{\pi(2)}, \lambda_{\pi(3)}) = t_j(\lambda_1, \lambda_2, \lambda_3)\,. \tag{5.47}$$

It is sufficient to show this for permutations which interchange two numbers. For the permutation $(1, 2, 3) \to (2, 1, 3)$ it follows with

$$Q = \begin{pmatrix} 0 & 1 & 0 \\ 1 & 0 & 0 \\ 0 & 0 & -1 \end{pmatrix}$$

and (5.44) that

$$\widehat{\sigma}\begin{pmatrix} \lambda_2 & 0 & 0 \\ 0 & \lambda_1 & 0 \\ 0 & 0 & \lambda_3 \end{pmatrix} = Q \begin{pmatrix} t_1(\lambda_2, \lambda_1, \lambda_3) & 0 & 0 \\ 0 & t_2(\lambda_2, \lambda_1, \lambda_3) & 0 \\ 0 & 0 & t_3(\lambda_2, \lambda_1, \lambda_3) \end{pmatrix} Q^{\mathsf{T}}$$

$$= \begin{pmatrix} t_2(\lambda_1, \lambda_2, \lambda_3) & 0 & 0 \\ 0 & t_1(\lambda_1, \lambda_2, \lambda_3) & 0 \\ 0 & 0 & t_3(\lambda_1, \lambda_2, \lambda_3) \end{pmatrix}\,.$$

Analogous one can show (5.47) for $(1, 2, 3) \to (3, 2, 1)$ and $(1, 2, 3) \to (1, 3, 2)$.

In the last step of the proof we show that (5.45) is true for appropriate functions $a_j = a_j(\lambda)$. This is equivalent to

$$t_j(\lambda) = a_0(\lambda) + a_1(\lambda)\lambda_j + a_2(\lambda)\lambda_j^2 \quad \text{for all } j = 1, 2, 3\,. \tag{5.48}$$

In the case that all eigenvalues are different we can determine the coefficients of $a_j$ with the help of polynomial interpolation. In case that two eigenvalues coincide, for example $\lambda_1 = \lambda_2 \neq \lambda_3$, we deduce from (5.47)

$$t_1(\lambda) = t_1(\lambda_1, \lambda_2, \lambda_3) = t_2(\lambda_2, \lambda_1, \lambda_3) = t_2(\lambda_1, \lambda_2, \lambda_3) = t_2(\lambda)\,.$$

In this case we obtain (5.48) with $a_2(\lambda) = 0$ and $a_0, a_1$ is given by

$$t_j(\lambda) = a_0(\lambda) + a_1(\lambda)\lambda_j \quad \text{for } j = 1, 3\,.$$

In the case that all eigenvalues are the same we have $t_1 = t_2 = t_3$ and

$$\widehat{\sigma}(A) = a_0(\lambda)I \ \text{ with } \ a_0(\lambda) = t_1(\lambda).$$

□

## 5.10   Modeling of Elastic Solids

Solids differ from fluids by the fact that the atoms and molecules are ordered in a lattice. A deformation of the solid deforms the lattice. However, the neighbor relations of the atoms and molecules are not effected. The restoring forces in the deformed state are generated by *displacements* of the molecules in the solid lattice and by the resulting intermolecular attractive and repulsive forces. This is true at least for elastic solids for which the exerted forces are not so large that they lead to *plastic* deformations or cracks and fractures. For elastic deformations it is hence meaningful to choose a description in *Lagrangian* coordinates. This is true since in Lagrangian coordinates it is easier to retrace the "original" neighbor relations in the solid lattice. In Fig. 5.11 we illustrate Lagrangian coordinates $x$ and Eulerian coordinates $y$. The notation is slightly changed in comparison to our earlier presentation as in elasticity theory it is common to use lower case letters for Lagrangian coordinates. The Lagrangian coordinates $x$ denote the position of a material point in the *reference configuration*. Typically this is the configuration a solid attains if no outer forces act. The Eulerian coordinates $y$ denote the position of the material point in the deformed state at time $t$.

The deformation of $\Omega$ at a time $t$ can be described by the *deformation field*

$$y : (t, x) \mapsto y(t, x)$$

or by the *displacement field*

$$u : (t, x) \mapsto u(t, x) = y(t, x) - x.$$

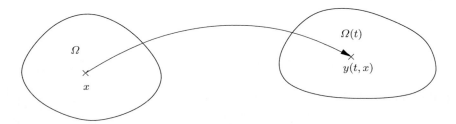

**Fig. 5.11**  Reference configuration, Lagrangian and Eulerian coordinates

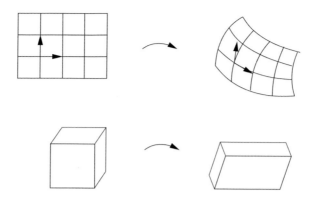

**Fig. 5.12** An illustration of the deformation gradient

A local measure for the deformation is given by the *deformation gradient*

$$Dy(t, x) = \left( \frac{\partial y_i}{\partial x_j}(t, x) \right)_{i,j=1}^d .$$

The columns of the deformation gradient consist of the tangential vectors at the image of the coordinate lines in the deformed state. This fact is visualized in Fig. 5.12.

The forces which act in the deformed elastic body depend on the length changes induced by the deformation. On the atomic scale these forces lead to displacements of the positions of atoms or molecules in the lattice of the solid. In order to describe the length changes we consider two points $x$ and $x + a$ with a small distance $a$. The distance before the deformation is $|a|$. The distance in the deformed state we can approximate with the help of a Taylor expansion as follows

$$|y(x + a) - y(x)| \approx |Dy(x)a| = \left( a^\top (Dy(x))^\top Dy(x) \, a \right)^{1/2} .$$

The matrix

$$C := (Dy)^\top Dy = (I + Du)^\top (I + Du)$$

is called *Cauchy–Green strain tensor*. This matrix describes local length changes. The matrix

$$G := \tfrac{1}{2}(C - I) = \tfrac{1}{2}\big((Dy)^\top Dy - I\big) = \tfrac{1}{2}\big(Du + (Du)^\top + (Du)^\top Du\big)$$

is called *Green* or *Green–St.Venant* strain tensor. A deformation *without* changes in the distance of points is characterized by

$$C = I \quad \text{and} \quad G = 0 .$$

Examples of such deformations are

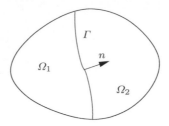

**Fig. 5.13** Illustration for the definition of the stress tensor

Translations: $y(x) = x + a$ or $u(x) = a$ with a translation vector $a \in \mathbb{R}^d$.
Rotations: $y(x) = x_M + Q(x - x_M)$ with a point $x_M$ on the rotation axis and a rotation matrix $Q$, which is an orthogonal matrix with determinant 1. The related displacement field is $u(x) = (Q - I)(x - x_M)$.

A general *rigid body motion* is given by

$$y(x) = Qx + a \ \text{ and } \ u(x) = (Q - I)x + a.$$

The deformation gradient of a rigid body motion is $Dy = Q$, the strain tensors are $C = I$ and $G = 0$.

### The Stress Tensor

The stress tensor describes how surface forces act on the deformed body. As discussed in Sect. 5.5 the stress tensor can be formulated in Eulerian coordinates. However, for applications in the context of elastic solids a description in Lagrangian coordinates is more appropriate. We consider a body which fills in the reference configuration a domain $\Omega \subset \mathbb{R}^d$. With the help of a surface $\Gamma$ we divide the domain in two parts $\Omega_1$ and $\Omega_2$, see Fig. 5.13. If the body deforms, forces act inside the body. In particular the *deformed* body $y(\Omega_2)$ exerts a force along the *deformed* surface $y(\Gamma)$ on the *deformed* subdomain $y(\Omega_1)$. This force is given by

$$F_{\Omega_2 \to \Omega_1} = \int_\Gamma \sigma n \, ds_x.$$

This formula looks at the first glance identical to the one in Eulerian coordinates. However, the formula has to be interpreted in a different way: The stress tensor $\sigma$ is defined on the reference configuration $\Omega$ and the vector $n$ is the normal vector of $\Gamma$ in the reference configuration. The vector $\sigma n$ is a force density. With this interpretation $\sigma$ is called the *first Piola–Kirchhoff stress tensor*. In the literature one finds also different stress tensors. In particular, also for elastic solids a formulation of the stress tensor in the deformed domain is considered. This stress tensor is called *Cauchy stress tensor*.

### The Equations of Elasticity

The differential equations for elastic solids follow from the *momentum balance law*. In the reference configuration the momentum balance is given as follows

$$\frac{d}{dt} \int_{\Omega} \varrho \, \partial_t u \, dx = \int_{\Omega} f \, dx + \int_{\partial \Omega} \sigma n \, ds_x \, .$$

Here $\varrho$ is the mass density with respect to the reference configuration and $f$ is a force density *per unit volume* in the reference configuration. We point out here that $\varrho$ does not depend on $t$. An application of the divergence theorem leads to

$$\int_{\Omega} \left( \varrho \, \partial_t^2 u - \nabla \cdot \sigma - f \right) dx = 0$$

and as this equation has to be true also for all subdomains of $\Omega$ we obtain the formulation

$$\varrho \, \partial_t^2 u - \nabla \cdot \sigma = f \, . \tag{5.49}$$

Since we considered a formulation in Lagrangian coordinates it was not necessary this time to use the Reynolds transport theorem. Therefore one also does not need a convective term in the differential equations. For a complete description of the elastic deformation we need a *constitutive law* to determine the stress tensor from the displacement field $u$, or alternatively from the deformation $y$. Characteristic for an *elastic* material is that the material deforms under stress and that this displacement disappears after the stress has vanished. The work performed in the deformation is stored in the deformed body as *elastic energy*. From the energy balance law we can now deduce a relation between the stress tensor and the elastic energy. Let $e$ be the density of the elastic energy per volume. Then the energy balance can be stated as

$$\frac{d}{dt} \int_{\Omega} \left( \tfrac{1}{2} \varrho |\partial_t u|^2 + e \right) dx = \int_{\Omega} f \cdot \partial_t u \, dx + \int_{\partial \Omega} \sigma n \cdot \partial_t u \, ds_x \, .$$

Here $\tfrac{1}{2} \varrho |\partial_t u|^2$ is the density of the kinetic energy per volume, $\int_{\Omega} f \cdot \partial_t u \, dx$ is the work of the volume forces and $\int_{\partial \Omega} \sigma n \cdot \partial_t u \, ds_x$ is the work of the surface forces. Since the domain $\Omega$ does not depend on time we can interchange time derivative and integration. An application of the divergence theorem yields

$$\int_{\Omega} \left( \varrho \, \partial_t u \cdot \partial_t^2 u + \partial_t e - f \cdot \partial_t u - (\nabla \cdot \sigma) \cdot \partial_t u - \sigma : \partial_t Du \right) dx = 0 \, .$$

Since the momentum balance equation holds all terms disappear with the exception of $\partial_t e - \sigma : \partial_t Du$. We hence obtain

$$\partial_t e - \sigma : \partial_t Du = 0 \, .$$

This equation has to hold for all possible evolutions. We can hence argue as in the context of the Coleman–Noll procedure and obtain that $e$ can only depend on $Du$. We now denote by $X = (X_{ij})_{i,j=1}^{d}$ the variables which are related to the matrix $Du$. We hence obtain

$$\sum_{i,j=1}^{d} \frac{\partial e}{\partial X_{ij}}(Du)\partial_t \frac{\partial u_i}{\partial x_j} - \sigma : \partial_t Du = 0.$$

Using again that one can choose arbitrary displacement fields we can conclude

$$\sigma_{ij}(u) = \frac{\partial e}{\partial X_{ij}}(Du).$$

We will call a material *elastic* if the stress tensor only depends on the space variable $x$ and on the displacement gradient $Du$,

$$\sigma = \sigma(x, Du(x)),$$

and if an energy function $e = e(x; Du)$ exists with

$$\sigma_{ij}(u) = \frac{\partial e}{\partial X_{ij}}(Du).$$

**Remark.** Here we considered besides mechanical forces no other phenomena which may lead to a deformation of the body or which could contribute to the energy balance. In particular, we did not consider a deformation of the body caused by temperature gradients. Strictly speaking this model is hence only suitable for *isothermal* situations.

### *Linear Elastic Materials*

In order to obtain a *linear* model for elastic materials one has to consider *two* linearizations. This is similar as in Sect. 2.2 in the case of an elastic frame. In the *geometric linearization* we linearize Green's strain tensor with respect to $Du$:

$$G = G(u) = \frac{1}{2}\left(Du + (Du)^\top + (Du)^\top Du\right) \approx \frac{1}{2}\left(Du + (Du)^\top\right).$$

The tensor

$$\varepsilon(u) = \frac{1}{2}\left(Du + (Du)^\top\right)$$

with components

$$\varepsilon_{ij}(u) = \tfrac{1}{2}\left(\partial_{x_i} u_j + \partial_{x_j} u_i\right)$$

is called *linearized strain tensor*. In practical applications it is often reasonable to assume small strains. In fact in many applications of the elasticity theory the deformation of the solid is so small that it cannot be observed with an "unaided eye". However, the forces acting can be relatively large in the same situation. Beside this geometric linearization one also needs a linear material law, i.e., a linear relation between $\sigma$ and $\varepsilon(u)$:

$$\sigma_{ij}(u) = \sum_{k,\ell=1}^{d} a_{ijk\ell}\, \varepsilon_{k\ell}(u) \ \text{ for } \ i,j = 1, \ldots, d\,. \tag{5.50}$$

This relation is called *Hooke's law*, and the tensor of fourth rank $A = (a_{ijk\ell})_{i,j,k,\ell=1}^{d} \in \mathbb{R}^{d,d,d,d}$ is called *Hooke's tensor*. This law follows from a quadratic ansatz function for the elastic energy density

$$e(u) = \sum_{i,j,k,\ell=1}^{d} \tfrac{1}{2}\, a_{ijk\ell}\, \varepsilon_{ij}(u)\, \varepsilon_{k\ell}(u)\,.$$

The coefficients $a_{ijk\ell}$ fulfill the symmetry conditions

$$a_{ijk\ell} = a_{jik\ell} = a_{k\ell ij}\,. \tag{5.51}$$

This is *no* restriction on the elastic energy density. In three dimensions $A$ has in general 21 different parameters. The linearized strain tensor has six independent components and a quadratic form in $\mathbb{R}^6$ is specified with the help of 21 independent coefficients. In two spatial dimensions $\varepsilon$ has three components and $A$ has six coefficients. To make sure that the energy density is positive one needs to require

$$\sum_{i,j,k,\ell=1}^{d} a_{ijk\ell}\xi_{ij}\xi_{k\ell} \geq a_0|\xi|^2 \ \text{ with } \ a_0 > 0 \tag{5.52}$$

for all symmetric matrices $\xi \in \mathbb{R}^{d,d}$ with $|\xi|^2 = \sum_{i,j=1}^{d} |\xi_{ij}|^2$. For an *isotropic* material we obtain from the principle of frame indifference and the Rivlin–Ericksen theorem, see (5.46),

$$\sigma_{ij}(u) = \lambda\,(\nabla \cdot u)\,\delta_{ij} + 2\mu\,\varepsilon_{ij}(u)$$

with the *Lamé constants* $\lambda$ and $\mu$. The corresponding Hooke's tensor is given by

$$a_{ijk\ell} = \lambda\,\delta_{ij}\delta_{k\ell} + \mu\,(\delta_{ik}\delta_{j\ell} + \delta_{i\ell}\delta_{jk})\,.$$

Instead of the Lamé constants one often also uses the *elastic modulus* (Young's modulus) $E$ and Poisson's ratio $\nu$:

$$E = \frac{\mu(3\lambda + 2\mu)}{\lambda + \mu}\,, \quad \nu = \frac{\lambda}{2(\lambda + \mu)}\,,$$

or respectively

$$\lambda = \frac{E\nu}{(1+\nu)(1-2\nu)}\,, \quad \mu = \frac{E}{2(1+\nu)}\,.$$

In the section on elastostatics we will understand the meaning of the quantities $E$ and $\nu$ a little better. The differential equations of linear elasticity are

$$\varrho\,\partial_t^2 u_i - \sum_{j,k,\ell=1}^{d} \partial_{x_j}(a_{ijk\ell}\varepsilon_{k\ell}) = f_i \text{ for } i = 1, \ldots, d.$$

A material is called *homogeneous* if all material parameters do not depend on the space variable $x$. For an isotropic, homogeneous material it holds

$$\varrho\,\partial_t^2 u - \mu\,\Delta u - (\lambda + \mu)\nabla\nabla \cdot u = f \tag{5.53}$$

in the formulation with the Lamé constants or

$$\varrho\,\partial_t^2 u - \frac{E}{2(1+\nu)}\Delta u - \frac{E}{2(1+\nu)(1-2\nu)}\nabla\nabla \cdot u = f \tag{5.54}$$

with the elastic modulus and Poisson's ratio.

### Elastostatics

In the static limit all time derivatives are zero, i.e., $\partial_t^2 u = 0$, and we obtain the equations of static elasticity theory

$$-\nabla \cdot \sigma(u) = f. \tag{5.55}$$

For a homogeneous, isotropic material we obtain the system

$$-\frac{E}{2(1+\nu)}\Delta u - \frac{E}{2(1+\nu)(1-2\nu)}\nabla\nabla \cdot u = f.$$

To illustrate the meaning of the constants $E$ and $\nu$, we now construct a solution of these equations for a cylinder

$$\Omega = \left\{x \in \mathbb{R}^3 \,|\, 0 < x_1 < L, \ \sqrt{x_2^2 + x_3^2} \leq R\right\}$$

with a force $f = 0$ and with the boundary conditions

$$\sigma(u)(-e_1) = -e_1 \text{ for } x_1 = 0, \quad \sigma(u)e_1 = e_1 \text{ for } x_1 = L$$

$$\text{and } \sigma(u)n = 0 \text{ for } \sqrt{x_2^2 + x_3^2} = R.$$

This means that at the left-hand side and right-hand side of the cylinder a unit force is pulling and on the curved sides no force is acting. We now use the ansatz

$$u(x) = \begin{pmatrix} a_1 x_1 \\ a_2 x_2 \\ a_2 x_3 \end{pmatrix}$$

with coefficients $a_1$ and $a_2$ to be determined. Since $\nabla \nabla \cdot u = 0$ and $\Delta u = 0$, we obtain that the differential equation is fulfilled. The stress tensor is

$$\sigma(u) = \frac{E}{2(1+\nu)}(Du + (Du)^\top) + \frac{E\nu}{(1+\nu)(1-2\nu)}\nabla \cdot u\, I$$

$$= \frac{E}{1+\nu}\left[\begin{pmatrix} a_1 & 0 & 0 \\ 0 & a_2 & 0 \\ 0 & 0 & a_2 \end{pmatrix} + \frac{\nu}{1-2\nu}\begin{pmatrix} a_1 + 2a_2 & 0 & 0 \\ 0 & a_1 + 2a_2 & 0 \\ 0 & 0 & a_1 + 2a_2 \end{pmatrix}\right].$$

From $\sigma(u)e_1 = e_1$ we deduce

$$a_1 + \frac{\nu}{1-2\nu}(a_1 + 2a_2) = \frac{1+\nu}{E},$$

and from $\sigma(u)e_2 = \sigma(u)e_3 = 0$ we obtain

$$a_2 + \frac{\nu}{1-2\nu}(a_1 + 2a_2) = 0.$$

The solution of this linear system can be computed as

$$a_1 = \frac{1}{E} \quad \text{and} \quad a_2 = -\frac{\nu}{E}.$$

Hence the displacement field is given as

$$u(x) = \frac{1}{E}\begin{pmatrix} x_1 \\ -\nu x_2 \\ -\nu x_3 \end{pmatrix}.$$

This means that the length of the cylinder is stretched by a factor of $\frac{1}{E}$ and at the same time the radius is reduced by a factor $\frac{\nu}{E}$, see Fig. 5.14. The force acting on the cylinder equals the stress on the boundary multiplied by the area of the boundary. The elastic modulus is hence the force per area with which one needs to pull on the sides in order to double the length of the cylinder. Poisson's ratio states by which fraction the radius is reduced at the same time. We hence observe that the elastic modulus in this section corresponds to the elastic modulus in the context of frames, see Sect. 2.2.

### Plane Strain Conditions and Plane Stress Conditions

In some applications one can use available *symmetries* to reduce the spatial dimension in which we need to solve the underlying differential equations. The most prominent

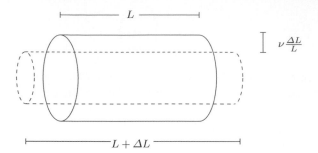

**Fig. 5.14** The meaning of the elastic modulus and of Poisson's ratio. Here we have $\Delta L = \frac{F}{EA}L$ with the force $F$ acting on the sides of the cylinder and the cross-sectional area $A$

examples in elasticity theory are plane stress and plane strain conditions for homogeneous, isotropic elastic materials. To obtain *plane strain conditions* one assumes that the strain tensor $\varepsilon$ is constant in one spatial dimension. Without loss of generality we choose the direction to be the $x_3$-direction and assume $\varepsilon_{i3}(u) = 0$ for $i = 1, 2, 3$. One obtains plane strain conditions in particular for infinitely long cylinders with a cylinder axis in the $x_3$-direction or for finitely long cylinders with boundary conditions $\sigma_{13} = \sigma_{23} = 0$ and $u_3 = 0$ on the "upper" and "lower" boundary surfaces, and in addition the acting volume and surface forces have to have components in the $x_1$- and $x_2$-directions. In practical situations one also uses the plane strain condition case for finitely long cylinders which are kept fixed at the top and the bottom. Plane strain conditions are in this case approximately fulfilled if the distance between the top and bottom boundary surfaces is large. The components of the stress tensor

$$\sigma = \lambda \operatorname{trace}(\varepsilon)I + 2\mu\varepsilon \tag{5.56}$$

in this case have the form

$$\sigma_{ij} = \lambda(\varepsilon_{11} + \varepsilon_{22})\delta_{ij} + 2\mu\,\varepsilon_{ij} \text{ for } i, j = 1, 2,$$
$$\sigma_{i3} = \sigma_{3i} = 0 \text{ for } i = 1, 2 \text{ and}$$
$$\sigma_{33} = \lambda(\varepsilon_{11} + \varepsilon_{22}).$$

Please note that here the component $\sigma_{33}$ in general will *not* be zero. Since $\varepsilon$ and hence also $\sigma$ does not depend on $x_3$, we obtain $\partial_{x_3}\sigma_{33} = 0$ and hence

$$(\nabla \cdot \sigma)_i = \sum_{j=1}^{2} \partial_{x_j}\sigma_{ij} = (\lambda + \mu)\partial_{x_i}(\partial_{x_1}u_1 + \partial_{x_2}u_2) + \mu\,\Delta u_i$$

for $i = 1, 2$. We now denote with $\nabla_2 = (\partial_{x_1}, \partial_{x_2})^\top$ the gradient and with $\Delta_2 = \partial_{x_1}^2 + \partial_{x_2}^2$ the Laplace operator with respect to the two dimensional spatial variables $(x_1, x_2)^\top$. With this notation we obtain for the plane strain condition the partial differential equations

$$\varrho\,\partial_t^2 \bar{u} - (\lambda + \mu)\nabla_2\nabla_2 \cdot \bar{u} - \mu\,\Delta_2\bar{u} = f$$

for the components $\bar{u} = (u_1, u_2)^\top$ of the displacement field. This corresponds to the two dimensional version of the elasticity equations (5.53) or (5.54) with the same values for the Lamé constants or respectively for the elastic modulus and the Poisson ratio as in three dimensions.

In *plane stress conditions* one assumes that the stress tensor does not depend on $x_3$ and that $\sigma_{i3} = 0$ for $i = 1, 2, 3$. From (5.56) we deduce first of all

$$\varepsilon_{13} = \varepsilon_{23} = 0 \ \text{ and } \ \varepsilon_{33} = -\frac{\lambda}{2\mu}\,\text{trace}(\varepsilon)$$

and

$$\varepsilon_{33} = -\frac{\lambda}{\lambda + 2\mu}(\varepsilon_{11} + \varepsilon_{22})\,. \tag{5.57}$$

In addition it follows

$$\sigma_{ij} = \frac{2\mu\lambda}{\lambda + 2\mu}(\varepsilon_{11} + \varepsilon_{22})\delta_{ij} + 2\mu\,\varepsilon_{ij} \ \text{ for } i, j = 1, 2\,,$$

which corresponds to the two dimensional version of Hooke's law with a modified Lamé constant

$$\tilde{\lambda} = \frac{2\mu\lambda}{\lambda + 2\mu}\,.$$

The equations of elasticity for plane stress conditions are given by

$$\varrho\,\partial_t^2 \bar{u} - \frac{2\mu^2 + 3\lambda\mu}{\lambda + 2\mu}\,\nabla_2\nabla_2 \cdot \bar{u} - \mu\,\Delta_2\bar{u} = f \tag{5.58}$$

or

$$\varrho\,\partial_t^2 \bar{u} - \frac{E}{2(1 - \nu)}\,\nabla_2\nabla_2 \cdot \bar{u} - \frac{E}{2(1 + \nu)}\,\Delta_2\bar{u} = f\,.$$

From these equations one can compute the components $u_1$ and $u_2$ of the displacement field. For the third component we obtain from (5.57)

$$\partial_{x_3}u_3 = -\frac{\lambda}{\lambda + 2\mu}(\varepsilon_{11} + \varepsilon_{22}) = -\frac{\nu}{1 - \nu}(\partial_{x_1}u_1 + \partial_{x_2}u_2)\,.$$

We obtain in particular that the displacement in the $x_3$-direction is different from zero. Plane stress conditions are appropriate for plates which are thin in the $x_3$-direction, and extended in the $x_1$- and $x_2$-directions, for which forces act only in the $x_1$- and $x_2$-direction. Actually Eq. (5.58) is part of *Kirchhoff's plate model*. This model not only describes loads acting along the $x_1$- and $x_2$-directions but also bending loads.

*Hyperelastic Materials*

Classical examples for nonlinear elasticity are *hyperelastic* materials. A material is called hyperelastic if an energy density function

$$W : \Omega \times M_+^d \to \mathbb{R}$$

exists. Here $M_+^d$ is the set of all regular matrices with positive determinant such that the elastic energy stored in the deformed body is given by

$$\int_\Omega W(x, Dy(x)) \, dx$$

and $y(x) = x + u(x)$ is the deformation field. The stress tensor in a hyperelastic material is given as (in the following we do not explicitly state the $x$-dependence)

$$\sigma_{ij} = \frac{\partial W}{\partial X_{ij}}(Dy),$$

where $\frac{\partial}{\partial X_{ij}}$ is the partial derivative of $W$ with respect to the component $X_{ij} = \partial_{x_j} y_i$ of the deformation gradient $X = Dy$. Meaningful properties of the energy functional are:

- The *frame invariance*
$$W(QF) = W(F)$$

  for all orthogonal matrices $Q \in \mathbb{R}^{d,d}$ with $\det Q = 1$ and all regular matrices $F \in \mathbb{R}^{d,d}$ with $\det F > 0$.
- The property
$$W(F) \to +\infty \quad \text{for } \det F \to 0.$$

  This condition means that a total compression of a material volume to the volume 0 would have an infinite energy density.
- An isotropic material is characterized by an energy functional which fulfills

$$W(FQ) = W(F)$$

  for all orthogonal matrices $Q \in \mathbb{R}^{d,d}$ with $\det Q = 1$ and all regular matrices $F \in \mathbb{R}^{d,d}$ with $\det F > 0$.

  Examples for hyperelastic materials are

- St. Venant–Kirchhoff materials:

$$W(F) = \frac{\lambda}{2}(\text{trace}(G))^2 + \mu \, \text{trace}(G^2) \text{ with } G = \frac{1}{2}\left(F^\top F - I\right),$$

and here $\lambda$ and $\mu$ are the Lamé constants. This corresponds to a linear material law without a geometric linearization. However this law does not fulfill the condition $W(F) \to \infty$ for $\det F \to 0$.

- Ogden materials:

$$W(F) = a \ \text{trace} \left(F^{\top} F\right)^{\delta/2} + b \ \text{trace} \left(\text{cof}(F^{\top} F)\right) + \gamma(\det F)$$

with coefficients $a, b > 0$, an exponent $\delta \geq 1$ and a convex function $\gamma : (0, \infty) \to \mathbb{R}$ with $\lim_{s \to 0} \gamma(s) = \infty$. Here cof is the cofactor matrix

$$(\text{cof} A)_{ij} = (-1)^{i+j} \frac{1}{\det A} \det \left(A^{(i,j)}\right),$$

and $A^{(i,j)}$ is the matrix $A$ without the $i^{\text{th}}$ row and the $j^{\text{th}}$ column.

### Bifurcation of Beams

If a body elastically deforms often the phenomenon occurs that the underlying equations are not uniquely solvable. For simplicity we consider the most simple model class, the *Euler–Bernoulli beam*.

In principle one can derive models for beams by an asymptotic analysis with respect to the thickness of the beam which is assumed to be small. Then one computes for the limit of vanishing beam thickness the deformation of the beam. Here one has to scale the load such that the deformation of the beam converges in an appropriate sense. This analysis is already in the linear case quite complicated. As we will see later for a meaningful analysis we will need a *geometrically nonlinear* model. Therefore we will now heuristically motivate the beam model we will use. We consider the *bending* of the beam which is loaded as illustrated in Fig. 5.15. The deformation of the beam essentially consists of a *stretching* of the "fibers" on top of centerline in Fig. 5.15 and a *compression* of the fibers below this line. Besides the bending the beam will also be compressed in the direction of the applied force. We will see that this last deformation is for a *small thickness* by an order lower and will hence be neglected.

For a geometrically nonlinear model it is meaningful to describe the centerline with the help of an *arc length parameter*:

$$\left\{(x(s), y(s))^{\top} \mid s \in (0, L)\right\}.$$

**Fig. 5.15** Bending of a beam

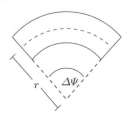

**Fig. 5.16** Derivation of the angular momentum of the beam

It is sufficient to consider the two coordinates $x(s)$ and $y(s)$ and we assume that the force only acts in the $x$-direction which is also the direction of the original centerline, and $y$ is the coordinate in the direction of the bending of the beam. The change of length of a fiber with distance $a$ to the centerline can be computed with the help of the curvature radius $r$, compare Fig. 5.16. The oriented distance $a$ is computed perpendicular to the "middle fiber". The "middle fiber" has the length $\ell_0 = 2\pi r \, \Delta\psi$, a fiber with distance $a$ has, before the deformation, also a length $\ell_0$ and after the deformation the length $2\pi(r + a) \, \Delta\psi$. The stretching is hence given by

$$\varepsilon(a) = \frac{\ell(a) - \ell_0}{\ell_0} = \frac{a}{r} \, .$$

We now choose the sign of the oriented distance $a$ and of the curvature $\kappa = \pm\frac{1}{r}$ in such a way that the fiber on top of the middle fiber has a positive distance and a beam which is curved to the top has a positive curvature. In this case we obtain

$$\varepsilon(a) = -\kappa a \, .$$

The stress is given by
$$E \, \varepsilon(a)$$

with the elastic modulus $E$. The angular momentum acting on a cross section $A(s)$ perpendicular to the middle fiber $(x(s), y(s))$ is given by

$$M(s) = \int_{A(s)} E \, \varepsilon(a) a \, da = -E \, J(s) \, \kappa(s) \tag{5.59}$$

with the second surface momentum

$$J(s) = \int_{A(s)} a^2 \, da \, . \tag{5.60}$$

The quantity $EJ(s)$ is called the bending stiffness of the beam. The force acting on the end of the beam leads to an angular momentum on the cross section perpendicular to $(x(s), y(s))$ which is given as
$$-Fy(s) \, .$$

Now the balance of angular momentum

$$M(s) - F y(s) = 0$$

leads to

$$\kappa(s) = -\lambda y(s) \quad \text{with} \quad \lambda = \frac{F}{EJ}. \tag{5.61}$$

We now assume that the beam has the thickness $\delta$. In this case the second surface momentum $J$ is proportional to $\delta^4$. This is true because the surface $A$ is proportional to $\delta^2$ and the integration variable $a$ is proportional to $\delta$. In order to obtain a vertical displacement of order 1 we need to scale the force proportional to $\delta^4$. The compression which is caused by this force acting on the beam is proportional to $\delta^2$. This we obtain from the formula

$$F = \sigma A = E\varepsilon A$$

with stress $\sigma$ and strain $\varepsilon$. The compression of the beam is hence two orders lower than the bending. This justifies that we neglect compressions in our model.

We will now use the *angle* $\varphi(s)$ as a variable instead of the displacement $(x(s), y(s))$ of the centerline, see Fig. 5.17. It holds

$$\kappa(s) = \varphi'(s), \quad x'(s) = \cos\varphi(s) \quad \text{and} \quad y'(s) = \sin\varphi(s).$$

Differentiating equation (5.61) with respect to $s$ gives the *beam equation*

$$\varphi''(s) + \lambda \sin\varphi(s) = 0. \tag{5.62}$$

In the situation sketched in Fig. 5.15 this equation has to be supplemented with the boundary conditions

$$\varphi'(0) = \varphi'(L) = 0.$$

This describes a fixed boundary on which no torque acts.

We now consider the *linearized* model

$$\varphi''(s) + \lambda\varphi(s) = 0, \quad \varphi'(0) = \varphi'(L) = 0. \tag{5.63}$$

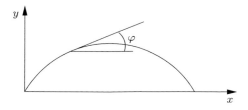

**Fig. 5.17** Coordinates in the beam model

The differential equation has the general solution

$$\varphi(s) = a\cos\left(\sqrt{\lambda}s\right) + b\sin\left(\sqrt{\lambda}s\right)$$

with coefficients $a, b \in \mathbb{R}$. There exists a non-trivial solution of the boundary value problem if and only if

$$\sqrt{\lambda}L = n\pi$$

with $n \in \mathbb{N}$. We hence have non-trivial solutions for the loads $F_n = EJ\lambda_n$, cf. (5.60), with

$$\lambda_n = \frac{n^2\pi^2}{L^2}.$$

The admissible values for $\lambda_n$ are the *eigenvalues* of the differential operator $\varphi \mapsto -\varphi''$ with domain of definition $\{\varphi \in C^2([0, L]) \mid \varphi'(0) = \varphi'(L) = 0\}$. The smallest eigenvalue $\lambda_1$ describes the *smallest* load for which a bending of the beam can appear. This load is called *Euler's buckling load*.

Below this load the beam will only be compressed. The outcome that a bending only occurs at certain fixed discrete values contradicts our experience. We will hence analyze the geometrically nonlinear model (5.62). Multiplying this equation with $\varphi'(s)$ and integration of the resulting equation leads to the following differential equation of first order

$$\tfrac{1}{2}(\varphi'(s))^2 - \lambda\cos\varphi(s) = -\lambda\cos\varphi_0$$

with a so far unknown *initial angle* $\varphi_0 = \varphi(0)$. This equation can be transformed to

$$\varphi'(s) = \pm\sqrt{2\lambda\big(\cos\varphi(s) - \cos\varphi_0\big)}\,. \tag{5.64}$$

We now assume $0 < \varphi_0 < \pi$. In this case we have $\varphi'(0) \le 0$ because otherwise the argument in the square root of (5.64) would be negative. We hence need to choose the negative sign on the right hand side of (5.64). By separation of variables we obtain

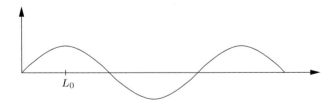

**Fig. 5.18** Bending of a beam

$$-\int_{\varphi_0}^{\varphi(s)} \frac{d\varphi}{\sqrt{\cos\varphi - \cos\varphi_0}} = \sqrt{2\lambda}\, s\,.$$

We now compute the position of $L_0$ on the beam as a function $\varphi_0$ for which the angle $\varphi$ attains the value 0 for the first time. One can then with the help of *symmetric extensions*, as illustrated in Fig. 5.18, compute the possible lengths $L_n = 2nL_0$ which will depend on the initial angle $\varphi_0$. The length $L_0$ is given by

$$L_0(\varphi_0) = \frac{1}{\sqrt{2\lambda}} \int_0^{\varphi_0} \frac{d\varphi}{\sqrt{\cos\varphi - \cos\varphi_0}}\,.$$

With the help of the identity $\cos\varphi = 1 - 2\sin^2\left(\frac{\varphi}{2}\right)$ we obtain

$$L(\varphi_0) = \frac{1}{\sqrt{4\lambda}} \int_0^{\varphi_0} \frac{d\varphi}{\sqrt{\sin^2\left(\frac{\varphi_0}{2}\right) - \sin^2\left(\frac{\varphi}{2}\right)}}\,.$$

The integral on the right hand side can be simplified with the help of the variable transformation $\varphi = \varphi(z)$ given by

$$\sin\left(\frac{\varphi(z)}{2}\right) = \sin\left(\frac{\varphi_0}{2}\right)\sin z\,.$$

Computing the derivative $\varphi'(z)$ with

$$\tfrac{1}{2}\cos\left(\frac{\varphi(z)}{2}\right)\varphi'(z) = \sin\left(\frac{\varphi_0}{2}\right)\cos z\,,$$

we obtain

$$\varphi'(z) = 2\sin\left(\frac{\varphi_0}{2}\right)\frac{\sqrt{1-\sin^2 z}}{\sqrt{1-\sin^2\left(\frac{\varphi(z)}{2}\right)}} = 2\sin\left(\frac{\varphi_0}{2}\right)\frac{\sqrt{1-\sin^2\left(\frac{\varphi(z)}{2}\right)/\sin^2\left(\frac{\varphi_0}{2}\right)}}{\sqrt{1-\sin^2\left(\frac{\varphi(z)}{2}\right)}}$$

$$= 2\frac{\sqrt{\sin^2\left(\frac{\varphi_0}{2}\right) - \sin^2\left(\frac{\varphi(z)}{2}\right)}}{\sqrt{1-\sin^2\left(\frac{\varphi(z)}{2}\right)}}\,.$$

With this identity we obtain

$$L_0(\varphi_0) = \frac{1}{\sqrt{\lambda}} \int_0^{\pi/2} \frac{dz}{\sqrt{1-\sin^2\left(\frac{\varphi_0}{2}\right)\sin^2 z}}\,. \tag{5.65}$$

From this representation of $L_0$ we can easily deduce the following properties of the function $\varphi_0 \mapsto L_0(\varphi_0)$:

• The function is strictly monotonically increasing on $(0, \pi)$.

- In the limit $\varphi_0 \to 0$ the function converges towards

$$L_0(0) = \frac{\pi}{2\sqrt{\lambda}}.$$

This means in the limit the admissible lengths are given by

$$L_n = \frac{n\pi}{\sqrt{\lambda}}$$

or that for a given length a non-trivial solution exists if

$$\lambda = \frac{n^2\pi^2}{L^2}.$$

These are exactly the eigenvalues of the linearized equation (5.63).
- For $\varphi_0 \to \pi$ the integral in (5.65) diverges to $+\infty$ and hence the length $L_0(\varphi_0)$ diverges to $+\infty$.

This means in particular that the mapping $L_0 : [0, \pi) \to \left[\frac{\pi}{2\sqrt{\lambda}}, \infty\right)$ is *bijective*. We hence obtain that for each length $L > \frac{n\pi}{\sqrt{\lambda}}$ we have an initial angle $\varphi_0$ with the related solution of problem (5.62) such that the deformed beam $\{(x(s), y(s)) \mid 0 < s < L\}$ has intersections with the $x$-axis.

Altogether one can describe the set of solutions in dependence of $\lambda$ as follows: For small $\lambda$, i.e., $\lambda \le \frac{\pi^2}{L^2}$, we have only the trivial solution $\varphi(s) = 0$. For $\frac{\pi^2}{L^2} < \lambda \le \frac{4\pi^2}{L^2}$ one has two solutions, namely the trivial solution and the solution characterized by (5.62) and initial conditions $\varphi(0) = \varphi_0$ and $\varphi'(0) = 0$ where $\varphi_0$ is the solution of the equation $L_0(\varphi_0) = L/2$. For each eigenvalue $\lambda_n = \frac{n^2\pi^2}{L^2}$ of the linearized equation we obtain an additional solution. The structure of the set of solutions is illustrated in Fig. 5.19. In our considerations above we assumed a positive initial angle. Obviously we can reflect all solutions with respect to the $\lambda$-axis. In this case we obtain the solutions with $\varphi_0 < 0$ sketched in Fig. 5.19. In three spatial dimensions the sketched

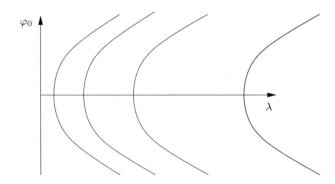

**Fig. 5.19** Bifurcation diagram

curves will become surfaces with rotational symmetry. This is because the beam can bend not only in the "top" and "bottom" directions but also in any other direction which is perpendicular to the direction of the load. The phenomenon of "branching" of solutions is called *bifurcation*. The points $\lambda_n = \frac{n^2 \pi^2}{L^2}$ are called bifurcation points or branching points. These points are typically characterized by the fact that the *linearized* equation has non-trivial solutions and that the trivial solution is *not stable*.

Bifurcations also appear when thin plates or so-called *shells*, which are curved thin plates, are loaded. These plates and shells buckle and the resulting deformations can be geometrically much more complicated as for beams.

A systematic introduction into bifurcation theory is given in the books Chow and Hale [21] and Kielhöfer [78]. Applications of bifurcation theory in the context of beams and other problems in elasticity theory can be found in Antman [7].

## 5.11  Electromagnetism

Another phenomenon which is important for many applications in science and technology is electromagnetism. In this section we will discuss models for electric and magnetic interactions in a continuum setting. In particular we will introduce Maxwell's equations, which are the essential basis for the mathematical description of electromagnetical processes in classical physics. Strictly speaking electromagnetism is not a part of mechanics and hence also not part of continuum mechanics. However electromagnetism is modeled with the help of a theory based on continuum fields and hence methodically it is close to continuum mechanics. This is the reason why we include this theory in this chapter.

### The Electric Field

An electric point charge $q_0$ at a position $x_0$ exerts a force on another point charge $q$ at a position $x$. The direction of this force is given by the direction of the line connecting the points and the strength decreases with the square of the distance. The force is attracting if $q_0$ and $q$ have a different sign and is repulsive if $q_0$ and $q$ have the same sign. Hence the force is given by

$$F = k \frac{q_0 q}{|x - x_0|^3} (x - x_0) \tag{5.66}$$

with a constant of proportionality $k$. This relation is called *Coulomb's law*. The effect of the point charge $q_0$ on other charges in space is described with the help of a force field which is the *electric field E*. This is defined in such a way that on a charge $q$ at $x$ the force

$$F(x) = q \, E(x)$$

acts. By comparing this expression with (5.66) one notices that the electric field induced by the point charge $q_0$ at the point $x_0$ is given by

$$E(x) = \frac{k q_0}{|x - x_0|^3}(x - x_0)\,. \tag{5.67}$$

For the electric fields induced by several point charges $q_1, \ldots, q_n$ at positions $x_1, \ldots, x_n$ one obtains by summing the fields of the individual charges

$$E(x) = \sum_{j=1}^{n} k \frac{q_j}{|x - x_j|^3}(x - x_j)\,.$$

The function $w(x) = \frac{1}{|x-x_0|^3}(x - x_0)$ has the following important properties:

(i)  The divergence of $w$ away from $x_0$ is zero,

$$\nabla \cdot \left( \frac{1}{|x - x_0|^3}(x - x_0) \right) = 0 \ \text{ for } \ x \neq x_0\,.$$

(ii)  For all balls $B_R(x_0)$ with radius $R$ centered at $x_0$ we obtain

$$\int_{\partial B_R(x_0)} w \cdot n \, ds_x = 4\pi\,,$$

and this is *independent* of the radius $R$. Because of (i) we obtain for *all* $C^1$-domains $\Omega$ with $x_0 \in \Omega$

$$\int_{\partial \Omega} w \cdot n \, ds_x = 4\pi\,.$$

This follows with the help of the divergence theorem. We take a ball $B_r(x_0)$ such that $B_r(x_0) \subset \Omega$ and compute

$$\int_{\partial \Omega} w \cdot n \, ds_x = \int_{\Omega \backslash B_r(x_0)} \nabla \cdot w \, dx + \int_{\partial B_r(x_0)} w \cdot n \, ds_x = 4\pi\,,$$

where the unit normal vector on $\partial B_r(x_0)$ points into the outside of $B_r(x_0)$ and hence to the inside of $\Omega \setminus B_r(x_0)$. For domains with $x_0 \notin \overline{\Omega}$ we have because of (i)

$$\int_{\partial \Omega} w \cdot n \, ds_x = 0\,.$$

(iii)  The curl of $w$ is zero,
$$\nabla \times w = 0 \ \text{ for } \ x \neq x_0\,.$$

In particular we obtain that $w$ has a potential

$$\varphi(x) = -\frac{1}{|x - x_0|} \ \text{ with } \ \nabla \varphi(x) = w(x) \ \text{ for all } \ x \neq x_0\,.$$

From (ii) we obtain for all ensembles of point charges at positions $x_1, \ldots, x_n$ and all smooth domains $\Omega$ whose boundary does not contain any of the $x_j$

$$\int_{\partial\Omega} E(x) \cdot n(x)\, ds_x = 4\pi k\, Q(\Omega) \tag{5.68}$$

with the charge $Q(\Omega)$ contained in $\Omega$ given as

$$Q(\Omega) = \sum_{\substack{j=1 \\ x_j \in \Omega}}^{n} q_j\,.$$

The Eq. (5.68) is called *Gauss' law*. Instead of the constant $k$ one uses the *permittivity* $\varepsilon = (4\pi k)^{-1}$. This constant depends on the material and for vacuum we have $\varepsilon = \varepsilon_0 \approx 8{,}854187817 \cdot 10^{-12}\, \mathrm{C^2 N^{-1} m^{-2}}$, where C stands for the electric charge unit "coulomb".

From (iii) it follows for all *closed* curves $\gamma$, which do not contain any of the points $x_j$,

$$\int_{\gamma} E \cdot dx = 0\,. \tag{5.69}$$

In a *continuum* we have instead of point charges a *charge density $\varrho$ per unit volume*. The right hand side (5.68) in this case has the form

$$Q(\Omega) = \int_{\Omega} \varrho\, dx,$$

which is the continuum formulation of Gauss's law

$$\int_{\partial\Omega} \varepsilon\, E(x) \cdot n(x)\, ds_x = \int_{\Omega} \varrho\, dx\,. \tag{5.70}$$

As we have no isolated charge sources anymore, this equation is true for *all* domains $\Omega$ which are smooth enough. With the help of the divergence theorem we obtain

$$\nabla \cdot (\varepsilon\, E) = \varrho\,. \tag{5.71}$$

Since (5.69) holds for *all* closed curves we obtain (see Exercise 5.35)

$$\nabla \times E = 0\,. \tag{5.72}$$

For a simply connected domain we now obtain the existence of a potential $V$ to $-E$,

$$E = -\nabla V\,. \tag{5.73}$$

Inserting this into (5.71) now gives

$$- \nabla \cdot (\varepsilon \nabla V) = \varrho\,. \tag{5.74}$$

In a homogeneous medium the permittivity does not depend on $x$ and we hence obtain Poisson's equation

$$-\Delta V = \varepsilon^{-1} \varrho\,.$$

These are the fundamental equations for the electric field in electrostatics. As an alternative to this approach which is based on (5.69) and (5.70) one can also derive the electric field generated by a continuum charge distribution

$$E(x) = \int_{\mathbb{R}^3} \frac{1}{4\pi\varepsilon} \frac{\varrho(y)}{|x - y|^3}(x - y)\,dy\,.$$

The associated potential is

$$V(x) = \int_{\mathbb{R}^3} \frac{1}{4\pi\varepsilon} \frac{\varrho(y)}{|x - y|}\,dy\,. \tag{5.75}$$

One can now show that this is a representation of a solution of Poisson's equation. One obtains this by a convolution of the right hand side $\varepsilon^{-1}\varrho$ with the singular function $\varphi(x) = \frac{1}{4\pi} \frac{1}{|x|}$. Formula (5.75) is the unique solution of Poisson's equation which is defined on the whole of $\mathbb{R}^3$ and whose values converge for $x \to +\infty$ to zero. The function $\varphi$ is a singular solution of Laplace's equation which is called *fundamental solution*. More on this we will learn in Sect. 6.1.

### Charge Balance

For the charge density $\varrho$ there holds a balance equation just as the balance equation for the mass density in continuum mechanics. Now let $j$ be the electric current density, i.e.,

$$j \cdot n = \lim_{\substack{\Delta A \to 0 \\ \Delta t \to 0}} \frac{\Delta q}{\Delta t\, \Delta A}\,,$$

where $\Delta q$ is the charge which flows through a piece of surface $\Delta A$ which has the unit normal $n$ in a time interval of length $\Delta t$. In this case the charge balance is given as

$$\frac{d}{dt} \int_{\Omega} \varrho\, dx = -\int_{\partial\Omega} j \cdot n\, ds_x\,.$$

The integral on the right hand side describes the charge which *leaves the domain*. An application of the divergence theorem gives

$$\int_{\Omega} (\partial_t \varrho + \nabla \cdot j)\, dx = 0\,.$$

Since this equation has to hold for all domains $\Omega$ we obtain

$$\partial_t \varrho + \nabla \cdot j = 0 . \tag{5.76}$$

This equation is often also called the *continuity equation*.

### Ohm's Law

An electric field acts as a force on charges. In the case that these charges can move they will be accelerated by this force. Materials which are electrically conductive oppose this movement by some resistance and charges will hence move with a velocity which depends on the electric field and the resistance. The resulting relation between electric field and electric current density can be formulated as

$$j = \sigma E . \tag{5.77}$$

The factor of proportionality $\sigma$ is called the *electric conductivity*. In so-called *ohmic materials* $\sigma$ is constant. In *non-ohmic materials* $\sigma$ can depend on $E$. Typically $\sigma$ depends also on other physical quantities, e.g., on the temperature.

### Electrostatics in Conductors, Surface Charges

We now consider processes with charge distributions in conducting materials which are *constant in time*. The continuity equation in this case is given as

$$\nabla \cdot j = 0 .$$

If a charge distribution is constant in time, this does not mean necessarily that no charge is transported. It just means that the charge density does not change in time which leads to the formula $\nabla \cdot j = 0$ for the charge transport. Together with Ohm's law $j = \sigma E$ we obtain for a constant conductivity $\sigma$

$$\nabla \cdot E = 0 .$$

Comparing this with (5.71) shows $\varrho = 0$. This means that the charge density in the interior of a conducting medium in the case of a stationary charge distribution is equal to zero. In an object made of conducting material, which is surrounded by a non-conducting material, we obtain in the stationary case that the charge will accumulate on the *surface*. In a continuum model this will be described with the help of charge densities per unit surface which are defined on the surface. Let $\Gamma$ be the surface of a conductor and $\varrho_\Gamma$ the charge density per unit surface on this surface. Gauss's law is in this case given by

$$\int_{\partial\Omega} \varepsilon E(x) \cdot n(x) \, ds_x = \int_{\Omega\cap\Gamma} \varrho_\Gamma \, ds_x . \tag{5.78}$$

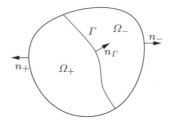

**Fig. 5.20** Illustration of the jump condition in the case of surface charges

This holds, if $\varrho_\Gamma$, $\Omega$ and $\Gamma$ are smooth enough and if the surface measure of the intersection $\partial\Omega \cap \Gamma$ is zero. Since (5.78) holds for general domains which have an empty intersection with $\Gamma$ we obtain first of all

$$\nabla \cdot E(x) = 0 \quad \text{for } x \notin \Gamma.$$

We now consider a domain $\Omega$ which is separated by $\Gamma$ in two parts $\Omega_+$ and $\Omega_-$, see Fig. 5.20 for an illustration. Let $n_+$ and $n_-$ be the outward pointing unit normal vectors of $\partial\Omega_+$ and $\partial\Omega_-$. On $\Gamma$ we have $n_+ = -n_- =: n_\Gamma$. We now assume that the one-sided limits $E_+(x) = \lim_{\substack{s \to 0 \\ s > 0}} E(x - sn_+)$ and $E_-(x) = \lim_{\substack{s \to 0 \\ s > 0}} E(x - sn_-)$ of $E$ for $x \in \Gamma$ exist and that they are integrable on the surface. An application of the divergence theorem gives

$$
\begin{aligned}
0 &= \int_{\Omega_+} \nabla \cdot (\varepsilon E)\, dx + \int_{\Omega_-} \nabla \cdot (\varepsilon E)\, dx \\
&= \int_{\partial\Omega_+} \varepsilon E \cdot n_+ \, ds_x + \int_{\partial\Omega_-} \varepsilon E \cdot n_- \, ds_x \\
&= \int_{\partial\Omega} \varepsilon E \cdot n \, ds_x + \int_{\Gamma \cap \Omega} (\varepsilon_+ E_+ \cdot n_+ + \varepsilon_- E_- \cdot n_-)\, ds_x.
\end{aligned}
$$

For the jump $[\varepsilon E \cdot n] = \varepsilon_+ E_+ \cdot n_+ + \varepsilon_- E_- \cdot n_-$ we have

$$\int_{\Omega \cap \Gamma} [\varepsilon E \cdot n]\, ds_x = -\int_{\Omega \cap \Gamma} \varrho_\Gamma \, ds_x.$$

As $\Omega$ is arbitrary, we obtain

$$[\varepsilon E \cdot n] = -\varrho_\Gamma. \tag{5.79}$$

Also in the case of charge densities per unit surface the formula (5.69) is still valid for all closed curves $\gamma$. From this we can deduce the existence of an electrostatic potential $V$ with $E = -\nabla V$. If $E$ is bounded, we obtain that $V$ is continuous and differentiable outside of $\Gamma$. Because of (5.79) the gradient of $V$ is not continuously extendible to all of $\Gamma$. The normal derivative of the potential fulfills the jump condition

$$[\varepsilon \nabla V \cdot n] = \varepsilon_+ \nabla V_+ \cdot n_+ + \varepsilon_- \nabla V_- \cdot n_- = -[\varepsilon E \cdot n] = \varrho_\Gamma.$$

**Fig. 5.21** Schematic image of a capacitor

The assumption that $E_+$ and $E_-$ are both continuously extendible onto $\Omega_+ \cup \Gamma$ and $\Omega_- \cup \Gamma$ implies that $V$ is on both sides of $\Gamma$ continuously differentiable and that the gradient can be extended continuously onto $\Gamma$. As $V$ is a potential of a bounded function we obtain that $V$ is continuous and hence also the tangential derivatives of $V$ are continuous. If a domain $\Omega$, which is filled with a charged medium, is in a *static* equilibrium, i.e., $j = -\sigma \nabla V = 0$ we obtain that $V$ is *constant* in $\Omega$. In particular the electric field is zero in the interior of the conductor.

**Example 1**: Capacitor. We consider two parallel plates with distance $a$, as shown in Fig. 5.21. We assume that the plates are oriented in the $x_1$–$x_2$ plane with plate 1 containing the origin and plate 2 containing the point $(0, 0, a)$. Let on plate 1 the potential $V_1$ and on plate 2 the potential $V_2$ be given. The electric field and the potential between the plates due to symmetry only depend on $x_3$. The equations

$$\Delta V = \partial_3^2 V = 0, \quad V(x_3 = 0) = V_1, \quad V(x_3 = a) = V_2$$

imply $V(x) = V_1 + x_3(V_2 - V_1)/a$. The electric field between the plates is

$$E = -\nabla V = \frac{(V_1 - V_2)}{a} e_3 .$$

Outside of the domain between the two plates we assume that $V$ is constant and hence $E = 0$. With the jump condition (5.79) we obtain that the charge density $\varrho_1$ on plate 1 is given by $\varrho_1 = \varepsilon E \cdot e_3 = \frac{\varepsilon}{a}(V_1 - V_2)$ and on plate 2 we obtain analogously $\varrho_2 = \frac{\varepsilon}{a}(V_2 - V_1)$. Here $\varepsilon$ is the permittivity for the medium *between* the plates.

These formulas also hold approximately for plates of finite size with surface area $A$ if the distance $a$ is very small. In this case the electric field deviates considerably only in the neighborhood of the boundaries of the plates. We then have the following relation between the potential difference $U = V_1 - V_2$ and the charge of the condensation plates:

$$Q = \frac{\varepsilon A}{a} U .$$

The quantity $C = \varepsilon A/a$ is the *capacitance* of the capacitor. The capacitance is proportional to the area and inversely proportional to the distance of the capacitor plates.

**Example 2**: Faraday cage. We now consider a domain $\Omega$ filled with a conducting material which encloses a non-conducting interior $\Omega_0$ as shown in Fig. 5.22. We assume that in the exterior $\mathbb{R}^3 \backslash \overline{\Omega \cup \Omega_0}$ there exists a stationary electric field $E$ which leads to a stationary distribution of surface charges on $\partial\Omega$. The potential in $\Omega$

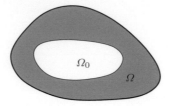

**Fig. 5.22** Schematic image of a Faraday cage

in this case is constant, i.e., $V = V_0$ with $V_0$ constant, and the electric field is $E = 0$.
In $\Omega_0$ we have to solve the potential equation

$$\Delta V = 0$$

with boundary condition $V = V_0$. The solution is $V = V_0$ and hence the electric field
in $\Omega_0$ is equal to zero. The conducting cover $\Omega$ insulates the exterior electric field
completely. Charges distribute themselves on the exterior surface of $\Omega$ such that
they compensate the exterior electric field. On the interior surface between $\Omega_0$ and
$\Omega$ there hence exist no surface charges.

**The Energy of the Electric Field**

In electrostatics an electric field is caused by charges. In an electric field the force
$F = qE = -q\nabla V$ acts on a charge $q$. The potential energy of the charge in the electric
field is hence $qV$. In a charge distribution in space there is hence an accumulated
potential energy of charges stored which is given by the electric fields of all other
charges. In order to compute this energy we consider a charge density $\varrho_0$ defined in
the whole space with compact support, i.e., it has to hold that $\varrho_0(x) = 0$ for all $|x| > R$
with a suitable $R$. We now assume that these charge distributions are built up slowly
and this fact will be described by a function $\varrho(t, x)$ with $\varrho(0, x) = 0$, $\varrho(1, x) = \varrho_0(x)$
and $\varrho(t, x) = 0$ for $|x| > R$. Let $E(t, x)$ be the electric field, $V(t, x)$ the electrostatic
potential and $W(t)$ the energy of the electric field at time $t$. It now holds $E(0, x) = 0$,
$W(0) = 0$ and

$$W'(t) = \int_{\mathbb{R}^3} V(t, x)\, \partial_t \varrho(t, x)\, dx\,.$$

The integrand here describes the power needed in order to obtain a rate of change
for the charge given as $\partial_t \varrho$ against the electric field $E$. From

$$\varrho = \nabla \cdot (\varepsilon E) = -\nabla \cdot (\varepsilon \nabla V)$$

we deduce

$$W'(t) = -\int_{\mathbb{R}^3} V(t, x)\, \partial_t \nabla \cdot (\varepsilon \nabla V(t, x))\, dx = \int_{\mathbb{R}^3} \frac{\varepsilon}{2} \partial_t |\nabla V(t, x)|^2\, dx\,.$$

The last equation follows from integration by parts with respect to $x$. The boundary terms at "infinity" vanish because $V$ grows like $\mathcal{O}(|x|^{-1})$ and $\nabla V$ like $\mathcal{O}(|x|^{-2})$ for $|x|$ large. This we can deduce for example from the formula (5.75). Integration gives

$$W(1) = \int_{\mathbb{R}^3} \frac{\varepsilon}{2}|E(1,x)|^2\, dx.$$

We can hence assign to an electric field $E$ the energy

$$W = \int_{\mathbb{R}^3} \frac{\varepsilon}{2}|E(x)|^2\, dx, \tag{5.80}$$

which is denoted as the *energy of the electric field*.

**Magnetostatic Interactions**

A current-carrying conductor exerts a force on magnetic poles close by. This is apparent for example in the phenomenon that iron filings on a plate which is perpendicular to the conductor will arrange in concentric circles. Another phenomenon is the fact that turnable magnetic needles will align perpendicular to the conductor. The reason for this is a force which acts with opposite signs on the two poles of a magnet and hence will lead to an angular momentum which leads to an alignment of the magnet. The strength of this force is proportional to the conducting current and to the reciprocal distance. The direction of the force is perpendicular to the conductor and to the connecting line between the magnet and the projection of this point on the conductor. We describe this by a field which we call the *magnetic induction B*. For a conductor in the form of a line with direction given by the vector $a$ with length 1, the magnetic induction is given by

$$B(x) = \frac{kI}{|x - Px|^2} a \times (x - Px),$$

where $k$ is a constant of proportionality, $I$ is the strength of the current and $Px$ is the projection of $x$ onto the conductor. The function $w(x) = \frac{1}{|x-Px|^2} a \times (x - Px)$ fulfills for $x \neq x_0$ the following important relations:

**Theorem 5.14** *Let $\gamma_0$ be a line with direction vector $a$, $Px$ the projection of $x \in \mathbb{R}^3$ onto $\gamma_0$ and $w(x) = \frac{1}{|x-Px|^2} a \times (x - Px)$ for $x \notin \gamma_0$. Let $\gamma$ be a smooth oriented closed curve with $\gamma \cap \gamma_0 = \emptyset$ which rotates around $\gamma_0$ once in positive mathematical orientation, and $\Omega$ is assumed to be a bounded domain with smooth boundary $\partial\Omega$. The intersection $\partial\Omega \cap \gamma_0$ is assumed to consist of isolated points in which the tangential plane of $\partial\Omega$ is not parallel to $\gamma_0$. In this case it holds*

$$\int_{\gamma} w(x) \cdot dx = 2\pi, \tag{5.81}$$

$$\int_{\partial\Omega} w(x) \cdot n(x)\, ds_x = 0. \tag{5.82}$$

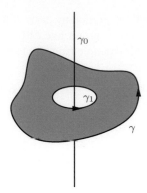

**Fig. 5.23** Illustration for the proof of (5.81)

*Proof* We choose a coordinate system with origin on $\gamma_0$ and an $x_3$-axis in the direction of $\gamma_0$. Then it holds for $(x_1, x_2) \neq (0, 0)$

$$w(x) = \frac{1}{x_1^2 + x_2^2} e_3 \times x = \frac{1}{x_1^2 + x_2^2} \begin{pmatrix} -x_2 \\ x_1 \\ 0 \end{pmatrix},$$

$$\nabla \times w = 0,$$

$$\nabla \cdot w = 0.$$

We first of all consider an oriented circle $\gamma_1$ in the $x_1$–$x_2$ plane with radius $r$ which rotates around $\gamma_0$ in positive direction. Using polar coordinates we obtain $x(\varphi) = \begin{pmatrix} r\cos\varphi \\ r\sin\varphi \\ 0 \end{pmatrix}$ and $x'(\varphi) = \begin{pmatrix} -r\sin\varphi \\ r\cos\varphi \\ 0 \end{pmatrix}$ and hence

$$\int_{\gamma_1} w(x) \cdot ds_x = \int_0^{2\pi} \frac{-x_2(\varphi)\, x_1'(\varphi) + x_1(\varphi)\, x_2'(\varphi)}{x_1^2(\varphi) + x_2^2(\varphi)}\, d\varphi = 2\pi.$$

For the curve $\gamma$ there exists a surface $\Gamma$ with boundary $\gamma \cup \gamma_1$, as shown in Fig. 5.23. An application of the Stokes theorem on this surface gives

$$0 = \int_\Gamma \nabla \times w \cdot n\, ds_x = \int_\gamma w \cdot ds - \int_{\gamma_1} w \cdot ds$$

and hence the relation (5.81) follows. For a domain $\Omega$ we define $\Omega_r := \{x \in \Omega \mid \operatorname{dist}(x, \gamma_0) > r\}$, see Fig. 5.24.

Since $\nabla \cdot w = 0$ in $\Omega_r$ it holds

$$0 = \int_{\Omega_r} \nabla \cdot w\, dx = \int_{\partial \Omega_r} w \cdot n\, ds_x.$$

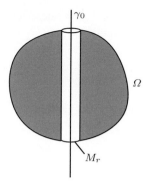

**Fig. 5.24** Illustration for the proof of (5.81)

On the lateral surface $M_r = \{x \in \partial\Omega_r \mid \text{dist}(x, \gamma_0) = r\}$ the normal $n(x)$ is parallel to $x - Px = (x_1, x_2, 0)^\top$, and hence it holds $w(x) \cdot n(x) = 0$. The remaining surface $\partial\Omega \setminus \partial\Omega_r$ consists of neighborhoods of the intersection points between $\partial\Omega$ and $\gamma_0$. These we can parametrize as a graph of the form $(x_1, x_2, h(x_1, x_2))$ with a smooth function $h$ for $x_1^2 + x_2^2 < r$. Let $c_1$ be an upper bound for $\sqrt{1 + |\nabla h|^2}$. We can then estimate

$$\left| \int_{\partial\Omega \setminus \partial\Omega_r} w \cdot n \, dx \right| \leq c_1 \ell \int_{x_1^2 + x_2^2 \leq r^2} \frac{1}{\left| \sqrt{x_1^2 + x_2^2} \right|} \, dx_1 \, dx_2 \to 0 \text{ for } r \to 0,$$

where $\ell$ is the number of intersection points. This shows (5.82). $\qquad\square$

The magnetic induction $B$ generated from a straight line which conducts a current of intensity $I$ fulfills due to Theorem 5.14 the following relations:

$$\int_\gamma B \cdot ds = \mu I, \tag{5.83}$$

$$\int_{\partial\Omega} B \cdot n \, ds_x = 0. \tag{5.84}$$

Here, $\mu := 2\pi k$ is the *magnetic permeability*. In a vacuum it holds $\mu = \mu_0 = 4\pi \, 10^{-7} \, \text{N/A}^2$. These relations are not only fulfilled for conductors which are straight but also for arbitrary conductor geometries. In a continuum formulation of the magnetic interactions we replace the current carrying conductor by a current density $j$. Instead of (5.83) we formulate the relation

$$\int_\gamma B \cdot ds = \int_\Gamma \mu j \cdot n \, ds_x, \tag{5.85}$$

where $\Gamma$ is an arbitrary surface with boundary $\gamma$. The integral on the right hand side describes the current enclosed by the curve $\gamma$. The integral does not depend on $\Gamma$ if $\nabla \cdot j = 0$ which is just the *stationary* case. This is because if $\Gamma_1$ and $\Gamma_2$ would be two different surfaces with the same boundary $\gamma$, these surfaces would enclose a domain $\Gamma$. The divergence theorem then implies

$$\int_{\Gamma_1} j \cdot n_1 \, ds_x - \int_{\Gamma_2} j \cdot n_2 \, ds_x = \int_{\partial\Omega} j \cdot n \, ds_x = \int_{\Omega} \nabla \cdot j \, dx = 0 \,.$$

An application of the Stokes theorem in (5.85) gives

$$\int_{\Gamma} (\nabla \times B - \mu j) \cdot n \, ds_x = 0 \,.$$

This is true for *each* bounded surface $\Gamma$ with smooth boundary. If $B$ and $j$ are smooth we can conclude

$$\nabla \times B = \mu j \,, \tag{5.86}$$

which is *Ampère's law*. From (5.84) we can deduce with the help of the divergence theorem and with the fact that $\Omega$ can be chosen arbitrarily that

$$\nabla \cdot B = 0 \,. \tag{5.87}$$

This completes the derivation of the basic equations of magnetostatics.

### The Lorentz Force

Magnetic fields exert forces on charges which move in time. The force on a point charge $q$ with velocity $v$ is perpendicular to the direction of movement and to the direction of the magnetic field. In addition it is proportional to the strength of the magnetic field and to the velocity of the charge and hence we obtain

$$F = q \, v \times B \,.$$

This force is called *Lorentz force*. A constant of proportionality does not appear here. The magnetic induction is just scaled by this relation. In a continuum formulation with spatially distributed charges, $qv$ is replaced by an electric charge density $j$ and $F$ is replaced by $f$ which is a force density per volume. The force $F_{\Omega}$ exerted on the charge in the domain $\Omega$ is given by

$$F_{\Omega} = \int_{\Omega} f \, dx = \int_{\Omega} j \times B \, dx \,.$$

This implies the equation

$$f = j \times B \,.$$

In the case of current carrying curves one can formulate an analogous variant. Let $\gamma = \{x(s) \mid s \in (0, 1)\}$ be a smooth curve through which the current flows. We can think of the current $I$ as a charge density $\varrho_{\gamma}$ *per unit length* which moves with velocity $v$, i.e., $I = \varrho_{\gamma} \, v$. The Lorentz force on a curve through which current flows is

$$F_\gamma = -\int_\gamma I\, B \times ds = -\int_0^1 I\, B(x(s)) \times x'(s)\, ds\,.$$

**Example**: We consider two parallel conductors with distance $a$. Currents flow through the conductors with strength $I_1$ and $I_2$. We choose a coordinate system such that the two conductors are oriented in the $x_3$-direction and such that conductor 1 contains the origin $(0,0,0)^\top$ and the conductor 2 contains the point $(a,0,0)^\top$. The magnetic induction generated by conductor 1 is given by

$$B_1(x) = \frac{\mu}{2\pi}\frac{I_1}{x_1^2 + x_2^2}\begin{pmatrix} -x_2 \\ x_1 \\ 0 \end{pmatrix}.$$

This magnetic field does not depend on $x_3$. The field exerts on a segment $\gamma_L$ of length $L$ of conductor 2 the force

$$F_{1\to 2}(I) = \int_{\gamma_L} I_2\, e_3 \times B_1\, dx = -\frac{\mu L I_1 I_2}{2\pi a}\, e_1\,.$$

This force is hence attracting if both currents have the same direction and repelling if they have different orientations.

**Maxwell's Equations**

A milestone in the development of electromagnetism was the insight that magnetic fields which change in time produce electric fields and vice versa, i.e., that also electric fields which change in time produce magnetic fields. The basis for this are experimental findings of the following kind:

- In a closed planar conductor loop in a spatially constant magnetic field, which changes in time, flows a current $I$. This current is proportional to the *change in time $\partial_t B$ of the magnetic field* and proportional to $A_\perp$, which is the area of surface enclosed by the loop

$$I \propto \partial_t B A_\perp\,.$$

- In a closed conductor loop which changes in time in a time independent magnetic field $B$ also current flows; this current is proportional to the *change* of $A_\perp$ and to $B$,

$$I \propto B\, \partial_t A_\perp\,.$$

As the constants of proportionality above are the same we can deduce that the current is proportional to the charge of the quantity $B A_\perp$ which is the *magnetic flux* through the conductor loop. According to Ohm's law the current will be produced by an electric field which is proportional to the strength of the current. For general conductor loops with geometry $\gamma$ and general magnetic fields which may be spatially varying one can postulate the following law for the relation between electric and magnetic fields:

$$\frac{d}{dt} \int_{\Gamma} B \cdot n \, ds_x \propto \int_{\gamma} E \cdot dx \,.$$

Here $\Gamma$ is a surface enclosed by $\gamma$ with unit normal $n$. An application of Stokes' theorem for a $\Gamma$ which is constant in time gives

$$\int_{\Gamma} \partial_t B \cdot n \, ds_x \propto \int_{\Gamma} \nabla \times E \cdot n \, ds_x \,.$$

We can hence deduce

$$\partial_t B \propto \nabla \times E \,.$$

The magnetic induction $B$ is scaled such that the constant of proportionality is equal to $-1$. Altogether we obtain the *Maxwell–Henry* equation

$$\partial_t B + \nabla \times E = 0 \,. \tag{5.88}$$

Without any experimental basis Maxwell guessed that an analogous relation between the time derivative of the electric field and the magnetic field has to hold. From Gauss's law (5.71) and the conservation of charge (5.76) we deduce

$$\nabla \cdot (\varepsilon \, \partial_t E + j) = 0 \,.$$

Hence $\varepsilon \, \partial_t E + j$ has a vector potential and it holds

$$\varepsilon \, \partial_t E + j = \nabla \times F$$

with a suitable vector field $F$. In the stationary case (5.86) implies

$$\nabla \times B = \mu j \,,$$

and one can hence conjecture $F = \mu^{-1} B$. We obtain the *Ampère–Maxwell* law

$$\varepsilon \mu \, \partial_t E - \nabla \times B + \mu j = 0 \,. \tag{5.89}$$

We now summarize the fundamental equations of electromagnetism

$$\nabla \cdot (\varepsilon E) = \varrho \,, \tag{5.90}$$
$$\nabla \cdot B = 0 \,, \tag{5.91}$$
$$\partial_t B + \nabla \times E = 0 \,, \tag{5.92}$$
$$\varepsilon \mu \, \partial_t E - \nabla \times B + \mu j = 0 \,. \tag{5.93}$$

These are *Maxwell's equations.*

**Electromagnetism in Materials**

In materials consisting of atoms and molecules there are electromagnetic interactions on the scale of atoms and molecules. These cannot be described directly in a continuous macroscopic model. Nevertheless they will influence the macroscopic electromagnetic properties of the material. More precisely we mean

- displacement of "bound charges" on the molecular scale,
- magnetic fields generated by currents on the molecular scale.

Bound charges are electrons which are bound to atoms or molecules. These charges are displaced within their molecules if an electric field is applied and this generates an electric dipole. This effect is called *polarization*. In addition, some materials can have a permanent polarization on the molecular scale which is due to their molecular composition. Macroscopically this polarization typically is not apparent as the available electric dipoles will orient themselves equally distributed. With an external electric field these electric dipoles will however orientate themselves into the direction of the field and will hence have a macroscopic effect which contributes to the polarization.

We distinguish between bound and free charges and modify Maxwell's equations such that the charge density $\varrho$ and the current density $j$ refer only to the *free* charges. The polarization macroscopically can be described by a current density $j_b$ of the bound charges which is proportional to the *variation of the electric field*

$$j_b = \chi \, \varepsilon_0 \, \partial_t E \, .$$

From the conservation of the bound charges

$$\partial_t \varrho_b + \nabla \cdot j_b = 0$$

we obtain

$$\partial_t \varrho_b + \partial_t \nabla \cdot (\chi \varepsilon_0 E) = 0$$

and after integration with the hypothetical initial conditions $\varrho_{b|t=0} = 0$, $E_{|t=0} = 0$ this leads to

$$\varrho_b = -\nabla \cdot (\chi \varepsilon_0 E) \, .$$

Inserting this into Gauss's law (5.90) with $\varepsilon = \varepsilon_0$, i.e.,

$$\nabla \cdot (\varepsilon_0 E) = \varrho + \varrho_b,$$

yields

$$\nabla \cdot (\varepsilon_0 (1 + \chi) E) = \varrho \, .$$

Motivated by this relation we define the field

$$D := \varepsilon_0(1 + \chi)E \,,$$

as the *electric displacement field*. It holds $D = \varepsilon E$ with the *permittivity* $\varepsilon = \varepsilon_0(1 + \chi)$. Often $\varepsilon$ is written as $\varepsilon = \varepsilon_0 \varepsilon_r$ with the *relative permittivity* $\varepsilon_r = 1 + \chi$.

Besides electric dipoles molecules can also have magnetic dipoles. These are generated on the molecular scale by currents generated by moving electrons. In most materials these magnetic dipoles are not apparent macroscopically because the dipoles are statistically equally distributed and therefore their impacts cancel each other. An exception are the so-called *ferromagnetic* substances in which the existing magnetic dipoles align due to mutual interactions. As an effect a permanent magnetic field will be generated. This effect gives rise to the existence of permanent magnets. In both cases the molecular magnetic dipoles are influenced by external magnetic fields and therefore experience a macroscopically apparent magnetization. This magnetization is described by a *density per unit volume of the magnetic moment* which is also called *magnetization vector* and will be denoted by $m$. We now distinguish between the external magnetic field and the field generated by the molecular magnetic dipoles. The external magnetic field is described by the *magnetic field strength* $H$ which in vacuum is given by $H = \mu_0^{-1}B$. The magnetization vector is proportional to the magnetic field strength

$$m = \psi H$$

with a material constant $\psi$. The magnetic induction $B$ is then given by

$$B = \mu_0(H + m) = \mu_0(1 + \psi)H = \mu H$$

with the *magnetic permeability* $\mu = \mu_0(1 + \psi)$. Often $\mu$ is written as $\mu = \mu_0 \mu_r$ with the *relative permeability* $\mu_r$.

The magnetization now generates currents of the bound electrons. In analogy to (5.93) we obtain a current density of the bound charges from

$$j_b = \chi \, \partial_t E + \nabla \times m \,.$$

Inserting this into the Ampère–Maxwell law (5.93) with current density $j + j_b$ yields

$$\mu_0 \, \partial_t(\varepsilon E) - \nabla \times (B - \mu_0 m) + \mu_0 j = 0$$

or, after dividing by $\mu_0$,

$$\partial_t D - \nabla \times H + j = 0 \,.$$

Altogether we obtain Maxwell's equations in the form

$$\nabla \cdot D = \varrho \,, \tag{5.94}$$

$$\nabla \cdot B = 0 \,, \tag{5.95}$$

$$\partial_t B + \nabla \times E = 0\,, \tag{5.96}$$

$$\partial_t D - \nabla \times H + j = 0\,. \tag{5.97}$$

These are supplemented by the equations $D = \varepsilon E$ and $B = \mu H$ as well as by the continuity equation (5.76) and by Ohm's law (5.77). Strictly speaking these are the same laws as in (5.90)–(5.93). In this representation we have just separated the processes on the microscopic scale from the macroscopic fields.

**The Energy of the Magnetic Field**

We now consider a magnetic field $H$ generated from a spatially distributed stationary electric field $j$. When the magnetic field is "switched on" it is not available instantaneously but will be created slowly in a certain period of time. This can be described with the help of "time-dependent" functions $H = H(t, x)$ and $j = j(t, x)$. The change in time of the magnetic field generates an electric field and its effect on the current density $j$ leads to a force. As a consequence energy will be needed to create the magnetic field and this energy will be set free if the magnetic field is removed. This energy is stored in the magnetic field. In order to compute this energy we consider a time-dependent solution of Maxwell's equations with charge density $\varrho = 0$ and a given time-dependent current density $j$. It holds $j(0, x) = 0$, $H(0, x) = 0$, $E(0, x) = 0$, $j(t, x) = 0$ for $|x| > R$ with a suitable $R > 0$ and $\nabla \cdot j(t, x) = 0$. The power needed at time $t$ to create the current density and the electric and magnetic fields is

$$P(t) = -\int_{\mathbb{R}^3} E \cdot j \, dx = -\int_{\mathbb{R}^3} E \cdot (\nabla \times H - \varepsilon \, \partial_t E) \, dx$$

$$= \frac{d}{dt} \left( \int_{\mathbb{R}^3} \frac{\varepsilon}{2} |E|^2 \, dx \right) - \int_{\mathbb{R}^3} E \cdot (\nabla \times H) \, dx\,.$$

It now holds $E \cdot (\nabla \times H) = \nabla \cdot (H \times E) + H \cdot (\nabla \times E)$ and $\nabla \times E = -\mu \, \partial_t H$. The integral over $\nabla \cdot (H \times E)$ vanishes, where we require that $E$ and $H$ decay rapidly for $|x| \rightarrow +\infty$. In addition, it holds

$$-\int_{\mathbb{R}^3} E \cdot (\nabla \times H) \, dx = \int_{\mathbb{R}^3} \mu H \cdot \partial_t H \, dx = \frac{d}{dt} \int_{\mathbb{R}^3} \frac{\mu}{2} |H|^2 \, dx\,.$$

Altogether we obtain

$$P(t) = \frac{d}{dt} \left( \int_{\mathbb{R}^3} \frac{\varepsilon}{2} |E|^2 \, dx + \int_{\mathbb{R}^3} \frac{\mu}{2} |H|^2 \, dx \right)\,.$$

According to (5.80) the energy of the electric field corresponds to the first integral on the right hand side of the above formula. In conclusion the magnetic field has the energy

$$W(t) = \int_{\mathbb{R}^3} \frac{\mu}{2} |H|^2 \, dx = \int_{\mathbb{R}^3} \frac{1}{2\mu} |B|^2 \, dx\,.$$

In the stationary limit after the magnetic field was "switched on" we obtain that the electric field disappears again. This is due to the fact that the corresponding potential $V$ solves $\nabla \cdot (\varepsilon \nabla V) = 0$ with boundary conditions $V \to 0$ for $|x| \to \infty$. Therefore, we obtain that the total energy is stored in the stationary magnetic field.

**Example**: We consider the turn-on of an electric coil. A current $I = I(t)$ flows through the coil. The magnetic field generated by this current is spatially inhomogeneous but proportional to the current strength $I$. For the energy stored in the magnetic field it hence holds

$$W(t) = \frac{L}{2} |I(t)|^2$$

with a constant $L$, called *inductance*, which depends on the coil. This constant is given by

$$L = \int_{\mathbb{R}^3} \mu |B_1(x)|^2 \, dx \,,$$

if $B_1$ is the magnetic induction generated by a current of strength 1. We now consider a change in the current which is so slow that the magnetic and electric fields are close to stationary states. The power supplied by the coil is then given by

$$P(t) = W'(t) = L I(t) \dot{I}(t) \,.$$

This power is provided by the voltage $U(t)$ which is applied to the coil and is given by the formula

$$P(t) = U(t) I(t) \,.$$

From this we obtain the formula for the voltage drop at the coil

$$U(t) = L \dot{I}(t) \,.$$

**Electromagnetic Waves**

Perhaps the most important application of electromagnetism are electromagnetic waves which are the basis of radiotelephones, of radio and television broadcast, of radars, of x-ray and of light. The existence of electromagnetic waves can be deduced from the Maxwell equations. Therefore, we consider the Eqs. (5.90)–(5.93) for $\varrho = 0$ and $j = 0$ and for constant $\varepsilon$, $\mu$. If we now add the time derivative of equation (5.93) and the curl of equation (5.92) we obtain

$$\varepsilon \mu \, \partial_t^2 E + \nabla \times \nabla \times E = 0 \,.$$

With the formulas $\nabla \times \nabla \times E = \nabla \nabla \cdot E - \Delta E$ and $\nabla \cdot E = 0$ we obtain the *wave equation*

$$\partial_t^2 E - c^2 \, \Delta E = 0 \tag{5.98}$$

with the *speed of light* $c = 1/\sqrt{\varepsilon\mu}$. Analogously we can deduce the corresponding version for the magnetic induction

$$\partial_t^2 B - c^2 \, \Delta B = 0 \,.$$

Equation (5.98) has travelling wave solutions of the form

$$E(t, x) = E_0 \sin(k \cdot x - t)$$

with a constant $E_0$ and a constant wave vector $k$, where $|k| = 1/c$ has to hold. From $\nabla \cdot E = 0$ we deduce $E_0 \cdot k = 0$. We hence obtain that the amplitude of the electric field is perpendicular to the direction of propagation. By using (5.92) we obtain that the corresponding magnetic induction is given as

$$B(t, x) = k \times E_0 \sin(k \cdot x - t) \,.$$

The magnetic induction is hence perpendicular to the direction of propagation and to the electric field. The existence of electromagnetic waves and the concrete determination of the direction of propagation have been important confirmations of Maxwell's laws.

## 5.12 Dispersion

The *dispersion* of a solution of a partial differential equation describes how harmonic waves of the form

$$u(t, x) = e^{i(k \cdot x - \omega t)} \tag{5.99}$$

with wave vector $k$ and circular frequency $\omega$ propagate. As an example we consider the one-dimensional convection diffusion equation

$$\partial_t u + c \, \partial_x u - a \, \partial_x^2 u = 0 \,.$$

Insertion of the ansatz $u(t, x) = e^{i(kx - \omega t)}$ gives

$$\left( -\omega i + cki + ak^2 \right) e^{i(kx - \omega t)} = 0 \,.$$

Between the wave number $k$ and the circular frequency $\omega$ the following relation has to hold

$$\omega = ck - a|k|^2 i \,.$$

The relation $k \to \omega = \omega(k)$ is called *dispersion relation*. From the dispersion relation one can read off the qualitative behavior of solutions of the differential equation under consideration. If $\omega(k)$ is real we obtain that the equation describes the propagation of a wave with wave number $2\pi/|k|$ and propagation velocity

$$v = \frac{\omega(k)}{|k|} .$$

If $\omega$ depends linearly on $|k|$ we obtain that the propagation velocity is for all wave numbers the same. This means that *every* wave packet will be transported with the same velocity. This is for example true for the wave equation. If $\omega(k)/|k|$ is *not* constant we obtain that waves with different wave lengths have different propagation velocities. This behavior is called *dispersion* and a corresponding differential equation is called *dispersive*. A wave packet or a "pulse" of an arbitrary form can be represented as a superposition of harmonic waves. For dispersive differential equations harmonic waves with different wave numbers will propagate with a different speed. This will lead to a *spread out* of a wave packet or a pulse of mixed wave lengths. If $\omega(k)$ is *not* real we call the differential equation *diffusive*. For $\omega = \omega_1 + i\,\omega_2$ the Eq. (5.99) has the form

$$u(t, x) = e^{\omega_2 t} e^{i(k \cdot x - \omega_1 t)} .$$

In the case $\omega_2 < 0$ this means that the *amplitude* of the harmonic wave will decrease. Oscillations of the wave will hence be *smoothed out* which is the case for so-called parabolic equations. In the case $\omega_2 > 0$ the equation is called *backwards* parabolic and this means that oscillations and existing gradients will be *strengthened*.

## 5.13  Literature

An extensive presentation of continuum mechanics can be found in [12, 59, 130, 132–134]. The mathematics of elastic and elastoplastic materials is treated in [7, 22, 94]. In particular one finds there models for beams, plates and shells. In [7] also bifurcation phenomena for such structures are studied. An introduction to bifurcation theory can be found in [21, 78]. Supplementing literature on electromagnetism are [53, 73]. A reference for Sect. 5.2 are the lecture notes [116].

## 5.14  Exercises

**Exercise 5.1** (*Divergence theorem, nabla calculus*) Let $\Omega \subset \mathbb{R}^3$ be a bounded domain with smooth boundary $\Gamma = \partial\Omega$. The exterior unit normal on $\Gamma$ we denote by $n$. Let $f, g$ be smooth functions and let $u = (u_1, u_2, u_3)$, $v = (v_1, v_2, v_3)$ be smooth vector fields on $\Omega$. All functions that appear are assumed to have continuous extensions onto $\Gamma$ and where needed the same is assumed for their derivatives. By $\nabla f$ we denote the gradient of $f$. We also use the notation $Du = (\partial_{x_j} u_i)_{i,j=1}^3$. The divergence and the curl of $u$ are given as

$$\nabla \cdot u = \partial_{x_1} u_1 + \partial_{x_2} u_2 + \partial_{x_3} u_3 \quad \text{and} \quad \nabla \times u = \begin{pmatrix} \partial_{x_2} u_3 - \partial_{x_3} u_2 \\ \partial_{x_3} u_1 - \partial_{x_1} u_3 \\ \partial_{x_1} u_2 - \partial_{x_2} u_1 \end{pmatrix}.$$

The Laplace operator is then given as $\Delta f = \nabla \cdot (\nabla f)$.

(a) Express $\nabla \times (\nabla \times u)$ and $\nabla \times (u \times v)$ by just using $\nabla$, $D$, $\nabla \cdot$, $+$ and $-$.
(b) Show with the help of the divergence theorem:

$$\int_\Gamma (\nabla \times u) \cdot n \, ds_x = 0.$$

Compute with the help of the divergence theorem the area of the surface of a ball with radius $a > 0$.

**Exercise 5.2** (*Tensor analysis*)

(a) Let $\phi$, $v$ and $S$ be smooth functions, $\phi$ scalar, $v \in \mathbb{R}^3$ and $S \in \mathbb{R}^{3,3}$. Show

$$D(\phi v) = \phi Dv + v(\nabla \phi)^\top,$$
$$\nabla \cdot (S^\top v) = S : Dv + v \cdot (\nabla \cdot S),$$
$$\nabla \cdot (\phi S) = \phi \nabla \cdot S + S \nabla \phi.$$

Here we define $A : B = \operatorname{trace}(A^\top B)$ as the inner product of two tensors $A, B \in \mathbb{R}^{3,3}$ and $\nabla \cdot A = \left( \sum_{j=1}^3 \partial_{x_j} a_{ij} \right)_{i=1}^3$ as the matrix divergence of $A \in \mathbb{R}^{3,3}$.

(b) Prove the following assertion: For $v \in C^2(\mathbb{R}^3; \mathbb{R}^3)$ it holds

$$\nabla \cdot (Dv^\top) = \nabla(\nabla \cdot v).$$

**Exercise 5.3** Show with the help of the divergence theorem the following formulas for a smooth domain $\Omega \subset \mathbb{R}^3$ with boundary $\Gamma$ and smooth functions $f, g : \Omega \to \mathbb{R}$, $u, v : \Omega \to \mathbb{R}^3$, $\sigma : \Omega \to \mathbb{R}^{3,3}$. Here we define the divergence of a matrix as $(\nabla \cdot \sigma)_i = \sum_{j=1}^3 \partial_{x_j} \sigma_{ij}$.

(a) $\displaystyle \int_\Omega f(x) \nabla \cdot u(x) \, dx = \int_\Gamma f(x) u(x) \cdot n(x) \, ds_x - \int_\Omega \nabla f(x) \cdot u(x) \, dx,$

(b) $\displaystyle \int_\Omega \nabla f(x) \cdot \nabla g(x) \, dx = \int_\Gamma f(x) \nabla g(x) \cdot n(x) \, ds_x - \int_\Omega f(x) \Delta g(x) \, dx,$

(c) $\displaystyle \int_\Omega f(x) \nabla g(x) \, dx = - \int_\Omega \nabla f(x) \, g(x) \, dx + \int_\Gamma f(x) \, g(x) \, n(x) \, ds_x,$

(d) $\displaystyle \int_\Omega \nabla \cdot \sigma(x) \, dx = \int_\Gamma \sigma(x) n(x) \, ds_x,$

(e) $\displaystyle \int_\Omega (\nabla \times u(x)) \cdot v(x) \, dx = \int_\Gamma (u(x) \times v(x)) \cdot n(x) \, ds_x +$

$\displaystyle \int_\Omega u(x) \cdot (\nabla \times v(x)) \, dx.$

**Exercise 5.4** We consider a mapping $x : \mathbb{R} \times \mathbb{R}^3 \to \mathbb{R}^3$ which maps Lagrangian coordinates onto Eulerian coordinates

$$x(t, X) = \begin{pmatrix} X_1 + t \\ e^t X_2 \\ X_3 + t X_1 \end{pmatrix}.$$

(a) Compute the corresponding velocity field $v(t, x)$ in Eulerian coordinates.
(b) Let the function

$$\varrho(t, x) = x_1 + x_2 \sin t$$

be given. Verify the formula

$$D_t \varrho(t, x) = \frac{d}{dt} \varrho(t, x(t, X))$$

with the material derivative $D_t$.
(c) We now consider the two-dimensional velocity field

$$v(t, x) = \begin{pmatrix} x_2 \\ x_1 \end{pmatrix}.$$

Show that the streamlines are given by hyperbolas and show that the streamlines coincide with the pathlines.

**Exercise 5.5** Let $\Omega \subset \mathbb{R}^d$ be a bounded domain and let $f : \Omega \to \mathbb{R}$ be a continuous function. Show the following fact: If

$$\int_U f(x) \, dx = 0$$

for all bounded subdomains $U \subset \Omega$ it holds $f(x) = 0$ for all $x \in \Omega$.

**Exercise 5.6** (*Observer invariance*) An observer notices the movement of a mass point with mass $m$ which in his coordinates $x = (x_1, x_2, x_3)$ is described by the momentum balance $m x''(t) = K$.

(a) Another observer has coordinates $y = (y_1, y_2, y_3)$, satisfying $y(t) = Q(t) x(t) + a(t)$ with smooth mappings $t \mapsto Q(t)$ and $t \mapsto a(t)$ from the coordinates of the first observer. Here $Q(t)$ is for all $t$ an orthogonal matrix with determinant 1. Show that this observer as well describes the movement with a momentum balance $m y''(t) = \widetilde{K}$. In order to do so you have to determine $\widetilde{K}$.
(b) Let $\{e_1, e_2, e_3\}$ be an orthonormal basis of the first observer. The basis of the second observer with vectors $\{b_1, b_2, b_3\}$ are computed from the $\{e_i\}_i$ by a rotation around the $e_3$-axis. We assume that the angular speed $\omega$ is constant. This means that the second observer rotates with a constant velocity around the $e_3$-axis.

Determine the matrix $Q(t)$ from a) and compute how the force $\widetilde{K}$ depends on $K$ and $\omega$.

**Exercise 5.7** (*Two body problem*) We want to describe the movement of a planet with the mass $m_P$ around a sun which is assumed to have mass $m_S$. By $x_P(t)$ and $x_S(t)$ we denote the positions of the planet and the sun and

$$x(t) = x_P(t) - x_S(t)$$

is the oriented distance between sun and planet.

(a) Derive the equations for the description of the movement of the sun and the planet. Compute the kinetic energy $E_{\text{kin}}$, the potential energy $E_{\text{pot}}$ and the angular momentum $L$ of the two body system and verify that the total energy $E = E_{\text{kin}} + E_{\text{pot}}$ and the angular momentum are conserved quantities.

(b) Show that the movement of the planet around the sun can be described by the equation

$$m_P\, x''(t) = -G\, m_P\, m\, \frac{x(t)}{|x(t)|^3}$$

with the gravitational constant $G$ and the total mass $m = m_S + m_P$.

(c) Prove Kepler's second law: The distance vector $x$ between planet and sun sweeps out the area $A_{\Delta t}$ in a given time interval $\Delta t$. Use and justify the fact that for $\Delta t = t_2 - t_1$ it holds:

$$A_{\Delta t} = \frac{1}{2} \int_{t_1}^{t_2} |x(t) \times x'(t)|\, dt\,.$$

**Exercise 5.8** (*Kepler's laws*) For the movement of a planet with mass $m_P$ around a sun with mass $m_S$ we already considered in Exercise 5.7 the following differential equation:

$$m_P\, x''(t) = -G\, m_P\, m\, \frac{x(t)}{|x(t)|^3}\,,$$

with $m = m_S + m_P$. The function $t \mapsto x(t)$ describes the orbit of the planet relative to the sun. We already computed that the energy and the angular momentum

$$E = \frac{-G\, m_P\, m}{|x|} + \frac{m_P}{2}|x'|^2 \quad \text{and} \quad L = m_P\, x \times x'\,,$$

are conserved quantities. If we assume $L \neq 0$, we obtain that the orbit of the planet lies in a plane. Let this plane be spanned by the orthonormal basis $\{e_1, e_2\}$. For an angle $\varphi$ we define

$$e_r := \cos\varphi\, e_1 + \sin\varphi\, e_2\,,$$
$$e_\varphi := -\sin\varphi\, e_1 + \cos\varphi\, e_2\,.$$

We now consider the movement in polar coordinates and set $x = r e_r$.

(a)  Derive from the conservation laws the following system:

$$r^2 \dot{\varphi} = \frac{|L|}{m_P} ,$$

(5.100)

$$(\dot{r})^2 + r^2 (\dot{\varphi})^2 - 2\frac{G\, m}{r} = 2\frac{E}{m_P} .$$

(5.101)

(b)  Consider $r$ as a function of $\varphi$ and show that

$$r = \frac{p}{1 + e \cos(\varphi)} \quad \text{with} \quad p = \frac{|L|^2}{G\, m\, m_P^2}, \quad e = \sqrt{1 + \frac{2E|L|^2}{G^2\, m^2\, m_P^3}}$$

fulfills the Eqs. (5.100) and (5.101).

We now consider the case $E < 0$ which is equivalent to $e < 1$.

(c)  Prove with the help of the coordinates $x_1 = r \cos \varphi$ and $x_2 = r \sin \varphi$ Kepler's first law: The planet moves on an elliptic orbit around the sun. In order to do so derive the equation of an ellipse

$$\frac{(x_1 + ea)^2}{a^2} + \frac{x_2^2}{b^2} = 1$$

with suitable constants $a, b$.

(d)  Derive the formula

$$t = \frac{m_P}{|L|} \int_{\varphi(0)}^{\varphi(t)} r^2(\varphi)\, d\varphi$$

for the time and derive for the orbital period $T$ the formula

$$\frac{T^2}{a^3} = \frac{4\pi^2}{G\, m} .$$

Neglecting the planet mass $m_P$ in comparison to the mass of the sun $m_S$ we obtain Kepler's third law $T^2/a^3 = \text{const}$ with a constant which does not depend on the particular planet.

**Exercise 5.9**  (*Piola identity*) The mapping $X \mapsto x(t, X)$ with $X \in E^3$ which maps Lagrangian coordinates to Eulerian coordinates is assumed to be smooth and invertible. We denote by $B(t, X) := \frac{\partial x}{\partial X}(t, X) = \left( \frac{\partial x_i}{\partial X_j} \right)_{i,j}$ the derivative and set $J(t, X) :=$ det $B(t, X)$.

Show: $\nabla_X \cdot \left( J B^{-\top} \right) = 0$ where $B^{-\top}$ is the transposed inverse of $B$.
Note: Let $V \subset E^3$ be open and arbitrary and let $U = x(V) \subset E^3$. For $\bar{\eta} \in C_0^\infty(U)$ let $\eta(x(X)) := \bar{\eta}(X)$. Show

$$\int_V \nabla_X \cdot (J(t, X)B^{-\top}(t, X))\, \overline{\eta}(X)\, dX = -\int_U \nabla_x \eta(x)\, dx = 0\,.$$

How does this help to show the assertion?

**Exercise 5.10** (*Conservation laws in Lagrangian coordinates*) In Sect. 5.5 we formulated the conservation laws for mass and momentum in Eulerian coordinates. With respect to Lagrangian coordinates $X$ and the Eulerian coordinates $x$ we use here the same notation as in Exercise 5.9 where also $J$ and $B$ have been defined.

Show that the conservation laws for mass and momentum in Lagrangian coordinates have the following form:

$$\partial_t(\overline{\varrho}J) = 0\,,$$
$$\varrho_0\, \partial_t V = \varrho_0 F + \nabla_X \cdot S\,,$$

where $\overline{\varrho}(t, X) = \varrho(t, x(t, X))$, $V(t, X) = v(t, x(t, X))$, $F(t, X) = f(t, x(t, X))$ and

$$\varrho_0(X) = (\overline{\varrho}J)(t, X) \quad \text{and} \quad S(t, X) = J(t, X)\sigma(t, x(t, X))B^{-\top}(t, X)\,.$$

Why is $\varrho_0$ well-defined?

**Exercise 5.11** (*Methods of characteristics*) In order to solve the mass balance equation

$$\partial_t \varrho + \partial_x(v\varrho) = 0 \ \text{ for } (t, x) \in (0, \infty) \times \mathbb{R}\,,$$
$$\varrho = \varrho_0 \ \text{ for } t = 0,\ x \in \mathbb{R}$$

one can use the *method of characteristics*. One computes the solutions of

$$\partial_t s(t, x_0) = v(t, s(t, x_0)), \quad s(0, x_0) = x_0 \tag{5.102}$$

for all $x_0 \in \mathbb{R}$ and writes $z(t, x_0) := \varrho(t, s(t, x_0))\,.$

(a) Show that $t \to z(t, x_0)$ then solves the following initial value problem:

$$\partial_t z(t, x_0) = -\partial_x v(t, x)_{|x=s(t, x_0)}\, z(t, x_0), \quad z(0, x_0) = \varrho_0(x_0) \tag{5.103}$$

(b) Let $v(t, x) = x/(1 + t)$ and $\varrho_0(x) = 1 - \tanh(x)$ be given. Solve the initial value problems (5.102) and (5.103) and sketch the characteristics $s(t, x_0)$ and the corresponding values $\varrho(t, s(t, x_0))$.

**Exercise 5.12** (*Methods of characteristics for multidimensional transport equations*) Consider the differential equation

$$\partial_t u(t, x) + v(t, x) \cdot \nabla u(t, x) = f(t, x) \tag{5.104}$$

with velocity field $v(t, x)$, source term $f(t, x)$ and initial values $u(0, x) = u_0(x)$.

(a) Let $s(t, y)$ be the solution of the equation for the characteristics

$$\frac{d}{dt}s(t, y) = v(t, s(t, y)), \quad s(0, y) = y$$

and $u(t, x)$ be a solution of (5.104). Show that $w(t, y) = u(t, s(t, y))$ solves the equation

$$\frac{d}{dt}w(t, y) = f(t, s(t, y))$$

and determine with the help of this an expression for the solution of the transport Eq. (5.104).
(b) Solve (5.104) for

$$v(t, x) = \begin{pmatrix} 1 \\ x_3 \\ -x_2 \end{pmatrix}, \quad f(t, x) = e^{-(t+x_1+x_2^2+x_3^2)} \text{ and } u_0(x) = 0.$$

**Exercise 5.13** Let $V$ be a tetrahedron in $\mathbb{R}^d$ with lateral surfaces $S_1, \ldots, S_d$ and $S$, where $S_j \subset \{x \in \mathbb{R}^d \mid x_j = 0\}$ and such that $S$ has the outer unit normal vector $n$. Furthermore $V$ is assumed to be such that $n_j > 0$ for $j = 1, \ldots, d$ and $x_j \geq 0$ for $x \in V$ and $j = 1, \ldots, d$.

(a) Show for $d = 3$ with geometric arguments that $\frac{|S_j|}{|S|} = n_j$.
(b) Show with the help of the divergence theorem that for all $d$ the equation $\frac{|S_j|}{|S|} = h_j$ holds.

**Exercise 5.14** Determine the physical dimensions of the following quantities: stress tensor $\sigma$, heat flux $q$, species flux $j_i$, heat conductivity $K$, diffusion constant $D$, viscosities $\mu$ and $\lambda$ of viscous flows.

**Exercise 5.15** Prove under the assumptions stated in Sect. 5.6 that the inequality (5.10) is true. Now let $L > 0$. For which initial data do we obtain equality in (5.10)?

**Exercise 5.16** (*Non-dimensionalization of the heat equation*) We assume that we know the solution of the problem

$$\partial_t u(t, x) - \partial_x^2 u(t, x) = 0 \text{ for all } t > 0, \ x \in (0, 1)$$

with $u(0, x) = 0$, $u(t, 0) = 0$, $u(t, 1) = 1$.
    Compute with the help of this solution the solution of

$$c_V \partial_t \tilde{u}(t, x) - \lambda \partial_x^2 \tilde{u}(t, x) = 0 \text{ for all } t > 0, \ x \in (0, L)$$

with $\tilde{u}(0, x) = u_0$, $\tilde{u}(t, 0) = u_0$, $\tilde{u}(t, L) = u_1$.
*Hint:* Use the ansatz $\tilde{u}(t, x) = a_0 + a_1 u(b_0 t, b_1 x)$ and determine the constants $a_0, a_1$, $b_0, b_1$ in a suitable way.

**Exercise 5.17** In the interior of a ball-shaped house with exterior radius $R = 10\,\mathrm{m}$ and wall thickness $a = 1\,\mathrm{m}$ the temperature is assumed to be $T = 20\,°\mathrm{C}$ and outside of the ball the temperature shall be given as $T = 0\,°\mathrm{C}$.

(a) Solve the stationary heat equation

$$\Delta T = 0$$

in the wall with boundary condition $T = 20\,°\mathrm{C}$ for $|x| = R - a$ and $T = 0\,°\mathrm{C}$ for $|x| = R$. In order to do so use the radially symmetric ansatz $T(x) = u(|x|)$.

(b) Compute the heat flux through the wall for the heat conductivity $\lambda = 1{,}0\,\mathrm{W/(mK)}$ (concrete) and compute the heating power needed to maintain the temperature.

**Exercise 5.18** We consider the heat equation

$$\partial_t u(t, x) - \Delta u(t, x) = 0 \text{ for } t > 0, \; x \in \Omega \tag{5.105}$$

with $u(0, x) = u_0(x)$ for $x \in \Omega$. The boundary conditions are either

(i) $u(t, x) = 0$ for $t > 0, x \in \Gamma = \partial\Omega$, or
(ii) $\nabla u(t, x) \cdot n(x) = 0$ for $t > 0, x \in \Gamma$.

The domain $\Omega \subset \mathbb{R}^d$ is assumed to be bounded and sufficiently smooth and the solution $u$ is also assumed to be sufficiently smooth.

(a) Show

$$\int_\Omega u^2(t, x)\, dx = \int_\Omega u_0^2(x)\, dx - \int_0^t \int_\Omega |\nabla u(s, x)|^2\, dx\, ds\,.$$

*Hint:* Multiply (5.105) by $u(t, x)$ and integrate over $\Omega$.

(b) Show for the case (ii):

$$\int_\Omega u(t, x)\, dx = \int_\Omega u_0(x)\, dx\,.$$

(c) Assume that the limits $\lim_{t \to +\infty} u(t, x) = w(x)$ and $\lim_{t \to +\infty} \nabla u(t, x) = \nabla w(x)$ exist where the convergence is assumed to be uniform with respect to $x$. Determine the limit $w(x)$ for the cases (i) and (ii).

(d) Now consider the solution of (5.105) with $u(0, x) = u_0(x)$ for $x \in \Omega$ and $u(t, x) = U(x)$ for $x \in \Gamma$. Assume that $\lim_{t \to +\infty} u(t, x) = w(x)$ and $\lim_{t \to +\infty} \nabla u(t, x) = \nabla w(x)$ exist uniformly with respect to $x \in \Omega$ and that $u, w$ are sufficiently smooth.
Show, that $w$ is a solution of

$$\Delta w = 0 \text{ for } x \in \Omega, \quad w(x) = U(x) \text{ for } x \in \Gamma\,.$$

*Hint:* Use (c) for a suitable variant of the problem.

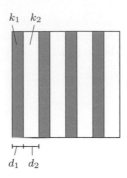

$k_1$  $k_2$

$d_1$  $d_2$

**Fig. 5.25**  Layered material

**Exercise 5.19**  (*Heat conduction through a layered material*) We consider a work-piece made up by thin layers of different materials as shown in Fig. 5.25. The two materials are assumed to have heat conductivities $k_1$ and $k_2$ and the layers are assumed to have thickness $d_1$ and $d_2$.

(a)  We want to compute the heat flux through a wall which is infinitely extended in two directions and which is composed of a composite material as above with total thickness $d = n(d_1 + d_2)$. The layers are assumed to be either (i) parallel to the wall or (ii) perpendicular to the wall. On the two exterior sides of the wall the two constant temperatures $T_I$ and $T_A$ are given. Compute for both cases the heat flux through the wall in dependence on the temperature difference $T_I - T_A$. *Hint:* Derive first coupling conditions for the interior surfaces between the layers. One of these conditions follows from the energy balance in integral form.

(b)  Compute with the help of part (a) an *effective heat conductivity* for the composite material for the two cases (i) parallel to the layers and (ii) perpendicular to the layers.

(c)  Compute with the help of (b) the *heat conductivity matrix*, which is the $3 \times 3$-matrix $K$ for the heat diffusion law

$$q = -K\nabla T$$

in a homogeneous substitute material for the layered material.

**Exercise 5.20**  In case that we neglect exterior forces and the diffusion of heat the Euler equations of gas dynamics have the form

$$\partial_t \varrho + \nabla \cdot (\varrho v) = 0,$$
$$\varrho \, \partial_t v + \varrho(v \cdot \nabla)v + \nabla p = 0,$$
$$\partial_t u + v \cdot \nabla u + \frac{p}{\varrho} \nabla \cdot v = 0$$

with a general pressure function $p = p(\varrho, u)$.

Show that these equations in $d$ spatial dimensions can be written as a *system of conservation laws of first order* for the vector field

$$w = (w_0, \ldots, w_{d+1}) = \left( \varrho, \varrho v, \varrho u + \tfrac{1}{2}\varrho |v|^2 \right).$$

This means the equations can be written in the form

$$\partial_t w_j + \nabla \cdot F_j(w) = 0 \quad \text{for } j = 0, \ldots, d+1.$$

How does one have to choose the functions $F_j(w)$? What is the physical meaning of the components of $w$?

**Exercise 5.21** We would like to describe the flow of an incompressible viscous flow through a pipe

$$\Omega = \left\{ x \in \mathbb{R}^3 \,\middle|\, 0 < x_1 < L, \; x_2^2 + x_3^2 < R^2 \right\}$$

with length $L$ and radius $R$.

(a) Solve the stationary Navier–Stokes equations

$$\nabla \cdot v = 0,$$
$$\varrho(v \cdot \nabla)v - \mu \, \Delta v = -\nabla p$$

in $\Omega$ with boundary conditions $p(x) = p_1$ for $x_1 = 0$, $p(x) = p_2$ for $x_1 = L$, $v(x) = 0$ for $x_2^2 + x_3^2 = R^2$. Use the ansatz

$$v(x) = w(r(x)) \, e_1 \text{ with } r(x) = \sqrt{x_2^2 + x_3^2} \text{ and } p(x) = q(x_1).$$

(b) Compute the flow rate through the pipe and verify the law of Hagen–Poiseuille from Exercise 2.16.

**Exercise 5.22** (*Curl-free flow, potential flow*)

(a) Let $v$ be a stationary, incompressible, curl-free flow and let the density $\varrho$ be constant. Show that $v$ is a solution of Euler's equations where the pressure $p$ is given as $p = -\tfrac{\varrho}{2}|v|^2$.
(b) In $\mathbb{R}^2 \backslash \{(0,0)\}$ we consider the velocity field

$$v(x_1, x_2) = \frac{1}{x_1^2 + x_2^2} \begin{pmatrix} -x_2 \\ x_1 \end{pmatrix}.$$

Show that $v$ fulfills the assumptions of part (a).
(c) Show that the velocity $v$ from part (b) does not describe a potential flow.

**Exercise 5.23** (*Couette flow*) Let $\Omega$ be the region between two concentric cylinders with radii $R_1$ and $R_2$ with $R_1 < R_2$. In cylindrical coordinates $(r, \varphi, z)$ we consider the velocity field

$$v = \left( -\left( \tfrac{A}{r} + Br \right) \sin \varphi, \; \left( \tfrac{A}{r} + Br \right) \cos \varphi, \; 0 \right),$$

where

$$A = -\frac{R_1^2 R_2^2 (\omega_2 - \omega_1)}{R_2^2 - R_1^2}, \quad B = -\frac{R_1^2 \omega_1 - R_2^2 \omega_2}{R_2^2 - R_1^2}.$$

(a)  Show that $v$ is a stationary solution of Euler's equations with $\varrho \equiv 1$. Which form has the pressure?
(b)  Compute $\nabla \times v$.
(c)  Which boundary conditions hold for $v$? How can we interpret $\omega_1$ and $\omega_2$?

**Exercise 5.24** (*Navier–Stokes operator, divergence theorem*) In a domain $\Omega$ let $(v, p)$ be a smooth solution of the homogeneous Navier–Stokes equations for incompressible fluids:

$$\varrho \, \partial_t v + \varrho \, (v \cdot \nabla) v - \mu \, \Delta v + \nabla p = 0 \quad \text{in } \Omega,$$
$$\nabla \cdot v = 0 \quad \text{in } \Omega,$$
$$v = 0 \quad \text{on } \Gamma = \partial\Omega.$$

We assume that the mass $\varrho$ and the dynamic viscosity $\mu$ are positive constants. Show

$$\frac{d}{dt}\left( \int_\Omega \frac{\varrho}{2} |v|^2 \, dx \right) + \int_\Omega \mu |Dv|^2 \, dx = 0.$$

Here $|Dv|^2 = \sum_{i=1}^{3} |\nabla v_i|^2$.

Remark: This identity shows that the kinetic energy in an incompressible fluid, on which no outer forces act, can never increase.

**Exercise 5.25** (*Theorem of Bernoulli*) We consider a flow described by the quantities $(v, \varrho, T)$ which fulfill the conservation laws for mass and momentum. The stress tensor is assumed to be given by $\sigma = -pI$ with a pressure $p$, the force in the conservation law for momentum is assumed to be given by $f = -\nabla\beta$. Here $\beta$ is the potential for the force. Derive the following assertions:

(a)  If a potential flow is given, i.e., $v = \nabla\varphi$ for a real function $\varphi$, we have

$$\nabla \left( \partial_t \varphi + \tfrac{1}{2}|v|^2 + \beta \right) + \tfrac{1}{\varrho}\nabla p = 0.$$

   *Hint:* In this case we have $\nabla \times v = 0$.
(b)  If the flow is stationary, i.e., if the partial derivatives with respect to time of $(v, \varrho, \sigma)$ vanish, we obtain

$$v \cdot \nabla \left( \tfrac{1}{2}|v|^2 + \beta \right) + \tfrac{1}{\varrho} v \cdot \nabla p = 0.$$

**Exercise 5.26** (*Gas dynamics*) The equations for a non-viscous (i.e., the stress tensor only contains a pressure part), non heat-conducting (i.e., $q = 0$) gas in one spatial dimension in Eulerian coordinates are given by

$$\partial_t \varrho + \partial_x(\varrho v) = 0\,,$$

$$\partial_t(\varrho\, v) + \partial_x\big(\varrho v^2 + p\big) = 0\,,$$

$$\partial_t\Big[\varrho\big(\tfrac{v^2}{2} + u\big)\Big] + \partial_x\Big[\varrho\, v\big(\tfrac{v^2}{2} + u\big) + p\, v\Big] = 0\,.$$

Show that under the assumption $\varrho = 1$ these equations are equivalent to the following formulation in Lagrangian coordinates:

$$\partial_t C - \partial_X V = 0\,,$$

$$\partial_t V + \partial_X P = 0\,,$$

$$\partial_t\Big(\tfrac{V^2}{2} + U\Big) + \partial_X(PV) = 0\,.$$

Here $c(t,x) = 1/\varrho(t,x)$ is the specific volume and the Lagrangian variables in capital letters correspond to small letters in Eulerian coordinates (e.g., $V(t,X) = v(t,x(t,X))$).

Show in addition, that the last equation is equivalent to $\partial_t S = 0$, where $S$ is the entropy in Lagrangian coordinates.

Remark: It is assumed that all functions are sufficiently smooth.

**Exercise 5.27**  (*Variable transformation*) We consider the mapping

$$\Phi : (\varrho, T) \mapsto (V, s) := \big(\tfrac{1}{\varrho}, -\partial_T f(\varrho, T)\big)\,,$$

where $\varrho$ is the mass, $T$ is the temperature, $V$ the specific volume, $s$ the entropy and $f = f(\varrho, T)$ is the density of the free energy. For the free energy density we assume:

$$\partial_T f(\varrho, \cdot) \text{ is strictly monotonically increasing.}$$

It is possible to consider the inverse of $\Phi$, which means that we can express $\varrho$ and $T$ as functions of $V$ and $s$. In doing so we obtain for the free energy density

$$u(V, s) = f(\varrho(V, s), T(V, s)) + T(V, s)s$$
$$= f(\Phi^{-1}(V, s)) + (\Phi^{-1})_2(V, s)s\,,$$

where $(\Phi^{-1})_2$ is the second component of $\Phi^{-1}$. By using the known thermodynamic relations show that the following identities are true

$$\partial_V u(V, s) = -p(V, s)\,, \quad \partial_s u(V, s) = T(V, s)\,.$$

**Exercise 5.28**  (*Gas dynamics*) For gases with a temperature close to room temperature and small densities, up to first order the following relations for the pressure $p$ and the specific heat capacity $c = \partial_T u$ are true:

$$p = r\varrho T\,, \quad c = \frac{\alpha r}{2}\,,$$

where $r$ is the gas constant and $\alpha$ is a natural number.

Show
$$u(T, \varrho) = c\,T + d\,, \quad s(T, \varrho) = c\ln(T\,\varrho^{1-\gamma})$$

with a constant $d$ and $\gamma = 1 + \frac{r}{c}$.

In addition, express $u$ and $p$ as functions of $s$, $V$, where $V = 1/\varrho$ is the specific volume.

**Exercise 5.29** (*Entropy identity*) We assume the conservation laws for mass, momentum and energy

$$\partial_t \varrho + \nabla \cdot (\varrho v) = 0\,,$$
$$\partial_t(\varrho v) + \nabla \cdot (\varrho v v^{\mathsf{T}} - \sigma) = \varrho \overline{f}\,,$$
$$\partial_t\left(\varrho\left(\frac{|v|^2}{2} + u\right)\right) + \nabla \cdot \left(\varrho v\left(\frac{|v|^2}{2} + u\right) + q - \sigma v\right) = \varrho \overline{f} \cdot v$$

with a force density $\overline{f}$ and the thermodynamic relations

$$s = -\partial_T f\,, \quad p = \varrho^2\partial_\varrho f\,, \quad u = f + Ts\,. \tag{5.106}$$

These relations are true if we write $s, p, u$ as functions of $T$ and $\varrho$. Show the following equation:

$$\partial_t(\varrho s) + \nabla \cdot \left(\frac{q}{T} + \varrho s v\right) = \frac{1}{T}(\sigma + pI) : Dv + q \cdot \nabla\left(\frac{1}{T}\right)\,.$$

**Exercise 5.30** (*Transport identity*) Let the stress tensor be given by $\sigma = -pI$ with a function $p = p(\varrho)$ for the pressure. In addition, let $q = 0$ (no heat flux) and $g = 0$ (no exterior heat sources). We consider the free energy $f$ and the internal energy $u$ as functions of density $\varrho$ and temperature $T$. In addition, we assume the thermodynamic relations (5.106).

(a) Show: $f(\varrho, T) = f_1(\varrho) + f_2(T)$ with suitable functions $f_1, f_2$ and $\partial_\varrho u = \partial_\varrho f$.
(b) Use the energy conservation law, the mass and momentum conservation and a suitable assumption on the heat capacity $c = \partial_T u$ to derive the following transport equation for the temperature:

$$\partial_t T + v \cdot \nabla T = 0\,.$$

(c) Assume that the velocity is given by $v(t, x) = x\,e^{-t}, x \in \mathbb{R}^3$. In addition, assume that the temperature at initial time $t = 0$ is given by a function $T_0$.
Solve the transport equation in (b) with the above velocity with the help of the methods of characteristics discussed in Exercise 5.12.
*Hint:* Take into account that here the equation has a simpler structure compared to Exercise 5.12.

**Exercise 5.31** (*Observer invariance*) In Sect. 5.5 we showed that for an isotropic material the stress tensor only depends on the symmetric part $\varepsilon(v) = \frac{1}{2}\big(Dv + (Dv)^\top\big)$ of $Dv$. We now assume $\sigma(\varrho, T, Dv) = -p(\varrho, T)I + S(\varepsilon(v))$.

With the help of the theorem of Rivlin–Ericksen we showed that under the assumption that $S$ is linear and observer invariant the following representation holds

$$S(E) = \lambda \, \mathrm{trace}(E)I + 2\mu E \, ,$$

where $\mu$ and $\lambda$ are real constants.

We now assume that the shear modulus $\mu$ only depends on $|E| = \sqrt{E : E}$, i.e., $\mu = \mu(|E|)$. Show that in this case $S$ is still observer independent and isotropic.

**Exercise 5.32** The mapping

$$\widehat{S} : \{A \in \mathbb{R}^{3,3} \mid A \text{ is symmetric }\} \to \mathbb{R}^{3,3}$$

is assumed to be linear and to fulfill one of the equivalent assertions in the Rivlin–Ericksen theorem 5.13. Show that there exists constants $\lambda, \mu \in \mathbb{R}$ such that

$$\widehat{S}(A) = 2\mu A + \lambda \, \mathrm{trace}(A)I \, .$$

**Exercise 5.33** We consider the elasticity system

$$\varrho \, \partial_t^2 u - \nabla \cdot \sigma(u) = f$$

in a bounded domain $\Omega \subset \mathbb{R}^d$ with initial conditions $u(t_0, x) = u_0(x)$ and $\partial_t u(t_0, x) = u_1(x)$ for $x \in \Omega$ and boundary conditions $\sigma(u; t, x)n(x) = b(t, x)$ for $t > 0$, $x \in \Gamma = \partial\Omega$.

(a)  (Linear elasticity) Let

$$\sigma_{ij}(u) = \sum_{k,\ell=1}^{d} a_{ijk\ell} \partial_{x_\ell} u_k$$

with coefficients $a_{ijk\ell} \in \mathbb{R}$, which fulfill the symmetry conditions $a_{ijk\ell} = a_{jik\ell} = a_{k\ell ij}$ be given. Derive the following *energy conservation identity*:

$$\int_{\Omega} \left[ \tfrac{\varrho}{2}|\partial_t u(t_1, x)|^2 + \tfrac{1}{2}\sigma(u; t_1, x) : Du(t_1, x) \right] dx$$

$$= \int_{\Omega} \left[ \tfrac{\varrho}{2}|u_1(x)|^2 + \tfrac{1}{2}\sigma(u_0; x) : Du_0(x) \right] dx$$

$$+ \int_{t_0}^{t_1} \left[ \int_{\Omega} f(t, x) \cdot \partial_t u(t, x) \, dx + \int_{\Gamma} b(t, x) \cdot \partial_t u(t, x) \, ds_x \right] dt$$

(b)  (Nonlinear elasticity) How does the energy conservation law look like for

$$\sigma_{ij}(u) = \frac{\partial W}{\partial X_{ij}}(Du)$$

with $W : \mathbb{R}^{d,d} \to \mathbb{R}$?

**Exercise 5.34**  (a)  Determine for an isotropic, linear elastic material the strain and the stress tensor for $u(x) = \gamma x_2 e_1$. Interpret $u$ geometrically and discuss why the Lamé constant $\mu$ is also called shear modulus.

(b)  Determine for an isotropic, linear elastic material the strain and the stress tensor for $u(x) = \delta x$. Why is the bulk modulus $K = \frac{2}{3}\mu + \lambda$ sometimes also called the compressive modulus?

**Exercise 5.35**  Use the classical Stokes theorem and the fact that (5.68) holds for all curves $\gamma$ in order to show that (5.71) is true.

**Exercise 5.36**  Compute the electric field and the potential for a charged ball with radius $R$ and charge $q$, where

(a)  the charge is homogeneously distributed in the volume of the ball,

or

(b)  the charge is distributed homogeneously on the surface of the ball.

**Exercise 5.37**  We consider a capacitor with two spherical shells with radii $0 < R - a < R$ and potential difference $U$. The medium between the two spherical shells is assumed to have a dielectric constant $\varepsilon$. Compute the electric field between the spherical shells and the charge densities on the spherical shells. With the help of this derive the capacitance of the capacitor.

# Chapter 6
# Partial Differential Equations

In this chapter we will discuss the partial differential equations which appear in continuum mechanics in more detail. The essential features of the analysis of these equations will enable us to understand properties of problems that appear in applications. This chapter cannot replace a textbook on partial differential equations. Often we will only sketch the analysis and justifications for assertions to serve as a motivation and we will not always be rigorous. For a deeper understanding of partial differential equations we refer to the vast literature on this theme. The main goal of the Chapter will be however to derive properties which are need to understand the underlying models.

## 6.1 Elliptic Equations

Elliptic equations often appear as *stationary* long-time limits of dynamical problems. Important examples are the stationary heat equation (5.14), the Stokes equations (5.16) or the equations of elastostatics (5.55). A simple example of an elliptic equation is

$$- \nabla \cdot (\lambda \nabla u) = f \quad \text{for } x \in \Omega \subset \mathbb{R}^d \tag{6.1}$$

with a function $\lambda : \Omega \to \mathbb{R}$ which may depend on the spatial variable $x$. A special case is Poisson's equation

$$-\Delta u = f .$$

To be able to obtain a unique solution one also has to formulate boundary conditions. Typically these are either

© Springer International Publishing AG 2017
C. Eck et al., *Mathematical Modeling*, Springer Undergraduate
Mathematics Series, DOI 10.1007/978-3-319-55161-6_6

*Dirichlet boundary conditions*, where the values of $u$ on the boundary are fixed,

$$u = u_0 \text{ on } \partial\Omega \tag{6.2}$$

or
*Neumann boundary conditions*, where the flux into the domain is fixed. This means $-\lambda \nabla u \cdot (-n)$ will be prescribed and we set

$$\lambda \nabla u \cdot n = g \text{ on } \partial\Omega, \tag{6.3}$$

where $g : \partial\Omega \to \mathbb{R}$ is the flux into the domain and $n$ is the outer unit normal on $\partial\Omega$.

One can understand easily the necessity of boundary conditions with the help of the stationary heat equation. The temperature in the interior of a body depends essentially on how much heat is supplied on the boundary. More specifically one has to know either the temperature on the boundary of the body or the heat flux on the boundary. Besides pure Dirichlet or Neumann conditions there are also boundary conditions of mixed forms. For example it is possible to specify Dirichlet conditions on parts of the boundary and Neumann conditions on the rest of the boundary. Another class of mixed form boundary conditions is the class of *third type boundary conditions* which are also often called *Robin boundary conditions*

$$- \lambda \nabla u \cdot n = a(u - b) \tag{6.4}$$

with functions $a, b : \partial\Omega \to \mathbb{R}$. For the example of the stationary heat equation this models the contact of a body with a heat conducting exterior medium with fixed temperature. In such a medium a diffusive boundary layer will occur close to the boundary in such a way that the temperature in the exterior medium close to the body under consideration will differ from the temperature further away in the medium. A sensible constitutive ansatz in this case is *Newton's law of cooling*. This law says that the heat loss on the boundary of the body under consideration is proportional to the difference between the temperature on the boundary and the boundary of the exterior medium. The function $b$ describes the temperature of the exterior medium, $a$ is the so-called *heat transfer coefficient*, which in particular will depend on the material data of the exterior medium.

### 6.1.1  Calculus of Variations

The classical interpretation of Eq. (6.1) demands that both sides of the equation are continuous functions and that the equation is fulfilled for all points $x$ in the domain $\Omega$. For this to hold one has to assume that $u$ is *two times continuously differentiable*, i.e., $u \in C^2(\Omega)$. In this case one speaks of a *classical solution*. We can often describe solutions of elliptic partial differential equations using an equivalent optimization problem. In the following theorem this will be formulated for the case of Dirichlet boundary conditions.

**Theorem 6.1** *Let $\Omega$ be a bounded domain with smooth boundary, the functions $f$ : $\overline{\Omega} \to \mathbb{R}$, $u_0 : \partial\Omega \to \mathbb{R}$ and $\lambda : \Omega \to \mathbb{R}$ are assumed to be smooth and $\lambda_0 \le \lambda(x) \le \lambda_1$ with $0 < \lambda_0 \le \lambda_1 < \infty$. Then a two times continuously differentiable function $u : \overline{\Omega} \to \mathbb{R}$ is a solution of the elliptic equation (6.1) with boundary condition $u(x) = u_0(x)$ for $x \in \partial\Omega$ if and only if $u$ is a solution of the optimization problem*

$$\min_{u \in V} \left\{ \int_\Omega \left( \tfrac{\lambda}{2} |\nabla u|^2 - fu \right) dx \,\Big|\, u \in V \right\}. \tag{6.5}$$

*Here we set $V = \{v \in C^2(\overline{\Omega}) \mid v(x) = u_0(x) \text{ for } x \in \partial\Omega\}$.*

*Proof* We show that (6.1) is a necessary criterion for an optimum in (6.5). Let $u$ be a solution of (6.5) and $v \in C^2(\overline{\Omega})$ with $v = 0$ on $\partial\Omega$. In this case $u + \varepsilon v \in V$ for $\varepsilon \in \mathbb{R}$ is an admissible comparison function in (6.5). As $u$ is a solution of the optimization problem we obtain

$$0 = \frac{d}{d\varepsilon} \int_\Omega \left( \lambda |\nabla(u + \varepsilon v)|^2 - 2f(u + \varepsilon v) \right) dx \Big|_{\varepsilon=0}$$

$$= \int_\Omega \left( 2\lambda \nabla u \cdot \nabla v - 2fv \right) dx = -2 \int_\Omega \left[ \nabla \cdot (\lambda \nabla u) + f \right] v \, dx.$$

Here the last identity follows from an application of the divergence theorem for the vector field $\lambda v \nabla u$. As this is fulfilled for all $v$ we obtain that (6.1) holds. If $u$ fulfills (6.1) and if $u = u_0$ on $\partial\Omega$ holds we obtain for all $w \in V$:

$$\int_\Omega \left( \lambda |\nabla w|^2 - 2fw \right) dx = \int_\Omega \lambda |\nabla(w - u)|^2 \, dx + \int_\Omega \left( \lambda |\nabla u|^2 - 2fu \right) dx$$

$$+ 2 \int_\Omega \left( \lambda \nabla u \cdot \nabla(w - u) - f(w - u) \right) dx$$

$$\ge \int_\Omega \left( \lambda |\nabla u|^2 - 2fu \right) dx - 2 \int_\Omega \left( \nabla \cdot (\lambda \nabla u) + f \right)(w - u) \, dx$$

$$= \int_\Omega \left( \lambda |\nabla u|^2 - 2fu \right) dx.$$

This shows that $u$ is a solution of the optimization problem (6.5). □

In the optimization problem (6.5) the condition $u \in C^2(\overline{\Omega})$ is not necessary. It is sufficient that the gradient is defined and that its square is integrable. This observation can be used to define the concept of a *weak solution* to the differential equation (6.1). Let

$$L_2(\Omega) := \left\{ [f] \mid f : \Omega \to \mathbb{R} \text{ measurable}, \int_\Omega f^2 \, dx \text{ exists} \right\},$$

where $[f]$ is the *equivalence class* of the function $f$ with respect to the equivalence relation

$$f \sim g \Leftrightarrow f - g = 0 \text{ almost everywhere}.$$

This function space is a *Hilbert space*, i.e., a complete space with inner product defined as

$$\langle f, g \rangle_{L_2(\Omega)} = \int_\Omega f \, g \, dx \,.$$

In addition we define

$$H^1(\Omega) := \left\{ f \in L_2(\Omega) \,|\, \partial_{x_i} f \in L_2(\Omega) \text{ for } i = 1, \ldots, d \right\} \,.$$

Here $\partial_{x_i} f$ is the $i^{\text{th}}$ *weak* partial derivative. The weak derivative $g = \partial_{x_i} f \in L_2(\Omega)$ exists, if

$$\int_\Omega f \, \partial_{x_i} \varphi \, dx = - \int_\Omega g \, \varphi \, dx \text{ for all } \varphi \in C_0^\infty(\Omega) \,.$$

Here $C_0^\infty(\Omega)$ denotes the set of all infinitely times differentiable functions with compact support $\text{supp}\, \varphi = \overline{\{x \in \Omega \,|\, \varphi(x) \neq 0\}}$ in $\Omega$. As $\Omega$ is open this means that $\varphi$ is zero close to the boundary. If $f$ is continuously differentiable we obtain that the weak derivative coincides with the "classical" derivative. This follows easily with the help of integration by parts. In this sense the weak derivative is a generalization of the classical derivative. The function space $H^1(\Omega)$ is also a Hilbert space with the inner product

$$\langle f, g \rangle_{H^1(\Omega)} = \int_\Omega \left( f \, g + \nabla f \cdot \nabla g \right) dx \,.$$

A suitable function space for the optimization problem (6.5) is obtained by replacing the function space $C^2(\overline{\Omega})$ by $H^1(\Omega)$ in the definition of $V$. The space $H^1(\Omega)$ contains all functions for which all first partial derivatives exist (in a weak sense) and are integrable. The space $H^2(\Omega)$ then contains functions for which all first derivatives possess square-integrable weak derivatives. For further details we refer to [4, 37].

In order to formulate (6.1) for functions in $H^1(\Omega)$ one can introduce a *weak formulation*. To do so one multiplies the differential equation with a "test function" $v \in H^1(\Omega)$, for which $v = 0$ on $\partial\Omega$ holds. After integration by parts one obtains for the left-hand side

$$-\int_\Omega \nabla \cdot (\lambda \nabla u) \, v \, dx = \int_\Omega \lambda \nabla u \cdot \nabla v \, dx \,.$$

In addition we choose a function $u_0 \in H^1(\Omega)$ whose values on $\partial\Omega$ specify the boundary conditions.

**Definition 6.2**  A function $u \in H^1(\Omega)$ with $u = u_0$ on $\partial\Omega$ and for which

$$\int_\Omega \lambda \nabla u \cdot \nabla v \, dx = \int_\Omega f \, v \, dx$$

holds for all $v \in H^1(\Omega)$ with $v = 0$ on $\partial\Omega$ is called a weak solution of the boundary value problem (6.1), (6.2).

The condition $u = u_0$ on $\partial\Omega$ means that $u$ and $u_0$ in a suitable sense have the same boundary values on $\partial\Omega$. To specify boundary values for a function $u \in H^1(\Omega)$ is a nontrivial task. This is due to the fact that values $u(x)$ of an element $u \in H^1(\Omega)$ are not well-defined, as the values of different *representatives* of $u$ can differ on a set of measure zero and the boundary of a domain is typically a set of measure zero (in particular if the domain has a smooth boundary). However, it turns out that for functions in $H^1(\Omega)$ it is possible to identify boundary values in a unique way. The resulting function on the boundary is called the *trace* of $u$, see, e.g., [4, 37]. In the following we will also use the functions space

$$H_0^1(\Omega) := \{v \in H^1(\Omega) \mid v = 0 \ \text{ on } \ \partial\Omega\}.$$

The differential equation (6.1) is called *elliptic* because the mapping

$$(u, v) \rightarrow a(u, v) = \int_\Omega \lambda \nabla u \cdot \nabla v \, dx$$

is *positive definite* in the following sense: It holds that

$$a(u, u) \geq 0, \quad a(u, u) = 0 \Leftrightarrow u = 0$$

for all $u \in H_0^1(\Omega)$. Elliptic differential equations can be interpreted as *infinite dimensional* generalizations of linear equations with *positive semi-definite* system matrix. There is a clear analogy between the stationary heat equation (6.1) and linear systems for electrical networks, space frames and pipeline systems which we considered in Chap. 2. The linear systems have a common structure which consists of the following components:

- A vector $x \in \mathbb{R}^n$ which is the "primal" variable. The vector $x$ contains the electric potentials in an electrical network, the displacements in elastic space frames or pressures in a pipeline system.
- A vector $e = -Ax \in \mathbb{R}^m$ of *driving forces*. This vector describes the voltages in an electrical network, the strains in an elastic space frame or the pressure differences in a pipeline system.
- A vector $y = Ce + b$ of "dual" variables which consists of electric currents in an electrical network, the elastic stresses in an elastic space frame or the flow rates in a pipeline system.

- The linear system $A^\top y = f$, which typically result from a conservation law or an equilibrium condition. For the electrical network this system describes the conservation of charge, in an elastic space frame it describes the force balance and in a pipeline system the conservation of mass.

For the stationary heat equation we obtain:

- The temperature $u : \Omega \to \mathbb{R}$ as the "primal variable", i.e., "$u$ corresponds to $x$".
- The negative temperature gradient $-\nabla u$ as "driving force" for the heat flux, i.e., "$\nabla u$ corresponds to $e$" and "$\nabla$ corresponds to $A$". This analogy makes sense: The matrix $A$ describes typically the difference of values at the endpoints of edges of the network and $\nabla u$ is the vector of the partial derivatives which is obtained as the limit of difference quotients of values of $u$.
- The heat flux $q = -\lambda \nabla T$ is the "dual variable", i.e., "$q$ corresponds to $y$" and "$\lambda$ corresponds to $C$".
- The energy conservation $-\nabla \cdot q = f$ is the relevant conservation law and hence "$-\nabla\cdot$ corresponds to $A^\top$". Indeed the negative divergence "$-\nabla\cdot$" can be interpreted as the adjoint operator "$\nabla$" with respect to the inner product in $L_2(\Omega)$. This is true because integration by parts leads to

$$\langle -\nabla \cdot q, u \rangle_{L_2(\Omega)} = - \int_\Omega \nabla \cdot q\, u\, dx = \int_\Omega q \cdot \nabla u\, dx = \langle q, \nabla u \rangle_{L_2(\Omega)},$$

if either $q \cdot n = 0$ or $u = 0$ on the boundary $\partial\Omega$.

With the help of the weak formulation and the condition $0 < \lambda_0 \le \lambda(x) \le \lambda_1$ which is assumed to hold for (almost) all $x \in \Omega$ one can easily derive a so-called *energy estimate*. For simplicity we assume that the boundary data $u_0$ is defined on the whole domain $\Omega$ and that $u_0$ is smooth. Using the test function $v = u - u_0$ in the weak formulation leads to

$$\int_\Omega \lambda \nabla u \cdot \nabla (u - u_0)\, dx = \int_\Omega f(u - u_0)\, dx.$$

With the help of the Cauchy–Schwarz inequality

$$\int_\Omega v\, w\, dx \le \left( \int_\Omega v^2\, dx \right)^{1/2} \left( \int_\Omega w^2\, dx \right)^{1/2}$$

and Young's inequality (see [37]) $|2ab| \le \eta|a|^2 + \eta^{-1}|b|^2$ which holds for arbitrary $\eta > 0$ we obtain

$$\lambda_0 \int_\Omega |\nabla u|^2\, dx \le \frac{\eta_1 \lambda_1}{2} \int_\Omega |\nabla u|^2\, dx + \frac{\lambda_1}{2\eta_1} \int_\Omega |\nabla u_0|^2\, dx$$
$$+ \frac{\eta_2}{2} \int_\Omega |f|^2\, dx + \frac{1}{2\eta_2} \int_\Omega |u - u_0|^2\, dx.$$

Using *Poincaré's inequality* (see [4, 37])

$$\int_\Omega |v|^2 \, dx \le c_P \int_\Omega |\nabla v|^2 \, dx \,,$$

which holds for all $v \in H_0^1(\Omega)$ with a constant $c_p$ which only depends on $\Omega$, and the inequality $|(u - u_0)|^2 \le 2|u|^2 + 2|u_0|^2$ we obtain

$$\left(\lambda_0 - \frac{\eta_1 \lambda_1}{2} - \frac{c_P}{\eta_2}\right) \int_\Omega |\nabla u|^2 \, dx \le \frac{\eta_2}{2} \int_\Omega |f|^2 \, dx$$
$$+ \left(\frac{\lambda_1}{2\eta_1} + \frac{c_P}{\eta_2}\right) \int_\Omega |\nabla u_0|^2 \, dx \,.$$

Choosing suitable $\eta_1$ and $\eta_2$ this can be simplified to

$$\int_\Omega |\nabla u|^2 \, dx \le c \int_\Omega \left(|f|^2 + |\nabla u_0|^2\right) dx \qquad (6.6)$$

with a constant $c$ which only depends on $\Omega$ and on the constants $\lambda_0$ and $\lambda_1$.

With the help of this inequality we can easily show uniqueness of a weak solution to the elliptic boundary value problem (6.1), (6.2). Let $u_1$ and $u_2$ be two solutions of (6.1) for the same right hand side $f$ and the same Dirichlet boundary data $u_0$ on $\partial\Omega$. In this case the difference $u_1 - u_2$ is a solution of

$$-\nabla \cdot (\lambda \nabla (u_1 - u_2)) = 0 \text{ in } \Omega,$$
$$u_1 - u_2 = 0 \text{ on } \partial\Omega \,.$$

Using the inequality (6.6) for this boundary value problem yields

$$\int_\Omega |\nabla(u_1 - u_2)|^2 \, dx \le 0 \,.$$

Hence, $u_1 - u_2$ has to be constant on each connected component of $\Omega$. Since $u_1 = u_2$ on $\partial\Omega$ we obtain $u_1 = u_2$ everywhere.

Under very general assumptions it is possible to show the existence of a weak solution to the boundary value problem (6.1), (6.2), see [4, 37]. Altogether we obtain the following theorem.

**Theorem 6.3** *Let $\Omega$ be a bounded domain with smooth boundary, and let the coefficient function $\lambda : \Omega \to \mathbb{R}$ be measurable such that $0 < \lambda_0 \le \lambda(x) \le \lambda_1$ for almost all $x \in \Omega$ with $0 < \lambda_0 \le \lambda_1 < \infty$. Furthermore, let $f \in L_2(\Omega)$ and $u_0 \in H^1(\Omega)$. Under these assumptions the problem (6.1) with the boundary condition $u = u_0$ on $\partial\Omega$ has a unique weak solution $u \in H^1(\Omega)$.*

In the formulation of this theorem we assume that the boundary data $u_0$ are defined on the whole domain $\Omega$.

Let us now consider (6.1) with *Neumann conditions* (6.3). A weak formulation of this boundary value problem is obtained after multiplying the differential equation (6.1) with a test function $v$ and integration over the domain $\Omega$. After integration by parts

$$-\int_{\Omega} \nabla \cdot (\lambda \nabla u)\, v\, dx = \int_{\Omega} \lambda \nabla u \cdot \nabla v\, dx - \int_{\partial\Omega} \lambda \nabla u \cdot n\, v\, ds_x,$$

and taking the boundary condition into account we obtain

$$\int_{\Omega} \lambda \nabla u \cdot \nabla v\, dx = \int_{\Omega} f\, v\, dx + \int_{\partial\Omega} g\, v\, ds_x. \tag{6.7}$$

If (6.7) is fulfilled for all $v \in H^1(\Omega)$ then we say that $u$ is a *weak solution* of the elliptic differential equation (6.1) with the Neumann boundary condition (6.3). Plugging the test function $v = 1$ into the weak formulation (6.7) we obtain

$$\int_{\Omega} f\, dx + \int_{\partial\Omega} g\, ds_x = 0. \tag{6.8}$$

The unknown function $u$ does not appear in this equation. Hence, (6.8) is a condition the data $f$ and $g$ have to fulfill in order to be able to obtain a solution. Since (6.1), (6.3) only depend on $u$ via $\nabla u$ we can add a constant $c$ to a solution $u$ to obtain a new solution $u + c$. In contrast to the Dirichlet problem, a solution to the Neumann problem may not exist for all right-hand sides, and if a solution exists the solution will not be unique. One can show that (6.8) together with additional assumptions on the smoothness of the domain $\Omega$ and on the data $f$ and $g$ will guarantee the existence of a solution to the Neumann problem. More precisely the following theorem is true.

**Theorem 6.4**  *Let $\Omega$ be a bounded domain with smooth boundary $\partial\Omega$, $f : \overline{\Omega} \to \mathbb{R}$ and $g : \partial\Omega \to \mathbb{R}$ are assumed to be smooth functions. Then the boundary value problem (6.1), (6.3) has a weak solution if condition (6.8) is fulfilled. The solution is unique up to adding a constant, i.e., if $u_1$ and $u_2$ are two solutions it holds that $u_1 - u_2$ is constant.*

The proof of this theorem can be found in books on partial differential equations, see, e.g., [37]. The solvability condition (6.8) is analogous to conditions for the solvability of systems of equations in linear algebra which are formulated with a symmetric matrix. This can be seen, e.g., by considering the weak formulation (6.7) of the boundary value problem. The left-hand side there can be interpreted as an *operator A* which maps an element $u \in V = H^1(\Omega)$ of the function space $V$ to an element $Au$ of the *dual space* $V^*$ of $V$. The dual space $V^*$ is the set of all linear, continuous maps from $V$ to $\mathbb{R}$ (these maps are the *linear functionals* on $V$). Indeed the mapping

$$v \mapsto \langle Au, v \rangle := \int_{\Omega} \lambda \nabla u \cdot \nabla v\, dx \tag{6.9}$$

defines a linear functional on $V$. The right-hand side of the weak formulation also defines a linear functional $b \in V^*$ as follows

$$v \mapsto \langle b, v \rangle := \int_\Omega f\, v\, dx + \int_\Gamma g\, v\, ds_x.$$

The weak formulation is then equivalent to the operator equation

$$Au = b$$

in the dual space $V^*$. It is known from linear algebra that a linear system $Ax = b$ with symmetric matrix $A \in \mathbb{R}^{n,n}$ has a solution if and only if $b$ is orthogonal to the kernel of the matrix $A$. The kernel of the operator $A$ defined in (6.9) consists of the *constant* functions $u(x) = c$. Indeed for all constant functions $u(x) = c$ and all $v \in V$ it holds that

$$\langle Au, v \rangle = \int_\Omega \lambda \nabla u \cdot \nabla v\, dx = 0,$$

and hence $Au$ is the "zero functional" $Au = 0 \in V^*$ and $u$ is an element of the kernel of $A$. Conversely, if $u$ is in the kernel of $A$ we obtain

$$0 = \langle Au, u \rangle = \int_\Omega \lambda \nabla u \cdot \nabla u\, dx,$$

and since $\Omega$ is connected it follows that $u = c$ with a constant $c$. The solvability condition (6.8) corresponds to the orthogonality relation

$$\langle b, c \rangle = 0$$

for all constants $c$.

We will now interpret the assertion of Theorem 6.4 for two different applications and we will observe that the solvability condition is related to obvious physical facts.

For the stationary heat equation $f$ is the volume density of the supplied heat and $g$ is the surface density of the heat flux across the boundary. The term

$$\int_\Omega f\, dx + \int_{\partial\Omega} g\, ds_x \tag{6.10}$$

describes the heat supplied per unit of time. The stationary problem describes the state towards which a dynamical problem converges for large times – this will be considered in detail in Sect. 6.2.2. If heat is continuously supplied or removed from a body the averaged temperature will either continuously increase or decrease and hence no stationary long-time limit will be obtained. Only if the balance of the heat supply (6.10) equals zero can one expect a solution to the stationary heat equation. The solution is only unique up to a constant because the total temperature

$$\int_\Omega u\,dx$$

is given by the temperature at the start of the dynamical process. The information on the initial condition is lost when we consider the stationary problem. The solutions to the stationary equation hence describe the stationary limit for *all* possible initial conditions.

For a *potential flow* we solve $\Delta u = 0$ and $u$ describes the potential of a velocity field $v = \nabla u$, $g$ is the normal component of the velocity field at the boundary and $f = 0$. In this case

$$\int_{\partial\Omega} g\,ds_x = \int_{\partial\Omega} v \cdot n\,ds_x$$

describes how much mass leaves or enters the domain (compare Sect. 5.6). In order to obtain a stationary limit in a time-dependent fluid model the total mass in the system is not allowed to change (which follows from mass conservation). The solution $u$ is unique only up to a constant. This is due to the fact that $u$ describes a potential and potentials are only unique up to a constant of integration. The application of Theorem 6.4 for potential flows also shows the assertion on page 233.

Also the minimization property of Theorem 6.1 can be transferred to the problem with Neumann boundary conditions.

**Theorem 6.5**  *Let $u$ be a solution of (6.1) with boundary condition $\lambda\nabla u \cdot n = g$ on $\partial\Omega$. The functions $f$ and $g$ are assumed to fulfill the solvability condition (6.8). Then $u$ is the solution of the optimization problem*

$$\min_{u\in H^1(\Omega)} \left( \int_\Omega \left(\tfrac{1}{2}\lambda|\nabla u|^2 - f\,u\right) dx - \int_{\partial\Omega} g\,u\,ds_x \right).$$

The minimum in the above minimization problem is not obtained in cases where the condition (6.8) is not fulfilled. If (6.8) is not fulfilled the expression $\int_\Omega \left(\tfrac{1}{2}\lambda|\nabla u|^2 - f\,u\right) dx - \int_{\partial\Omega} gu\,ds_x$ is not bounded from below and hence no minimum exists. This follows by choosing constant functions $u$ with $u \mapsto \pm\infty$, where the sign of $u$ is the same as the sign of $\int_\Omega f\,dx + \int_{\partial\Omega} g\,ds_x$.

For stationary incompressible flows one can show that for a given normal part of the velocity on the boundary of the domain the minimizer of the kinetic energy will be a curl-free velocity field. Under appropriate assumptions on the domain this will be a potential flow. This means that vortices will only increase the kinetic energy.

**Theorem 6.6**  *Let $\Omega$ be a bounded domain with smooth boundary $\partial\Omega$, $g : \partial\Omega \to \mathbb{R}$ is a given smooth function and $v = \nabla\varphi$ is assumed to be smooth with $\Delta\varphi = 0$ in $\Omega$ and $v \cdot n = g$ on $\partial\Omega$. In this case $v$ is a solution of the optimization problem*

$$\min \left\{ \int_\Omega \tfrac{1}{2}|v|^2\,dx \,\middle|\, v \in L_2(\Omega)^d,\ \nabla \cdot v = 0 \text{ in } \Omega,\ v \cdot n = g \text{ on } \partial\Omega \right\}.$$

The conditions $\nabla \cdot v = 0$ in $\Omega$ and $v \cdot n = g$ on $\partial\Omega$ for a function $v \in L_2(\Omega)^d$ have to be interpreted in a generalized so-called *distributional* sense. Both conditions are fulfilled for a smooth function $v$ if and only if

$$\int_\Omega v \cdot \nabla\psi \, dx = \int_{\partial\Omega} g\psi \, ds_x \quad \text{for all } \psi \in H^1(\Omega). \tag{6.11}$$

To understand this we observe that for a sufficiently smooth function $v$ it holds that

$$\int_\Omega v \cdot \nabla\psi \, dx = -\int_\Omega \nabla \cdot v\,\psi \, dx + \int_{\partial\Omega} v \cdot n\,\psi \, ds_x \, .$$

We say $v \in L_2(\Omega)^d$ fulfills $\nabla \cdot v = 0$ in $\Omega$ and $v \cdot n = g$ on $\partial\Omega$ in a weak sense if and only if (6.11) is fulfilled.

*Proof of the Theorem 6.6* We define

$$J(w) := \int_\Omega \frac{1}{2}|w|^2 \, dx$$

which is the functional we aim to minimize. It is sufficient to show that

$$J(v + w) \geq J(v)$$

for a given solution $v$ and all functions $w \in L_2(\Omega)^d$ which fulfill $\nabla \cdot w = 0$ in $\Omega$ and $w \cdot n = 0$ on $\partial\Omega$ in the weak sense. It holds that

$$J(v + w) = \frac{1}{2}\int_\Omega |v + w|^2 \, dx = \frac{1}{2}\int_\Omega \left(|v|^2 + 2\,v \cdot w + |w|^2\right) dx$$

$$\geq J(v) + \int_\Omega v \cdot w \, dx \, .$$

With integration by parts we deduce from $v = \nabla\varphi$ and (6.11)

$$\int_\Omega v \cdot w \, dx = \int_\Omega \nabla\varphi \cdot w \, dx = 0 \, .$$

This proves the assertion. $\qquad\qquad\qquad\qquad\qquad\qquad\qquad\qquad\qquad\qquad\qquad\quad\square$

In bounded domains $\Omega$ it holds that the only incompressible, stationary flow with $v \cdot n = 0$ on $\partial\Omega$ is given by $v \equiv 0$. This follows from the fact that $v \equiv 0$ minimizes the kinetic energy.

## 6.1.2   The Fundamental Solution

We will now consider an important special solution of Laplace's equation

$$\Delta u = 0 \ \text{in} \ \mathbb{R}^d$$

for $d \geq 2$. We make the *radially symmetric* ansatz

$$u(x) = w(|x|) = w(r) \ \text{with} \ r = |x|.$$

Using $\partial_{x_i} |x| = x_i/|x|$ we obtain

$$\nabla u(x) = \frac{w'(|x|)}{|x|}x \ \text{and} \ \Delta u(x) = w''(|x|) + \frac{d-1}{|x|}w'(|x|). \tag{6.12}$$

Using the technique of separation of variables we obtain

$$w'(r) = c\,r^{-(d-1)}$$

and after integration

$$w(r) = \begin{cases} a\,r^{2-d} + b & \text{if } d > 2, \\ a\ln r + b & \text{if } d = 2, \end{cases}$$

for suitable constants $a, b$. We are especially interested in solutions which vanish as $r \to +\infty$ and hence at least for $d > 2$ we can achieve this by setting $b = 0$. We now consider the special solutions

$$u(x) = c_d \ln r \ \text{if} \ d = 2 \ \text{and} \ u(x) = c_d |x|^{2-d} \ \text{if} \ d > 2.$$

These solutions have a singularity at $r = 0$.

In order to fix the factor $c_d$ we consider the condition

$$1 = -\lim_{r \to 0} \int_{|x|=r} \nabla u \cdot n\,ds_x, \tag{6.13}$$

where $n$ is the exterior unit normal vector on the surface of the sphere $\{|x| = r\}$. The meaning of this condition will become clear in the proof of Theorem 6.7. With $n = x/|x|$ and $\nabla u(x) = w'(|x|)x/|x|$ it follows that

$$\int_{|x|=r} \nabla u \cdot n\,ds_x = \omega_d\,r^{d-1}w'(r) = \begin{cases} c_2\,\omega_2 & \text{if } d = 2, \\ c_d\,\omega_d(2-d) & \text{if } d \geq 3, \end{cases}$$

where $\omega_d$ is the surface area of the unit sphere in $\mathbb{R}^d$. We hence obtain

$$c_d = \begin{cases} -1/\omega_2 & \text{if } d = 2, \\ -1/((2-d)\omega_d) & \text{if } d \geq 3, \end{cases}$$

and the fundamental solution is given by

$$\Phi(x) = \begin{cases} -\frac{1}{2\pi} \ln |x| & \text{if } d = 2, \\ \frac{1}{(d-2)\omega_d} |x|^{2-d} & \text{if } d \geq 3. \end{cases}$$

The importance of the fundamental solution lies in the fact that with its help one can obtain a representation formula for solutions of *Poisson's equation*

$$-\Delta u = f \quad \text{in } \mathbb{R}^d. \tag{6.14}$$

**Theorem 6.7** *Let* $f : \mathbb{R}^d \to \mathbb{R}$ *be two times continuously differentiable and assume that* $f = 0$ *holds outside of a ball with radius R. In this case*

$$u(x) = \int_{\mathbb{R}^d} \Phi(y - x) f(y) \, dy \tag{6.15}$$

*is a solution of* (6.14).

*Proof* A simple variable transformation and using $\Phi(y) = \Phi(-y)$ gives

$$u(x) = \int_{\mathbb{R}^d} \Phi(y - x) f(y) \, dy = \int_{\mathbb{R}^d} \Phi(y) f(x - y) \, dy.$$

Applying the Laplace operator to this identity gives

$$-\Delta u(x) = -\int_{\mathbb{R}^d} \Phi(y) \Delta_x f(x - y) \, dy = \int_{\mathbb{R}^d} \Phi(y) \Delta_y f(x - y) \, dy.$$

One can show that it is possible to interchange integration and differentiation if one considers the definition of derivatives as limits of difference quotients:

$$\partial_{x_i}^2 u(x) = \lim_{h \to 0} \frac{u(x - he_i) - 2u(x) + u(x + he_i)}{h^2}$$

$$= \lim_{h \to 0} \int_{\mathbb{R}^d} \Phi(y) \frac{f(x - he_i - y) - 2f(x - y) + f(x + he_i - y)}{h^2} \, dy$$

$$= \int_{\mathbb{R}^d} \Phi(y) \lim_{h \to 0} \frac{f(x - he_i - y) - 2f(x - y) + f(x + he_i - y)}{h^2} \, dy$$

$$= \int_{\mathbb{R}^d} \Phi(y) \partial_{x_i}^2 f(x - y) \, dy.$$

It is possible to interchange the limit process and the integral using the fact that the convergence $h^{-2}(f(x - he_i) - 2f(x) + f(x + he_i)) \to \partial_{x_i}^2 f$ is uniform. We now cut out the ball $B_\varepsilon(0)$ around the singularity $y = 0$ and consider

$$I_\varepsilon(x) = \int_{B_\varepsilon(0)} \Phi(y)\,\Delta_y f(x - y)\,dy \ \text{ and } \ J_\varepsilon(x) = \int_{\mathbb{R}^d \setminus B_\varepsilon(0)} \Phi(y)\,\Delta_y f(x - y)\,dy\,.$$

As $\Delta_y f$ is bounded it follows from the special form of the fundamental solution that $\Phi(x) = \widetilde{\Phi}(r)$ with $\widetilde{\Phi}(r) \sim \ln r$ for $d = 2$ and $\widetilde{\Phi}(r) \sim r^{2-d}$ for $d \geq 3$, thus

$$|I_\varepsilon(x)| \leq C_1 \int_0^\varepsilon |\widetilde{\Phi}(r)|\,r^{d-1}\,dr \to 0 \ \text{ as } \ \varepsilon \to 0\,.$$

Using Green's formula we obtain

$$J_\varepsilon(x) = \int_{\mathbb{R}^d \setminus B_\varepsilon(0)} \Delta_y \Phi(y)\,f(x - y)\,dy$$

$$+ \int_{\partial B_\varepsilon(0)} \big(\Phi(y)\,\nabla_y f(x - y) \cdot n - \nabla_y \Phi(y) \cdot n\,f(x - y)\big)\,ds_y\,.$$

Here $n$ is the outer unit normal on $\partial B_\varepsilon(0)$. The first integral on the right-hand side vanishes because $\Delta\Phi(y) = 0$ for $y \neq 0$. The first part in the second integral can be estimated as follows

$$\left| \int_{\partial B_\varepsilon(0)} \Phi(y)\,\nabla_y f(x - y) \cdot n\,ds_y \right| \leq C\,|\widetilde{\Phi}(\varepsilon)|\,\varepsilon^{d-1} \to 0 \ \text{ as } \ \varepsilon \to 0\,.$$

For the remaining term it follows that

$$-\int_{\partial B_\varepsilon(0)} \nabla_y \Phi(y) \cdot n\,f(x - y)\,ds_y = -f(x) \int_{\partial B_\varepsilon(0)} \nabla_y \Phi(y) \cdot n\,ds_y$$

$$-\int_{\partial B_\varepsilon(0)} \nabla_y \Phi(y) \cdot n\,(f(x - y) - f(x))\,ds_y\,.$$

For the last integral we use $|f(x - y) - f(x)| \leq C\varepsilon$ which holds for $y \in \partial B_\varepsilon(0)$ to estimate

$$\left| \int_{\partial B_\varepsilon(0)} \nabla_y \Phi(y) \cdot n\,(f(x - y) - f(x))\,ds_y \right|$$

$$\leq C\varepsilon \int_{\partial B_\varepsilon(0)} |\nabla_y \Phi(y) \cdot n|\,ds_y \to 0 \ \text{ for } \ \varepsilon \to 0\,.$$

Altogether we obtain with (6.13)

$$\lim_{\varepsilon \to 0} J_\varepsilon(x) = f(x)$$

and this shows the assertion.                                                                    □

**Remarks**

1. The decay property $\lim_{|x| \to +\infty} \Phi(x) = 0$ for $d \geq 3$ implies that the solution (6.15) also decays for $|x| \to +\infty$. Formula (6.15) hence yields a solution with boundary condition $\lim_{|x| \to +\infty} u(x) = 0$ which is attained at "infinity". One can show that only one solution of $-\Delta u = f$ with this property exists.
2. One can show the formula (6.15) also under much weaker assumptions on the function $f$, see, e.g., [49].
3. The formula (6.15) expresses how function values of $f$ influence the solution of Poisson's equation. The fundamental solution can be interpreted as the *distributional* solution to

$$-\Delta \Phi = \delta$$

where $\delta$ is the *Dirac distribution*. The function $f$ can formally be written as

$$f(x) = \langle \delta(\cdot - x), f \rangle = \int_{\mathbb{R}^d} \delta(y - x) \, f(y) \, dy \, ,$$

where the integral here strictly speaking is just a formal way to express the application of a distribution to a function. This is in some sense a generalization of the inner product in $L_2(\mathbb{R}^3)$. The formula (6.15) is then obtained as a "superposition" of the solutions to "all" values $f(x)$,

$$u(x) = \int_{\mathbb{R}^d} \Phi(y - x) \, f(y) \, dx \, .$$

### 6.1.3  Mean Value Theorem and Maximum Principles

Solutions $u$ of Laplace's equation
$$\Delta u = 0$$

have the interesting property that $u(x)$ equals the mean value of $u$ over all balls with center $x$. It holds:

**Theorem 6.8** *Let $u$ be a two times continuously differentiable solution of $\Delta u = 0$ on a domain $\Omega$ and let $B_R(x) \subset \Omega$ be a ball with radius $R$. Then it holds that*

$$\frac{1}{|B_R(x)|} \int_{B_R(x)} u(y)\,dy = \frac{1}{|\partial B_R(x)|} \int_{\partial B_R(x)} u(y)\,ds_y = u(x),$$

where $|B_R(x)|$ is the $d$–dimensional volume and $|\partial B_R(x)|$ is the $(d-1)$–dimensional surface area.

*Proof* Let $\omega_d(r) := |\partial B_r(0)|$ and

$$\varphi(r) := \frac{1}{\omega_d(r)} \int_{\partial B_r(x)} u(y)\,ds_y = \frac{1}{\omega_d(1)} \int_{\partial B_1(0)} u(x+rz)\,ds_z.$$

With the help of the divergence theorem we obtain

$$\varphi'(r) = \frac{1}{\omega_d(1)} \int_{\partial B_1(0)} \nabla u(x+rz)\cdot z\,ds_z = \frac{1}{\omega_d(r)} \int_{\partial B_r(x)} \nabla u(y)\cdot n\,ds_y$$

$$= \frac{1}{\omega_d(r)} \int_{B_r(x)} \Delta u(y)\,dy = 0.$$

This implies that the function $r \mapsto \varphi(r)$ is constant. In the limit $r \to 0$ it follows that

$$\varphi(r) = u(x) \quad \text{for all } r < R.$$

With generalized polar coordinates we compute

$$\int_{B_R(x)} u(y)\,dy = \int_0^R \int_{\partial B_r(x)} u(y)\,ds_y\,dr = \int_0^R \omega_d(r)\,\varphi(r)\,dr = u(x)|B_R(x)|.$$

$\square$

With the help of this property of solutions of Laplace's equation one can easily deduce the *maximum principle* for Laplace's equation. This property says that solutions attain its maximum (and its minimum) at the *boundary* of each domain under consideration.

**Theorem 6.9** *Let $\Omega \subset \mathbb{R}^d$ be a bounded domain and $u : \overline{\Omega} \to \mathbb{R}$ a two times continuously differentiable solution of Laplace's equation in $\Omega$. Then it holds:*

*(i)  u attains its maximum at the boundary, $\max_{x\in\Omega} u(x) = \max_{x\in\partial\Omega} u(x)$.*
*(ii) If an $x \in \Omega$ exists such that $u(x) = \max_{y\in\Omega} u(y)$, then u is constant.*

*Proof* It is sufficient to prove *(ii)*. If an $x \in \Omega$ with $u(x) = \max_{y\in\Omega} u(y)$ exists, the mean property implies that for all $r > 0$ with $B_r(x) \subset \Omega$,

$$u(x) = \frac{1}{|B_r(x)|} \int_{B_r(x)} u(y)\,dy.$$

This identity together with the inequality $u(y) \leq u(x)$ yields

$$u(y) = u(x) \quad \text{for all } y \in B_r(x). \tag{6.16}$$

We now consider the set

$$M := \{y \in \Omega \mid u(y) = u(x)\}.$$

This set is open as for all $y \in M$ a ball $B_r(y)$ exists such that $B_r(y) \subset \Omega$ and (6.16) yields $B_r(y) \subset M$. On the other hand $M$ is relatively closed with respect to $\Omega$. This can be seen as follows. For all $y_n \in M$ with $\lim_{n \to +\infty} y_n = y \in \Omega$, the continuity of $u$ implies $y \in M$. As $\Omega$ is connected it follows that $M = \Omega$. $\qquad\qquad\square$

### 6.1.4 Plane Potential Flows, the Method of Complex Variables

An important application of Laplace's equation is potential flows. For two-dimensional flows one can derive with the help of complex analysis some interesting conclusions. In two dimensions the conditions $\nabla \times v = 0$, $\nabla \cdot v = 0$ for curl-free, incompressible flows are given as

$$\partial_{x_2} v_1 - \partial_{x_1} v_2 = 0,$$
$$\partial_{x_1} v_1 + \partial_{x_2} v_2 = 0.$$

The vector field $(v_1, -v_2)$ hence solves the *Cauchy–Riemann* differential equations

$$\partial_{x_1} v_1 = \partial_{x_2}(-v_2),$$
$$\partial_{x_2} v_1 = -\partial_{x_1}(-v_2).$$

This implies that

$$w : \mathbb{C} \mapsto \mathbb{C}, \quad w(x_1 + i x_2) := v_1(x_1, x_2) - i \, v_2(x_1, x_2)$$

is a *holomorphic* function.

**Definition 6.10** The function $w$ is called the *complex velocity.*

On the other hand it holds that for all holomorphic functions $w$, the velocity field $v = (v_1, v_2)$ with $v_1 = \operatorname{Re} w$ and $v_2 = -\operatorname{Im} w$ is an incompressible, stationary potential flow. One can hence derive with the help of complex analysis some interesting conclusions on incompressible potential flows.

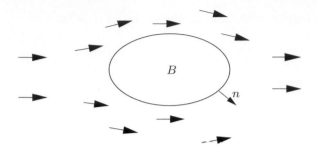

**Fig. 6.1**  Flow past a plane obstacle

We now consider an *inviscid* flow past an obstacle $B$ as shown in Fig. 6.1. In the following we will now always assume that $\partial B$ can be parameterized by a smooth, closed curve. The force $f$ acting on the obstacle is given as the integral over the normal stresses on the boundary

$$f = \int_{\partial B} \sigma(x)\, n(x)\, ds_x = -\int_{\partial B} p(x)\, n(x)\, ds_x \,. \tag{6.17}$$

In the following we will identify vectors $\begin{pmatrix} a_1 \\ a_2 \end{pmatrix} \in \mathbb{R}^2$ with complex numbers $a_1 + i\, a_2$. In particular $f$ will be interpreted as a complex number. The following theorem states a possibility to compute the force $f$ with the help of the complex velocity $w$.

**Theorem 6.11**  (Blasius Theorem) *Let $B$ be a bounded domain with smooth boundary. In addition, let $v$ be a potential flow fulfilling Euler's equations with constant density $\varrho$ and the conditions $\nabla \cdot v = 0$ in the exterior of $B$, and $v \cdot n = 0$ on $\partial B$ are assumed to hold. Under these assumptions it holds that*

$$f = \frac{-i\varrho}{2} \overline{\int_{\partial B} w^2\, dz} \,.$$

*Proof*  Let $\gamma : [0, L] \to \partial B$ be an arc-length parameterization of the boundary $\partial B$. After identifying vectors in $\mathbb{R}^2$ with complex numbers we can express the unit tangent $t(s)$ and the unit normal $n(s)$ on $\partial B$ as follows

$$t(s) = \gamma'(s) = \gamma_1'(s) + i\,\gamma_2'(s)\,,$$
$$n(s) = \gamma_2'(s) - i\,\gamma_1'(s)\,.$$

Here, the orientation of $\gamma$ is chosen such that $n(s)$ is the exterior normal to $B$. Bernoulli's theorem now implies that $p$ up to an additive constant is given as $p = -\frac{\varrho}{2}\left(v_1^2 + v_2^2\right)$ (compare p. 233). With this we obtain

$$f = -\int_{\partial B} p(n_1 + i\, n_2)\, ds_x = \int_0^L p(\gamma(s))(-\gamma_2'(s) + i\, \gamma_1'(s))\, ds$$

$$= i \int_0^L p(\gamma(s))(\gamma_1'(s) + i\, \gamma_2'(s))\, ds \qquad\qquad (6.18)$$

$$= -\frac{i\varrho}{2} \int_0^L \left(v_1^2 + v_2^2\right)(\gamma_1'(s) + i\, \gamma_2'(s))\, ds\,.$$

The boundary condition $v \cdot n = 0$ implies

$$0 = v_1 \gamma_2'(s) - v_2 \gamma_1'(s) \quad \text{on} \quad \partial B$$

and hence

$$w^2 = (v_1 - i\, v_2)^2 = v_1^2 - v_2^2 - 2i\, v_1 v_2$$

implies

$$w^2(\gamma_1'(s) + i\, \gamma_2'(s)) = \left(v_1^2 - v_2^2 - 2i\, v_1 v_2\right)(\gamma_1'(s) + i\, \gamma_2'(s))$$
$$= \left(v_1^2 + v_2^2\right)(\gamma_1'(s) - i\, \gamma_2'(s))\,.$$

With (6.18) it follows that

$$f = -\frac{i\varrho}{2} \overline{\int_0^L w^2\, \gamma'(s)\, ds} = -\frac{i\varrho}{2} \overline{\int_{\partial R} w^2\, dz}\,.$$

This shows the assertion.                                                                     $\square$

We now consider a situation in which the flow for $|x| \to \infty$ approaches a constant value $V$, see Fig. 6.2. This is related to the situation in which a body moves with constant velocity through a resting fluid. This is the case for example for the flow past an airplane where the coordinate system moves with the body.

The following theorem states that the force acting on $B$ is always orthogonal to the direction $V$.

**Theorem 6.12** (Kutta–Joukowski Theorem) *Let an incompressible potential flow $v$ be given outside of a domain $B$ with smooth boundary. In addition, $v \cdot n = 0$ on $\partial B$ and*

$$v(x) \to V \quad as \quad |x| \to \infty$$

*with $V \in \mathbb{R}^2$ given. In this case the force acting on the body $B$ is given as*

$$f = \varrho \Gamma_{\partial B} |V| \widehat{n}$$

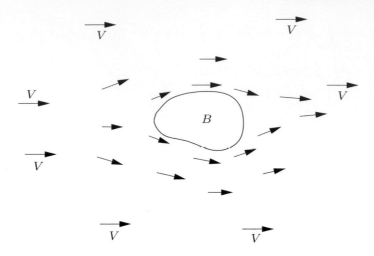

**Fig. 6.2** Flow past a plane obstacle

*with the* circulation

$$\Gamma_{\partial B} = \int_{\partial B} v \cdot t \, ds_x$$

*of the vector field $v$ around $B$ and the unit normal $\widehat{n} = \frac{(V_2, -V_1)}{|V|}$ related to $V$.*

*Proof* The complex velocity $w = v_1 - i\, v_2$ is analytic outside of $B$. Therefore, it is possible to express $w$ outside of each ball which contains $B$ in a Laurent series

$$w(z) = \sum_{k=-\infty}^{\infty} a_k z^k, \ z = x_1 + i\, x_2$$

with $a_k \in \mathbb{C}$. From

$$w \to V_1 - i\, V_2 \ \text{as} \ |z| \to \infty$$

we deduce

$$a_k = 0 \ \text{for all} \ k > 0$$

and

$$a_0 = V_1 - i\, V_2.$$

Integrating the series $\displaystyle\sum_{k=-\infty}^{0} a_k z^k$ over the boundary of a large ball $B_R(0)$ we obtain, using the fact that we are allowed to integrate term by term,

$$\int_{\partial B_R(0)} w(z)\, dz = \sum_{k=-\infty}^{0} a_k \int_{\partial B_R(0)} z^k\, dz$$

$$= \sum_{k=-\infty}^{0} a_k \int_{0}^{2\pi} R^k e^{iks}\, Ri\, e^{is}\, ds = a_{-1} 2\pi i\,.$$

Here we used the parameterization $\gamma : [0, 2\pi] \to \mathbb{C}, s \to R\, e^{is}$ and the fact that for $\ell \neq 0$

$$\int_{0}^{2\pi} e^{i\ell s}\, ds = 0$$

holds. Cauchy's integral theorem implies that integration over $\partial B_R(0)$ and $\partial B$ yields the same result. Hence it follows that

$$\int_{\partial B} w(z)\, dz = 2\pi i\, a_{-1}\,.$$

As $v_1 \gamma'_2 = v_2 \gamma'_1$ on $\partial B$ it holds that

$$\int_{\partial B} w\, dz = \int_{0}^{2\pi} (v_1 - i\, v_2)(\gamma'_1 + i\, \gamma'_2)\, ds$$

$$= \int_{0}^{2\pi} (v_1 \gamma'_1 + v_2 \gamma'_2 + i\, v_1 \gamma'_2 - i\, v_2 \gamma'_1)\, ds$$

$$= \int_{0}^{2\pi} (v_1 \gamma'_1 + v_2 \gamma'_2)\, ds = \int_{\partial B} v \cdot t\, ds_x = \Gamma_{\partial B}\,.$$

This implies

$$a_{-1} = \frac{\Gamma_{\partial B}}{2\pi i}\,.$$

Squaring $w$ leads to the Laurent series

$$w^2 = a_0^2 + \frac{2\, a_0\, a_{-1}}{z} + \frac{2\, a_0\, a_{-2} + a_{-1}^2}{z^2} + \cdots\,.$$

Theorem 6.11 then implies

$$f = -\frac{i\varrho}{2} \overline{\int_{\partial B} w^2\, dz} = -\frac{i\varrho}{2}\, \overline{2\pi i\, 2\, a_0\, a_{-1}} = \varrho \Gamma_{\partial B}(V_2 - i V_1)\,.$$

This shows the theorem.                                                                                           □

**Remark.** The Kutta–Joukowski theorem describes the so-called *d'Alembert'* *paradox* in two dimensions. It says that for incompressible and inviscid potential flows no force in the direction of the velocity $V$ is acting on the body $B$. Hence, there are no drag forces $\frac{1}{|V|} f \cdot V$ acting on the body $B$ in two-dimensions. The d'Alembert paradox in three-dimensions says that the force exerted on a bounded body by an incompressible and inviscid potential flow is equal to zero. Hence, there are no drag forces and also no lift forces as the components perpendicular to the fluid velocity are also zero. This in contrast to our experience for example if a strong wind blows against our body when we ride our bike. We refer to Chorin and Marsden [20], Chap. 2, for more details. More realistic models like the Navier–Stokes equations take viscous effects into account. The influence of viscous effects on a flow past a body will be considered in Sect. 6.6 where we study the boundary layers which appear close to the body $B$.

**Complex Potentials and Stream Functions**

We now assume that the complex velocity $w = v_1 - i\, v_2$ of an incompressible, curl-free flow $v = \begin{pmatrix} v_1 \\ v_2 \end{pmatrix}$ has a primitive $W$, i.e.,

$$ w = \frac{dW}{dz}. $$

This antiderivative is called the *complex velocity potential* for $w$. Defining $\varphi$ and $\psi$ via

$$ W = \varphi + i\,\psi, $$

we obtain

$$ v_1 = \partial_{x_1}\varphi = \partial_{x_2}\psi, $$
$$ v_2 = \partial_{x_2}\varphi = -\partial_{x_1}\psi. $$

This implies that

$$ v = \nabla\varphi, $$

and it can be shown that $\psi$ is a so-called *stream function*. A function $\psi$ is called stream function if the streamlines of the flow are the level sets of $\psi$. We now state a simple example and refer to the exercises for more interesting examples.

**Example**: Let $U = \begin{pmatrix} U_1 \\ U_2 \end{pmatrix} \in \mathbb{R}^2$ be a *constant* velocity. Setting $w = U_1 - i\, U_2$, we obtain

$$ W(z) = wz = (w_1 + i\, w_2)(x_1 + i\, x_2) $$
$$ = (w_1 x_1 - w_2 x_2) + i\,(w_2 x_1 + w_1 x_2). $$

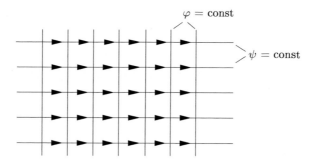

**Fig. 6.3** Potential and stream function for a simple flow

With this we get that

$$\varphi = w_1 x_1 - w_2 x_2 \quad \text{is a potential and}$$
$$\psi = w_2 x_1 + w_1 x_2 \quad \text{is a stream function.}$$

For $U = \binom{1}{0}$ we obtain the situation represented in Fig. 6.3.

### 6.1.5 The Stokes Equations

We now consider a Dirichlet boundary value problem for the Stokes equations (of Sect. 5.6),

$$\nabla \cdot v = 0 \,,$$
$$-\mu \, \Delta v + \nabla p = \varrho f, \tag{6.19}$$

in a bounded domain $\Omega \subset \mathbb{R}^d$ with given boundary data $v = b$ on the boundary $\partial \Omega$ of $\Omega$. We will now show that this system are the necessary conditions for the solution of the optimization problem

$$\min \left\{ J(v) \,\middle|\, v \in H^1(\Omega)^d, \ \nabla \cdot v = 0 \text{ in } \Omega, \ v = b \text{ on } \partial \Omega \right\} \tag{6.20}$$

with the functional

$$J(v) := \int_\Omega \left( \frac{\mu}{2} |Dv|^2 - \varrho f \cdot v \right) dx \,.$$

This is an optimization problem in an *infinite-dimensional* space. It is possible to extend the method of Lagrange multipliers to infinite-dimensional Hilbert spaces if one uses appropriate inner products. Here the constraint $\nabla \cdot v = 0$ can be coupled with the help of a Lagrange multiplier $-p \in L_2(\Omega)$ if one uses the inner product in $L_2(\Omega)$. The corresponding Lagrangian is

$$L(v, p) = J(v) + \langle -p, \nabla \cdot v \rangle_{L_2(\Omega)} = \int_\Omega \left( \frac{\mu}{2} |Dv|^2 - \varrho f \cdot v - p \nabla \cdot v \right) dx \,.$$

The $L_2(\Omega)$ inner product here replaces the inner product of two vectors in $\mathbb{R}^m$ which is used for optimization problems with $m$ constraints. The optimality criteria have to be formulated with the help of *directional derivatives*:

$$\left\langle \frac{\delta L(v, p)}{\delta v}, w \right\rangle := \frac{d}{d\varepsilon} L(v + \varepsilon w, p) \Big|_{\varepsilon=0} = 0 \text{ and}$$

$$\left\langle \frac{\delta L(v, p)}{\delta p}, q \right\rangle := \frac{d}{d\varepsilon} L(v, p + \varepsilon q) \Big|_{\varepsilon=0} = 0 \,.$$

This has to hold for all $w \in H^1(\Omega)^d$ with $w = 0$ on $\partial\Omega$ and all $q \in L_2(\Omega)$. The derivative $\langle \frac{\delta L(v)}{\delta v}, w \rangle$ can be formulated with the help of the formula

$$\frac{1}{2} \frac{d}{d\varepsilon} |D(v + \varepsilon w)|^2 \Big|_{\varepsilon=0} = \frac{1}{2} \frac{d}{d\varepsilon} \sum_{i,j=1}^d |\partial_{x_i}(v_j + \varepsilon w_j)|^2 \Big|_{\varepsilon=0} = \sum_{i,j=1}^d \partial_{x_i} v_j \, \partial_{x_i} w_j$$

$$= \sum_{i,j=1}^d \left[ \partial_{x_i}(\partial_{x_i} v_j \, w_j) - \partial_{x_i}^2 v_j \, w_j \right]$$

and by using the identity

$$-\int_\Omega \Delta v \cdot w \, dx = -\int_\Omega \sum_{i,j=1}^d \partial_{x_i}^2 v_j \, w_j \, dx = \int_\Omega \sum_{i,j=1}^d \partial_{x_i} v_j \, \partial_{x_i} w_j \, dx$$

$$= \int_\Omega Dv : Dw \, dx \tag{6.21}$$

as follows

$$\frac{d}{d\varepsilon} \int_\Omega \left( \frac{\mu}{2} |D(v + \varepsilon w)|^2 - \varrho f \cdot (v + \varepsilon w) - p \nabla \cdot (v + \varepsilon w) \right) dx \Big|_{\varepsilon=0}$$

$$= \int_\Omega (-\mu \Delta v - \varrho f + \nabla p) \cdot w \, dx \overset{!}{=} 0 \,.$$

As this has to hold for *all* suitable test functions $w$ we obtain

$$-\mu \Delta v = -\nabla p + \varrho f \,.$$

The directional derivative with respect to $p$ is given as

$$\frac{d}{d\varepsilon} \int_\Omega \left( \frac{\mu}{2} |Dv|^2 - \varrho f \cdot v - (p + \varepsilon q) \nabla \cdot v \right) dx = -\int_\Omega q \nabla \cdot v \, dx = 0 \,,$$

and as this also has to hold for all test functions $q$ it follows that

$$\nabla \cdot v = 0 \,.$$

This shows that the Stokes equations are necessary conditions for a critical point of (6.20). Here the pressure can be interpreted as a *Lagrange parameter* of the constraint $\nabla \cdot v = 0$.

This property indicates that the Stokes equations are also an elliptic system. This is in fact the case. In the derivation of the weak formulation it is appropriate to incorporate the constraint $\nabla \cdot v = 0$ into the definition of the underlying function space. We hence define

$$V = \left\{ v \in H^1(\Omega)^d \mid \nabla \cdot v = 0 \text{ in } \Omega, \ v = 0 \text{ on } \partial\Omega \right\} \,.$$

Multiplying $-\mu \Delta v + \nabla p = \varrho f$ with a test function $w \in V$, integrating and using integration by parts yields with the help of (6.21) and

$$\int_\Omega \nabla p \cdot w \, dx = -\int_\Omega p \nabla \cdot w \, dx = 0$$

the identity

$$\int_\Omega \mu \, Dv : Dw \, dx = \int_\Omega \varrho f \cdot w \, dx \,.$$

The pressure disappears in this weak formulation as the constraint is incorporated into the function space $V$ and hence the Lagrange parameter is not needed any longer. One can now easily see that the bilinear form

$$a(v, w) = \int_\Omega Dv : Dw \, dx$$

is symmetric and positive definite on $V$ and in this sense the Stokes equations are elliptic.

### 6.1.6 Homogenization

In many applications media with a complex structure appear in which for example diffusion constants can vary drastically. A simple example is a medium which is composed of different materials, e.g., a layered medium. In such a situation different length scales play a role. The aim of homogenization is to derive laws on large scales for processes in which oscillations on small scales play a role. With the help of this typically simple equations arise as an averaging over microscopic information takes place. We will now discuss a simple example in which two scales appear, one microscopic scale of order $\varepsilon$ and a macroscopic scale of order 1. We will assume that the dependence on the microscopic variable is periodic.

Let $\Omega \subset \mathbb{R}^d$ be a smooth domain and $f : \Omega \to \mathbb{R}$ is assumed to be smooth. We now consider

$$(P^\varepsilon) \quad \begin{cases} -\nabla \cdot (\lambda(\tfrac{x}{\varepsilon})\nabla u^\varepsilon) = f & \text{if } x \in \Omega, \\ u^\varepsilon = 0 & \text{if } x \in \partial\Omega. \end{cases}$$

Here $\lambda : \mathbb{R}^d \to \mathbb{R}$ is assumed to be *periodic* with respect to all coordinate directions with period 1 and it is assumed that $\lambda_0, \lambda_1 \in \mathbb{R}$ exist such that

$$0 < \lambda_0 < \lambda(y) \le \lambda_1 \quad \text{for all} \quad y \in \mathbb{R}^d.$$

The diffusion coefficient in $(P^\varepsilon)$ hence varies for small $\varepsilon$ and we would like to identify a limit problem in the limit $\varepsilon \to 0$.

In order to do so we consider the two-scale ansatz (as we have done in Exercise 1.10)

$$u^\varepsilon(x) = u_0\left(x, \tfrac{x}{\varepsilon}\right) + \varepsilon\, u_1\left(x, \tfrac{x}{\varepsilon}\right) + \varepsilon^2\, u_2\left(x, \tfrac{x}{\varepsilon}\right) + \cdots \tag{6.22}$$

with

$$u_i : \Omega \times \mathbb{R}^d \to \mathbb{R}, \quad i = 0, 1, 2, \ldots$$

where $u_i(x, y)$ for all $x \in \Omega$ is periodic with period 1 with respect to all coordinates in $y$. We plug this ansatz into $(P^\varepsilon)$ and since

$$\nabla\left[u_i\left(x, \tfrac{x}{\varepsilon}\right)\right] = \frac{1}{\varepsilon}\nabla_y u_i\left(x, \tfrac{x}{\varepsilon}\right) + \nabla_x u_i\left(x, \tfrac{x}{\varepsilon}\right),$$

we obtain at the orders $\mathcal{O}(\varepsilon^{-2})$, $\mathcal{O}(\varepsilon^{-1})$, and $\mathcal{O}(1)$

$$0 = -\nabla_y \cdot (\lambda(y)\nabla_y u_0(x, y)), \tag{6.23}$$

$$\begin{aligned} 0 = -\nabla_y \cdot (\lambda(y)\nabla_y u_1(x, y)) - \nabla_y \cdot (\lambda(y)\nabla_x u_0(x, y)) \\ - \nabla_x \cdot (\lambda(y)\nabla_y u_0(x, y)), \end{aligned} \tag{6.24}$$

$$\begin{aligned} f(x) = -\nabla_y \cdot (\lambda(y)\nabla_y u_2(x, y)) - \nabla_y \cdot (\lambda(y)\nabla_x u_1(x, y)) \\ - \nabla_x \cdot (\lambda(y)\nabla_y u_1(x, y)) - \nabla_x \cdot (\lambda(y)\nabla_x u_0(x, y)). \end{aligned} \tag{6.25}$$

Since $u_0(x, \cdot)$ is periodic we obtain that (6.23) only has solutions which are constant with respect o $y$. This follows for example with the help of an energy argument: Multiplying (6.23) with $u_0(x, \cdot)$ we obtain after integration in $Y = [0, 1]^d$ with the help of the divergence theorem

$$\int_Y \lambda(y)|\nabla_y u_0(x, y)|^2 dy = 0 \quad \text{for all} \quad x \in \Omega. \tag{6.26}$$

There are no boundary integrals appearing from the integration by parts due to the fact that we have periodic boundary conditions with respect to the $y$-variable. The fact that $\lambda$ is positive implies

$$u_0(x, y) = u(x) \quad \text{for all } x \in \Omega$$

with a function $u : \Omega \to \mathbb{R}$. Our goal now is to derive an equation for $u$.

The Eq. (6.24) now simplifies to

$$-\nabla_y \cdot (\lambda(y)\nabla_y u_1(x, y)) = \nabla_y \lambda(y) \cdot \nabla_x u(x) = \sum_{i=1}^{d} \partial_{y_i} \lambda(y)\, \partial_{x_i} u(x)\,.$$

For all $x \in \Omega$ we have to solve a linear partial differential equation to determine $u_1(x, \cdot)$. As the right-hand side of the above differential equation is given as a sum we make the following ansatz for $u_1$

$$u_1(x, y) = \sum_{i=1}^{d} v^{(i)}(x, y)\,,$$

where $v^{(i)}$ is for all $x \in \Omega$ a solution of

$$-\nabla_y \cdot \left(\lambda(y)\nabla_y v^{(i)}(x, y)\right) = \partial_{y_i} \lambda(y)\, \partial_{x_i} u(x) \tag{6.27}$$

in $Y$ with periodic boundary conditions. The product structure of the right-hand side suggests to choose $v^{(i)}$ as follows

$$v^{(i)}(x, y) = g^{(i)}(x)\, h^{(i)}(y) + f^{(i)}(x)$$

with some arbitrary function $f^{(i)}(x)$,

$$g^{(i)}(x) = -\partial_{x_i} u(x)\,,$$

and $h^{(i)}$, $i = 1, \ldots, d$, such that the differential equations

$$-\nabla_y \cdot \left(\lambda(y)\nabla_y h^{(i)}(y)\right) = -\partial_{y_i} \lambda(y) \tag{6.28}$$

are fulfilled. The differential operator on the left-hand side only has constant functions as periodic homogeneous solutions. The Fredholm alternative, (see Alt [4] and Evans [37]) now implies that a periodic solution to (6.28) exists if and only if the right-hand side is orthogonal with respect to the $L_2$–inner product to all homogeneous solutions. This implies the requirement

$$\int_Y \partial_{y_i} \lambda(y)\, dy = 0 \quad \text{with} \quad Y = [0, 1]^d\,.$$

The fact that $\lambda$ is periodic implies that the integral vanishes and the existence of a solution is guaranteed. We hence obtain a solution

$$u_1(x, y) = -\sum_{i=1}^{d} h^{(i)}(y) \, \partial_{x_i} u(x) + \tilde{u}_1(x)$$

with an arbitrary function $\tilde{u}_1 : \Omega \to \mathbb{R}$.

Now Eq. (6.25) yields

$$-\nabla_y \cdot (\lambda(y)\nabla_y u_2(x, y)) = f(x) + \nabla_y \cdot (\lambda(y)\nabla_x u_1(x, y))$$
$$+ \nabla_x \cdot (\lambda(y)\nabla_y u_1(x, y)) + \nabla_x \cdot (\lambda(y)\nabla_x u(x)).$$

As solvability condition we obtain again that the integral with respect to $Y$ of the right-hand side has to vanish. Using the periodicity of $\lambda$ and $u_1(x, \cdot)$ we now obtain

$$f(x) = \int_Y f(x) \, dy$$
$$= -\int_Y \nabla_x \cdot (\lambda(y)\nabla_y u_1(x, y)) \, dy - \int_Y \nabla_x \cdot (\lambda(y)\nabla_x u(x)) \, dy$$
$$= \sum_{i=1}^{d} \nabla_x \cdot \left( \partial_{x_i} u(x) \int_Y \lambda(y)\nabla_y h^{(i)}(y) \, dy \right) - \int_Y \lambda(y) \, dy \, \Delta u(x).$$

Defining a matrix $K = (k_{ij})$ with

$$k_{ij} = \int_Y \lambda(y) \left( \delta_{ij} - \partial_{y_i} h^j(y) \right) dy$$

where $\delta_{ij}$ is the Kronecker symbol, we obtain that $u$ solves

$$(P^0) \quad \begin{cases} -\nabla \cdot (K \nabla u(x)) = f(x) & \text{in } \Omega, \\ \qquad\qquad\qquad u = 0 & \text{on } \partial\Omega. \end{cases}$$

$(P^0)$ is called *homogenized problem* related to $(P^\varepsilon)$. The solution $u = u_0$ of the homogenized problem is for small $\varepsilon$ a good approximation of the solution $u^\varepsilon$ of the problem $(P^\varepsilon)$. The elliptic differential equation (6.28) together with periodic boundary condition is called the corresponding *corrector problem*. With the help of solutions to the corrector problem one can construct an improved approximation

$$u_1^\varepsilon(x) = u_0(x) + \varepsilon \, u_1(x, x/\varepsilon)$$

of the solution to the original problem. In particular first partial derivatives of $u_1^\varepsilon$ approximate the corresponding partial derivatives of $u^\varepsilon$,

$$\partial_{x_j} u_1^\varepsilon(x) \sim \partial_{x_j} u^\varepsilon(x) \, ,$$

which in general cannot be true for $u_0$. This is suggested already by the ansatz (6.22): From

$$u^\varepsilon(x) \sim u_0(x) + \varepsilon u_1(x, x/\varepsilon)$$

we obtain to order $\mathcal{O}(1)$,

$$\partial_{x_i} u^\varepsilon(x) \sim \partial_{x_i} u_0(x) + \partial_{y_i} u_1(x, y)_{|y=x/\varepsilon} \, ,$$

and so $u_1$ is essential for an approximation of the derivatives of $u^\varepsilon$.

**Remarks**

1. $K$ does not depend on $x$ and in general is not diagonal.
2. If we consider media which only vary in one direction, for example $\lambda(y)$ is given by

$$\lambda(y) = \overline{\lambda}(y_1) \quad \text{for all} \quad y = (y_1, \dots, y_d) \in \mathbb{R}^d \, ,$$

we obtain non-constant solutions $h^1$ of the corrector problem (6.28) of the form $h^1(y_1, \dots, y_d) = h(y_1)$, where $h$ is a periodic solution of

$$-(\overline{\lambda} h')' = -\overline{\lambda}' \, .$$

It follows that

$$h(y_1) = y_1 - c \int_0^{y_1} \frac{1}{\overline{\lambda}(z)} \, dz \quad \text{with} \quad c = \left( \int_0^1 \frac{1}{\overline{\lambda}(z)} \, dz \right)^{-1} \, .$$

Hence $K$ is a diagonal matrix with

$$K_{11} = \left( \int_0^1 \frac{1}{\overline{\lambda}(z)} \, dz \right)^{-1} \quad \text{and} \quad K_{ii} = \int_0^1 \overline{\lambda}(z) \, dz \quad \text{for} \quad i = 2, \dots, d \, .$$

In the $x_1$-direction we hence do not obtain the arithmetic mean but the harmonic mean. This result can also be obtained with slightly different arguments with the help of Exercise 5.19. In fact, the diffusion constant $K_{11}$ in general is *smaller* than the arithmetic mean. This follows with the help of the Cauchy–Schwarz inequality as follows

$$1 = \int_0^1 \sqrt{\overline{\lambda}(z)} \frac{1}{\sqrt{\overline{\lambda}(z)}} \, dz \leq \left( \int_0^1 \overline{\lambda}(z) \, dz \right)^{1/2} \left( \int_0^1 \frac{1}{\overline{\lambda}(z)} \, dz \right)^{1/2} \, .$$

In the Cauchy–Schwarz inequality we have equality if and only if the two factors are linearly dependent. One hence obtains in the case of a non-constant $\overline{\lambda}$

$$K_{11} = \left( \int_0^1 \frac{1}{\overline{\lambda}(z)}\, dz \right)^{-1} < \int_0^1 \overline{\lambda}(z)\, dz = K_{ii} \ \text{ for } \ i = 2, \dots, d\,.$$

Hence, the diffusion in $x_1$-direction is *slower* than the diffusion in all other coordinate directions.

### 6.1.7   Optimal Control of Elliptic Differential Equations

In Sect. 4.8 we considered the optimal control of ordinary differential equations. We saw in Chap. 5 that many applications lead to models which are formulated with the help of partial differential equations. Therefore, the control of processes which are modeled with the help of partial differential equations are of great practical interest. In this book we will only discuss very few aspects and we refer to the books of Tröltzsch [131], Hinze et al. [67] and Lions [90] on optimal control of partial differential equations.

Let us consider a simple control problem in which we try to control the temperature in a set $\Omega \subset \mathbb{R}^d$. The temperature is described with the help of a suitable scaled variable $y : \Omega \to \mathbb{R}$, and we want to control the temperature in $\Omega$ which can be achieved for example by electromagnetic induction or with the help of microwaves. A well-directed control of the temperature with the help of microwaves is, e.g., a possibility in the heat treatment (hyperthermia) of tumors. In a simple situation we would like to achieve a given temperature distribution $z : \Omega \to \mathbb{R}$. It is assumed that we can produce heat sources in the body where the heat source is given as a function $u : \Omega \to \mathbb{R}$. Our goal is to solve the following problem.

(P$_1$) Minimize

$$\mathcal{F}(y, u) := \tfrac{1}{2} \int_\Omega (y(x) - z(x))^2\, dx + \tfrac{\alpha}{2} \int_\Omega u^2(x)\, dx$$

under the constraints

$$-\Delta y = u \ \text{ in } \ \Omega\,, \tag{6.29}$$

$$y = 0 \ \text{ on } \ \partial\Omega\,, \tag{6.30}$$

$$u_-(x) \le u(x) \le u_+(x) \ \text{ for all } x \in \Omega\,.$$

Here $u_-, u_+ : \Omega \to \mathbb{R} \cup \{-\infty, \infty\}$ with $u_- \le u_+$ are given functions. The term $\tfrac{\alpha}{2} \int_\Omega u^2(x)\, dx$ takes the energy costs for heating and cooling into account. The inequality constraints on the control $u : \Omega \to \mathbb{R}$ reflect constraints on the ability to heat or to cool. In practice also inequality constraints on the state $y$ are meaningful

– e.g., in situations in which a cancer patient has a heat treatment. These constraints lead to measure-valued Lagrange multipliers and will not be discussed here any further. For information on this aspect we refer, e.g., to Casas [18].

Often it is more realistic to control the temperature with the help of the temperature at the boundary. In this case we assume a given temperature in the surrounding space and that the heat loss at the boundary is proportional to the temperature difference at the boundary. Let $u : \partial \Omega \to \mathbb{R}$ be the temperature of the surrounding medium at the boundary of $\Omega$ and $a : \partial \Omega \to \mathbb{R}$ is the heat transfer coefficient, compare (6.4). We consider $u : \partial \Omega \to \mathbb{R}$ as control and want to solve the following problem.

(P$_2$) Minimize

$$\mathcal{F}(y, u) := \tfrac{1}{2} \int_{\Omega} (y(x) - z(x))^2 \, dx + \tfrac{\alpha}{2} \int_{\partial \Omega} u^2(x) \, ds_x$$

under the constraints

$$
\begin{array}{lll}
-\Delta y = 0 & \text{in } \Omega , & (6.31) \\
-\nabla y \cdot n = a(y - u) & \text{on } \partial \Omega , & (6.32) \\
u_-(x) \le u(x) \le u_+(x) & \text{for all } x \in \partial \Omega .
\end{array}
$$

Here $u_-, u_+ : \partial \Omega \to \mathbb{R} \cup \{-\infty, \infty\}$ with $u_- \le u_+$ are given functions on $\partial \Omega$.

As in the case of optimal control of ordinary differential equations we would like to derive necessary optimality conditions for solutions of the above control problem. We will argue mostly formally and refer to the book of Tröltzsch [131] for a rigorous treatment.

First of all we will derive the necessary conditions for the problem (P$_1$). As in the case of optimal control of ordinary differential equations we multiply the constraint $\Delta y + u = 0$ with a Lagrange multiplier $p$, then we integrate and add the resulting term to the functional $\mathcal{F}$. We obtain the Lagrangian

$$\mathcal{L}(y, u, p) = \mathcal{F}(y, u) - \int_{\Omega} (-\Delta y - u) \, p \, dx .$$

Assuming now that $p$ is continuously differentiable and in addition vanishes on the boundary, after integration by parts we obtain

$$\mathcal{L}(y, u, p) = \mathcal{F}(y, u) - \int_{\Omega} (\nabla y \cdot \nabla p - up) \, dx .$$

Here $\mathcal{L}$ can be written as a functional on the set $H_0^1(\Omega) \times L_2(\Omega) \times H_0^1(\Omega)$. Motivated from our considerations in the case of ordinary differential equations we conjecture that the necessary condition of a solution $(\bar{y}, \bar{u})$ of (P$_1$) can be formulated with the help of an adjoint state (the Lagrange multiplier) $\bar{p}$ and by taking the first variation of the Lagrangian.

As in the case of ordinary differential equations we obtain:

$$\frac{\delta \mathcal{L}}{\delta y}(\overline{y}, \overline{u}, \overline{p})(y) = 0 \quad \text{for all } y \in H_0^1(\Omega), \tag{6.33}$$

$$\frac{\delta \mathcal{L}}{\delta u}(\overline{y}, \overline{u}, \overline{p})(u - \overline{u}) \geq 0 \quad \text{for all } u \in L_2(\Omega) \text{ with } u_- \leq u \leq u_+, \tag{6.34}$$

$$\frac{\delta \mathcal{L}}{\delta p}(\overline{y}, \overline{u}, \overline{p})(p) = 0 \quad \text{for all } p \in H_0^1(\Omega). \tag{6.35}$$

It follows that

$$\frac{\delta \mathcal{L}}{\delta y}(\overline{y}, \overline{u}, \overline{p})(y) = \int_\Omega \big[(\overline{y}(x) - z(x))\, y(x) - \nabla \overline{p} \cdot \nabla y\big] dx = 0$$

for all $y \in H_0^1(\Omega)$. As we require $\overline{p} \in H_0^1(\Omega)$, it follows that $\overline{p}$ is a weak solution of the boundary value problem

$$-\Delta \overline{p} = \overline{y} - z \quad \text{in } \Omega, \tag{6.36}$$

$$\overline{p} = 0 \quad \text{on } \partial\Omega. \tag{6.37}$$

The condition $\frac{\delta \mathcal{L}}{\delta p}(\overline{y}, \overline{u}, \overline{p}) = 0$ leads to

$$\int_\Omega (\nabla \overline{y} \cdot \nabla p - \overline{u}\, p) = 0$$

for all $p \in H_0^1(\Omega)$ and hence $\overline{y}$ is a weak solution of (6.29), (6.30).
From (6.34) we deduce

$$\int_\Omega (\alpha \overline{u}(x) + \overline{p}(x))(u(x) - \overline{u}(x))\, dx \geq 0 \tag{6.38}$$

for all $u$ with $u_-(x) \leq u(x) \leq u_+(x)$, $x \in \Omega$. This variational inequality says that the control $\overline{u}$ is just the $L_2$ projection of $-\frac{1}{\alpha}\overline{p}$ onto the set of all admissible controls, compare Exercise 6.7. It can be shown (see Exercise 6.9) that (6.38) holds if and only if the following pointwise variational inequality holds

$$(\alpha \overline{u}(x) + \overline{p}(x))(v - \overline{u}(x)) \geq 0 \quad \text{for all} \quad v \in [u_-(x), u_+(x)]. \tag{6.39}$$

The last assertion can be interpreted as follows (Exercise 6.8). The mapping

$$v \mapsto \overline{p}(x)v + \tfrac{\alpha}{2}v^2 \tag{6.40}$$

attains its minimum at the point $\overline{u}(x)$. In the case of a maximization problem for an optimal control in the context of ordinary differential equations we discussed

*Pontryagin's maximum principle.* If we had searched for a maximum instead of a minimum, we could have formulated a corresponding minimum principle. In the case of the control problem ($P_1$) it now holds analogously a minimum principle. In order to formulate the analogy we consider the Hamiltonian $H(u, p) = pu + \frac{\alpha}{2}u^2$.

**Theorem 6.13** *A weak solution* $(\overline{y}, \overline{u}) \in H_0^1(\Omega) \times U$ *of* (6.29), (6.30) *with* $U = \{u \in L_2(\Omega) \mid u_-(x) \le u(x) \le u_+(x)$ *for almost all* $x \in \Omega\}$ *solves* ($P_1$) *if and only if an adjoint state* $\overline{p} \in H_0^1(\Omega)$ *exists which solves the boundary value problem*

$$-\Delta\overline{p} = \overline{y} - z \quad in \ \Omega\,, $$
$$\overline{p} = 0 \quad on \ \partial\Omega$$

*in a weak sense and the minimum principle*

$$H(\overline{u}(x), \overline{p}(x)) = \min_{v \in [u_-(x), u_+(x)]} H(v, \overline{p}(x))$$

*holds.*

For a proof of this theorem we refer to the book of Tröltzsch [131]. In this case the conditions formulated are also sufficient. The problem ($P_1$) interpreted as a minimization problem for $u$ is strictly convex and this is responsible for the fact that the above characterization of critical points only allows for one solution which then has to be a minimum. The control problems for ordinary differential equations which we considered in Chap. 4 are typically not convex and the set of critical points can be very large. In particular many (local) minima and maxima can occur.

At the end we would like to briefly discuss how we can deal with the problem ($P_2$). First of all we want to incorporate the constraints (6.31) and (6.32) into $\mathcal{F}_2$ with the help of Lagrange multipliers. In order to do so we choose a function $p_1 : \Omega \to \mathbb{R}$ and compute

$$\int_\Omega \Delta y\, p_1 \, dx = -\int_\Omega \nabla y \cdot \nabla p_1 \, dx + \int_{\partial\Omega} (\nabla y \cdot n)\, p_1 \, ds_x\,.$$

The second constraint is incorporated with the help of a function $p_2 : \partial\Omega \to \mathbb{R}$ and the expression

$$\int_{\partial\Omega} (-\nabla y \cdot n - a(y - u)) p_2 \, ds_x\,.$$

As Lagrangian one obtains

$$\mathcal{L}(y, u, p_1, p_2) = \mathcal{F}(y, u) - \int_\Omega \nabla y \cdot \nabla p_1 \, dx + \int_{\partial\Omega} (\nabla y \cdot n)(p_1 - p_2)\, ds_x$$
$$- \int_{\partial\Omega} a(y - u)p_2 \, ds_x\,.$$

The condition $\frac{\delta \mathcal{L}}{\delta y}(\bar{y}, \bar{u}, \bar{p}_1, \bar{p}_2)(y) = 0$ for $y \in C_0^\infty(\Omega)$ yields

$$\int_\Omega (\bar{y}(x) - z(x))\, y(x)\, dx - \int_\Omega \nabla p_1 \cdot \nabla y\, dx = 0 .$$

This implies that $p_1$ is a weak solution of the elliptic differential equation

$$-\Delta p_1 = \bar{y} - z \quad \text{in} \quad \Omega .$$

If we consider $\frac{\delta \mathcal{L}}{\delta y}(\bar{y}, \bar{u}, \bar{p}_1, \bar{p}_2)(y) = 0$ for functions $y$, for which $y \neq 0$ or $\nabla y \cdot n \neq 0$ on $\partial\Omega$ holds, then we obtain for a $p_1$ which is sufficiently smooth that

$$\int_\Omega \big(\bar{y}(x) - z(x) + \Delta p_1(x)\big) y(x)\, dx - \int_{\partial\Omega} (\nabla p_1 \cdot n)\, y\, ds_x$$
$$+ \int_{\partial\Omega} (\nabla y \cdot n)(p_1 - p_2)\, ds_x - \int_{\partial\Omega} a\, y\, p_2\, ds_x = 0 , \tag{6.41}$$

where we used integration by parts to obtain this expression. The first integral vanishes because the integrand is zero. If we choose test functions $y$ with $y = 0$ on $\partial\Omega$ we can still choose $\nabla y \cdot n$ arbitrarily and it follows that

$$p_1 = p_2 \quad \text{on} \quad \partial\Omega .$$

We now define $p := p_1$ and it holds that $p = p_2$ on $\partial\Omega$. As $y$ on $\partial\Omega$ can be chosen arbitrarily we obtain from (6.41)

$$-\nabla p \cdot n = ap \quad \text{on} \quad \partial\Omega .$$

Hence, the adjoint state is a solution of the boundary value problem

$$-\Delta p = \bar{y} - z \text{ in } \Omega ,$$
$$-\nabla p \cdot n = ap \quad \text{on } \partial\Omega .$$

For problem (P$_2$) one can now formulate a minimum principle as in Theorem 6.13. The details have to be worked out in Exercise 6.10.


## 6.1.8  Parameter Identification and Inverse Problems

In the mathematical models discussed in this book always one or more parameters appear. In many applications these parameters cannot be determined quantitatively or qualitatively in a satisfactory way. For example the detailed structure of a medium, which could be the soil, or the human body or a metallic workpiece formed from the melt, is not known. It is hence necessary to identify the parameters in a model or at least to estimate these values. In this subsection we will discuss some relevant questions that appear in this context and we also want to point out some problems in the mathematical treatment of these questions. We consider a stationary heat conduction process in a body which occupies a domain $\Omega \subset \mathbb{R}^d$. A heat source $f : \Omega \to \mathbb{R}$ acts on this body and the temperature on the boundary is given by a function $u_0 : \partial\Omega \to \mathbb{R}$. This is characterized by the following boundary value problem:

$$-\nabla \cdot (\lambda(x)\nabla u(x)) = f(x) \text{ if } x \in \Omega ,$$
$$u(x) = u_0(x) \text{ if } x \in \partial\Omega .$$

We now assume that the heat conduction coefficient $\lambda : \Omega \to \mathbb{R}$ is not known. As the body under consideration in general will not be spatially homogeneous we have to assume that $\lambda$ depends on the spatial variable $x$. If we know for a given $f$ and $u_0$ a temperature distribution $z$ via measurements the question arises whether the heat conduction coefficient $\lambda$ can be determined. In general the answer is negative as the example $u_0 \equiv 0$, $f \equiv 0$ shows. In this case $u \equiv 0$ is a solution for all heat conduction coefficients $\lambda$.

The question of how to determine the data of a problem from the knowledge of one or more solutions is a typical *inverse problem*. Determining the solution of the heat equation is the *direct problem*. This is typically the original problem. The converse, which is determining the data from the solution, is called the inverse problem. Inverse problems are often *ill-posed problems*. This means that one of the following conditions is not fulfilled:

- there exists a solution,
- the solution is unique,
- the solution depends continuously on the data.

These three conditions were requested by Hadamard for a *well-posed problem*. In this context we would like to mention one aspect. Let us assume that we are in the case where always a solution exists. If this solution depends discontinuously on the data then severe consequences will follow. In such a case an attempt to compute the solution by approximation, e.g., with the help of a numerical method, will typically lead to a failure. This is due to the fact that slightly different data lead to completely different solutions. One says the problem is unstable. In such a case one typically applies regularization techniques. These change the problem "slightly" such that stability is guaranteed.

We now illustrate the unstable dependence on the data with an example. For this purpose we consider the following one-dimensional diffusion problem in $\Omega = (0, 1)$ and recollect the considerations in Sect. 6.1.4 on homogenization:

$$\tfrac{d}{dx}(\lambda(x)u'(x)) = 0 \quad \text{for} \quad x \in (0, 1) ,$$
$$u(0) = 0 , \, u(1) = 1 .$$

The flux $q(x) = -\lambda(x)u'(x)$ is hence constant in $\Omega$ and we denote this constant by $\bar{q}$. We obtain

$$u'(x) = -\frac{\bar{q}}{\lambda(x)}$$

and using the boundary condition for $x = 0$ we get

$$u(x) = -\bar{q}\int_0^x \frac{1}{\lambda(y)}\,dy.$$

As we require $u(1) = 1$, it follows that

$$\bar{q} = -\left(\int_0^1 \frac{1}{\lambda(y)}\,dy\right)^{-1}.$$

As in the section on homogenization we consider $\varepsilon = \frac{1}{n}$, $n \in \mathbb{N}$, for the functions

$$\lambda^\varepsilon(x) = \lambda\left(\frac{x}{\varepsilon}\right) \quad \text{with} \quad \lambda : \mathbb{R} \to \mathbb{R} \quad 1\text{-periodic}.$$

In this case we obtain as solution of

$$\frac{d}{dx}\left(\lambda^\varepsilon(x)\frac{d}{dx}u^\varepsilon(x)\right) = 0 \quad \text{for} \quad x \in (0,1),$$
$$u^\varepsilon(0) = 0,\, u^\varepsilon(1) = 1,$$

the function

$$u^\varepsilon(x) = \frac{\int_0^x \frac{1}{\lambda^\varepsilon(y)}\,dy}{\int_0^1 \frac{1}{\lambda^\varepsilon(y)}\,dy} = \frac{\int_0^{nx} \frac{1}{\lambda(y)}\,dy}{\int_0^n \frac{1}{\lambda(y)}\,dy} = x + \frac{\int_0^{nx} \frac{1}{\lambda(y)}\,dy - xn\int_0^1 \frac{1}{\lambda(y)}\,dy}{n\int_0^1 \frac{1}{\lambda(y)}\,dy}.$$

As $\lambda$ is periodic we obtain that $\int_0^{nx} \frac{1}{\lambda(y)}\,dy - xn\int_0^1 \frac{1}{\lambda(y)}\,dy$ is bounded and we obtain with $u(x) = x$

$$u^\varepsilon \to u \quad \text{as} \quad \varepsilon \to 0 \quad \text{uniformly in the interval} \quad [0,1].$$

This implies that $u^\varepsilon$ converges uniformly despite the fact that $\lambda^\varepsilon$, in cases where $\lambda$ is not constant, does not even converges pointwise almost everywhere.

Whether a problem is well-posed or not can depend on the choice of the norm and on the function spaces. More generally speaking well-posedness depends on the topology which is used to study the continuity. In the above example it holds that $\lim_{\varepsilon \to 0} u^\varepsilon = u$ in the space $L_\infty(0,1)$ but not in $H^1(0,1)$. Furthermore, one can show that although $\lambda^\varepsilon$ does not converge in $L_2(\Omega)$ it converges in the *dual space* of $H^1(0,1)$. Inverse problems typically have the problem that the norms for which the problem is well-posed are far too strong in order to take errors that appear in the data into account. These errors are typically at best controlled in the $L_2$– or $L_\infty$–norm.

How can we deal with the parameter identification problem? One variant leads to problems which we considered similarly in the section on optimal control. A first proposal would be to solve the following problem.

(Q) Minimize

$$\mathcal{F}(u, \lambda) = \tfrac{1}{2} \int_\Omega (u(x) - z(x))^2 \, dx$$

subject to the constraints

$$-\nabla \cdot (\lambda(x)\nabla u(x)) = f(x) \text{ for } x \in \Omega ,$$
$$u(x) = u_0(x) \text{ for } x \in \partial\Omega ,$$

with respect to all $\lambda \in \{\mu \in L_\infty(\Omega) \mid \mu \text{ uniformly positive}\}$. Here $\lambda$ has the role of a control and $z$ is a given function.

As this is just an equivalent reformulation of the original problem it turns out that also problem (Q) in general is ill-posed. One strategy to obtain a solution is to move to an "approximate" problem which is well-posed. This is called *regularization*. By this the modeling or approximation error increases. However, the influence of errors in the data decreases. It will hence be important to find an optimal "balance" between the sources for errors. Regularization for (Q) can be done by introducing constraints for the admissible set for $\lambda$. What is helpful here for example is the knowledge of bounds $0 < \underline{\lambda}, \overline{\lambda}$ such that

$$\underline{\lambda} \le \lambda(x) \le \overline{\lambda} \text{ for almost all } x \in \Omega \tag{6.42}$$

or further qualitative *a priori information*. A regularization is also possible by discretization, i.e., we restrict $\lambda$ to a finite-dimensional vector space. Related to the restriction of $\lambda$ by $L_\infty$–bounds is the so-called *Tikhonov regularization*. In this regularization a quadratic term is added to the functional $\mathcal{F}$. We obtain for example

$$\mathcal{F}_r(u, \lambda) = \tfrac{1}{2} \int_\Omega (u(x) - z(x))^2 dx + \tfrac{\alpha}{2}\|\lambda\|^2 ,$$

where $\| \cdot \|$ is a suitable norm. In the case of the above diffusion problem the norm has to be chosen strong enough to control the supremum norm of $\lambda$. A possible norm is the $L_\infty$–norm and in three spatial dimensions it is possible to choose the Sobolev norm $\|.\|_{H^2}$. The norm $\|u\|_{H^2}$ is equivalent to the sum of the $L_2$–norms of $u, \nabla u, D^2 u$. In fact the regularization with such high Sobolev norms in general is not used in practice. It is important to guarantee the positivity of the heat conduction coefficient. This can be achieved for example with the constraint (6.42). In this case it is sufficient to choose the $L_2$–norm in the Tikhonov regularization. This is true because the constraints in (6.42) already guarantee the boundedness. As one approximates an ill-posed problem typically there will be no convergence for $\alpha \to 0$.

However, often there will be an optimal regularization parameter $\alpha$. For further information on parameter identification we refer to the book of Banks and Kunisch [10]. Regularization techniques are dealt with in detail in the book Engl, Hanke and Neubauer [34].

### 6.1.9  Linear Elasticity Theory

We will now discuss some aspects of the analysis of the basic equations in linear elastostatics. We consider the boundary value problem for

$$ -\nabla \cdot \sigma(u) = f \tag{6.43} $$

in a bounded domain $\Omega \subset \mathbb{R}^d$ with boundary conditions $u = u_0$ on the boundary $\partial\Omega$. Here, $u$ is the displacement, $\sigma$ the stress tensor and $f$ are the outer forces per unit volume. The stress tensor is given by Hooke's law,

$$ \sigma_{ij}(u) = \sum_{k,\ell=1}^{d} a_{ijk\ell}\, \varepsilon_{k\ell}(u)\,, $$

where $\varepsilon_{ij}(u) = \frac{1}{2}(\partial_i u_j + \partial_j u_i)$ are the components of the linearized strain tensor. The coefficients $a_{ijk\ell}$ are assumed to be bounded functions which fulfill the symmetry relations $a_{ijk\ell} = a_{jik\ell} = a_{k\ell ij}$. More information on the model can be found in Sect. 5.10.

We now take the product of the Eq. (6.43) with a test function $v$, integrate over $\Omega$ and perform an integration by parts. This gives the *weak formulation*:

*Find $u \in H^1(\Omega)^d$ with $u = u_0$ on $\partial\Omega$ such that for all $v \in H^1(\Omega)^d$ with $v = 0$ on $\partial\Omega$ it holds that*

$$ \int_\Omega \sigma(u) : Dv\, dx = \int_\Omega f \cdot v\, dx\,. $$

The left-hand side is for a linear elastic material is given by

$$ a(u,v) = \int_\Omega \sigma(u) : Dv\, dx = \int_\Omega \sum_{i,j,k,\ell=1}^{d} a_{ijk\ell}\, \varepsilon_{k\ell}(u)\partial_{x_j} v_i\, dx $$

$$ = \int_\Omega \sum_{i,j,k,\ell=1}^{d} a_{ijk\ell}\, \varepsilon_{k\ell}(u)\, \varepsilon_{ij}(v)\, dx = \int_\Omega \sum_{i,j,k,\ell=1}^{d} a_{ijk\ell}\, \partial_{x_\ell} u_k\, \partial_{x_j} v_i\, dx\,. $$

The last two identities follow from the symmetry $a_{ijk\ell} = a_{jik\ell} = a_{ij\ell k}$. The mapping $(u,v) \mapsto a(u,v)$ is a *bilinear form* on the space $H^1(\Omega)^d$. A bilinear form is a function which maps two elements in a vector space $V$ to a real number and is linear in both arguments, i.e.,

$$a(\alpha u + \beta v, w) = \alpha\, a(u, w) + \beta\, a(v, w) \quad \text{and} \quad a(u, \alpha v + \beta w) = \alpha\, a(u, v) + \beta\, a(u, w)$$

for all $u, v, w \in V$ and all $\alpha, \beta \in \mathbb{R}$. Using the symmetry condition $a_{ijk\ell} = a_{k\ell ij}$ one can show that the bilinear form of linear elasticity is *symmetric*, i.e., it holds that $a(u, v) = a(v, u)$.

We already discussed that positive definiteness is a possible property of a bilinear form, i.e.,

$$a(u, u) \geq 0, \quad a(u, u) = 0 \Leftrightarrow u = 0.$$

A symmetric, positive definite bilinear form defines an *inner product*. Being positive definite in its weak formulation is a characteristic property of an *elliptic* system of partial differential equations: A differential operator

$$u \to \left( -\sum_{j,k,\ell=1}^{d} \partial_{x_j}\left(b_{ijk\ell}\partial_{x_\ell}u_k\right) \right)_{i=1}^{d}$$

is elliptic if and only if the corresponding bilinear form

$$b(u, v) = \int_{\Omega} \sum_{i,j,k,\ell=1}^{d} b_{ijk\ell}\, \partial_{x_\ell}u_k\, \partial_{x_j}v_i\, dx$$

is positive definite. The bilinear form of elastostatics is positive definite if the coefficients $a_{ijk\ell}$ of Hooke' tensor fulfill condition (5.52). Then

$$a(u, u) \geq a_0 \int_{\Omega} \sum_{i,j=1}^{d} |\varepsilon_{ij}(u)|^2\, dx$$

follows, and the positive definiteness of $a(\cdot, \cdot)$ is then equivalent to the property

$$\int_{\Omega} \sum_{i,j=1}^{d} |\varepsilon_{ij}(u)|^2\, dx = 0 \quad \text{if and only if} \quad u = 0.$$

The condition on the left-hand side is equivalent to $\varepsilon_{ij}(u) = 0$ for all $i, j = 1, \ldots, d$. This is now equivalent to the fact that $u$ is an element of the set of all *linearized rigid body displacements*

$$\mathcal{R} = \left\{ x \mapsto Bx + a \,\middle|\, B \in \mathbb{R}^{d,d},\ B^\top = -B,\ a \in \mathbb{R}^d \right\}.$$

If $u \in \mathcal{R}$ then $\varepsilon_{ij}(u) = \frac{1}{2}(b_{ij} + b_{ji}) = 0$, where $(B)_{ij} = b_{ij}$. On the other hand if $\varepsilon_{ij}(u) = 0$ for all $i, j$, then we obtain $\partial_i u_j = -\partial_j u_i$ and hence

$$\partial_i \partial_j u_k = -\partial_i \partial_k u_j = \partial_k \partial_j u_i = -\partial_j \partial_i u_k .$$

It hence follows that $\partial_i \partial_j u_k = -\partial_i \partial_j u_k$ and $\partial_i \partial_j u_k = 0$ for all $i, j, k$. If $\Omega$ is connected then we obtain that $u$ is an *affine* function,

$$u(x) = Bx + a \quad \text{with} \quad B \in \mathbb{R}^{d,d}, \ a \in \mathbb{R}^d .$$

The identities $\varepsilon_{ij}(u) = \frac{1}{2}(b_{ij} + b_{ji}) = 0$ for all $i, j$ then imply that $B$ is skew symmetric, i.e., $B^\top = -B$. Using these considerations and some facts from functional analysis one can deduce the following important result:

**Theorem 6.14** (Korn's inequality) *Let $\Omega \subset \mathbb{R}^d$ be a bounded domain with smooth boundary $\partial\Omega$. Under these assumptions there exist positive constants $c_0, c_1, c_2$, such that*

*(i)* $\displaystyle \int_\Omega \sum_{i,j=1}^d |\varepsilon_{ij}(u)|^2 \, dx \geq c_0 \int_\Omega \left( |u|^2 + |Du|^2 \right) dx$ *for all $u \in H^1(\Omega)^d$ with $u = 0$ on $\partial\Omega$,*

*(ii)* $\displaystyle \int_\Omega \sum_{i,j=1}^d |\varepsilon_{ij}(u)|^2 \, dx + c_1 \int_\Omega |u|^2 \, dx \geq c_2 \int_\Omega |Du|^2 \, dx$ *for all $u \in H^1(\Omega)^d$.*

*The constants $c_0, c_1, c_2$ depend on the domain $\Omega$.*

*Proof* We show Korn's inequality for the case (i) under the assumption that $u$ in $H^2(\Omega)$ holds. We refer to Zeidler [140] for a proof of the generalization. Using integration by parts twice, and the boundary condition $u = 0$ on $\partial\Omega$ one easily deduces

$$\int_\Omega \partial_i u_j \, \partial_j u_i \, dx = \int_\Omega \partial_i u_i \, \partial_j u_j \, dx .$$

This then implies

$$\int_\Omega \sum_{i,j=1}^d |\varepsilon_{ij}(u)|^2 \, dx = \int_\Omega \frac{1}{4} \sum_{i,j=1}^d \left( (\partial_i u_j)^2 + (\partial_j u_i)^2 + 2\,\partial_i u_j \, \partial_j u_i \right) dx$$

$$= \int_\Omega \frac{1}{2} \left[ \sum_{i,j=1}^d (\partial_i u_j)^2 + \left( \sum_{i=1}^d \partial_i u_i \right) \left( \sum_{j=1}^d \partial_j u_j \right) \right] dx$$

$$= \tfrac{1}{2} \left( \|Du\|_{L_2(\Omega)}^2 + \|\nabla \cdot u\|_{L_2(\Omega)}^2 \right) \geq \tfrac{1}{2} \|Du\|_{L_2(\Omega)}^2 .$$

Using the *Poincaré inequality*

$$\|\varphi\|_{L_2(\Omega)} \leq c_P \|D\varphi\|_{L_2(\Omega)} \quad \text{for all} \quad \varphi \in H_0^1(\Omega)$$

with a constant $c_P$, depending on the domain $\Omega$, we obtain

$$\|\varepsilon(u)\|^2_{L_2(\Omega)} \geq \frac{1}{2}\|Du\|^2_{L_2(\Omega)} \geq \frac{1}{2(1 + c_P^2)}\left(\|u\|^2_{L_2(\Omega)} + \|Du\|^2_{L_2(\Omega)}\right).$$

This is Korn's inequality with $c_0 = \sqrt{1/(2(1 + c_P^2))}$. $\qquad\qquad\qquad\square$

As the name already indicates, the set $\mathcal{R}$ of all linearized rigid body displacements is the set of all linearizations of "true" rigid body displacements

$$x \mapsto (Q - I)x,$$

where $Q$ is an orthogonal matrix with determinant $\det Q = 1$. To demonstrate the relation between the rotation matrix $Q$ and a skew symmetric matrix $B$ we consider a one-parameter family of orthogonal matrices $Q_\delta, \delta \in \mathbb{R}$, which converge as $\delta \to 0$ to the identity matrix $I$ and have the asymptotic behavior $Q_\delta = I + \delta B + O(\delta^2)$. The orthogonality condition $Q_\delta^\top Q_\delta = I$ yields

$$0 = \lim_{\delta \to 0} \frac{1}{\delta}(Q_\delta^\top Q_\delta - I) = \lim_{\delta \to 0} \frac{1}{\delta}(\delta B^\top I + \delta I B + O(\delta^2)) = B^\top + B.$$

In this sense we can consider skew symmetric matrices as linearizations of orthogonal matrices.

The boundary value problem of elastostatics is equivalent to the optimization problem

$$\min\left\{J(u) \,\middle|\, u \in H^1(\Omega)^d,\ u = u_0 \text{ on } \partial\Omega\right\}$$

with the functional

$$J(u) = \int_\Omega \left[\frac{1}{2} \sum_{i,j,k,\ell=1}^{d} a_{ijk\ell}\, \varepsilon_{k\ell}(u)\, \varepsilon_{ij}(u) - f \cdot u\right] dx.$$

Here

$$\int_\Omega \frac{1}{2} \sum_{i,j,k,\ell=1}^{d} a_{ijk\ell}\, \varepsilon_{k\ell}(u)\, \varepsilon_{ij}(u)\, dx$$

is the stored elastic energy in the deformed body and

$$\int_\Omega f \cdot u\, dx$$

is the work performed by the volume forces.

## 6.2  Parabolic Equations

In this section we analyze as a model example for a parabolic equation the *heat equation*

$$\partial_t u - D\Delta u = f \tag{6.44}$$

with a diffuse constant $D > 0$. This equation also describes *diffusion processes* in general. These are characterized by the fact that a physical quantity described by $u$, which can be the temperature or the concentration of a substance, flows from region where it attains high values to regions where small values are attained. The special form (6.44) is obtained from the constitutive law

$$q = -D\nabla u$$

for the *flux q*. One hence assumes that the flux is directly proportional to the gradient of the quantity, compare (5.9). Here also other physical laws, in particular nonlinear laws, are possible. These lead to nonlinear variants of (6.44) and the resulting equations are also called *parabolic* as long as the direction of the flow is in the direction of the negative gradient. The term $f$ can describe a heat source for example given by electromagnetic radiation (like in a microwave), by electric currents or by chemical reactions. For a species diffusion $f$ represents a source term for the diffusing substance, which could be due to a chemical reaction.

For the elliptic equations in the previous section suitable *boundary conditions* were necessary in order to obtain a meaningful problem. For parabolic equations boundary conditions are also necessary. In addition, we also need *initial conditions* which describe the system at a fixed given time. In order to determine the evolution of the temperature in a given domain, we need to know the temperature distribution at initial time and also data which describe the heat supply across the boundary during the process. The initial conditions fix the values of the solution $u$ at a fixed time instance $t_0$, i.e.,

$$u(t_0, x) = u_0(x), \ x \in \Omega .$$

As boundary conditions the same conditions which we considered for elliptic equations are possible. Namely

- Dirichlet boundary conditions

$$u(t, x) = u_{\partial\Omega}(t, x) \ \text{ for all } \ x \in \partial\Omega , \ t > t_0$$

with a given function $u_{\partial\Omega}$. These describe situations in which values for temperature and concentration at the boundary of the domain under consideration are prescribed.
- Neumann boundary conditions

$$-D\nabla u \cdot n = g \ \text{ for all } \ x \in \partial\Omega , \ t > t_0 .$$

These prescribe values for the *flux* of the temperature or the species. Often one has $g = 0$ and in this case the domain is *isolated* at the boundary.

- Mixed forms, in particular boundary conditions of the third kind

$$-D\nabla u \cdot n = a(u - b) \quad \text{for all } x \in \partial\Omega \text{ and } t > t_0 \,.$$

For the heat equation this condition means that the heat loss on the boundary is proportional to the difference between the temperature at the boundary and a given exterior temperature $b$. This corresponds to *Newton's law of cooling*.

Now the question arises whether the above boundary conditions guarantee the *existence and uniqueness of a solution*. It could be that additional conditions have to be imposed. To answer such questions is an important task in the context of mathematical modeling as in real world applications exactly one state appears.

It is possible to show existence and uniqueness of a solution for all three boundary conditions if also initial conditions are prescribed. To show existence of a solution is more elaborate and we will discuss here only a uniqueness result.

## 6.2.1  Uniqueness of Solutions, the Energy Method

**Theorem 6.15** *Let $\Omega \subset \mathbb{R}^d$ be a bounded domain with smooth boundary, the functions $f : \mathbb{R}_+ \times \Omega \to \mathbb{R}$, $u_0 : \Omega \to \mathbb{R}$, $a, b : \mathbb{R}_+ \times \partial\Omega \to \mathbb{R}$ are assumed to be smooth and let $a \geq 0$. Then there exists at most one solution of the heat equation*

$$\partial_t u - D\Delta u = f$$

*with initial condition*

$$u(0, x) = u_0(x) \quad \text{for all } x \in \Omega$$

*and with one of the boundary conditions*

$$u(t, x) = b(t, x) \quad \text{for all } x \in \partial\Omega \,, \ t > 0, \ \text{or}$$
$$-D\nabla u(t, x) \cdot n(x) = b(t, x) \quad \text{for all } x \in \partial\Omega \,, \ t > 0, \ \text{or}$$
$$-D\nabla u(t, x) \cdot n(x) = a(t, x)(u(t, x) - b(t, x)) \quad \text{for all } x \in \partial\Omega \,, \ t > 0 \,.$$

**Remark**. Here we did not specify in which sense the heat equation and the boundary conditions are assumed to hold. Also the specific assumptions on the given data $f$, $u_0$, $a$, $b$ have not been made precise. To simplify matters we assume that the solution is classical. This means that all derivatives that appear exist and that the differential equation and the boundary condition are fulfilled pointwise. However, also a uniqueness result for generalized (so-called weak) solutions exists in a similar sense as we formulated it for elliptic equations in the previous section.

*Proof*  The proof uses the so-called *energy method*, however, the quantities appearing need not be related to a physical energy.

We assume that two solutions $u_1$ and $u_2$ exist. Their difference

$$w = u_1 - u_2$$

then, due to the linearity of the problem, also solves a parabolic boundary value problem

$$\partial_t w = D \Delta w \qquad \text{for all } x \in \Omega, \ t > 0,$$
$$w(0, x) = 0 \qquad \text{for all } x \in \Omega,$$

as well as one of the following conditions

$$w = 0 \qquad \text{for all } x \in \partial\Omega, \ t > 0, \text{ or} \qquad (6.45)$$
$$\nabla w \cdot n = 0 \qquad \text{for all } x \in \partial\Omega, \ t > 0, \text{ or} \qquad (6.46)$$
$$-D\nabla w \cdot n = a\, w \qquad \text{for all } x \in \partial\Omega, \ t > 0. \qquad (6.47)$$

We now multiply the equation $\partial_t w - D \Delta w = 0$ with $w$ and integrate over $Q_s = (0, s) \times \Omega$. This gives

$$\int_{Q_s} \left( \partial_t w\, w - D \Delta w\, w \right) dx\, dt = 0.$$

With the help of the identity $\partial_t w\, w = \frac{1}{2}\partial_t w^2$, the divergence theorem and the identity $\nabla \cdot (\nabla w\, w) = \Delta w\, w + |\nabla w|^2$ it follows that

$$\int_{Q_s} \left( \frac{1}{2}\partial_t (w^2) + D|\nabla w|^2 \right) dx\, dt - \int_0^s \int_{\partial\Omega} w\, D\nabla w \cdot n\, ds_x\, dt = 0.$$

The fundamental theorem of calculus and the identity $w(0, x) \equiv 0$ imply

$$\frac{1}{2}\int_\Omega |w(s, x)|^2\, dx + D \int_{Q_s} |\nabla w|^2\, dx\, dt$$

$$= \begin{cases} 0 & \text{for the condition (6.45) or (6.46),} \\ -\displaystyle\int_0^s \int_{\partial\Omega} a w^2\, ds_x\, dt & \text{for condition (6.47).} \end{cases}$$

It hence holds for all $s \in [0, T]$:

$$\frac{1}{2}\int_\Omega |w(s, x)|^2\, dx \leq 0$$

and hence

$$w \equiv 0.$$

This implies

$$u_1 = u_2$$

which yields the uniqueness of a solution. □

**Remark**. The condition $a \geq 0$ was needed for the way we argued. By a refinement of the technique it is possible to prove uniqueness without this condition.

## 6.2.2 Large Time Behavior

If we prescribe for the diffusion equation data $a, b, f$ which are constant in time one obtains that the solution converges as $t \to \infty$ to a solution of the *stationary* diffusion equation

$$-D\Delta u = f$$

with the corresponding boundary condition. We now aim for a more rigorous mathematical derivation of this property. In doing so we will also gain information on the *convergence speed* towards the equilibrium in dependence on $\Omega$. As an example we study a Dirichlet problem on a bounded domain $\Omega$ with given initial data $u(0, x) = u_0(x)$ for $x \in \Omega$ and boundary data $u(t, x) = b(t, x)$ for $x \in \partial\Omega, t > 0$. The time-dependent parabolic problem is then given by

$$
\begin{align}
\partial_t u(t, x) - D\Delta u(t, x) &= f(x) & \text{for } x \in \Omega, \, t > 0, & \quad (6.48) \\
u(0, x) &= u_0(x) & \text{for } x \in \Omega, & \quad (6.49) \\
u(t, x) &= b(x) & \text{for } x \in \partial\Omega, & \quad (6.50)
\end{align}
$$

and the stationary limit solves the boundary value problem

$$
\begin{align}
-D\Delta\bar{u}(x) &= f(x) & \text{for } x \in \Omega, & \quad (6.51) \\
\bar{u}(x) &= b(x) & \text{for } x \in \partial\Omega. & \quad (6.52)
\end{align}
$$

The difference $w(t, x) = u(t, x) - \bar{u}(x)$ is a solution of the parabolic problem

$$
\begin{align}
\partial_t w(t, x) - D\Delta w(t, x) &= 0 & \text{for } x \in \Omega, \, t > 0, & \quad (6.53) \\
w(0, x) &= u_0(x) - \bar{u}(x) & \text{for } x \in \Omega, & \quad (6.54) \\
w(t, x) &= 0 & \text{for } x \in \partial\Omega. & \quad (6.55)
\end{align}
$$

After multiplication of the differential equation with $w$ we obtain after integration over $\Omega$, similar as in the proof of the uniqueness theorem,

$$\frac{d}{dt}\frac{1}{2}\int_{\Omega} w^2\,dx = -D\int_{\Omega} |\nabla w|^2\,dx\,.$$

As $w = 0$ on $\partial\Omega$ we can use the Poincaré inequality

$$\int_{\Omega} w^2\,dx \le c_P \int_{\Omega} |\nabla w|^2\,dx$$

with a constant $c_P$ that depends on the domain $\Omega$, see, e.g., [4]. Here $c_P$ is the *smallest* constant for which the above inequality is true for all functions which are zero on the boundary. It then follows that

$$\frac{d}{dt}\int_{\Omega} w^2\,dx \le -\frac{2D}{c_P}\int_{\Omega} w^2\,dx\,,$$

or with $y(t) = \int_{\Omega} w^2(t, x)\,dx$,

$$y'(t) \le -\frac{2D}{c_P}y(t)\,.$$

Multiplication with $e^{2Dt/c_P}$ yields

$$\frac{d}{dt}\left(y(t)\,e^{2Dt/c_P}\right) \le 0\,.$$

After integration with respect to $t$ it follows that

$$\int_{\Omega} w^2(t, x)\,dx \le \left(\int_{\Omega} w^2(0, x)\,dx\right)e^{-2Dt/c_P}\,. \qquad (6.56)$$

**Remark**. It can be shown that equality in the Poincaré inequality

$$\int_{\Omega} w^2\,dx \le c_P \int_{\Omega} |\nabla w|^2\,dx$$

is true only for *eigenfunctions* $w$ of the *smallest eigenvalue* $\mu_1$ of the Laplace operator. Here, $w \ne 0$ is eigenfunction with eigenvalue $\mu$ if

$$\begin{aligned}-\Delta w &= \mu\,w \quad &&\text{in } \Omega\,,\\ w &= 0 \quad &&\text{on } \partial\Omega\,.\end{aligned}$$

Let $\mu_1$ be the smallest number for which the above problem has a solution $\overline{w} \ne 0$. Then it holds that

$$\int_{\Omega} \overline{w}^2\,dx = -\frac{1}{\mu_1}\int_{\Omega} \overline{w}\,\Delta\overline{w}\,dx = \frac{1}{\mu_1}\int_{\Omega} |\nabla\overline{w}|^2\,dx$$

and hence

$$c_P = \frac{1}{\mu_1}.$$

With

$$w(0, x) = \overline{w}(x),$$

one can show for

$$w(t, x) = e^{-D\mu_1 t}\, \overline{w}(x)$$

the identities

$$\partial_t w = -D\mu_1 w,$$
$$-D\Delta w = D\mu_1 w,$$

and hence

$$\partial_t w - D\Delta w = 0.$$

In addition, it holds that

$$\int_\Omega w^2(t, x)\, dx = e^{-2D\mu_1 t} \int_\Omega w^2(0, x)\, dx.$$

As $c_P = \frac{1}{\mu_1}$ this shows that the estimate (6.56) is sharp. Hence, in general the convergence to zero is not faster than (6.56).

Altogether this shows that the heat equation in fact balances out temperature differences and that solutions converge for large times to an equilibrium profile. In fact the times do not need to be "very large" due to the fact that the convergence is exponentially fast.

We now want to discuss how the convergence velocity depends on $D$ and the size of $\Omega$. In this context we remark that $c_P$ has a *dimension*. In the inequality

$$\int_\Omega w^2\, dx \le c_P \int_\Omega |\nabla w|^2\, dx$$

the quantity $w^2$ has dimension $[w]^2$ and $|\nabla w|^2$ the dimension $[w]^2/[L]^2$. This implies that $c_P$ has the dimension $[L]^2$. A more mathematical view uses a *dilation* of the domain $\Omega$ to $\Omega_L = L\Omega$. This can be realized by a variable transformation $x^{\text{old}} = x^{\text{new}}/L$. In this case it holds that

$$w_L(x) = w\left(\tfrac{x}{L}\right),$$
$$\nabla w_L(x) = \tfrac{1}{L} \nabla w\left(\tfrac{x}{L}\right)$$

and the inequality

$$\int_{\Omega} w^2(x)\,dx \le c_P \int_{\Omega} |\nabla w(x)|^2\,dx$$

transforms to

$$\int_{\Omega_L} w_L^2(x)\,dx \le c_P L^2 \int_{\Omega_L} |\nabla w_L(x)|^2\,dx\,.$$

It follows that $c_{PL} = c_P L^2$, where $c_{PL}$ is the Poincaré constant of the domain $\Omega_L$. The rate with which a solution with homogeneous boundary values converges to zero is hence given by

$$-\frac{2D}{c_P}$$

with dimension $1/[T]$. After scaling of the domain $L$ we obtain the new convergence rate

$$-\frac{2D}{c_P L^2}\,.$$

The convergence rate hence decreases *quadratically* with the diameter of the domain. With a larger domain the convergence rate decreases drastically. The rate is also *proportional* to the diffusion constant.

**Nondimensionalization** The fact that $\frac{D}{L^2}$ is important for the temporal behavior can be seen also with the help of dimensional analysis. Let $u(t, x)$ be a solution of

$$\begin{aligned}
\partial_t u - D\Delta u &= 0 && \text{for } x \in \Omega\,,\ t > 0\,, \\
u(t, x) &= 0 && \text{for } x \in \partial\Omega\,,\ t > 0\,, \\
u(0, x) &= u_0(x) && \text{for } x \in \Omega\,.
\end{aligned}$$

We now introduce nondimensional quantities:

$$\tilde{x} = \frac{x}{L}\,,$$

$$\tilde{t} = \frac{t}{\tau}\,,$$

where $L$ is a *characteristic length*, which can be chosen for example as the diameter of the domain $\Omega$. In addition, let $\tau$ be a characteristic time. As $D$ has the dimension $[L^2]/[T]$ it makes sense to choose $\tau = \frac{L^2}{D}$. In addition we set

$$w = \frac{u}{\bar{u}} \quad \text{with a reference value } \bar{u}\,.$$

It now holds that

$$\bar{u}\,w(\tilde{t}, \tilde{x}) = u(t, x)\,.$$

The equation

$$\partial_t u - D \Delta u = 0$$

implies

$$\frac{\bar{u}}{\tau} \partial_{\tilde{t}} w - \frac{D \bar{u}}{L^2} \Delta_{\tilde{x}} w = 0$$

and hence

$$\partial_{\tilde{t}} w - \frac{\tau}{L^2} D \Delta_{\tilde{x}} w = 0.$$

The choice $\tau = \frac{L^2}{D}$ yields the nondimensional heat equation

$$\partial_{\tilde{t}} w - \Delta_{\tilde{x}} w = 0.$$

The assertions for the nondimensional equations also hold for the original equation with dimensions when the time scale is modified by $\frac{L^2}{D}$.

### 6.2.3 Separation of Variables and Eigenfunctions

We now consider an elementary technique to compute solutions of the heat equation on bounded domains $\Omega$. The same technique is helpful also for many other linear equations of different types. We demonstrate this technique with the example of a simple initial value problem with Dirichlet boundary data

$$\begin{aligned}
\partial_t u - \Delta u &= 0 && \text{for } x \in \Omega, \; t > 0, \\
u(t, x) &= 0 && \text{for } x \in \partial\Omega, \; t > 0, \\
u(0, x) &= u_0(x) && \text{for } x \in \Omega.
\end{aligned} \tag{6.57}$$

Separation of variables is based on the following ansatz for the solution

$$u(t, x) = v(t) w(x)$$

with functions $v : \mathbb{R}_+ \to \mathbb{R}$, $w : \Omega \to \mathbb{R}$. This gives

$$v'(t) w(x) - v(t) \Delta w(x) = 0,$$

or after dividing by $v(t) w(x)$,

$$\frac{v'(t)}{v(t)} = \frac{\Delta w(x)}{w(x)}.$$

As the left-hand side only depends on $t$ and the right-hand side only depends on $x$ we deduce that both sides are equal to a fixed constant which will be denoted by $-\mu$ in what follows. The equation then splits into two parts

$$v' = -\mu\,v\,,$$
$$\Delta w = -\mu\,w\,.$$

For $w$ in addition the boundary condition has to be fulfilled, i.e.,

$$w(x) = 0 \ \text{ for } \ x \in \partial\Omega\,.$$

Functions $w \neq 0$ with

$$-\Delta w = \mu\,w \ \text{ in } \ \Omega \ \text{ and } \ w(x) = 0 \ \text{ on } \ \partial\Omega \tag{6.58}$$

are just the *eigenfunctions* of the Laplace operator with Dirichlet boundary data 0. With the help of the *spectral theory for compact, self-adjoint operators* which is studied in functional analysis, see, e.g., [4, 138], one can show that sequences

$$\{w_n\}_{n\in\mathbb{N}},\ \{\mu_n\}_{n\in\mathbb{N}} \ \text{ with } 0 < \mu_1 \leq \mu_2 \leq \mu_3 \leq \cdots,\ \mu_n \to +\infty \text{ for } n \to +\infty$$

of eigenfunctions $w_n$ with eigenvalues $\mu_n$ exist. The linear span of the eigenfunctions

$$\{w_1, w_2, w_3, \dots\}$$

is *dense* in $L_2(\Omega)$ (and also in $H_0^1(\Omega)$), i.e., functions in $L_2(\Omega)$ or in $H_0^1(\Omega)$ are approximated arbitrarily well by finite linear combinations of eigenfunctions. For $a_n \in \mathbb{R}$ the function

$$a_n e^{-\mu_n t} w_n(x)$$

is a solution of

$$\partial_t u - \Delta u = 0\,. \tag{6.59}$$

This means in particular that initial data $u(0, x) = w_n(x)$ will decrease exponentially with the rate $\mu_n$. In addition every finite linear combination

$$\sum_{n=1}^{N} a_n\, e^{-\mu_n t} w_n(x)$$

is also a solution of (6.59). In general one can show that for all $u_0 \in L_2(\Omega)$ a representation

$$u_0(x) = \sum_{n=1}^{\infty} a_n\, w_n(x)$$

exists. The solution of the initial boundary value problem (6.57) is then given by

$$u(t, x) = \sum_{n=1}^{\infty} a_n e^{-\mu_n t} w_n(x) . \tag{6.60}$$

This is a generalization of a method which was developed by Fourier to solve the heat equation. This approach leads to *Fourier analysis* which deals with Fourier series and Fourier transformation.

The individual terms in the series (6.60) decay differently. The term $a_1 e^{-\mu_1 t} w_1(x)$ has the slowest decay. We already saw earlier that the smallest eigenvalue of the Laplace operator was responsible for the slowest decay.

**Remark**. If one can show that

$$\sum_{n=1}^{\infty} a_n e^{-\mu_n t} w_n(x)$$

is a solution of $\partial_t u - \Delta u = 0$ one has solved the heat equation with boundary condition $u = 0$ for general initial data $u_0 \in L_2(\Omega)$. To justify this approach rigorously one needs some functional analysis, see, e.g., Evans [37].

In order to state the solutions explicitly one needs to compute the eigenfunctions of the Laplace operator. For some specific domains this is possible.

**Example**: Let $d = 1$ and $\Omega = (0, 1)$. Then it holds that

$$w_n(x) - \sin(n\pi x) \quad \text{and} \quad \mu_n - (n\pi)^2 .$$

The solution of the heat equation is given as

$$u(t, x) = \sum_{n=1}^{\infty} a_n e^{-(n\pi)^2 t} \sin(n\pi x) ,$$

where the coefficients $a_n$ stem from the Fourier series of the initial data

$$u(0, x) = \sum_{n=1}^{\infty} a_n \sin(n\pi x) .$$

The lower the "frequency" $n$ of a term the slower the corresponding part in the solution decays. Terms with many oscillations are damped rapidly.

## 6.2.4   The Maximum Principle

The characteristic property of the diffusion equation is that the (heat or species) flux points in the direction of the negative (temperature or species) *gradient*. This leads to the property that for a vanishing right-hand side, i.e., $f = 0$, the maximum of the temperature or concentration will always lie on the "parabolic boundary"

$$\Gamma = ([0, T] \times \partial\Omega) \cup (\{0\} \times \Omega)$$

of the space-time domain $Q_T = (0, T) \times \Omega$. This means the maximum is either attained at initial time or at the boundary of the spatial domain ("there are no hot spots in the interior").

**Theorem 6.16** *Let $\Omega$ be a bounded domain and u be a classical solution of the heat equation*

$$\partial_t u = D \Delta u \ \ for \ x \in \Omega, \ 0 < t \leq T. \tag{6.61}$$

*Then u attains its maximum (and its minimum) either at time $t = 0$ or on the set $[0, T] \times \partial\Omega$.*

*Proof* If $u$ does not attain its maximum on the parabolic boundary there exists a point $(t_0, x_0) \in Q_T \setminus \Gamma$ such that

$$B = u(t_0, x_0) = \max_{(t,x)\in\overline{Q}_T} u(t, x) > \max_{(t,x)\in\Gamma} u(t, x) = A,$$

and it follows that

$$\nabla u(t_0, x_0) = 0,$$
$$D^2 u(t_0, x_0) \text{ is negative semidefinite and hence}$$
$$\Delta u(t_0, x_0) = \text{trace } D^2 u(t_0, x_0) \leq 0.$$

In addition from $t_0 > 0$ and $u(t, x) \leq u(t_0, x)$ for $t < t_0$, it follows that

$$\partial_t u(t_0, x_0) \geq 0.$$

If

$$\Delta u(t_0, x_0) < 0 \ \ \text{or} \ \ \partial_t u(t_0, x_0) > 0 \tag{6.62}$$

would hold this would contradict equation (6.61). However, from what we know so far we cannot deduce (6.62) for $u$. Therefore, we consider

$$w(t, x) = u(t, x) - \varepsilon(t - t_0).$$

It now holds for $t < t_0$,

$$\max_{(t,x)\in Q_{t_0}} w(t, x) \geq u(t_0, x_0) = B$$

and

$$\max_{(t,x)\in\Gamma} w(t, x) = \max_{(t,x)\in\Gamma} (u(t, x) - \varepsilon(t - t_0)) \leq \max_{(t,x)\in\Gamma} u(t, x) + \varepsilon t_0 = A + \varepsilon t_0 .$$

Choosing $\varepsilon$ so small that

$$B > A + \varepsilon t_0 , \text{ which means } \frac{B - A}{t_0} > \varepsilon ,$$

we obtain that $w$ attains its maximum not on $\Gamma$ but at some point $(t_1, x_1)$ in the interior. This implies that for $\varepsilon$ sufficiently small

$$\partial_t w(t_1, x_1) \geq 0 ,$$
$$\Delta w(t_1, x_1) \leq 0 ,$$

and hence
$$\partial_t w(t_1, x_1) - D\Delta w(t_1, x_1) \geq 0 .$$

This leads to a contradiction to

$$\partial_t w(t_1, x_1) - D\Delta w(t_1, x_1) = \partial_t u(t_1, x_1) - c - D\Delta u(t_1, x_1) = \quad c < 0 .$$

This implies the assertion of the theorem.                                                                        □

## 6.2.5  The Fundamental Solution

We will now consider a situation in which heat is released to a system at one fixed point in space and time. We hence search for a solution of

$$\partial_t u = D \Delta u \quad \text{for} \quad t > 0 , \ x \in \mathbb{R}^d ,$$

with the properties

$$u(0, x) = 0 \quad \text{for} \quad x \neq 0 ,$$
$$\int_{\mathbb{R}^d} u(t, x) = Q \quad \text{for} \quad t > 0 .$$

Here $Q$ is a constant which is proportional to the total amount of heat released. Such a solution can be used to construct solutions to problems with general heat supplies and general initial data. One can do this by a convolution of all data which in some sense "adds" all these influences.

We now want to find an ansatz for the solution with the help of dimensional analysis. A solution $u$ can depend on $(t, x)$ and on the parameters $D$ and $Q$. We now search for dimensionless combinations and if possible for typical length and time scales. As the domain on which we search for the solution is the whole of $\mathbb{R}^d$ we have no parameter which provides information about the size of the domain that can be used as a length scale. The appearing quantities have the following dimensions

| quantity | $x$ | $t$ | $D$ | $Q$ | $u$ |
|---|---|---|---|---|---|
| dimension | $L$ | $T$ | $L^2/T$ | $KL^d$ | $K$ |

where $K$ has the dimension of temperature. For a dimensionless quantity

$$\Pi = x^a t^b D^c Q^e, \quad a, b, c, e \in \mathbb{Z}$$

of the independent "variables" it holds that

$$\text{dimension of } \Pi = L^a T^b L^{2c} T^{-c} K^e L^{de}$$

and hence it follows that $e = 0$, $a + 2c = 0$, $b = c$. We therefore obtain

$$\Pi = \left(\frac{x}{\sqrt{Dt}}\right)^a$$

and up to powers $\frac{x}{\sqrt{Dt}}$ is the only dimensionless combination of the independent "variables". A dimensionless combination of the *independent* "variables" in the ansatz for the solution

$$u(t, x, D, Q)$$

hence cannot contain $Q$. What is a good scaling of the dependent variable $u$? As only the quantity $Q$ contains the unit Kelvin it makes sense to use the scalings

$$\frac{Q}{x^d} \quad \text{or} \quad \frac{Q}{(Dt)^{d/2}}.$$

The heat propagation of the point source should have a representation of the solution which is invariant under a change of the dimension. Therefore, we make the ansatz

$$u(t, x) = \frac{Q}{(Dt)^{d/2}} U\left(\frac{x}{\sqrt{Dt}}\right). \tag{6.63}$$

We call such a solution *self-similar*. The shape of $u(t, \cdot)$ does not change if we change the time under consideration. If we know the solution at a fixed time instance we know the solution for all times after performing a simple scaling (similarity transformation). In particular we observe that the problem is reduced by one dimension.

We now want to derive an equation for $U$. Setting $y = \frac{x}{\sqrt{Dt}}$ and denoting by $D_y$ the first and by $D_y^2$ the second derivative with respect to $y$ we obtain for a function $u$ of the form (6.63)

$$\nabla u = \frac{Q}{(Dt)^{d/2}} \frac{1}{\sqrt{Dt}} \nabla_y U \left( \frac{x}{\sqrt{Dt}} \right),$$

$$D^2 u = \frac{Q}{(Dt)^{d/2}} \frac{1}{Dt} D_y^2 U \left( \frac{x}{\sqrt{Dt}} \right),$$

$$\Delta u = \operatorname{trace} D^2 u = \frac{Q}{(Dt)^{(d+2)/2}} \Delta_y U \left( \frac{x}{\sqrt{Dt}} \right),$$

$$\partial_t u = -\frac{d \, Q \, U}{2 \, t^{d/2+1} \, D^{d/2}} - \frac{Q}{2 \, D^{(d+1)/2} \, t^{(d+3)/2}} \, x \cdot \nabla_y U.$$

The equation $-\partial_t u + D \Delta u = 0$ implies

$$\frac{d \, Q}{2 \, (Dt)^{d/2} \, t} U + \frac{Q}{2 \, (Dt)^{d/2} \, t} \, y \cdot \nabla_y U + \frac{Q}{(Dt)^{d/2} \, t} \Delta_y U = 0$$

and hence

$$\tfrac{d}{2} U + \tfrac{1}{2} y \cdot \nabla_y U + \Delta_y U = 0. \tag{6.64}$$

Moreover, we observe that the total amount of heat at all times is given by the total amount of heat at the initial time. We obtain from

$$\int_{\mathbb{R}^d} u(t, x) \, dx = Q$$

by setting $Dt = 1$

$$\int_{\mathbb{R}^d} U(y) \, dy = 1.$$

As the equation, or physically speaking since the body under consideration, is isotropic we conjecture that the heat propagates equally in all directions and we hence make the ansatz

$$U(y) = v(|y|).$$

From (6.64) it follows with $r = |y|$ and (6.12)

$$\tfrac{d}{2} v + \tfrac{1}{2} v' r + v'' + \tfrac{d-1}{r} v' = 0.$$

Multiplying this identity with $r^{d-1}$ yields

$$(r^{d-1}v')' + \tfrac{1}{2}(r^d v)' = 0.$$

Integration gives

$$r^{d-1}v' + \tfrac{1}{2}r^d v = a \quad \text{with} \quad a \in \mathbb{R}. \tag{6.65}$$

In order to obtain a solution $u$ which is smooth at the origin we require $v'(0) = 0$. Setting $r = 0$ in (6.65) we obtain $a = 0$ and hence

$$v' + \tfrac{1}{2}rv = 0.$$

As solution we obtain for some $b \in \mathbb{R}$,

$$v(r) = b\,e^{-\frac{r^2}{4}}.$$

The identity

$$1 = \int_{\mathbb{R}^d} U(y)\,dy = b \int_{\mathbb{R}^d} e^{-|y|^2/4}dy$$

determines $b$ and we obtain with $\int_{\mathbb{R}^d} e^{-|y|^2/4}dy = (4\pi)^{d/2}$ the solution

$$u(t,x) = \frac{Q}{(4\pi Dt)^{d/2}}e^{-|x|^2/(4Dt)} \quad \text{for} \quad t > 0,\ x \in \mathbb{R}^d.$$

It holds that

$$\lim_{t\searrow 0} u(t,x) = 0 \qquad\qquad \text{for} \quad x \neq 0,$$

$$\int_{\mathbb{R}^d} u(t,x)\,dx = Q \qquad\qquad \text{for} \quad t > 0.$$

In fact in the distributional sense it holds that

$$\partial_t u - D\Delta u = Q\delta_{(0,0)},$$

where $\delta_{(0,0)}$ is the Dirac distribution concentrated in the point $(0,0) \in \mathbb{R} \times \mathbb{R}^d$.

### 6.2.6   Diffusion Times

In this section we want to estimate the time the diffusion process needs to diffuse heat or a substance over a given distance. We now want to derive an assertion which is called a "fundamental fact concerning diffusion" in the classical book of Lin and Segel [89].

**Remark**. In the following we analyze very particular initial data. By using a representation formula for the solution, which is derived in Sect. 6.2.8, one can interpret the result in a more general context.

Now let $u$ be the self-similar solution from the previous subsection. We now want to know how much time is needed such that 50% of the heat has diffused out of a ball with radius $L$. We will denote this period of time by $t_{L,D}$. We know that a self-similar solution has the form

$$u(t, x) = \frac{1}{(Dt)^{d/2}} U\left(\frac{x}{\sqrt{Dt}}\right) = \frac{1}{(4\pi Dt)^{d/2}} e^{-|x|^2/(4Dt)}.$$

Now we choose $L_{\text{ref}}$ such that

$$\int_{\mathbb{R}^d \setminus B_{L_{\text{ref}}}(0)} U(y)\, dy = \tfrac{1}{2}.$$

We now want to determine $t_{L,D}$ such that

$$\int_{\mathbb{R}^d \setminus B_L(0)} \frac{1}{(Dt_{L,D})^{d/2}} U\left(\frac{x}{\sqrt{Dt_{L,D}}}\right) dx = \frac{1}{2}$$

holds. With the variable transformation $y = \frac{x}{\sqrt{Dt_{L,D}}}$ it follows that

$$\int_{\mathbb{R}^d \setminus B_{L/\sqrt{Dt_{L,D}}}(0)} U(y)\, dy = \frac{1}{2}.$$

With this we obtain

$$L = L_{\text{ref}} \sqrt{Dt_{L,D}} \quad \text{and} \quad t_{L,D} = \frac{1}{L_{\text{ref}}^2} \frac{L^2}{D}.$$

We now summarize the result:

1. Diffusion over a distance $L$ needs $\sim \frac{L^2}{D}$ time units.
2. In a period of time $t$ heat diffuses $\sim \sqrt{Dt}$ space units. These assertions result as simple conclusions from the scaling behavior, i.e., from the invariance transformations, of the heat equation. In the following we want to analyze the scaling behavior of the heat equation in more detail.

### 6.2.7  Invariant Transformations

An additional approach which leads to the fundamental solution uses invariant transformations. We now ask ourselves in which situations a simple transformation in the variables $u$, $x$, and $t$ leaves the problem

$$\partial_t u = D \Delta u \quad \text{for} \quad x \in \mathbb{R}^d , \ t > 0 ,$$
$$u(0, x) = 0 \quad \text{for} \quad x \in \mathbb{R}^d , \ x \neq 0 ,$$
$$\int_{\mathbb{R}^d} u(t, x) \, dx = Q$$

invariant. We set

$$u^*(t^*, x^*) = \gamma u(t, x) , \quad t^* = \alpha t, \ x^* = \beta x .$$

In the new variables we obtain

$$\partial_{t^*} u^* = \frac{\gamma}{\alpha} \partial_t u = \frac{\gamma}{\alpha} D \Delta_x u = \frac{\beta^2}{\alpha} D \Delta_{x^*} u^* .$$

It follows that

$$u^*(0, x^*) = 0 \quad \text{for} \quad x^* \neq 0 ,$$
$$\int_{\mathbb{R}^d} u^*(t^*, x^*) \, dx^* = \int_{\mathbb{R}^d} \gamma u \left( \frac{t^*}{\alpha}, \frac{x^*}{\beta} \right) dx^* = \int_{\mathbb{R}^d} \beta^d \gamma u(t, x) \, dx = \beta^d \gamma Q .$$

It hence follows that the equation $\partial_t u = D \Delta u$ remains invariant if $\alpha = \beta^2$ holds and if the integral constraint remains invariant, i.e., if $\beta^d \gamma = 1$ holds. Possible invariant transformations are hence

$$u^* = \alpha^{-d/2} u , \quad t^* = \alpha t , \quad x^* = \sqrt{\alpha} x \ \text{with} \ \alpha \in \mathbb{R}_+ .$$

Assuming that the solution is unique one obtains

$$u(t, x) = \gamma u \left( \frac{t}{\alpha}, \frac{x}{\beta} \right)$$

with $\alpha, \beta, \gamma$ such that

$$\beta = \sqrt{\alpha} \quad \text{and} \quad \gamma = \alpha^{-d/2} .$$

Setting $t = \alpha$ we obtain

$$u(t, x) = t^{-d/2} u \left( 1, \frac{x}{\sqrt{t}} \right) .$$

We reduced the problem by one dimension and observe that $U(y) = u(1, y)$ corresponds to the function $U$ of Sect. 6.2.5.

## 6.2.8 General Initial Data

Let the initial temperature be now given by a function $g$ and we consider for simplicity the case $d = 1$. This means we consider heat propagation in a rod. We search for a function $u$ such that

$$\partial_t u - \partial_{xx} u = 0 \,,$$
$$u(0, x) = g(x) \,.$$

To find such a function we consider heat sources at the points

$$y_i = ih, \quad i = 0, \pm 1, \cdots, \pm M \,,$$

which release the heat $g(y_i)\, h$. Here $h > 0$ is a given real number. These finitely many heat sources are chosen as an approximation of the continuous distribution $g$. If we consider only heat source of strength 1 in $y_i$ we obtain the solution

$$v(t, x - y_i) = \frac{1}{\sqrt{4\pi t}}\, e^{-(x-y_i)^2/(4t)} \,.$$

We can now take a linear combination of those solutions and obtain

$$\sum_{i=-M}^{M} v(t, x - y_i)\, g(y_i)\, h \tag{6.66}$$

as solution for the initial data with finitely many heat sources described above. The representation (6.66) can be considered as an approximation of an integral. In the limit $M \to \infty$, $h \to 0$ we obtain, if $g$ is continuous and decays for $|x| \to \infty$ sufficiently fast, the function

$$u(t, x) = \int_{-\infty}^{\infty} v(t, x - y)g(y)\, dy \,. \tag{6.67}$$

The finitely many heat sources given above converge in the sense of measures or in the sense of distributions to the function $g$. We hence expect that $u$ solves the heat equation with initial value $g$. For a proof of this fact we refer to the book of Evans [37]. The representation formula (6.67) demonstrates that the heat equation has the property of *infinite speed of propagation*. If we take continuous initial data $g$ with $g \geq 0$, $g \not\equiv 0$, which has compact support we obtain

$$u(t, x) > 0 \quad \text{for all} \quad t > 0, \; x \in \mathbb{R}.$$

A heat source which is localized for $t = 0$ close to the origin will immediately make the solution positive everywhere in space.

## 6.2.9   Brownian Motion

The apparently completely irregular motion of small particles suspended in a liquid or gas is called *Brownian molecular motion*. If the particles are concentrated in specific regions initially they will spread due to the disordered motion. This so-called diffusion process is a microscopic process. The macroscopic quantity "density of the particles" under suitable assumptions satisfies a diffusion equation.

Here we want to consider a very simplified model for a stochastic microscopic process which allows us to derive a deterministic macroscopic description. We make the following assumptions:

- The particle jumps randomly back and forth on the real line with a step size $h$.
- After a time interval $\tau$ a new step will be performed.
- Jumps to the left or right are equally likely.

After the time $n\tau$ a particle which started at the point $x = 0$ will lie at one of the points

$$-nh, \ldots, -h, 0, h, \ldots, nh.$$

However, the probability to lie at these points is different. Let $p(n, m)$ be the probability that a particle which started at $x = 0$ will lie at $x = mh$ after $n$ steps. A particle will reach the point $mh$ by $a$ jumps to the right and $b$ jumps to the left where

$$\left. \begin{array}{l} m = a - b \\ n = a + b \end{array} \right\} \quad \text{and hence} \quad \left\{ \begin{array}{l} a = \frac{n+m}{2}, \\ b = n - a. \end{array} \right.$$

The total number of possible paths to reach the point $x = mh$ can be computed with the help of a little combinatorics as

$$\binom{n}{a} = \frac{n!}{a!\,(n-a)!} = \frac{n!}{a!\,b!} = \frac{n!}{b!\,(n-b)!} = \binom{n}{b}.$$

The total number of paths is $2^n$. It follows that

$$p(n, m) = \frac{1}{2^n} \binom{n}{a}$$

with $a = \frac{n+m}{2}$ if $n + m$ is even. The quantity $p$ describes the probabilities of the *binomial distribution* for fixed $n$ as a function of $a$. For large $n$ Stirling's formula states

$$n! \simeq (2\pi n)^{1/2} n^n e^{-n} \quad \text{as} \quad n \to \infty,$$

i.e., $n!/(\sqrt{2\pi n}\, n^n e^{-n}) \to 1$ as $n \to \infty$. The proof uses the integral representation of the $\Gamma$-function. More precisely one obtains: For each $n \in \mathbb{N}$ there exists a $\vartheta(n) \in (0, 1)$ such that

$$n! = (2\pi n)^{1/2} n^n e^{-n} e^{\vartheta(n)/(12n)}.$$

For $n, n + m, n - m$ large and $m + n$ even it follows that

$$p(n, m) = \frac{1}{2^n} \binom{n}{\frac{m+n}{2}} \simeq \left(\tfrac{2}{\pi n}\right)^{1/2} e^{-m^2/(2n)}.$$

This is part of the de Moivre–Laplace theorem and can be shown with the help of Stirling's formula. We now set

$$n\tau = t, \quad mh = x.$$

We formally want to consider the limit $\tau, h \to 0$. For this we introduce the density

$$u_{\tau,h}(n\tau, mh) = \frac{p(n, m)}{2h}$$

with $m + n$ even. Then the probability that a particle can be found at time $t = n\tau$ in the interval $[ih, kh]$ is given as

$$\sum_{m=i}^{k}{}'' u_{\tau,h}(n\tau, mh)\, 2h.$$

Here $\sum''$ only sums over all indices $m$ for which $m + n$ is even and $i$ and $k$ are assumed to be both even if $n$ is even and both odd if $n$ is odd. For $t = n\tau$, $x = mh$ and $m + n$ even it holds that

$$u_{\tau,h}(t, x) = \frac{p(\frac{t}{\tau}, \frac{x}{h})}{2h} \approx \frac{1}{2h}\left(\frac{2\tau}{\pi t}\right)^{1/2} e^{-\frac{x^2}{2t}\frac{\tau}{h^2}} = \left(\frac{\tau}{h^2}\frac{1}{2\pi t}\right)^{1/2} e^{-\frac{x^2}{2t}\frac{\tau}{h^2}}.$$

We now consider the limit $\tau, h \to 0$, where $\tau$ and $h$ are chosen such that

$$\lim_{\tau,h\to 0} \frac{h^2}{2\tau} = D \neq 0 \tag{6.68}$$

holds. Then it follows that

$$u_{\tau,h} \to \left(\frac{1}{4\pi D t}\right)^{1/2} e^{-x^2/(4Dt)}.$$

As limit we obtain the fundamental solution of the heat equation. The fundamental solution hence can be interpreted as a probability density function for the probability that particles which start in the origin are at point $x$ at time $t$. The diffusion coefficient $D$ with unit $\frac{L^2}{T}$ specifies how efficient particles diffuse. If the time steps after which particles jump are small compared to the square of the space intervals then $D$ is large.

In the following we heuristically justify why we require the condition (6.68) in the limit $\tau, h \to 0$. In order to apply Stirling's formula we need that $n, n+m$ and $n - m$ are large. Hence $\frac{m}{n}$ has to be small and for fixed $(t, x) = (n\tau, mh)$ we obtain, if we require $\frac{m}{n} \to 0$,

$$\frac{m}{n} = \frac{m h \tau}{n \tau h} = \frac{x \tau}{t h} \to 0 \qquad \text{as } n \to \infty. \tag{6.69}$$

Choosing $\frac{h^2}{2\tau} = D$ with fixed $D > 0$ gives

$$\frac{\tau}{h} = \frac{1}{D}\frac{h}{2} \to 0,$$

and as $\frac{x}{t}$ is fixed the limit in (6.69) follows. The shift $x = mh$ from the origin has to be small in comparison to the total path of the particle. This assumption allows for enough stochastic variations. This consideration can be justified more precisely with the help of statistics. The variance of the displacement at time $t = n\tau$ is given as

$$nh^2 = t \frac{h^2}{\tau}$$

and the assumption $h^2/(2\tau) \to D$ says that the variance and hence the standard deviation $\sqrt{n}\, h$ has a well-defined limit. If the variance converges to $0$ or $\infty$ we would not obtain a well-defined probability density function.

Finally we want to consider another possibility to derive the diffusion equation heuristically from a simple Brownian motion model. We again consider grid points $0, \pm h, \pm 2h, \ldots$ and denote by $u(t, x)$ the probability of finding a particle at the point $x$ at time $t$. Now let $\alpha$ be the probability of jumping in a time step $\tau$ one point to left or to the right. We hence assume that both directions are equally likely. In addition, $1 - 2\alpha$ is the probability of staying at the same point. After a time step $\tau$ we then obtain

$$u(t + \tau, x) = \alpha u(t, x - h) + (1 - 2\alpha) u(t, x) + \alpha u(t, x + h)$$

for all $x = 0, \pm h, \pm 2h, \dots$. This gives

$$u(t + \tau, x) - u(t, x) = \alpha(u(t, x - h) - 2u(t, x) + u(t, x + h)).$$

We now chose $\alpha = \frac{\tau}{h^2} D$ and assume that $u$ can be extended to a smooth function. With the help of the Taylor expansion in the identity

$$\frac{u(t + \tau, x) - u(t, x)}{\tau} = D \frac{u(t, x - h) - 2u(t, x) + u(t, x + h)}{h^2}$$

we obtain

$$\partial_t u = D \partial_x^2 u + \mathcal{O}\big(|\tau| + |h|^2\big).$$

This yields that $u$ solves to leading order the heat equation.

### 6.2.10 Traveling Waves

Many nonlinear parabolic equations have traveling waves as solutions. Such solutions have the form

$$u(t, x) = U(x \cdot n - Vt), \quad U : \mathbb{R} \to \mathbb{R}$$

where $V$ is the velocity of the wave and $n \in \mathbb{R}^d$ with $|n| = 1$ is the direction in which the wave "travels". Such solutions are characterized by a profile which is given by the function $U$. This profile moves with velocity $V$. Such solutions appear in particular in many biological and chemical applications. In mathematical biology for example traveling waves describe the propagation of infections or the healing of wounds. The function $u$ in this context is, e.g., a chemical concentration, an electrical signal, the density of a population or a mechanical deformation. Traveling waves are of major importance as often solutions to quite general initial data converge for large times to a traveling wave solution.

We now consider traveling waves for the equation

$$\partial_t u + \partial_x f(u) = \eta \, \partial_{xx} u, \quad \eta > 0. \tag{6.70}$$

In the case $f(u) = u^2/2$ this is the *viscous Burgers' equation*, which appears in fluid dynamics as a simple one-dimensional model. In this case $u$ is the velocity, $\partial_x (u^2/2)$ is the convection term and $\eta \, \partial_{xx} u$ is the viscous term.

We now search for a solution of the form

$$u(t, x) = U(x - Vt), \quad V \in \mathbb{R} \tag{6.71}$$

with

$$\lim_{x \to -\infty} u(t, x) = u_- \quad \text{and} \quad \lim_{x \to \infty} u(t, x) = u_+. \tag{6.72}$$

This traveling wave describes a transition between the values $u_-$ and $u_+$. Inserting the ansatz (6.71) into (6.70) yields for $U(z)$ with $z = x - Vt$:

$$-VU' + f(U)' = \eta U''. \tag{6.73}$$

Integration gives

$$\eta U' = -VU + f(U) + Vu_- - f(u_-) \tag{6.74}$$

where we used that a solution of (6.73) can only fulfill $\lim\limits_{z \to -\infty} U(z) = u_-$ if $\lim\limits_{z \to -\infty}$ $U'(z) = 0$.

A necessary condition for the existence of a solution to the differential equation (6.74) which in addition has the property $\lim\limits_{z \to \infty} U(z) = u_+$ is the validity of $\lim\limits_{z \to \infty} U'(z) = 0$ and hence

$$0 = -Vu_+ + f(u_+) + Vu_- - f(u_-).$$

If $u_- \neq u_+$ this implies that

$$V = \frac{f(u_+) - f(u_-)}{u_+ - u_-}.$$

If we consider the case

$$u_- > u_+,$$

we obtain that a solution of (6.72), (6.74) necessarily has to fulfill

$$U'(z) < 0, \quad z \in \mathbb{R}.$$

If $U'(z) = 0$ would hold for a $z \in \mathbb{R}$ then the fact that the initial value problem for (6.74) is uniquely solvable would imply that $U$ necessarily would have to be constant. This is a contradiction to the given boundary values. We hence obtain the necessary condition

$$-Vu + f(u) + Vu_- - f(u_-) < 0 \quad \text{for all} \quad u \in (u_+, u_-).$$

This is equivalent to

$$\frac{f(u) - f(u_-)}{u - u_-} > V = \frac{f(u_+) - f(u_-)}{u_+ - u_-} \quad \text{for all} \quad u \in (u_+, u_-). \tag{6.75}$$

If the condition (6.75) is fulfilled we can easily deduce the existence of a traveling wave by solving the initial boundary value problem

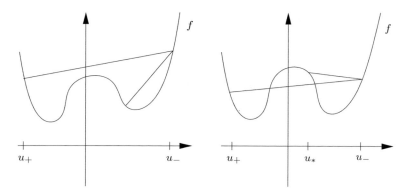

**Fig. 6.4** Illustrating situations in which traveling waves exist (*left*) or do not exist (*right*)

$$\eta U' = -VU + f(U) + Vu_- - f(u_-), \quad U(0) = \frac{u_- + u_+}{2}.$$

The right-hand side of the differential equation is positive for $u \in (u_+, u_-)$ and this implies that $U$ is strongly monotone. We can then deduce $\lim_{z \to \pm\infty} U(z) = u_\pm$ which follows since $U$ is monotone and from the fact that the right-hand side in the above differential equation has no zero in the interval $(u_+, u_-)$.

In the case $u_- \geq u_+$ the condition (6.75) characterizes cases in which a monotone traveling wave connecting $u_-$ and $u_+$ exists. Such a solution will be called *wave front* in what follows. The secant of $f$ connecting the points $(u_-, f(u_-))$ and $(u_+, f(u_+))$ hence has to lie above the graph of $f$ in order to obtain a traveling wave solution. A necessary condition for this is the strict convexity of $f$.

In Fig. 6.4 on the left we see a situation in which a traveling wave exists. In the situation to the right we cannot connect the points $u_-$ and $u_+$ with a traveling wave. However, there exists a traveling wave which connects $u_-$ and $u_*$. An analogous discussion can be performed for the case $u_+ > u_-$ and we obtain traveling wave solutions if the secant lies below of the graph.

### 6.2.11   Reaction Diffusion Equations and Traveling Waves

In Chap. 3 we considered chemical reactions in multi-species systems and we saw that the change of the mole number can be described by ordinary differential equations. If we now consider concentrations $u(t, x) \in \mathbb{R}^N$ of chemical substances which react with each other and diffuse in space, we have to combine the considerations on reaction kinetics and on diffusion and we obtain the following system of reaction diffusion equations

$$\partial_t u = D\Delta u + F(u), \tag{6.76}$$

where $D$ is a $N \times N$-matrix of diffusion coefficients and $F : \mathbb{R}^N \to \mathbb{R}^N$ is a function which describes the occurring chemical reactions as in Chap. 3. Setting $D = 0$ gives the reaction kinetics of Chap. 3 and $F \equiv 0$ leads to simple diffusion equations.

Similar equations appear in mathematical biology. If one considers a random movement of the population in the population model with limited resource, or in the predator-prey model, one obtains for the density $u$ of the population equations of the form (6.76). We do not state more details on the modeling of this aspect and refer to the excellent book by Murray [105] for more details. Many other situations are also considered there in which traveling waves for reaction-diffusion systems play an important role in applications.

We now want to consider the existence of traveling waves in the case of *Fisher's equation*

$$\partial_t u = D \partial_{xx} u + q u (1 - u) , \quad D, q, t \in \mathbb{R}_+, u, x \in \mathbb{R} . \tag{6.77}$$

Fisher's equation originally was introduced as a model for the spread of mutated genes. Furthermore, this equation has applications in neurophysiology, for autocatalytic chemical reactions and as already mentioned as a population model with limited resources. We obtain for $D = 0$ the equation of limited growth which is also called the logistic differential equation. In the case $q = 0$ we obtain the diffusion equation. For simplicity we rescale (6.77) by

$$t^* = qt , \quad x^* = x \left( \tfrac{q}{D} \right)^{1/2}$$

and obtain

$$\partial_t u = \partial_{xx} u + u(1 - u) \tag{6.78}$$

where we omitted the star for simplicity. We know from Chap. 1 that the stationary spatially constant solutions are

$$u \equiv 0 \quad \text{and} \quad u \equiv 1 .$$

We now search for a solution with

$$u(t, x) = U(z) , \quad z = x - Vt$$

and

$$\lim_{z \to -\infty} U(z) = 0 , \quad \lim_{z \to \infty} U(z) = 1$$

with the wave velocity $V$. In applications such a solution $u$ describes with which velocity a front, for example the domain with already mutated genes, will spread. Inserting the traveling wave ansatz in (6.78) one obtains

$$U'' + V U' + U(1 - U) = 0 .$$

With $W = U'$ one obtains the following system of ordinary differential equations

$$U' = W, \tag{6.79}$$
$$W' = -VW - U(1 - U). \tag{6.80}$$

The problem of finding a traveling wave which connects the stationary states is equivalent to the problem of finding a curve in the $(U, W)$ plane which connects the points $(0, 0)$ and $(1, 0)$ and solves (6.79), (6.80). In the language of dynamical systems one says: We search for a *heteroclinic orbit*, which connects the stationary points $(0, 0)$ and $(1, 0)$, compare also [6]. We now want to consider the phase portrait for (6.79), (6.80), see also Sect. 4.4. To do so we linearize the equations in the neighborhood of the stationary states.

Considering the point $(0, 0)$, we obtain

$$\begin{pmatrix} U' \\ W' \end{pmatrix} = \begin{pmatrix} 0 & 1 \\ -1 & -V \end{pmatrix} \begin{pmatrix} U \\ W \end{pmatrix}. \tag{6.81}$$

The eigenvalues of the above matrix are

$$\lambda_{1,2} = \tfrac{1}{2}\left(-V \pm \sqrt{V^2 - 4}\right).$$

If the linearization at a stationary point only has eigenvalues with a real part different from zero one speaks of a hyperbolic stationary point. The qualitative theory of ordinary differential equations now says that the phase portrait of the nonlinear problem close to stationary hyperbolic points has the same structure as its linearization. We now consider the phase portrait of (6.81) for $V \neq 0$. In Exercise 6.12 one shows that in the case $V = 0$ no traveling wave exists.

**Case 1**: $0 < V^2 < 4$.

In this case all complex solutions of the linearized equation are given by (see Sect. 4.6)

$$e^{\lambda_1 z}\begin{pmatrix} \widetilde{u}_1 \\ \widetilde{w}_1 \end{pmatrix} + e^{\lambda_2 z}\begin{pmatrix} \widetilde{u}_2 \\ \widetilde{w}_2 \end{pmatrix} = e^{-\frac{1}{2}Vz}\left(e^{\frac{1}{2}i\sqrt{4-V^2}z}\begin{pmatrix} \widetilde{u}_1 \\ \widetilde{w}_1 \end{pmatrix} + e^{-\frac{1}{2}i\sqrt{4-V^2}z}\begin{pmatrix} \widetilde{u}_2 \\ \widetilde{w}_2 \end{pmatrix}\right),$$

where $(\widetilde{u}_1, \widetilde{w}_1)^\top, (\widetilde{u}_2, \widetilde{w}_2)^\top \in \mathbb{C}^2$ are eigenvectors to the eigenvalues $\lambda_1$ and $\lambda_2$. One now easily computes that all real solutions are given by

$$e^{-\frac{1}{2}Vz}\left[\alpha\begin{pmatrix} \cos(\omega z) \\ \sin(\omega z) \end{pmatrix} + \beta\begin{pmatrix} -\sin(\omega z) \\ \cos(\omega z) \end{pmatrix}\right] \quad \text{with} \quad \omega = \tfrac{1}{2}\sqrt{4 - V^2}.$$

An initial data $(\alpha, \beta)^\top$ will hence be stretched by the factor $e^{-\frac{1}{2}Vz}$ and rotated by the angle $\omega z$ with $\omega = \tfrac{1}{2}\sqrt{4 - V^2}$. For $V > 0$ we hence obtain a phase portrait as given

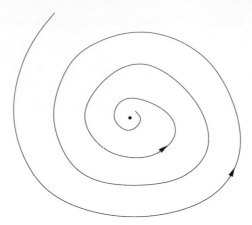

**Fig. 6.5** Phase portrait of a stable spiral

in Fig. 6.5. A hyperbolic point with eigenvalues of the linearization which complex conjugate is called a *stable spiral* or respectively an *unstable spiral*, depending on whether the solutions converge to zero or to infinity for $z$ going to infinity

**Case 2**: $V^2 \geq 4$.

In this case we have two real eigenvalues with the same sign and one speaks of a knot. The representation formula from Sect. 4.6 yields that all solutions converge to zero if the eigenvalues are negative. We note that the case $V^2 = 4$ has to be considered separately. If the eigenvalues are positive all solutions to non-zero initial data converge to zero for $z$ tending to $-\infty$, see Fig. 6.6. Since we search for solutions with $\lim_{z \to -\infty} U(z) = 0$ and $U(z) \geq 0$, $z \in \mathbb{R}$ we consider the case

$$V \leq -2 .$$

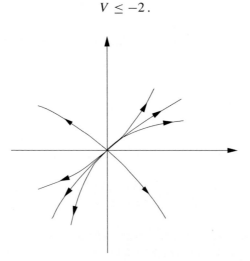

**Fig. 6.6** Phase portrait of a knot

If $V^2 \in (0, 4)$, the solutions oscillate around zero and we cannot expect non-negative solutions. For $V > 2$ we cannot satisfy the condition $\lim\limits_{z \to -\infty} U(z) = 0$ which is due to the fact that all solutions evolve away from the origin for $z \to -\infty$.

Linearization around $(U, W) = (1, 0)$ yields

$$\begin{pmatrix} U' \\ W' \end{pmatrix} = \begin{pmatrix} 0 & 1 \\ 1 & -V \end{pmatrix} \begin{pmatrix} U \\ W \end{pmatrix}$$

and we obtain the eigenvalues

$$\lambda_{1,2} = \tfrac{1}{2} \left( -V \pm \sqrt{V^2 + 4} \right).$$

This implies that the two eigenvalues have different signs and an analysis of the representation formula from Sect. 4.6 yields the phase portrait in Fig. 6.7. A stationary point for which the eigenvalues have different signs is called a *saddle*.

For a saddle point in the plane a one-dimensional curve $M_s$ exists such that all solutions to initial data on this curve converge exponentially for large time to the stationary point (see [6, 58]). This curve is called the *stable manifold* of the stationary point. The stable manifold reaches the stationary point with a tangent whose direction is given by an eigenvector of the negative eigenvalue. In our case we obtain in the point $(1, 0)$,

$$(1, \lambda_2) \quad \text{is an eigenvector of} \quad \lambda_2 = \tfrac{1}{2} \left( -V - \sqrt{V^2 + 4} \right) < 0.$$

Analogously there exists an unstable manifold which consists of initial data of solutions which converge to the stationary point as $t \to -\infty$. Our goal is now to find a solution curve which connects the knot $(0, 0)$ with the hyperbolic point $(1, 0)$. An

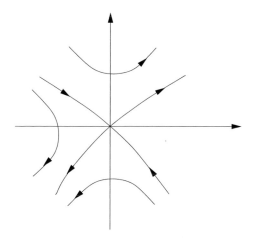

**Fig. 6.7** Phase portrait of a saddle

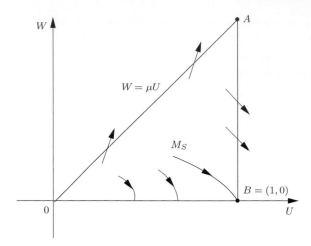

**Fig. 6.8** Phase portrait for the construction of the traveling wave solution

inspection of the phase portrait in the neighborhood of the point $(1, 0)$ yields that the solution curve has to be part of the stable manifold of the point $(1, 0)$.

As the eigenspace to the negative eigenvalue $\lambda_2$ is spanned by the vector $(1, \lambda_2)$ we obtain that the stable manifold reaches the point $(1, 0)$ from the "upper left" as sketched in Fig. 6.8.

We now want to show that the unstable manifold meets the origin $(0, 0)$. If we retrace the stable manifold in Fig. 6.8 back there are two possibilities. Either the stable manifold converges to a stationary point in the closed triangle $\Delta(0, A, B)$ or the curve $M_s$ leaves the triangle. Other possibilities cannot occur as $U'(z) = W(z) \geq 0$ as long as the solution $(U, W)$ is contained in the triangle. In this context we refer to the Poincaré–Bendixon theorem, see [6].

We now show that a $\mu > 0$ exists such that the stable manifold to the stationary point $B$ cannot leave the triangle $\Delta(0, A, B)$. Now let $(U, W)(z)$ be a solution of (6.79), (6.80) for which the image lies on the stable manifold. As $U'(z) \geq 0$ in $\Delta(0, A, B)$ the stable manifold cannot leave the triangle via the segment $\overline{AB}$. As long as $U'(z) = W(z) > 0$ we can write $W$ as function of $U$ and the other way around

$$\frac{dW}{dU} = \frac{W'}{U'} = -V - \frac{U(1 - U)}{W} < 0 \quad \text{for} \quad W \text{ small}, \ U \in (0, 1).$$

Solution curves in the $(U, W)$–plane therefore can approach points $(U, 0)$ with $U \in (0, 1)$ only from the left and hence we can conclude that the stable manifold cannot leave the segment $\overline{0B}$ through a point $U > 0$.

On the segment $\overline{0A}$ it holds that

$$\frac{dW}{dU} = -V - \frac{U(1 - U)}{W} = -V - \frac{1}{\mu}(1 - U).$$

We now want to choose $\mu$ such that on the segment $\overline{0A}$

$$\frac{dW}{dU} > \mu$$

holds. This condition is equivalent to

$$-V - \frac{1}{\mu}(1 - U) > \mu \text{ for all } U \in (0, 1)$$

and

$$U > \mu^2 + V\mu + 1 \text{ for all } U \in (0, 1).$$

Since $V \leq -2$ we can choose, e.g., $\mu = 1$. For this $\mu$ we can conclude that the stable manifold for $B$ cannot leave the triangle $\Delta(0, A, B)$ via a point $(U, W)$ on $\overline{0A}$ with $U > 0$.

The only remaining possibility is that the unstable manifold meets the point $(0, 0)$ and this shows the existence of a traveling wave for all $V \leq -2$. The existence of a traveling wave can also be shown without using knowledge about stable manifolds, see Exercise 6.14.

The solution with $V = -2$ has a special significance. If we start with non-negative initial data with compact support one obtains that the solution asymptotically for large times spreads like two traveling waves with velocity $-2$ and $2$ (see Fig. 6.9). One wave spreads to $-\infty$ and the other to $\infty$ and both waves asymptotically have the traveling wave profile of the above discussed solution with $V = -2$. The wave traveling to the right is obtained by a reflection of the solution for $V = -2$. This example demonstrates that the waves discussed above are relevant also for general initial data. For further details we refer to Murray [105], Vol. 1, page 443.

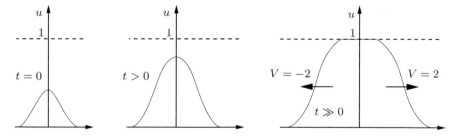

**Fig. 6.9** For large times solutions of Fisher's equation with compact initial data lead to waves which propagate with velocities $-2$ and $2$. The profile of the individual waves corresponds to the traveling wave solution considered in Sect. 6.2.11

## 6.2.12   Turing Instability and Pattern Formation

We first consider linear systems of ordinary differential equations of the form

$$x'(t) = Ax(t), \quad x'(t) = Bx(t) \quad \text{with} \quad x(t) \in \mathbb{R}^2, \; A, B \in \mathbb{R}^{2 \times 2}. \qquad (6.82)$$

If $x \equiv 0$ is a stable solution of both differential equations one would naively expect that $x \equiv 0$ is also a stable solution of

$$x' = (A + B)x.$$

This is however not the case as we will see in what follows. If the matrices $A$ and $B$ have eigenvalues with negative real parts we obtain that $x \equiv 0$ is a stable solution of the systems (6.82). However if an eigenvalue with positive real part exists one obtains that $x \equiv 0$ is unstable. These facts have been discussed in Sect. 4.6.

We can characterize the case where the eigenvalues $\lambda_1$, $\lambda_2$ have negative real parts with the conditions

$$\lambda_1 + \lambda_2 = \text{trace } B < 0 \quad \text{and} \quad \lambda_1 \lambda_2 = \det B > 0.$$

We now want to demonstrate that the following case is possible

$$\left. \begin{array}{l} \text{trace } A < 0, \; \det A > 0 \\ \text{trace } B < 0, \; \det B > 0 \end{array} \right\} \quad \text{but} \quad \det (A + B) < 0. \qquad (6.83)$$

In this case one of the eigenvalues of $A + B$ has to have a positive real part. For simplicity we consider

$$A = \begin{pmatrix} -1 & 0 \\ 0 & -d \end{pmatrix} \quad \text{with} \quad d > 0 \quad \text{and} \quad B = \begin{pmatrix} a & e \\ c & b \end{pmatrix}, \; a, b, c, e \in \mathbb{R}.$$

To fulfill (6.83), we demand

$$a + b < 0, \; ab > ce \quad \text{and} \quad (a - 1)(b - d) < ce.$$

These conditions can be fulfilled in many ways. For example for the case

$$A = \begin{pmatrix} -1 & 0 \\ 0 & -8 \end{pmatrix}, \; B = \begin{pmatrix} 2 & 3 \\ -3 & -3 \end{pmatrix},$$

we obtain that $x \equiv 0$ is unstable for $x' = (A + B)x$.

The observation that systems which are stable for themselves become unstable if they are brought together is the essence of an idea by the British mathematician Alan Turing from the year 1952. Turing observed a similar phenomenon for reaction-diffusion systems and derived a multitude of pattern formation scenarios. Instabilities of the kind discussed above are called Turing instabilities.

In the following we consider a reaction-diffusion system in a domain $\Omega$ which is already nondimensionalized and has the form

$$\partial_t u = \Delta u + \gamma f(u, v) \quad \text{for } x \in \Omega, \ t \geq 0, \tag{6.84}$$

$$\partial_t v = d \Delta v + \gamma g(u, v) \quad \text{for } x \in \Omega, \ t \geq 0 \tag{6.85}$$

together with Neumann boundary conditions

$$\nabla u \cdot n = 0, \quad \nabla v \cdot n = 0 \quad \text{for } x \in \partial\Omega, \ t \geq 0.$$

The diffusion constant $d$ is assumed to be positive and the functions $f$ and $g$ describe the reaction kinetics or other interactions between $u$ and $v$. The constant $\gamma > 0$ describes the relative strength of the reaction term in comparison to the diffusion terms.

In the case $\gamma = 0$ the two equations decouple and solutions converge as $t \to \infty$ to a constant state (compare Exercise 6.15). The constant state obtained for $t \to \infty$ is the average of the initial data. This is expected as diffusion balances out differences in the concentration. We obtain that the stationary states $(u, v) \equiv (u_0, v_0) \in \mathbb{R}^2$ are stable.

Turing observed that states $(u_0, v_0) \in \mathbb{R}^2$ exist which are unstable for the full system (6.84), (6.85) although these states are stable solutions of the system of differential equations

$$\partial_t u = \gamma f(u, v), \quad \partial_t v = \gamma g(u, v). \tag{6.86}$$

We hence have a similar situation to the beginning of this subsection. The state $(u_0, v_0)$ is stable for the individual systems and after combining of reaction and diffusion $(u_0, v_0)$ becomes unstable. This *diffusion driven instability* leads to spatially inhomogeneous patterns (see, e.g., Murray [105]). One speaks of *self-organization*, as there is no influence from the outside acting on the system. We will now discuss in detail when such instabilities can occur.

We consider $(u_0, v_0) \in \mathbb{R}^2$ with

$$f(u_0, v_0) = 0, \quad g(u_0, v_0) = 0.$$

To ensure that $(u_0, v_0)$ is a stable solution of (6.86) we require that

$$A = \begin{pmatrix} f_{,u}(u_0, v_0) & f_{,v}(u_0, v_0) \\ g_{,u}(u_0, v_0) & g_{,v}(u_0, v_0) \end{pmatrix}$$

has only eigenvalues with negative real part. Here $f_{,u}, f_{,v}, g_{,u}, g_{,v}$ denote partial derivatives of $f$ and $g$ with respect to the variables $u, v$. After our discussions above we know that stability is guaranteed if

$$\text{trace } A = f_{,u} + g_{,v} < 0, \quad \det A = f_{,u} g_{,v} - f_{,v} g_{,u} > 0.$$

Linearization of the full system (6.84), (6.85) at $(u_0, v_0)$ yields the following system of linear partial differential equations

$$\partial_t W = D \Delta W + \gamma A W \quad \text{with } D = \begin{pmatrix} 1 & 0 \\ 0 & d \end{pmatrix}$$

for

$$W = \begin{pmatrix} u - u_0 \\ v - v_0 \end{pmatrix}.$$

For simplicity we only consider in the following the spatially one-dimensional case $\Omega = (0, a), a > 0$. With the ansatz

$$W(t, x) = h(t) g(x) c \quad \text{with } c \in \mathbb{R}^2 \tag{6.87}$$

and real valued functions $h$ and $g$ we obtain a solution of the linearized equation if and only if

$$h' gc = h g'' Dc + \gamma h g A c. \tag{6.88}$$

For $h(t) \neq 0$, $g(x) \neq 0$ this identity is equivalent to

$$\frac{h'(t)}{h(t)} c = \frac{g''(x)}{g(x)} Dc + \gamma A c.$$

As the left-hand side only depends on $t$ and as the right-hand side only depends on $x$ there have to exist constants $\lambda, \mu \in \mathbb{R}$ such that

$$h' = \lambda h,$$
$$g'' = -\mu g, \quad g'(0) = g'(a) = 0. \tag{6.89}$$

The last equation has solutions if

$$\mu = \mu_k = \left( \frac{k\pi}{a} \right)^2, \quad k = 0, 1, 2, 3, \dots,$$

and the solutions are multiples of

$$g_k(x) = \cos \left( \frac{k\pi x}{a} \right).$$

Equation (6.88) has solutions if for $k \in \mathbb{N}$, there exist values $\lambda \in \mathbb{R}$ and vectors $c \in \mathbb{R}^2$ such that

$$(-\mu_k D + \gamma A) c = \lambda c. \tag{6.90}$$

To make sure that a solution $W$ in the representation (6.87) grows in time one needs solutions of the Eq. (6.90) with Re $\lambda > 0$. Equation (6.90) has nontrivial solutions if $\lambda$ solves the equation

$$\det(\lambda \operatorname{Id} - \gamma A + \mu_k D) = 0.$$

This corresponds to the quadratic equation

$$\lambda^2 + a_1(\mu_k)\,\lambda + a_0(\mu_k) = 0 \qquad (6.91)$$

with coefficients

$$a_0(s) = d\,s^2 - \gamma(d\,f_{,u} + g_{,v})s + \gamma^2 \det A,$$
$$a_1(s) = s(1+d) - \gamma(f_{,u} + g_{,v}),$$

where $f_{,u}, f_{,v}, g_{,u}, g_{,v}$ are evaluated at $(u_0, v_0)$. As solutions we obtain

$$\lambda_{1,2} = -\frac{a_1(\mu_K)}{2} \pm \tfrac{1}{2}\sqrt{(a_1(\mu_K))^2 - 4a_0(\mu_K)}.$$

The conditions $f_{,u} + g_{,v} < 0$ and $d > 0$ imply

$$a_1(\mu_K) \geq 0 \quad \text{for} \quad \mu_K \geq 0.$$

A solution of (6.91) with positive real part can only exist if $a_0(\mu_K) < 0$ holds. As $\det A > 0$, $\mu_K \geq 0$, and $d > 0$ this case can only occur if

$$d\,f_{,u} + g_{,v} > 0 \qquad (6.92)$$

holds. As $f_{,u} + g_{,v} < 0$ we obtain the necessary condition

$$d \neq 1.$$

If

$$f_{,u} > 0 \quad \text{and} \quad g_{,v} < 0$$

hold we can hope that (6.92) is fulfilled if $d > 1$.

To make sure that we obtain solutions with positive real part we need to make sure that $a_0(s)$ is negative for at least one positive $s$. As minimum of $a_0$ we obtain

$$s_{\min} = \gamma\frac{d\,f_{,u} + g_{,v}}{2d} \quad \text{with} \quad a_0(s_{\min}) = \gamma^2\left[\det A - \frac{(d\,f_{,u} + g_{,v})^2}{4d}\right],$$

where the last term has to become negative to make sure that we obtain solutions $\lambda$ with Re $\lambda > 0$. As $\det A = f_{,u}g_{,v} - f_{,v}g_{,u}$ holds we altogether have to fulfill the following conditions to make sure that we obtain a Turing instability

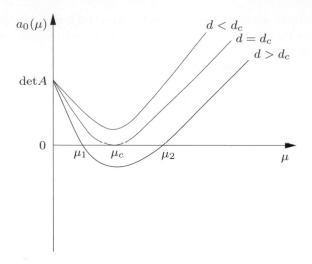

**Fig. 6.10** Turing instability: If the diffusion coefficient $d$ is larger than $d_c$ instabilities can arise

$$f_{,u} + g_{,v} < 0, \qquad\qquad f_{,u}\, g_{,v} - f_{,v}\, g_{,u} > 0,$$
$$d\, f_{,u} + g_{,v} > 0, \; (d\, f_{,u} + g_{,v})^2 - 4d\, (f_{,u}\, g_{,v} - f_{,v}\, g_{,u}) > 0.$$

If the conditions in the first line are fulfilled the conditions in the second line hold in the case $f_{,u} > 0$ and $g_{,v} < 0$ if and only if $d$ is large enough. A further investigation yields the existence of a critical diffusion coefficient $d_c$ such that for all $d > d_c$ instabilities occur. In Fig. 6.10 we will show how $a_0$ looks like for different $d$. In Fig. 6.11 we sketch the larger of the two real parts of the solution of (6.91) as a function of $\mu$ and we see that for $d > d_c$ a whole interval of unstable wavenumbers exist. For a detailed analysis we refer to Murray [105].

A simple reaction-diffusion system with a Turing instability is the *Schnakenberg system*

$$\partial_t u = \Delta u + \gamma(\alpha - u + u^2 v)\,, \qquad\qquad\qquad (6.93)$$
$$\partial_t v = d\,\Delta v + \gamma(\beta - u^2 v)\,, \; \alpha, \beta \in \mathbb{R}\,, \; d > 0\,, \qquad (6.94)$$

which is discussed in Exercise 6.16. There the question regarding how large $\Omega$ has to be to ensure that a Turing instability can be observed is also discussed. It has to be guaranteed that a $k \in \mathbb{N}$ exists such that $\mu_k = \left(\frac{k\pi}{a}\right)^2$ lies in the unstable interval, see Exercise 6.12. More general solutions of the linearized system can be obtained by an infinite linear combination of solutions of the form (6.87). After a certain time the parts of the solution which correspond to a $\mu$ for which Re $\lambda > 0$ is large are particularly enhanced whereas the parts with Re $\lambda < 0$ are damped. The parts which are particularly enhanced "imprint" their pattern on the general solution. The patterns arise due to the fact that the wavelengths which are enhanced the most dominate the solution, as shown in Fig. 6.12. We will discuss this phenomenon in the following section with the help of a scalar equation.

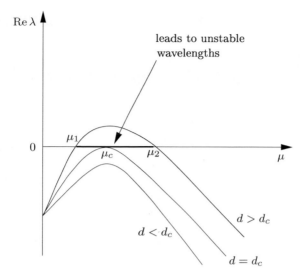

**Fig. 6.11** Turing instability: For large $d$ eigenvalues $\lambda$ in (6.90) with positive real parts can appear in such a way that unstable wavelengths can occur

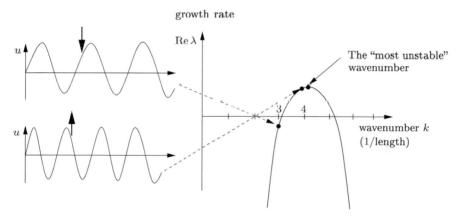

**Fig. 6.12** This figure shows a typical result of a linear stability analysis. Some wavelengths are damped and others are amplified

In the book of Murray [105] also the two-dimensional case is discussed and it is demonstrated that Turing instabilities can lead to a multitude of patterns. As examples the formation of animal coat patterns (zebras, leopards,...), butterfly wings, chemical reaction patterns and patterns on seashells are discussed (Fig. 6.13). Complex patterns arise for example by the fact that on complex domains the eigenfunctions of the boundary value problem, see (6.89),

$$\Delta g = -\mu g \ \text{in} \ \Omega, \quad \nabla g \cdot n = 0 \ \text{on} \ \partial\Omega$$

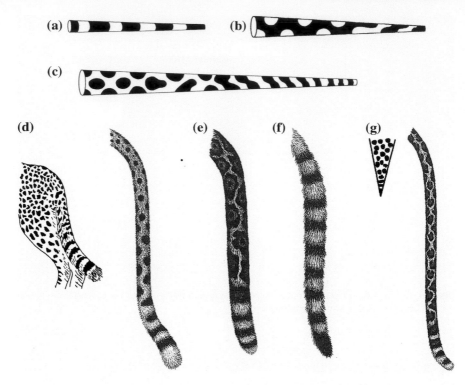

**Fig. 6.13** (From Murray [105]) **a–c** show numerical simulations for the system (6.95)–(6.97) with Neumann boundary conditions. **d–g** show typical animal coats: **d** adult cheetah **e** adult jaguar **f** common genet (prenatal) **g** adult leopard

have diverse zero level lines (so-called nodal lines). These can have stripe or point patterns or combinations of these. For eigenfunctions corresponding to an eigenvalue $\mu$ which is particularly enhanced we will see that the pattern is given by the nodal lines as a result of the reaction-diffusion process. Figure 6.13 shows in (a)–(c) numerical simulations of Murray [105] for the reaction-diffusion system

$$\partial_t u = \Delta u + \gamma f(u, v), \quad \partial_t v = d\Delta v + \gamma g(u, v), \qquad (6.95)$$

$$f(u, v) = a - u - h(u, v), \quad g(u, v) = \alpha(b - v) - h(u, v), \qquad (6.96)$$

$$h(u, v) = \frac{\varrho u v}{1 + u + K u^2} \qquad (6.97)$$

with positive constants $a, b, \alpha, \varrho, K$ and $d > 1$.

The system is solved with Neumann boundary conditions for the parameters $\alpha = 1{,}5$, $K = 0{,}1$, $\varrho = 18{,}5$, $a = 92$, $b = 64$ and $d = 10$. With these values $(u_0, v_0) = (10, 9)$ is a stationary solution. As initial data we choose a perturbation of the stationary state $(u_0, v_0)$ and for positive times the areas with $u > u_0$ (in the Fig. 6.13a and b) and the areas with $u < u_0$ (in Fig. 6.13c) have been shown in dark.

It is apparent that depending on the size and shape of the domain different patterns arise which resemble animal coats of different big cats.

We refer to the article of Fiedler [38] in which questions related to the Turing instability and the self-organization are brought in the context of Romeo and Juliet. The work of Gierer and Meinhardt [48] and Meinhardt [95] present several interesting and beautiful computer simulations of Turing systems and show that the patterns can be compared with patterns on seashells.

### 6.2.13  Cahn–Hilliard Equation and Pattern Formation Processes

In this section we analyze pattern formation processes in the context of a scalar equation of fourth order. The *Cahn–Hilliard equation* describes diffusion processes in two-component mixtures. Let $c_1$ and $c_2$ be the concentrations of the two components and we assume that the system under consideration is isothermal and isobaric and occupies a domain $\Omega \subset \mathbb{R}^d$. Then, as in Sect. 5.5, the conservation laws

$$\partial_t c_i + \nabla \cdot j_i = 0 , \quad i = 1, 2 , \tag{6.98}$$

have to hold where for simplicity we set $\varrho \equiv 1$. In (5.8) we assume $M = 2$, $v = 0$, $r_1 = r_2 = 0$, and $\varrho = 1$. The concentrations $c_1, c_2$ describe the local portion of the concentrations and hence we require $c_1 + c_2 = 1$. To ensure $\partial_t(c_1 + c_2) = 0$ we assume that $j_1 + j_2 = 0$. We can then reduce the system by using the independent variables

$$c = c_1 - c_2 \quad \text{and} \quad j = j_1 - j_2 .$$

In Sect. 5.6 we already discussed that the chemical potential is the driving force for the evolution. Considering now free energies of the form $\int_\Omega f(c)\, dx$ we obtain the flux $j = -L\nabla\mu$ with $\mu = f'(c)$ and a mobility $L \geq 0$. If $f''(c) < 0$ holds we obtain a mass flux from areas with a low concentration to regions with a high concentration ("uphill"–diffusion). It can be shown that solutions to the resulting partial differential equations $\partial_t c = \nabla \cdot (L f''(c)\nabla c)$ do not depend continuously on the initial data, see, e.g., Exercise 6.20. For non-convex energy densities $f$ the equation is not parabolic any longer and in general cannot be solved, see [37, 75, 110]. In many cases, and always if phase transitions occur, non-convex free energies appear. Cahn and Hilliard proposed to model the total free energy in such situations by

$$\mathcal{F}(c) = \int_\Omega \left( f(c) + \frac{\gamma}{2}|\nabla c|^2 \right) dx \quad \text{with} \quad \gamma > 0 \quad \text{constant} .$$

It turns out that the additional gradient term can be interpreted as an energy for phase boundaries, compare Sect. 7.9. In this case we define the chemical potential as variational derivative as follows

$$\langle \mu, v \rangle_{L^2(\Omega)} := \delta \mathcal{F}(c)(v) = \frac{d}{d\varepsilon} \mathcal{F}(c + \varepsilon v)|_{\varepsilon=0} \quad \text{for all} \quad v \in C_0^1(\Omega).$$

We obtain

$$\langle \mu, v \rangle_{L^2(\Omega)} = \int_\Omega \left( f'(c)v + \gamma \nabla c \cdot \nabla v \right) dx$$

$$= \int_\Omega \left( f'(c) - \gamma \Delta c \right) v \, dx$$

where the last identity follows from the divergence theorem. The above identity is true for all $v \in C_0^\infty(\Omega)$ if and only if

$$\mu = -\gamma \Delta c + f'(c).$$

In the following $f$ is assumed to be a non-convex potential. An example is the so-called *double well potential*

$$f(c) = \alpha \left( c^2 - a^2 \right)^2, \quad \alpha, a \in \mathbb{R}_+.$$

Altogether we obtain the Cahn–Hilliard equation

$$\partial_t c = L\Delta(-\gamma \Delta c + f'(c)). \tag{6.99}$$

We now consider the Cahn–Hilliard equation in the rectangular parallelepiped

$$\Omega = [0, \ell]^d, \quad \ell > 0,$$

with periodic boundary conditions. We then obtain after some computations (Exercise 6.17)

$$\frac{d}{dt} \int_\Omega c \, dx = 0, \tag{6.100}$$

$$\frac{d}{dt} \int_\Omega \left( \frac{\gamma}{2} |\nabla c|^2 + f(c) \right) dx \leq 0. \tag{6.101}$$

The integral in (6.101) is the free energy and (6.101) shows that the total energy cannot increase. We now consider homogeneous stationary solutions

$$c \equiv c_m, \quad c_m \in \mathbb{R}$$

with

$$f''(c_m) < 0$$

and we will see that these solutions can be unstable. Actually a small perturbation can lead to the generation of patterns in the Cahn–Hilliard model. As the total mass has to be preserved, we consider perturbations

$$c = c_m + u$$

with

$$\int_\Omega u\,dx = 0\,. \tag{6.102}$$

Linearization of (6.99) around $c_m$ gives for $L = 1$

$$\partial_t u = (-\Delta)\big(\gamma\Delta u - f''(c_m)u\big)\,. \tag{6.103}$$

The eigenfunctions of the operator

$$u \mapsto (-\Delta)\big(\gamma\Delta u - f''(c_m)u\big)$$

for periodic boundary conditions are given by

$$\varphi_k(x) = e^{ik\cdot x}$$

with $k \in \frac{2\pi}{\ell}\mathbb{Z}^d \setminus \{0\}$, see, e.g., Courant and Hilbert [25]. The corresponding eigenvalues are

$$\lambda_k = |k|^2\big(-\gamma|k|^2 - f''(c_m)\big)\,. \tag{6.104}$$

General solutions of (6.103) with the property (6.102) can be constructed as infinite linear combinations as follows

$$u(t,x) = \sum_{k \in \frac{2\pi}{\ell}\mathbb{Z}^d \setminus \{0\}} \alpha_k e^{\lambda_k t} e^{ik\cdot x}, \quad \alpha_k \in \mathbb{C}\,. \tag{6.105}$$

The function $u \equiv 0$ is an unstable solution of (6.103) if the largest eigenvalue is positive. We see that this cannot happen if $f''(c_m) > 0$ holds.

From (6.104) it follows that

$$\lambda_k = -\gamma\left[|k|^2 + \frac{f''(c_m)}{2\gamma}\right]^2 + \frac{f''(c_m)^2}{4\gamma} \tag{6.106}$$

and we obtain from (6.104) and (6.105) that the most unstable wavelength $\bar{\ell} = 2\pi/|k|$ in the case $f''(c_m) < 0$ is given as

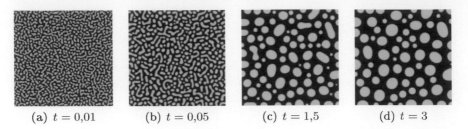

(a) $t = 0{,}01$      (b) $t = 0{,}05$      (c) $t = 1{,}5$      (d) $t = 3$

**Fig. 6.14** Solutions of the Cahn–Hilliard equation with Neumann boundary conditions for $c$ and $\Delta c$

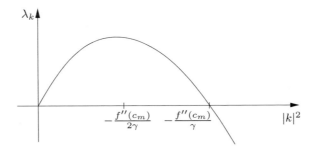

**Fig. 6.15** Linear stability analysis of the Cahn–Hilliard equation: dependency of the eigenvalues on the wavenumber

$$\bar{\ell} = 2\pi \sqrt{-\frac{2\gamma}{f''(c_m)}}. \tag{6.107}$$

This length scale can actually been observed in numerical simulations. In Fig. 6.14 we show a numerical simulation of the Cahn–Hilliard equation. At initial time $t = 0$ the evolution started with a small random perturbation of a constant unstable solution as initial data and we show the solution for different positive times. The values of $c$ are shown on a grey scale. The length scale on which the solution oscillates can be seen in the first image (a) and follows from formula (6.107). In Fig. 6.15 we show $\lambda_k$ as a function of $|k|^2$. In particular, we see that an interval of values of $|k|^2$ exists for which instabilities occur. We also see that $\lambda_k$ is stable if and only if

$$-\frac{f''(c_m)}{\gamma} < |k|^2.$$

From this we see that for $\ell$ such that

$$-\frac{f''(c_m)}{\gamma} < \frac{4\pi^2}{\ell^2} \quad \text{or} \quad \ell < 2\pi \sqrt{-\frac{\gamma}{f''(c_m)}},$$

we will not observe instabilities. In particular, it follows that the domain $\Omega$ has to be large enough in order to observe instabilities.

## 6.3 Hyperbolic Conservation Laws

The third important class of partial differential equations are *hyperbolic* differential equations. In Chap. 5 we considered two slightly different types of hyperbolic equations: The Euler equations are an example of a hyperbolic system of *first* order and the wave equation is a hyperbolic equation of *second* order. The analysis of both equations is quite different. In this section we only consider hyperbolic equations of first order. We restrict ourselves to the simplest case of a scalar equation.

We consider the nonlinear scalar hyperbolic conservation law

$$\partial_t u(t, x) + \partial_x\big(F(u(t, x))\big) = 0 \qquad \text{for } t > 0, \ x \in \mathbb{R},$$
$$u(0, x) = g(x) \quad \text{for } x \in \mathbb{R}, \tag{6.108}$$

with a function $F : \mathbb{R} \to \mathbb{R}$ which in general will be *nonlinear*. For $F(u) = \frac{1}{2}u^2$, i.e., $\partial_x F(u) = u \, \partial_x u$, this is called Burgers' equation. This equation can be considered as a highly simplified scalar version of Euler's equations

$$\partial_t \varrho + \nabla \cdot (\varrho\, v) = 0,$$
$$\partial_t v + v \cdot \nabla v = -\nabla(p(\varrho)),$$

for isothermal, incompressible gases.

**Method of Characteristics**

Partial differential equations of first order can be solved with the help of the method of characteristics. A *characteristic* $s \mapsto x(s)$ in this case is a curve along which a solution $u$ of a partial differential equation can be described as a solution of an ordinary differential equation for the function $s \mapsto z(s) = u(s, x(s))$. In the case of the hyperbolic equation (6.108) we obtain

$$z'(s) = \partial_t u\big(s, x(s)\big) + \partial_x u\big(s, x(s)\big)\, x'(s).$$

Choosing $s \mapsto x(s)$ as solution of

$$x'(s) = F'(z(s)), \tag{6.109}$$

we obtain $z'(s) = 0$. It hence holds that

$$z(s) = u(s, x(s)) = z(0) = g(x(0)).$$

We now obtain the solution formula

$$u(t, x(t)) = g(x(0)) \quad \text{for all } t > 0.$$

This means that the solution is *constant* along the characteristic $s \mapsto x(s)$. In this case the characteristic curves are affine linear.

**Example 1: The transport equation.** For $F(u) = cu$ it holds that $F'(u) = c$ and hence

$$x'(s) = c, \quad \text{and so } x(s) = cs + x_0$$

with $x_0 = x(0)$. The solution of the equation is

$$u(t, ct + x_0) = g(x_0)$$

or after the transformation $x_0 = x - ct$,

$$u(t, x) = g(x - ct). \tag{6.110}$$

**Example 2: Burgers' equation.** For $F(u) = \frac{1}{2}u^2$ we obtain the characteristic equation

$$x'(s) = z(s) = z(0) = g(x_0),$$

and hence $x(s) = x_0 + s \, g(x_0)$ which then implies

$$u(t, x_0 + t \, g(x_0)) = g(x_0). \tag{6.111}$$

The characteristics are lines in the $(x, t)$–plane with slope $1/g(x_0)$. In order to derive a solution formula for $(t, x) \mapsto u(t, x)$ we need to compute $x_0 = x_0(t, x)$ by solving $x = x_0 + t \, g(x_0)$. In the following we consider a so-called *Riemann problem*, which is the initial value problem (6.108) with piecewise constant initial data

$$g(x) = \begin{cases} a & \text{for } x < 0, \\ b & \text{for } x > 0. \end{cases}$$

The corresponding characteristics are sketched in Fig. 6.16. In the case $a > b$ there exists an area in which characteristics intersect each other. For each point in this area the formula (6.111) proposes two *different* values for the solution. In the case $a < b$ there is an area in which formula (6.111) does not provide a value for the solution as no characteristic connects points in this area with the initial data. For *nonlinear* hyperbolic conservation laws the characteristics only provide an incomplete picture of the solution. The reason for this is that the equation

$$x = x_0 + t \, g(x_0)$$

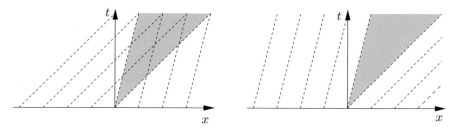

**Fig. 6.16** Characteristics of the Riemann problem for Burgers' equation for initial data $a > b$ (*left*) and $a < b$ (*right*). The slope of the characteristic is the inverse of the corresponding value at initial time

in general has no solution $x_0$. The problems that arise due to characteristics which intersect also appear for *smooth* initial data. For example for the initial data

$$g(x) = \begin{cases} 1 & \text{for } x < -1, \\ -1 & \text{for } x > 1 \end{cases}$$

we obtain that a characteristic with initial point $(0, x_1)$ for $x_1 < -1$ and a characteristic with initial point $(0, x_2)$ for $x_2 > 1$ will *always* intersect each other independently on how the initial data are defined in the interval $(-1, 1)$.

This example shows that for nonlinear hyperbolic equations the classical definition of a solution is not sufficient.

**Weak Solutions, the Rankine–Hugoniot Condition**

One obtains a weaker solution concept if one multiplies Eq. (6.108) with a smooth *test function*, integrates with respect to space and time and performs an integration by parts with respect to all derivatives. The assumptions on the solution in general can be reduced to the requirement that the solution is locally integrable. We hence require that the solution lies in the space $L_{1,\text{loc}}(\mathbb{R}_+ \times \mathbb{R})$.

**Definition 6.17** A function $u \in L_{1,\text{loc}}(\mathbb{R}_+ \times \mathbb{R})$ is called a *weak solution* of (6.108) if for all $w \in C_0^1([0, \infty) \times \mathbb{R})$ the following identity holds

$$\int_{\mathbb{R}_+} \int_{\mathbb{R}} \left( u \, \partial_t w + F(u) \, \partial_x w \right) dx \, dt + \int_{\mathbb{R}} g \, w(0, \cdot) \, dx = 0.$$

This formulation allows for *discontinuities* in the solution. We will show now that discontinuities need to satisfy a special condition which is called the *Rankine–Hugoniot condition*. To this end, we consider a weak solution $u$ which can be discontinuous along a curve $\Gamma$ but is smooth away from $\Gamma$. We consider a curve $\Gamma$ of the form

$$\Gamma = \{(t, s(t)) \mid t \in \mathbb{R}_+\}$$

with a smooth function $t \mapsto s(t)$. The curve divides the domain $\mathbb{R}_+ \times \mathbb{R}$ into three sets: $\Gamma$,

$$Q_- = \{(t, x) \mid x < s(t), t \in \mathbb{R}_+\}, \text{ and } Q_+ = \{(t, x) \mid x > s(t), t \in \mathbb{R}_+\}.$$

We assume that $u_{|Q_-}$ and $u_{|Q_+}$ can be extended as continuously differentiable functions on $\overline{Q}_-$ and on $\overline{Q}_+$, respectively. From the weak formulation with a test function $w$ with $w(0, x) = 0$ we obtain via integration by parts in $Q_-$ and in $Q_+$

$$
\begin{aligned}
0 &= \int_{\mathbb{R}_+} \int_{\mathbb{R}} \left( u\, \partial_t w + F(u)\, \partial_x w \right) dx\, dt \\
&= \int_{Q_-} \left( u\, \partial_t w + F(u)\, \partial_x w \right) dx\, dt + \int_{Q_+} \left( u\, \partial_t w + F(u)\, \partial_x w \right) dx\, dt \\
&= -\int_{Q_-} \left( \partial_t u + \partial_x F(u) \right) w\, dx\, dt + \int_{\Gamma} \left( u_- n_{t,-} + F(u_-) n_{x,-} \right) ds_{(t,x)} \\
&\quad - \int_{Q_+} \left( \partial_t u + \partial_x F(u) \right) w\, dx\, dt + \int_{\Gamma} \left( u_+ n_{t,+} + F(u_+) n_{x,+} \right) ds_{(t,x)}.
\end{aligned}
$$

Here $u_-$ and $u_+$ are the limits of $u$ from $Q_-$ and $Q_+$ onto $\Gamma$ and $n_{\pm} = (n_{t,\pm}, n_{x,\pm})$ are the normal vectors on $\Gamma$ which are oriented to the exterior of $Q_{\pm}$. These normal vectors can be represented as

$$n_- = \frac{1}{\sqrt{1 + \dot{s}^2(t)}} \begin{pmatrix} -\dot{s}(t) \\ 1 \end{pmatrix} \quad \text{and} \quad n_+ = -n_-.$$

Choosing test functions which vanish on $\Gamma$ we obtain that the differential equation $\partial_t u + \partial_x F(u) = 0$ has to hold in $Q_-$ and $Q_+$. As the volume integrals in the above formula are zero we obtain

$$0 = \int_{\Gamma} \left( (u_- - u_+) n_{t,-} + (F(u_-) - F(u_+)) n_{x,-} \right) w\, ds_{(t,x)}$$

for *all* test functions $w$. We hence obtain that pointwise on $\Gamma$ the following identity has to hold

$$0 = (u_- - u_+) n_{t,-} + (F(u_-) - F(u_+)) n_{x,-}.$$

We now denote by

$$V := \dot{s}$$

the *velocity* of the discontinuity curve $\Gamma$ and obtain the condition

$$F(u_+) - F(u_-) = V(u_+ - u_-).\qquad(6.112)$$

This is the *Rankine–Hugoniot condition*. This condition states that the jump in the flux $F(u)$ is given as the jump of the solution times the velocity of the discontinuity curve. In this context we refer to (7.50) for a variant of the Rankine–Hugoniot condition in higher space dimensions.

Condition (6.112) was already formulated in (6.75). There this condition was the necessary condition in order to be able to connect the limits $\lim_{x\to\pm\infty} u(t, x)$ with a traveling wave solution with velocity $V$ satisfying

$$\partial_t u + \partial_x (f(u)) = \eta\, \partial_x^2 u.\qquad(6.113)$$

From (6.113) we formally obtain (6.108) in the limit $\eta \to 0$. In this sense the hyperbolic conservation law can be interpreted as the *singular* limit of a family of equations of parabolic type. As we will see also later most difficulties in the analysis of hyperbolic conservation laws arise from the fact that the viscosity term $\eta\, \partial_x^2 u$ is not taken into account. For more details we refer to the book of Smoller [119].

With the help of the Rankine–Hugoniot condition we can now find a unique solution to the Riemann problem for Burgers' equation in the case $a > b$. We expect a discontinuity of the solution along a curve which connects the values $u_- = a$ and $u_+ = b$. From the Rankine–Hugoniot condition the *velocity* of the discontinuity is computed as

$$V = \frac{F(u_+)\quad F(u_-)}{u_+ - u_-} = \frac{1}{2}\frac{b^2 - a^2}{b - a} = \frac{a+b}{2}.$$

The corresponding solution is

$$u(t, x) = \begin{cases} a & \text{for } x < \frac{a+b}{2}t, \\ b & \text{for } x > \frac{a+b}{2}t. \end{cases}\qquad(6.114)$$

The discontinuity of such a solution is called a *shock wave*.

**The Entropy Condition**

We now consider Burgers' equation with initial data $u(0, x) = a$ for $x < 0$ and $u(0, x) = b$ for $x > 0$ with $a < b$. In this case we can construct with the help of the Rankine–Hugoniot condition a discontinuous solution. The velocity of the shock is given by $V = \frac{a+b}{2}$ and we have

$$u(t, x) = \begin{cases} a & \text{for } x < \frac{a+b}{2}t, \\ b & \text{for } x > \frac{a+b}{2}t. \end{cases}$$

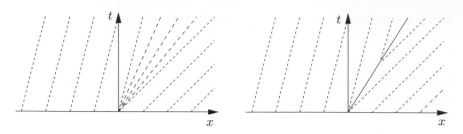

**Fig. 6.17** Rarefaction wave (*left*) and unphysical shock (*right*). We sketch the shocks and some chosen characteristics. The unphysical shock does not fulfil the entropy condition

The characteristics of this solution are sketched in Fig. 6.17 to the right. Besides this discontinuous solution there also exists a continuous solution which can be obtained by interpolating the values $u(t, at) = a$ and $u(t, bt) = b$ for $x \in (at, bt)$:

$$u(t, x) = \begin{cases} a & \text{for } x < at\,, \\ \frac{x}{t} & \text{for } at < x < bt\,, \\ b & \text{for } x > bt\,. \end{cases} \qquad (6.115)$$

One easily computes that this solution solves for all $x \notin \{at, bt\}$ the differential equation and on the lines $x = at$ and $x = bt$ we have $u_+ = u_-$ and $F(u_+) = F(u_-)$, which implies that the Rankine–Hugoniot conditions holds. The characteristics of the solution are sketched in Fig. 6.17 to the left. The solution has a *rarefaction wave* in the area given by $x \in (at, bt)$. There are also other solutions which can be obtained by choosing a value $c \in (a, b)$ and introducing two discontinuities; one connecting $a$ and $c$ and one connecting $c$ and $b$. These shocks bound a domain where $u(t, x) = c$ holds. The Rankine–Hugoniot condition for the shocks gives the shock velocities as

$$V_{a,c} = \frac{a + c}{2} \quad \text{and} \quad V_{c,b} = \frac{c + b}{2}\,.$$

The corresponding solution is hence given as

$$u(t, x) = \begin{cases} a & \text{for } x < \frac{a+c}{2}t\,, \\ c & \text{for } \frac{a+c}{2}t < x < \frac{c+b}{2}t\,, \\ b & \text{for } x > \frac{c+b}{2}t\,. \end{cases} \qquad (6.116)$$

By analogous construction principles one can obtain solutions by combining an arbitrary number of shocks and rarefaction waves.

Now the question arises: how to choose from this multitude of solutions one which is physically *meaningful*? One obvious conjecture is that the *continuous* solution (6.115) is the most meaningful. We will now give some arguments in favor of this solution.

- The characteristics are the curves along which *information* about the initial data is transported. For the discontinuous solution (6.114) with $a > b$ the characteristics *end* in the shock and the velocity of the shock is given by information about the initial data which are transported to the shock by the characteristics. For all discontinuous solutions for $a < b$ some characteristics leave the shock. This means that the shock does not obtain information and in contrast generates information virtually out of "nothing".
- We can approximate the initial data by a continuous function as sketched in Fig. 6.18. A continuous approximation is given by

$$u_0(x) = \begin{cases} a & \text{for } x < -\frac{\varepsilon}{2}, \\ \frac{a+b}{2} + \frac{b-a}{\varepsilon}x & \text{for } -\frac{\varepsilon}{2} < x < \frac{\varepsilon}{2}, \\ b & \text{for } x > \frac{\varepsilon}{2}. \end{cases}$$

With the method of characteristics we obtain the unique continuous solution

$$u(t, x) = \begin{cases} a & \text{for } x < at - \frac{\varepsilon}{2}, \\ \frac{a+b}{2}\left(1 - \frac{t}{t+\varepsilon/(b-a)}\right) + \frac{x}{t+\varepsilon/(b-a)} & \text{for } at - \frac{\varepsilon}{2} < x < bt + \frac{\varepsilon}{2}, \\ b & \text{for } x > bt + \frac{\varepsilon}{2}. \end{cases}$$

This solution converges as $\varepsilon \to 0$ to the continuous solution (6.115). This also leads to an interpretation for the multitude of characteristics which in the rarefaction wave leave the origin: These characteristics transport the information about the discontinuous initial data into the domain $\{(t, x) \mid t > 0, \ at < x < bt\}$ or in other words the information about the values between $a$ and $b$.

We hence now require for a meaningful solution that no characteristic originates out of a shock. In the case of the scalar hyperbolic equation (6.108) this condition can be made precise quite easily. A characteristic along which the value $u$ is transported

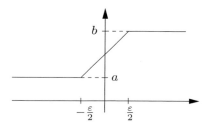

**Fig. 6.18** Approximation of discontinuous initial data

due to Eq. (6.109) has the form $\{(t, F'(u(t))) \mid t > 0\}$. For a shock with velocity $V$ and values $u_-$ and $u_+$ to the left and right of the shock we hence require

$$F'(u_-) \geq V \geq F'(u_+) . \tag{6.117}$$

This condition is called the *entropy condition*. One can show that a hyperbolic conservation law with convex and sufficiently smooth flux function possesses precisely one weak solution which fulfills the entropy condition along each shock wave.

There is another important argument for the entropy condition. In deriving the Euler equations, which is the physically most important example of a system of hyperbolic conservation laws, we neglected the *viscosity* of the gas. The viscosity of a gas is often very small but never zero. In the case of the Navier–Stokes equations the viscosity leads to an additional term with *second* derivatives. If we interpret this in the case of our scalar equation in one spatial dimension we obtain the differential equation

$$\partial_t u + \partial_x (F(u)) - \varepsilon \, \partial_x^2 u = 0 \tag{6.118}$$

with a *small* parameter $\varepsilon$. The parameter $\varepsilon$ in (6.118) multiplies the term with the *highest order of derivatives*. Neglecting this term changes the *type* of the partial differential equation from parabolic to hyperbolic and we obtain the equation

$$\partial_t u + \partial_x (F(u)) = 0 . \tag{6.119}$$

The solutions of Eq. (6.118) are *continuous* and also *unique* for meaningful boundary- and initial data. This obviously is not the case for solutions of (6.119). Motivated by this we can require for solutions of (6.119) that they are obtained as *limits* of the Eq. (6.118) when $\varepsilon$ tends to zero. Such a solution of (6.119) is called a *viscosity solution*. One can show that every viscosity solution has to fulfill an *entropy condition* which in the case of a scalar equation has the form (6.117).

A discontinuity for which characteristics on at least one side of the discontinuity have the same slope as the discontinuity itself is called a *contact discontinuity*. The simplest example of a contact discontinuity appears in the case of a transport equation with discontinuous initial data. We can take for example $u(0, x) = a$ for $x < 0$ and $u(0, x) = b$ for $x > 0$. The representation formula (6.110) then gives

$$u(t, x) = \begin{cases} a & \text{for } x < ct , \\ b & \text{for } x > ct . \end{cases}$$

The flux function is $F(u) = cu$ and the solution hence fulfills the Rankine–Hugoniot condition $F(u_+) - F(u_-) = c(u_+ - u_-)$ and the entropy condition in the form $F'(u_-) = c = F'(u_+)$. The characteristics here have on *both* sides of the shock curve the same velocity as the discontinuity. The contact discontinuity hence transports the discontinuous initial data. Contact discontinuities with different slopes for the

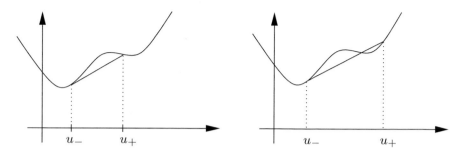

**Fig. 6.19** Non-convex flux function and initial data for an admissible (*left*) and a non-admissible (*right*) contact discontinuity

characteristics on both sides can be obtained for nonlinear equations with non-convex flux functions. A criterion for this can be obtained from the Rankine–Hugoniot condition and the entropy condition. For a contact discontinuity for which the characteristics left to the shock have the same slope as the discontinuity itself it has to hold

$$F'(u_-) = V = \frac{F(u_+) - F(u_-)}{u_+ - u_-} \geq F'(u_+).$$

For $F'(u_+) < V$ we obtain from this formula and the mean value theorem that $F$ has to be non-convex. In Fig. 6.19 the flux functions and the value of a solution to the left and the right of a discontinuity are shown for an admissible (left) and non-admissible (right) contact discontinuity. The non admissible contact discontinuity violates the entropy condition.

## 6.4 The Wave Equation

Wave phenomena often appear in applications, for example in compressible flows which describe the propagation of sound in gases. Other examples are electromagnetic waves or elastic waves in solids. Waves are often modeled with the help of *systems* of differential equations. To discuss the properties of wave type equations we now consider the mathematically simplest scalar version

$$\partial_t^2 u - \Delta u = f. \tag{6.120}$$

This equation for example describes

- the propagation of a longitudinal wave in a vibrating string or of a planar transversal wave in an elastic rod or the wave of an air column in a pipe (all examples are for $d = 1$).
- a vibrating membrane for space dimension $d = 2$.

### The One-Dimensional Cauchy Problem

We first consider the wave equation in one spatial dimension with initial conditions on the complete real axis:

$$\partial_t^2 u - \partial_x^2 u = 0 \quad \text{for } (t, x) \in (0, \infty) \times \mathbb{R}, \tag{6.121}$$

$$u = g \text{ and } \partial_t u = h \quad \text{for } t = 0, \ x \in \mathbb{R}. \tag{6.122}$$

A solution of (6.121) can be obtained from the following observation: The differential operator $\partial_t^2 - \partial_x^2$ can be factorized into

$$\partial_t^2 - \partial_x^2 = (\partial_t + \partial_x)(\partial_t - \partial_x).$$

If $u$ is a solution of (6.121) we obtain that

$$v(t, x) = (\partial_t - \partial_x)u(t, x)$$

is a solution of the *transport equation*

$$\partial_t v(t, x) + \partial_x v(t, x) = 0$$

with the initial condition

$$v(0, x) = \partial_t u(0, x)|_{t=0} - \partial_x u(0, x) = h(x) - \partial_x g(x) =: a(x).$$

The transport equation was already solved in (6.110). It holds that

$$v(t, x) = w(x - t)$$

with an arbitrary function $w : \mathbb{R} \to \mathbb{R}$. From the initial condition it follows that $v(t, x) = a(x - t)$ and hence

$$\partial_t u(t, x) - \partial_x u(t, x) = a(x - t).$$

This is an *inhomogeneous* transport equation with the solution

$$u(t, x) = u(0, x + t) + \int_0^t a(x + t - 2s)\, ds = g(x + t) + \frac{1}{2} \int_{x-t}^{x+t} a(y)\, dy.$$

Integrating $a$ gives

$$u(t, x) = g(x + t) + \frac{1}{2} \int_{x-t}^{x+t} h(y)\, dy - \frac{1}{2}\big(g(x + t) - g(x - t)\big),$$

and hence

$$u(t, x) = \frac{1}{2}\big(g(x + t) + g(x - t)\big) + \frac{1}{2}\int_{x-t}^{x+t} h(y)\, dy.$$  (6.123)

This is *d'Alembert's formula*.

### Uniqueness of the Solution

The uniqueness of the solution can be obtained with the help of energy estimates. This is similar to our considerations for parabolic equations.

**Theorem 6.18** *Let $\Omega \subset \mathbb{R}^d$ be a bounded domain with smooth boundary. In addition, $f : \mathbb{R}_+ \times \Omega \to \mathbb{R}$, $g, h : \Omega \to \mathbb{R}$ and $u_{\partial\Omega} : (0, T) \times \partial\Omega \to \mathbb{R}$ are assumed to be smooth. Then there exists at most one solution $u \in C^2\big([0, T] \times \overline{\Omega}\big)$ of the initial boundary value problem*

$$\begin{aligned}
\partial_t^2 u - \Delta u &= f && \text{in } (0, T) \times \Omega, \\
u &= u_{\partial\Omega} && \text{on } (0, T) \times \partial\Omega, \\
u = g \text{ and } \partial_t u &= h && \text{on } \{0\} \times \Omega.
\end{aligned}$$  (6.124)

*Proof* Assume that $u_1$ and $u_2$ solve (6.124). Then it follows that $w := u_1 - u_2$ solves

$$\begin{aligned}
\partial_t^2 w - \Delta w &= 0 && \text{in } (0, T) \times \Omega, \\
w &= 0 && \text{on } (0, T) \times \partial\Omega, \\
w &= 0 && \text{on } \{0\} \times \Omega.
\end{aligned}$$

For the "energy"

$$E(t) := \frac{1}{2}\int_{\Omega} \big(|\partial_t w|^2 + |\nabla w|^2\big) dx$$

it follows with the help of integration by parts and by using the fact that $\partial_t w = 0$ on $(0, T) \times \partial\Omega$,

$$\begin{aligned}
\frac{d}{dt} E(t) &= \int_{\Omega} \big(\partial_t w \cdot \partial_t^2 w + \nabla w \cdot \partial_t \nabla w\big) dx \\
&= \int_{\Omega} \partial_t w \big(\partial_t^2 w - \Delta w\big) dx + \int_{\partial\Omega} \partial_t w\, \nabla w \cdot n\, ds_x = 0.
\end{aligned}$$

Hence

$$E(t) = E(0) = 0$$

for all $t > 0$. This implies $\nabla w = 0$, $\partial_t w = 0$ and as $w = 0$ on $(0, T) \times \partial\Omega$ we obtain $w = 0$ and $u_1 = u_2$. $\qquad\square$

### Remarks

1. In some applications the quantity $E(t)$ can be a physical energy. This is in contrast to the diffusion equation where the corresponding energy typically is not a

physically meaningful energy. The term $\frac{1}{2}\int_\Omega |\partial_t w|^2\, dx$ is a kinetic energy and $\frac{1}{2}\int_\Omega |\nabla w|^2\, dx$ describes a "deformation energy".

2. The condition $u \in C^2([0, T] \times \overline{\Omega})$ is an assumption on the regularity of the solution. Strictly speaking we only showed the uniqueness of a solution with sufficient regularity. One can also show uniqueness under the weaker condition that the energy is bounded for almost all times.

**Domain of Dependence**

A characteristic feature of the wave equation is the property that "information" has a *finite speed of propagation*.

**Theorem 6.19** *Let $u \in C^2(\mathbb{R}_+ \times \mathbb{R}^d)$ be a solution of*

$$\partial_t^2 u - \Delta u = 0 \quad in\ \mathbb{R}_+ \times \mathbb{R}^d$$

*with the initial condition*

$$u(0, x) = \partial_t u(0, x) = 0 \quad for\ x \in B_{t_0}(x_0),$$

*where $B_{t_0}(x_0)$ is a ball with radius $t_0$ and center $x_0$. Then it holds that*

$$u(t, x) = 0 \ for\ (t, x) \in C := \left\{ (t, x) \,|\, 0 \leq t \leq t_0,\ |x - x_0| < t_0 - t \right\}.$$

*Proof* For

$$E(t) = \frac{1}{2} \int_{B_{t_0 - t}(x_0)} \left( |\partial_t u(t, x)|^2 + |\nabla u(t, x)|^2 \right) dx$$

it holds that

$$\frac{d}{dt} E(t) = \int_{B_{t_0 - t}(x_0)} \left( \partial_t u\, \partial_t^2 u + \nabla u \cdot \nabla \partial_t u \right) dx$$

$$- \frac{1}{2} \int_{\partial B_{t_0 - t}(x_0)} \left( |\partial_t u|^2 + |\nabla u|^2 \right) ds_x$$

$$= \int_{B_{t_0 - t}(x_0)} \partial_t u\, (\partial_t^2 u - \Delta u)\, dx$$

$$+ \int_{\partial B_{t_0 - t}(x_0)} \left( \partial_t u\, \nabla u \cdot n - \frac{1}{2}(|\partial_t u|^2 + |\nabla u|^2) \right) ds_x$$

$$\leq 0.$$

Here, the last inequality follows with the Cauchy-Schwarz inequality

$$|\partial_t u||\nabla u \cdot n| \leq |\partial_t u||\nabla u| \leq \tfrac{1}{2} \left( |\partial_t u|^2 + |\nabla u|^2 \right).$$

It hence holds that $E(t) \leq E(0) = 0$ and because of $E(t) \geq 0$ it follows that $E(t) = 0$ and hence $u = 0$ in $C$.                                                                          □

The wave equation is a simple example for a hyperbolic differential equation of second order. A characteristic feature of *hyperbolic* equations is that, solutions are *not* smoother than its initial and boundary data (cf. (6.123)). This is in contrast to parabolic equations where the solution becomes more regular. Another characteristic feature of hyperbolic equations is that the "information" given by the initial and boundary data spreads with *finite* speed.

### Acoustic Approximation of the Euler Equations

The propagation of sound in gases shows that waves can also appear in gases. In fact a suitable linearization of the Euler equations can lead to a wave equation. In order to derive this we assume that the *velocity* and the *deviation of the density* around its mean $\varrho_0$ are very small. The Euler equations for isothermal flows are

$$\partial_t \varrho + \nabla \cdot (\varrho v) = 0,$$
$$\partial_t v + (v \cdot \nabla)v + \frac{1}{\varrho}\nabla(p(\varrho)) = 0.$$

The ansatz

$$\varrho = \varrho_0(1 + g)$$

with a small dimensionless quantity $g$ yields

$$\partial_t g + (1 + g)\nabla \cdot v \mid v \cdot \nabla g - 0,$$
$$\partial_t v + (v \cdot \nabla)v + \frac{1}{\varrho_0(1 + g)}\nabla(p(\varrho)) = 0.$$

We assume that $g$, $\nabla g$, $v$, and $\nabla v$ are small and now linearize the system. The terms $g\nabla \cdot v$, $v \cdot \nabla g$, and $(v \cdot \nabla)v$ are of second order and do not appear in the linearized system. The gradient of the pressure is linearized via $\nabla p(\varrho) = p'(\varrho_0)\nabla g$ and as this term is of second order we obtain from the prefactor $\frac{1}{\varrho_0(1+g)}$ only the term $\frac{1}{\varrho_0}$ of zeroth order. We hence obtain

$$\partial_t g + \nabla \cdot v = 0,$$
$$\partial_t v + c^2 \nabla g = 0$$

with the *speed of sound*

$$c = \sqrt{\frac{p'(\varrho_0)}{\varrho_0}}.$$

Subtracting the divergence of the second equation from the time derivative of the first equation yields the *acoustic approximation* of the Euler equations

$$\partial_t^2 g - c^2 \Delta g = 0\,.$$

In the case of an ideal gas under adiabatic conditions the equation of state

$$p(\varrho) = K \varrho^\gamma$$

holds. The speed of sound is then given as

$$c^2 = \frac{p'(\varrho_0)}{\varrho_0} = \frac{\gamma K \varrho_0^{\gamma-1}}{\varrho_0} = \frac{\gamma\, p(\varrho_0)}{\varrho_0^2}\,.$$

**Spherical Waves**

The propagation of waves which are generated from a point source can be computed from the three-dimensional wave equation

$$\partial_t^2 u - c^2 \Delta u = 0$$

by using a radially symmetric ansatz. Let

$$u(t,x) = w(t,|x|) \quad \text{where} \quad w : \mathbb{R}_+ \times \mathbb{R}_+ \to \mathbb{R}$$

and using (6.12) yields

$$\Delta u = \partial_r^2 w + \tfrac{2}{r}\partial_r w$$

and hence

$$\partial_t^2 w - c^2 \left(\partial_r^2 w + \tfrac{2}{r}\partial_r w\right) = 0\,.$$

This equation can be transformed to the one-dimensional wave equation by using the ansatz

$$w(t,r) = \frac{z(t,r)}{r}\,.$$

It then follows that

$$\partial_t^2 z - c^2 \partial_r^2 z = 0\,.$$

The general solution is given by

$$z(t,r) = f(r - ct) - g(r + ct)$$

with arbitrary functions $f$ and $g$. Here $f$ describes an outgoing wave and $g$ an incoming wave. Transforming back yields

$$u(t, x) = \frac{f(|x| - ct)}{|x|} + \frac{g(|x| + ct)}{|x|}.$$

This solution has a singularity at the origin $|x| = 0$. We now consider an incoming wave and observe that the energy density which is distributed on the surface of a ball will be concentrated on a point when it reaches the origin. In the case of an outgoing wave the singularity can be avoided by requiring $f(r) = 0$ for $r \leq 0$.

**Scattering of Waves**

We now consider a one-dimensional wave which travels from a homogeneous medium of density $\varrho_-$ and elasticity constant $a_-$ to another medium with density $\varrho_+$ and elasticity constant $a_+$. At the interface between both media, which is assumed to lie at $x = 0$, the *displacements*, the acting *forces*, and the *stresses* have to be equal. This yields the differential equations

$$\partial_t^2 u - c_-^2 \, \partial_x^2 u = 0 \ \text{for } x < 0, \quad \partial_t^2 u - c_+^2 \partial_x^2 u = 0 \ \text{for } x > 0$$

and the sound velocities $c_- = \sqrt{\frac{a_-}{\varrho_-}}$ and $c_+ = \sqrt{\frac{a_+}{\varrho_+}}$ and the *coupling conditions*

$$u(t, 0-) = u(t, 0+) \ \text{and} \ a_- \, \partial_x u(t, 0-) = a_+ \, \partial_x u(t, 0+).$$

Here $u(t, 0\pm)$ is the limit $\lim\limits_{\substack{x \to 0 \\ x > 0}} u(t, \pm x)$. A wave coming from the left is then given by

$$u(t, x) = f(x - c_- t) \ \text{for } t < 0, \ x \in \mathbb{R},$$

where $f(x) = 0$ is assumed for $x > 0$. The corresponding initial conditions are

$$u(0, x) = f(x), \ \partial_t u(0, x) = -c_- f'(x).$$

A meaningful ansatz is

$$u(t, x) = \begin{cases} f(x - c_- t) + g(x + c_- t) & \text{for } x < 0, \\ h(x - c_+ t) & \text{for } x > 0. \end{cases}$$

Here the function $g$ describes a wave which is obtained by a reflection at the interface. The function $h$ describes the part of the wave which propagates to the right. The differential equations are fulfilled by this ansatz. The coupling conditions lead to the following linear system

$$\begin{aligned} f(-c_- t) + g(c_- t) &= h(-c_+ t), \\ a_- f'(-c_- t) + a_- g'(c_- t) &= a_+ h'(-c_+ t). \end{aligned} \tag{6.125}$$

Differentiating the first equation with respect to $t$ yields after a multiplication with $a_-$ and after subtracting $c_-$ times the second equation

$$2\,a_-\,c_-\,f'(-c_-t) = (a_-\,c_+ + a_+\,c_-)\,h'(-c_+t)\,.$$

Because of $f(0) = h(0) = 0$ we obtain after integration

$$h(x) = \frac{2\,a_-\,c_+}{a_-\,c_+ + a_+\,c_-}\,f\left(\frac{c_-}{c_+}x\right)\,.$$

The first equation in (6.125) then yields

$$g(x) = \frac{a_-\,c_+ - a_+\,c_-}{a_-\,c_+ + a_+\,c_-}\,f(-x)\,.$$

The factors in the expressions above determine which parts of the wave are reflected and which pass through the interface.

## 6.5   The Navier–Stokes Equations

The Navier–Stokes equations describe flows with viscous friction. They are based on the conservation of mass and momentum by assuming a constitutive law for the stress tensor which besides the pressure also models viscous forces. We now consider the variant for *incompressible* flows

$$\nabla \cdot v = 0\,, \tag{6.126}$$
$$\partial_t v + (v \cdot \nabla)v - \eta\,\Delta v = -\nabla p + f\,. \tag{6.127}$$

Comparing this with the original version (5.15) of the Navier–Stokes equations one observes that we included the factor $1/\varrho$ in front of $\nabla p$ in the definition of the pressure. This equation has two characteristic features: The constraint that the divergence of the velocity is zero and the *convective* term $(v \cdot \nabla)v$. As in the Stokes equations the pressure is also related to the condition $\nabla \cdot v = 0$ and will be discussed later. The convective term leads to problems in the analysis of the system. The term is *nonlinear* and has the *strongest growth* for large velocities, which is the source of the main difficulties in the analysis. Scaling the velocity with a constant $\alpha$ leads to a *quadratic* growth in the convective term with all other terms being still linear.

   The momentum balance (6.127) can be regarded as an evolution equation for the velocity field,

$$\partial_t v = -(v \cdot \nabla)v + \eta\,\Delta v + f - \nabla p\,. \tag{6.128}$$

If the velocity field is divergence free at a time $t_0$, i.e., $\nabla \cdot v(t_0, \cdot) = 0$ holds, one has to ensure that $\nabla \cdot \partial_t v = 0$ holds during the evolution. This has to hold to ensure the constraint $\nabla \cdot v = 0$ also holds for $t > t_0$. This observation leads to *Poisson's equation* for the pressure

$$-\Delta p = \nabla \cdot \left((v \cdot \nabla)v - \eta \, \Delta v - f\right).$$

With suitable boundary conditions for the pressure $p$, one can derive from this equation the "corresponding" pressure for a given velocity field. The pressure typically is only defined up to an additive constant. The gradient of the pressure is related to a *projection* on the set of all divergence-free functions. The mathematical background is given by the following theorem.

**Theorem 6.20** (Helmholtz–Hodge decomposition) *Let $\Omega \subset \mathbb{R}^d$ be a bounded domain with smooth boundary. Every vector field*

$$w \in C^2\left(\overline{\Omega}, \mathbb{R}^d\right)$$

*has a unique orthogonal decomposition in $L_2\left(\Omega; \mathbb{R}^d\right)$ of the form*

$$w = u + \nabla p,$$

*where $\nabla \cdot u = 0$ in $\Omega$ and $u \cdot n = 0$ on $\partial\Omega$. This decomposition is called the Helmholtz–Hodge decomposition.*

*Proof* We first show that $u$ and $\nabla p$ have to be orthogonal to each other. It holds that

$$\nabla \cdot (p\, u) = p \nabla \cdot u + u \cdot \nabla p = u \cdot \nabla p.$$

The divergence theorem now yields

$$\int_\Omega u \cdot \nabla p \, dx = \int_\Omega \nabla \cdot (p\, u) \, dx = \int_{\partial\Omega} p\, u \cdot n \, ds_x = 0.$$

We now show the uniqueness of the decomposition and assume that $w$ has two decompositions

$$w = u_1 + \nabla p_1 = u_2 + \nabla p_2$$

with the properties from the Helmholtz–Hodge decomposition. It then holds that

$$0 = u_1 - u_2 + \nabla(p_1 - p_2).$$

This implies after multiplication with $u_1 - u_2$ and integration over $\Omega$

$$0 = \int_\Omega |u_1 - u_2|^2 \, dx + \int_\Omega (u_1 - u_2) \cdot \nabla(p_1 - p_2) \, dx.$$

The second integral is zero due to the orthogonality of $u_1 - u_2$ and $\nabla(p_1 - p_2)$. It follows that $u_1 = u_2$ and hence $\nabla p_1 = \nabla p_2$. It remains to show the existence of the decomposition. Taking the divergence of $w = u + \nabla p$ yields Poisson's equation

$$\Delta p = \nabla \cdot w \ \text{ in } \Omega .$$

Together with the boundary condition

$$\nabla p \cdot n = w \cdot n \ \text{ on } \partial\Omega$$

this equation has a unique solution $p$, compare Theorem 6.4. If one assumes that $\Omega$ is connected one obtains that this solution is unique up to a constant. The necessary solvability condition

$$\int_\Omega \nabla \cdot w \, dx - \int_{\partial\Omega} w \cdot n \, ds_x = 0$$

is fulfilled due to the choice of the boundary data. Regularity theory [49] yields $p \in C^2(\overline{\Omega})$. Defining $u = w - \nabla p$ gives

$$\nabla \cdot u = 0 \ \text{ in } \Omega ,$$
$$u \cdot n = 0 \ \text{ on } \partial\Omega .$$

This shows the assertion of the theorem.                                                    □

We are now in a position to show an energy estimate for the kinetic energy in the Navier–Stokes equations.

**Theorem 6.21** *Let $\Omega$ be a bounded domain with smooth boundary and let $v$ be a sufficiently smooth solution of the Navier–Stokes equations in $\Omega$ with Dirichlet boundary conditions $v = 0$ on $\partial\Omega$. Then the estimate*

$$\int_\Omega \tfrac{1}{2}|v(t,x)|^2 \, dx + \int_0^t \int_\Omega \eta \, |Dv(s,x)|^2 \, dx \, ds$$
$$\leq C \left( \int_\Omega \tfrac{1}{2}|v(0,x)|^2 dx + \int_0^t \int_\Omega |f(s,x)|^2 \, dx \, ds \right)$$

*holds with a constant $C$ which only depends on $t$.*

*Proof* We multiply Eq. (6.127) with $v$ and integrate over the domain $\Omega$. Using the identity

$$\int_{\Omega} (v \cdot \nabla) v \cdot v \, dx = \int_{\Omega} \sum_{i,j=1}^{d} v_j \, \partial_{x_j} v_i \, v_i \, dx = \int_{\Omega} \sum_{i,j=1}^{d} \frac{1}{2} v_j \, \partial_{x_j} (v_i^2) \, dx$$

$$= -\int_{\Omega} \sum_{i,j=1}^{d} \frac{1}{2} \partial_{x_j} v_j \, v_i^2 \, dx = -\int_{\Omega} (\nabla \cdot v) \frac{1}{2} |v|^2 \, dx = 0$$

and integration by parts

$$-\int_{\Omega} \eta \, \Delta v \, v \, dx = -\int_{\Omega} \eta \sum_{i,j=1}^{d} \partial_{x_i}^2 v_j \, v_j \, dx = \int_{\Omega} \eta \sum_{i,j=1}^{d} \partial_{x_i} v_j \, \partial_{x_i} v_j \, dx$$

$$= \int_{\Omega} \eta \, |Dv|^2 \, dx \,,$$

it follows that

$$\frac{1}{2} \frac{d}{dt} \int_{\Omega} |v|^2 \, dx + \int_{\Omega} \eta \, |Dv|^2 \, dx = \int_{\Omega} f \cdot v \, dx \,. \tag{6.129}$$

This equation expresses the conservation of energy. The rate of change in time of the energy together with the energy which is dissipated by viscous friction at time $t$ equals the power brought into the system. Applying the elementary inequality

$$\int_{\Omega} f \cdot v \, dx \leq \frac{1}{2} \left( \int_{\Omega} |f|^2 \, dx + \int_{\Omega} |v|^2 \, dx \right) \tag{6.130}$$

yields for $y(t) = \frac{1}{2} \int_{\Omega} |v|^2 dx$

$$y' \leq y + \frac{1}{2} \int_{\Omega} |f|^2 dx \,.$$

With Gronwall's inequality, see Lemma 4.5, it follows that

$$\frac{1}{2} \int_{\Omega} |v(t, x)|^2 \, dx \leq e^t \left( \int_{\Omega} |v(0, x)|^2 \, dx + \int_0^t \int_{\Omega} \frac{1}{2} |f(s, x)|^2 \, dx \, ds \right) \,.$$

The estimate for $\int_0^t \int_{\Omega} \eta |Dv(s, x)|^2 \, dx \, ds$ now follows after integrating (6.129) and using (6.130), and the above inequality. □

This inequality yields the essential prerequisite for the proof of the existence of weak solutions to the Navier–Stokes equations. A proof would then use methods of functional analysis and cannot be given here, see [44, 118, 126] for a proof. For *sufficiently smooth* solutions one can show uniqueness of a solution.

**Theorem 6.22** *Let $\Omega$ be a bounded domain with smooth boundary. Then there exists at most one two times continuously differentiable solution of the Navier–Stokes equations to the same initial data and the same Dirichlet boundary data.*

*Proof* We now assume that $v^{(1)}$ and $v^{(2)}$ are two solutions. We now multiply the equation for $v^{(1)}$ with $v^{(1)} - v^{(2)}$, the equation for $v^{(2)}$ with $v^{(2)} - v^{(1)}$ and add both equations. After integration over $\Omega$ and integration by parts of the viscous term we obtain

$$
\int_\Omega \Big[ \partial_t \big(v^{(1)} - v^{(2)}\big) \cdot \big(v^{(1)} - v^{(2)}\big)
$$
$$
+ \Big( \big(v^{(1)} \cdot \nabla\big)v^{(1)} - \big(v^{(2)} \cdot \nabla\big)v^{(2)} \Big) \cdot \big(v^{(1)} - v^{(2)}\big)
$$
$$
+ \eta\, D\big(v^{(1)} - v^{(2)}\big) : D\big(v^{(1)} - v^{(2)}\big) \Big]\, dx = 0 \,.
$$

For the convective term the decomposition

$$
\big(v^{(1)} \cdot \nabla\big)v^{(1)} - \big(v^{(2)} \cdot \nabla\big)v^{(2)} = \big(v^{(1)} \cdot \nabla\big)\big(v^{(1)} - v^{(2)}\big)
$$
$$
+ \big((v^{(1)} - v^{(2)}) \cdot \nabla\big)v^{(2)}
$$

yields the estimate

$$
\big|(v^{(1)} \cdot \nabla)v^{(1)} - (v^{(2)} \cdot \nabla)v^{(2)}\big|
$$
$$
\leq \big|v^{(1)}\big|\big|D(v^{(1)} - v^{(2)})\big| + \big|Dv^{(2)}\big|\big|v^{(1)} - v^{(2)}\big| \,.
$$

We then obtain

$$
\left| \int_\Omega \Big( (v^{(1)} \cdot \nabla)v^{(1)} - (v^{(2)} \cdot \nabla)v^{(2)} \Big) \cdot \big(v^{(1)} - v^{(2)}\big)\, dx \right|
$$
$$
\leq C_1 \int_\Omega \big( |D(v^{(1)} - v^{(2)})| + |v^{(1)} - v^{(2)}| \big) |v^{(1)} - v^{(2)}|\, dx
$$
$$
\leq \frac{\eta}{2} \int_\Omega |D(v^{(1)} - v^{(2)})|^2\, dx + \frac{C_1^2}{2\eta} \int_\Omega |v^{(1)} - v^{(2)}|^2\, dx \,.
$$

Altogether it follows that

$$
\frac{1}{2} \frac{d}{dt} \int_\Omega |v^{(1)} - v^{(2)}|^2\, dx + \int_\Omega \frac{\eta}{2} |D(v^{(1)} - v^{(2)})|^2\, dx
$$
$$
\leq C \int_\Omega |v^{(1)} - v^{(2)}|^2\, dx \,.
$$

Applying Gronwall's inequality with $y(0) = 0$ and $\beta(t) = 0$ yields $v^{(1)} = v^{(2)}$.

□

The regularity assumptions for the solution required in the uniqueness theorem can be weakened considerably. In particular it can be shown that only *one* solution has to be smooth in order to exclude the existence of another weak solution. Nevertheless nobody has succeeded so far in showing the regularity required in the uniqueness theorem for the relevant space dimension 3. The reason for this is the convective term $(v \cdot \nabla)v$ which grows for large $v$ much stronger than the other terms. For more details we refer to the book of Temam [129].

## 6.6 Boundary Layers

In Chap. 5 we studied two models for flows: The *Euler equations* and the *Navier–Stokes equations*. The models differ from each other by the fact that the "internal friction" of the flow has been neglected in case of the Euler equations. In the Navier–Stokes equations however viscosity is taken into account. In cases where the viscosity becomes smaller and smaller we expect that solutions to the Navier–Stokes equations will converge to solutions of Euler's equations. This is what we want to analyze in more detail in what follows.

We consider the Navier–Stokes equations in nondimensional form

$$\left. \begin{array}{r} \partial_t v + (v \cdot \nabla)v - \frac{1}{\mathrm{Re}}\, \Delta v + \nabla p = 0 \\ \nabla \cdot v = 0 \end{array} \right\} \text{ in } \Omega \,,$$

$$v = 0 \qquad\qquad \text{on } \partial\Omega \,,$$

and Euler's equations for incompressible flows also in its nondimensional form

$$\left. \begin{array}{r} \partial_t v + (v \cdot \nabla)v + \nabla p = 0 \\ \nabla \cdot v = 0 \end{array} \right\} \text{ in } \Omega \,,$$

$$v \cdot n = 0 \qquad\quad \text{on } \partial\Omega \,.$$

It will become apparent that solutions to the Navier–Stokes equations will have *boundary layers*. At the boundary of the domain under consideration the velocity profile will change drastically in order to fulfill the boundary conditions. The reason for this different behavior is due to the fact that in the Navier–Stokes equations terms with a higher differentiability order appear when compared to the Euler equations. The Navier–Stokes equations are in fact a *singular perturbation* of the Euler equations.

The consequences of the different differentiability order on the qualitative behavior of solutions will now be discussed for a simple example.

**Example 1**: We consider the boundary value problem

$$y'(x) = a \,, \quad y(1) = 1 \,,$$

where $a \in \mathbb{R}$ is constant and $x \in (0, 1)$. The solution is

$$y(x) = a(x - 1) + 1 \, .$$

We now add the "small" term $\varepsilon y''$ with $\varepsilon > 0$ to the above differential equation and require a second condition at the boundary. We now consider

$$\varepsilon \, y'' + y' = a \ \text{ with } \ y(0) = 0 \, , \ y(1) = 1 \, .$$

It is straightforward to compute that

$$y_\varepsilon(x) = \frac{1 - a}{1 - e^{-1/\varepsilon}} \left(1 - e^{-x/\varepsilon}\right) + ax$$

is a solution of this problem. For $0 < a < 1$ the solution is sketched in Fig. 6.20.

The solution $y_\varepsilon$ of the perturbed problem differs from $y$ mainly in a small strip with thickness $\mathcal{O}(\varepsilon)$ around the point $x = 0$.

**Example 2**: Flow over a plate. In the following example we compute a boundary layer for the Navier–Stokes equations. We consider a two-dimensional flow in the upper half plane $\{x \in \mathbb{R}^2 \,|\, x_2 \geq 0\}$ and assume that the lower boundary, the "plate", is fixed, see Fig. 6.21. In addition, we assume that

$$v(t, x_1, x_2) \to \begin{pmatrix} U_1 \\ 0 \end{pmatrix} \ \text{ as } \ x_2 \to \infty$$

holds.

We search for a solution with the following properties

$$v(t, x_1, x_2) = \begin{pmatrix} u(t, x_2) \\ 0 \end{pmatrix} , \ \nabla p(t, x_1, x_2) = 0 \, .$$

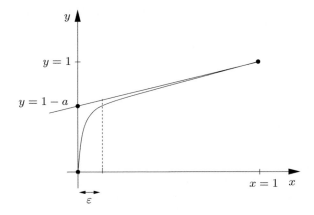

**Fig. 6.20**  A boundary layer in the solution of a boundary value problem

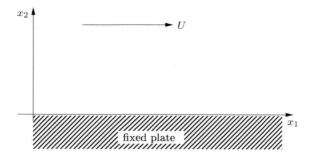

**Fig. 6.21** Flow over a fixed plate

Then it holds that $(v \cdot \nabla)v = v_1 \partial_{x_1} v + v_2 \partial_{x_2} v = 0$ and $\nabla p = 0$. This implies that

$$\partial_t u = \frac{1}{\text{Re}} \partial_{x_2}^2 u \,,$$
$$u(t, 0) = 0 \,, \tag{6.131}$$
$$u(t, x_2) \to U_1 \text{ as } x_2 \to \infty \,,$$

hold. This problem has the following scaling property: If $u$ is a solution we obtain that

$$w(t, x_2) = u\left(\tfrac{t}{T}, \tfrac{x_2}{L}\right) \text{ with fixed } T, L \in \mathbb{R}$$

solves the differential equation

$$\partial_t w = \frac{\partial_t u}{T} = \frac{1}{\text{Re}} \frac{1}{T} \partial_{x_2}^2 u = \frac{1}{\text{Re}} \frac{L^2}{T} \partial_{x_2}^2 w \,.$$

If $L^2 = T$ holds, we obtain that $w$ solves the system (6.131). In particular the same boundary conditions are fulfilled.

We now assume that (6.131) has a unique solution. In this case $w$ and $u$ have to coincide. This implies

$$u\left(\tfrac{t}{T}, \tfrac{x_2}{L}\right) = u(t, x_2) \,, \text{ if } L^2 = T \,.$$

Choosing for $t$ fixed the values $T = t$ and $L = \sqrt{t}$, we obtain

$$u\left(1, \tfrac{x_2}{\sqrt{t}}\right) = u(t, x_2) \,.$$

This implies that the solution has to be *self-similar*. In particular, this means that we know the solution at all times if we know the solution at one particular time, as we can use a simple rescaling of the $x$–variable.

We now introduce the variable $\eta = \frac{\sqrt{\text{Re}}}{2} \frac{x_2}{\sqrt{t}}$ and define

$$f(\eta) = \frac{1}{U_1} u\left(1, \frac{2}{\sqrt{\mathrm{Re}}} \eta\right) = \frac{1}{U_1} u\left(t, \frac{2\sqrt{t}\,\eta}{\sqrt{\mathrm{Re}}}\right).$$

It now follows that

$$f(0) = 0, \quad f(\infty) = 1.$$

Which equation has to be fulfilled by $f$? With

$$\frac{\partial}{\partial t}\eta = -\frac{1}{4}\sqrt{\mathrm{Re}}\,\frac{x_2}{t^{3/2}} = -\frac{1}{2}\frac{\eta}{t}$$

and

$$\frac{\partial}{\partial x_2}\eta = \frac{\sqrt{\mathrm{Re}}}{2\sqrt{t}}$$

we obtain

$$0 = \frac{1}{U_1}\left(\partial_t u - \frac{1}{\mathrm{Re}}\partial_{x_2}^2 u\right) = f'(\eta)\left(-\frac{\eta}{2t}\right) - \frac{1}{\mathrm{Re}}f''(\eta)\frac{\mathrm{Re}}{4t}.$$

It hence holds that

$$0 = 2\eta\,f'(\eta) + f''(\eta), \quad f(0) = 0, \quad f(\infty) = 1.$$

This implies

$$f'(\eta) = c\,e^{-\eta^2}$$

and integration using $f(\infty) = 1$ yields

$$f(\eta) = \mathrm{erf}(\eta) = \frac{2}{\sqrt{\pi}}\int_0^\eta e^{-s^2}\,ds.$$

Here we used

$$\int_0^\infty e^{-s^2}\,ds = \frac{\sqrt{\pi}}{2}.$$

For $u$ we obtain (Fig. 6.22)

$$u(t, x_2) = U_1\,\mathrm{erf}\left(\frac{\sqrt{\mathrm{Re}}}{2}\frac{x_2}{\sqrt{t}}\right) = U_1\frac{2}{\sqrt{\pi}}\int_0^{\frac{\sqrt{\mathrm{Re}}}{2}\frac{x_2}{\sqrt{t}}} e^{-s^2}\,ds.$$

We say, we are in the boundary layer if

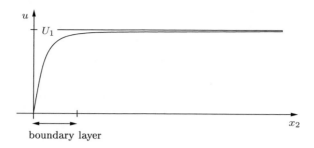

**Fig. 6.22** Boundary layer for a flow over a fixed plate

$$u < (1 - \varepsilon)U_1 \text{ for a fixed chosen } \varepsilon.$$

There is a unique $\eta_0$ with
$$\operatorname{erf}(\eta_0) = 1 - \varepsilon.$$

It then follows that the boundary layer is given by all $x_2$ such that
$$\frac{\sqrt{\operatorname{Re}}}{2} \frac{x_2}{\sqrt{t}} < \eta_0$$

or equivalently
$$x_2 < 2\eta_0 \sqrt{t/\operatorname{Re}}.$$

This implies that the boundary layer thickness is proportional to the square root of the "small" parameter $\frac{1}{\operatorname{Re}}$.

**Asymptotic Expansions for Boundary Layers**

We now want to study for a simple example a method which allows us to compute boundary layers for very complex problems. We consider the problem

$$\varepsilon y'' + 2y' + 2y = 0 \text{ for } x \in (0, 1) \text{ with } y(0) = 0, \ y(1) = 1. \tag{6.132}$$

Here $\varepsilon > 0$ is a small parameter. The solution has the form

$$y(x) = a\, e^{\alpha x/\varepsilon} + b\, e^{\beta x/\varepsilon}$$

with

$$\alpha = -1 + \sqrt{1 - 2\varepsilon} \approx -\varepsilon \text{ and}$$
$$\beta = -1 - \sqrt{1 - 2\varepsilon} \approx -2 + \varepsilon.$$

The boundary conditions yield

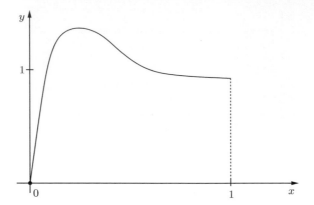

**Fig. 6.23** Solution of the boundary value problem (6.132)

$$0 = a + b,$$
$$1 = a \left( e^{\alpha/\varepsilon} - e^{\beta/\varepsilon} \right) \approx a \left( e^{-1} - e^{-2/\varepsilon} \right) \approx a\, e^{-1}.$$

This implies

$$y(x) \approx e^{1-x} - e^{1-2x/\varepsilon}.$$

This approximate solution is sketched in Fig. 6.23.

In this simple case one can recognize the structure of the solution without much effort. By means of this example we will now perform an asymptotic analysis which involves expansions in the boundary layer and in areas away from the boundary layer. In doing so we will use the fact that the parameter $\varepsilon$ is small.

**Remark.** For $\varepsilon = 0$ we obtain

$$y_0' + y_0 = 0 \quad \text{and hence} \quad y_0(x) = a\, e^{-x}.$$

We cannot fulfill both boundary conditions. This is due to the fact that the governing equations for $\varepsilon > 0$ are a singular perturbation of the above problem. We will not get any further unless we take care of the boundary layers in the asymptotic expansion.

Our ansatz for an asymptotic expansion will be as follows: We will make separate expansions for $x \geq \delta(\varepsilon) > 0$ and for $x \in [0, \delta(\varepsilon)]$. These have to be suitably put together. We will proceed in several steps.

**Step 1**: Outer expansion.
We assume that the solution in large parts of the interval $[0, 1]$ allows for an expansion of the form

$$y(x) = y_0(x) + \varepsilon\, y_1(x) + \cdots.$$

Plugging this ansatz into the differential equation (6.132) yields

$$2(y_0' + y_0) + \varepsilon(\cdots) + \varepsilon^2(\cdots) + \cdots = 0.$$

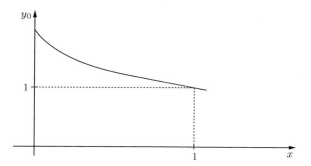

**Fig. 6.24** The outer solution to lowest order

At the order $\mathcal{O}(1)$ the equation

$$y_0' + y_0 = 0$$

has to be fulfilled together with the boundary conditions

$$y_0(0) = 0 \quad \text{and} \quad y_0(1) = 1.$$

We can only fulfill one boundary condition and opt for the boundary condition at the point $x = 1$. For the differential equation we obtain the solution

$$y_0(x) = a\,e^{-x}.$$

One can show that the following approach does not yield a meaningful solution if we require a boundary layer at $x = 1$. It hence follows that $y_0(1) = 1$ and

$$y_0(x) = e^{1-x}.$$

This solution is sketched in Fig. 6.24.

**Step 2**: Inner expansion.
We now introduce new coordinates $z = \frac{x}{\varepsilon^\alpha}$ for an $\alpha > 0$ and set $Y(z) = y(x)$. Plugging this into the differential equation yields

$$\varepsilon^{1-2\alpha}\frac{d^2}{dz^2}Y + 2\varepsilon^{-\alpha}\frac{d}{dz}Y + 2Y = 0 \quad \text{and} \quad Y(0) = 0.$$

The ansatz

$$Y(z) = Y_0(z) + \varepsilon^\gamma Y_1(z) + \cdots \quad \text{with} \quad \gamma > 0$$

yields

$$\varepsilon^{1-2\alpha}\left(Y_0'' + \varepsilon^\gamma Y_1'' + \cdots\right) + 2\varepsilon^{-\alpha}\left(Y_0' + \varepsilon^\gamma Y_1' + \cdots\right)$$
$$+ 2\left(Y_0 + c^\gamma Y_1 + \cdots\right) = 0.$$

(6.133)

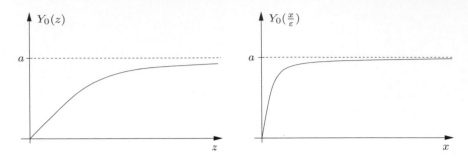

**Fig. 6.25** The inner solution at lowest order in the inner variable (*left*) and in the outer variable (*right*)

For terms to lowest order in $\varepsilon$ we obtain $\varepsilon^{1-2\alpha}Y_0''$ and/or $2\varepsilon^{-\alpha}Y_0'$. A balance of these terms is only possible for $\alpha = 1$. We hence obtain to the order $\varepsilon^{-1}$ in (6.133)

$$Y_0'' + 2Y_0' = 0 \quad \text{for} \ \ 0 < z < \infty$$

and

$$Y_0(0) = 0 \, .$$

This implies

$$Y_0(z) = a\big(1 - e^{-2z}\big) \, ,$$

where $a$ still has to be determined (Fig. 6.25).

**Step 3**: Matching of the inner and outer expansions.
We now search for an $a$, such that $Y_0$ and $y_0$ match, as sketched in Fig. 6.26.
In order to do so we introduce an intermediate variable $x_\eta = \frac{x}{\varepsilon^\beta} = z\varepsilon^{1-\beta}$ with $0 < \beta < \alpha = 1$. We consider for a fixed but arbitrary $x_\eta$ the limit $\varepsilon \to 0$, i.e., $z \to \infty$ and $x \to 0$. The inner and outer expansions have to coincide if they are expressed by the variable $x_\eta$. We define

$$\widetilde{y}_0(x_\eta) = y_0\big(\varepsilon^\beta x_\eta\big) \quad \text{and}$$
$$\widetilde{Y}_0(x_\eta) = Y_0\big(\varepsilon^{\beta-1}x_\eta\big) \, .$$

It then holds that

$$\widetilde{y}_0(x_\eta) = e^{1-\varepsilon^\beta x_\eta} \to e^1 \quad \text{for} \ \varepsilon \to 0 \quad \text{and}$$
$$\widetilde{Y}_0(x_\eta) = a\big(1 - e^{-2\varepsilon^{\beta-1}x_\eta}\big) \to a \quad \text{for} \ \varepsilon \to 0 \, .$$

Hence

$$a = e \, .$$

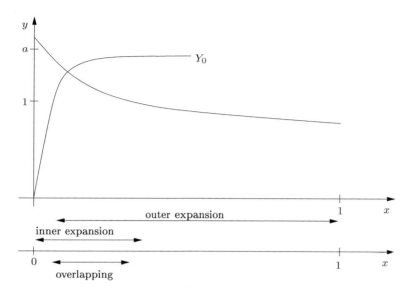

**Fig. 6.26**  In a transition zone both expansion have to coincide

**Step 4**: Composite expansion.
We now want to put both parts together. This is done by adding both solutions and subtracting a suitable common part. As $y_0(\varepsilon^\beta x_\eta) \to e$ and $Y_0(\varepsilon^{\beta-1} x_\eta) \to e$ as $\varepsilon \to 0$ a suitable common part is $e = y_0(0)$. We obtain

$$y(x) \approx y_0(x) + Y_0\left(\tfrac{x}{\varepsilon}\right) - y_0(0)$$
$$= e^{1-x} + e\left(1 - e^{-2x/\varepsilon}\right) - e$$
$$= e^{1-x} - e^{1-2x/\varepsilon}.$$

We add solutions which we obtained in different parts of the interval. It holds that $y_0$ is close to $y_0(0) = e$ in the proximity of the boundary. This part we subtract. Further away from the boundary $Y_0$ is close to $e = y_0(0)$. But also this part we subtract. It now happens that the constructed approximate solution is very close to the exact solution.

**Step 5**: Second term in the expansion.
Assuming that $y_0' + y_0 = 0$ we obtain for the outer expansion

$$\varepsilon(y_0'' + 2y_1' + 2y_1) + \varepsilon^2(\cdots) = 0.$$

At the order $\mathcal{O}(\varepsilon)$ we obtain

$$y_1' + y_1 = -\frac{1}{2}y_0'' = -\frac{1}{2}e^{1-x}, \tag{6.134}$$

as well as the boundary condition $y_1(1) = 0$. To obtain a solution one can use the "variation of constants" formula. A solution of the homogeneous equation is $x \mapsto e^{-x}$. With the ansatz

$$y_1(x) = e^{-x}h(x)$$

from (6.134) we can derive the identity

$$-e^{-x}h(x) + e^{-x}h'(x) + e^{-x}h(x) = -\frac{1}{2}e^{1-x}.$$

It follows that

$$h'(x) = -\frac{1}{2}e,$$

and with $h(1) = 0$ we obtain

$$h(x) = \frac{1}{2}e(1-x).$$

Hence Eq. (6.134) has the solution

$$y_1(x) = \tfrac{1}{2}(1-x)\,e^{1-x}.$$

In the *inner expansion* we obtain

$$\varepsilon^{-1}\big(Y_0'' + \varepsilon^\gamma Y_1'' + \cdots\big) + 2\varepsilon^{-1}\big(Y_0' + \varepsilon^\gamma Y_1' + \cdots\big) + 2\big(Y_0 + \varepsilon^\gamma Y_1 + \cdots\big) = 0.$$

As $Y_0$ has already been determined such that the terms of order $\mathcal{O}(\varepsilon^{-1})$ vanish we obtain that the term to lowest order now is either of order $\mathcal{O}(\varepsilon^{\gamma-1})$ or of order $\mathcal{O}(1)$. These two terms have to balance each other as otherwise it would follow that $Y_1 \equiv 0$. This implies $\gamma = 1$ and we obtain

$$Y_1'' + 2Y_1' = -2Y_0 = -2e\big(1 - e^{-2z}\big) \quad \text{and} \quad Y_1(0) = 0.$$

It follows that

$$Y_1(z) = c\big(1 - e^{-2z}\big) - z\,e\big(1 + e^{-2z}\big).$$

The constant $c$ has to be determined by a matching procedure. It holds that

$$\begin{aligned}
\tilde{y}_0(x_\eta) + \varepsilon\,\tilde{y}_1(x_\eta) &= y_0\big(\varepsilon^\beta x_\eta\big) + \varepsilon\,y_1\big(\varepsilon^\beta x_\eta\big) \\
&= e\big(1 - \varepsilon^\beta x_\eta\big) + \frac{\varepsilon}{2}\big(1 - \varepsilon^\beta x_\eta\big)e\big(1 - \varepsilon^\beta x_\eta\big) + \cdots \\
&= e - e\,\varepsilon^\beta x_\eta + \frac{\varepsilon}{2}e - \varepsilon^{1+\beta}e\,x_\eta + \frac{1}{2}\varepsilon^{1+2\beta}e\,x_\eta^2 + \cdots
\end{aligned}$$

and

$$\tilde{Y}_0(x_\eta) + \varepsilon\,\tilde{Y}_1(x_\eta) = Y_0\big(\varepsilon^{-1+\beta}x_\eta\big) + \varepsilon\,Y_1\big(\varepsilon^{-1+\beta}x_\eta\big)$$

$$= e\left(1 - e^{-2\varepsilon^{\beta-1}x_\eta}\right) + \varepsilon\left[c\left(1 - e^{-2\varepsilon^{\beta-1}x_\eta}\right) - \frac{x_\eta e}{\varepsilon^{1-\beta}}\left(1 + e^{-2\varepsilon^{\beta-1}x_\eta}\right)\right]$$

$$\approx e + \varepsilon c - \varepsilon^\beta e\,x_\eta\,.$$

It follows that

$$c = \frac{e}{2}\,.$$

We can put these expansions together to obtain

$$y(x) \approx y_0(x) + \varepsilon\,y_1(x) + Y_0\big(\tfrac{x}{\varepsilon}\big) + \varepsilon\,Y_1\big(\tfrac{x}{\varepsilon}\big) - \big[e - \varepsilon^\beta e\,x_\eta + \tfrac{\varepsilon}{2}e\big]$$

$$\approx e^{1-x} - (1+x)e^{1-2x/\varepsilon} + \frac{\varepsilon}{2}\big((1-x)e^{1-x} - e^{1-2x/\varepsilon}\big)\,.$$

Here, the term in the square brackets is the common part of both expansions.

**Prandtl's Boundary Layer Equations**

We now want to perform asymptotic expansions for the Navier–Stokes equations in the neighborhood of a solid body. To simplify the situation we consider a flat body. In addition we restrict ourselves to the spatially two-dimensional case. However, a generalization to three dimensions is possible. In this section the velocity vector is given by $(u, v)^\top \in \mathbb{R}^2$ and the spatial variables $(x, y)^\top \in \mathbb{R}^2$. In this case the Navier–Stokes equations in nondimensional form are given as

$$\partial_t u + u\,\partial_x u + v\,\partial_y u = -\partial_x p + \frac{1}{\mathrm{Re}}\Delta u\,,$$

$$\partial_t v + u\,\partial_x v + v\,\partial_y v = -\partial_y p + \frac{1}{\mathrm{Re}}\Delta v\,,$$

$$\partial_x u + \partial_y v = 0$$

for $y > 0$ and

$$u = v = 0 \quad\text{for}\quad y = 0\,.$$

We consider flows for large Reynolds numbers Re. In most practical applications in gas flows the Reynolds number in fact is large. The parameter

$$\varepsilon = \frac{1}{\mathrm{Re}}$$

in the following will be our small parameter.

**Step 1**: Outer expansion.
In areas far away from the boundary layer we consider an outer expansion for $(u, v)^\top$ of the form

$$
\begin{aligned}
u &= u_0 + \varepsilon\, u_1 + \varepsilon^2 u_2 + \cdots, \\
v &= v_0 + \varepsilon\, v_1 + \varepsilon^2 v_2 + \cdots, \\
p &= p_0 + \varepsilon\, p_1 + \varepsilon^2 p_2 + \cdots.
\end{aligned}
$$

At lowest order we obtain for $(u_0, v_0)$ the *Euler equations*

$$
\begin{aligned}
\partial_t u_0 + u_0\,\partial_x u_0 + v_0\,\partial_y u_0 &= -\partial_x p_0, \\
\partial_t v_0 + u_0\,\partial_x v_0 + v_0\,\partial_y v_0 &= -\partial_y p_0, \\
\partial_x u_0 + \partial_y v_0 &= 0.
\end{aligned}
\tag{6.135}
$$

**Step 2**: Inner expansion.
We expect the solution to change drastically in the $y$-direction. The velocity $(u, v)^\top$ vanishes for $y = 0$ and changes its value quickly in $y$-direction. Hence for the inner expansion near $\{y = 0\}$ we make the ansatz

$$
t = t, \ X = x, \ Y = \frac{y}{\varepsilon^k}
$$

and

$$
\begin{aligned}
U(t, X, Y) &= u\big(t, X, \varepsilon^k Y\big), \\
V(t, X, Y) &= v\big(t, X, \varepsilon^k Y\big), \\
P(t, X, Y) &= p\big(t, X, \varepsilon^k Y\big).
\end{aligned}
$$

For the unknown functions $U, V, P$ we assume an expansion of the form:

$$
\begin{aligned}
U &= U_0 + \varepsilon^m U_1 + \varepsilon^{2m} U_2 + \cdots, \\
V &= V_0 + \varepsilon^m V_1 + \varepsilon^{2m} V_2 + \cdots, \\
P &= P_0 + \varepsilon^m P_1 + \varepsilon^{2m} P_2 + \cdots.
\end{aligned}
$$

The divergence equation $\partial_x u + \partial_y v = 0$ yields

$$
\partial_X U_0 + \varepsilon^m \partial_X U_1 + \cdots + \varepsilon^{-k}\big(\partial_Y V_0 + \varepsilon^m \partial_Y V_1 + \cdots\big) = 0.
$$

This implies

$$
\partial_Y V_0 = 0,
$$

and as $V_0(t, x, 0) = 0$ holds, it follows that

$$V_0 \equiv 0 \,.$$

We only obtain a meaningful equation to the next order if we choose $m = k$. Only with this choice we obtain a meaningful balance between $\partial_X U_0$ and other terms in the equation. The resulting equation in this case is

$$\partial_X U_0 + \partial_Y V_1 = 0 \,. \tag{6.136}$$

The momentum equation yields

$$\partial_t U_0 + U_0 \, \partial_X U_0 + \varepsilon^k V_1 \, \partial_Y U_0 \, \varepsilon^{-k} + \cdots$$
$$= -\partial_X P_0 + \varepsilon \, \partial_X^2 U_0 + \varepsilon^{1-2k} \partial_Y^2 U_0 + \cdots ,$$
$$\varepsilon^k \partial_t V_1 + \varepsilon^k U_0 \, \partial_X V_1 + \varepsilon^{2k} V_1 \, \partial_Y V_1 \, \varepsilon^{-k} + \cdots$$
$$= -\varepsilon^{-k} \partial_Y P_0 + \varepsilon^{k+1} \partial_X^2 V_1 + \varepsilon^{k+1-2k} \partial_Y^2 V_1 + \cdots .$$

In the first equation the term $\varepsilon^{1-2k} \partial_Y^2 U_0$ can only be balanced if we set $k = \frac{1}{2}$. This then implies that viscosity terms appear to lowest order. We already expected beforehand that friction at the solid body is responsible for the boundary layer. It is hence meaningful to scale the viscosity term in such a way that the term remains to lowest order. A further motivation follows from our earlier thoughts about boundary layers in the Navier–Stokes equations. We already observed that the term $\varepsilon^{1/2} = \left(\frac{1}{\mathrm{Re}}\right)^{1/2}$ determines the thickness of the boundary layer. If we would choose $k > \frac{1}{2}$ the term $\partial_Y^2 U_0$ would dominate and we would obtain

$$\partial_Y^2 U_0 = 0 \,.$$

In this case we would only take viscous terms into account and the acceleration terms would be ignored. We would consider the flow too close to the solid body and the friction terms would dominate too much. In fact, the power $k$ determines via

$$Y = \varepsilon^k y \,,$$

how far we are away from the boundary. For $k < \frac{1}{2}$ we would ignore the viscous term and in this case we would be too far away from the boundary to "feel" the friction. Altogether, choosing $k = \frac{1}{2}$ yields to lowest order

$$\partial_t U_0 + U_0 \, \partial_X U_0 + V_1 \, \partial_Y U_0 = -\partial_X P_0 + \partial_Y^2 U_0 \,,$$
$$\partial_Y P_0 = 0 \,.$$

These equations have to be solved together with Eq. (6.136). As boundary conditions at the lower boundary we obtain

$$U_0 = V_1 = 0 \text{ for } Y = 0.$$

In particular, it holds that

$$P_0 = P_0(t, x).$$

**Step 3**: Matching.
Let

$$y_\eta = \frac{y}{\varepsilon^\beta} = \varepsilon^{1/2-\beta} Y$$

with $y = \varepsilon^{1/2} Y$ and

$$0 < \beta < \frac{1}{2}.$$

In the following we do not express the dependence on $t$ and $x$ explicitly. We define

$$\widetilde{u}_0(y_\eta) = u_0(\varepsilon^\beta y_\eta),$$
$$\widetilde{v}_0(y_\eta) = v_0(\varepsilon^\beta y_\eta),$$
$$\widetilde{p}_0(y_\eta) = p_0(\varepsilon^\beta y_\eta)$$

and

$$\widetilde{U}_0(y_\eta) = U_0(\varepsilon^{\beta-1/2} y_\eta),$$
$$\widetilde{V}_0(y_\eta) = V_0(\varepsilon^{\beta-1/2} y_\eta) \equiv 0,$$
$$\widetilde{P}_0(y_\eta) = P_0(\varepsilon^{\beta-1/2} y_\eta).$$

We require, that $(\widetilde{u}_0, \widetilde{v}_0, \widetilde{p}_0)$ and $(\widetilde{U}_0, \widetilde{V}_0, \widetilde{P}_0)$ coincide in the limit as $\varepsilon \to 0$. It follows that

$$\lim_{y \to 0} v_0(y) = 0 \text{ for all } t \text{ and } x.$$

From (6.135) it then follows that

$$\partial_t u_0(t, x, 0) + u_0(t, x, 0) \partial_x u_0(t, x, 0) = -\partial_x p_0(t, x, 0).$$

In addition it follows that

$$u_0(t, x, 0) = U_0(t, x, \infty),$$
$$p_0(t, x, 0) = P_0(t, x, \infty).$$

Because of

$$\partial_Y P_0 = 0$$

it follows that

$$p_0(t, x, 0) = P_0(t, x, Y) \text{ for all } Y > 0.$$

In the boundary layer we hence obtain *Prandtl's boundary layer equations*

$$\partial_t U_0 + U_0\,\partial_X U_0 + V_1\,\partial_Y U_0 = \partial_t u_0 + u_0\,\partial_x u_0 + \partial_Y^2 U_0\,,$$
$$\partial_X U_0 + \partial_Y V_1 = 0\,.$$

As boundary conditions we have

$$U_0 = V_1 = 0 \text{ for } Y = 0\,,$$

and

$$U_0(t, x, Y) = u_0(t, x, 0) \text{ for } Y = \infty\,.$$

This is a system of equations for $U_0$, $V_1$. For $V_1$ we only have one boundary condition. This is however sufficient as only first derivatives with respect to $V_1$ appear and our experience says that one boundary condition is sufficient in this case. For the outer expansion we solve the Euler equations with the boundary condition

$$v_0 = 0\,.$$

This is the usual boundary condition for the Euler equations, as this condition just says that the *normal component of the velocity* is zero.

**Summary**: Outside of the boundary layer we have to solve the Euler equations

$$\partial_t u_0 + u_0\,\partial_x u_0 + v_0\,\partial_y u_0 = -\partial_x p_0\,,$$
$$\partial_t v_0 + u_0\,\partial_x v_0 + v_0\,\partial_y v_0 = -\partial_y p_0\,,$$
$$\partial_x u_0 + \partial_y v_0 = 0$$

with the typical boundary condition

$$\begin{pmatrix} u_0 \\ v_0 \end{pmatrix} \cdot n = 0 \text{ for } y = 0\,.$$

In the boundary layer *Prandtl's boundary layer equations*

$$\partial_t U_0 + U_0\,\partial_X U_0 + V_1\,\partial_Y U_0 = \partial_Y^2 U_0 + \partial_t u_0 + u_0\,\partial_x u_0\,,$$
$$\partial_X U_0 + \partial_Y V_1 = 0$$

have to hold for $t, Y > 0$ with the boundary conditions

$$U_0 = V_1 = 0 \text{ for } Y = 0$$

and
$$U_0(t, X, Y = \infty) = u_0(t, x, 0) \quad \text{for all} \ t, \ X = x \,.$$

**Remarks**

1. Small curvatures at the boundary of the body can be neglected. For large curvatures additional terms would appear in the equations.
2. The error between the Navier–Stokes equations and the boundary layer equations in the boundary layer is, measured in a suitable norm, of order $\mathcal{O}(\varepsilon^{1/2})$.

## 6.7  Literature

A good survey on the analysis of partial differential equations is given by [37, 75]. For different classes of partial differential equations there exists specialized literature, like [49] for elliptic equations and [44, 120, 129] for the analysis of the Navier–Stokes equations. A kaleidoscope of applications of partial differential equations can be found in [93]. Applications of asymptotic analysis on partial differential equations can be found in [68, 77]. The homogenization of partial differential equations is discussed in detail in [23, 24, 71, 141]. Numerical methods to solve partial differential equations are introduced and analyzed in [56, 62, 74, 81, 97, 109]. For some equations specific numerical methods exist as the ones in [52] for the Navier–Stokes equations and the books [84, 87] for hyperbolic problems.

## 6.8  Exercises

**Exercise 6.1** We consider the following elliptic differential equation with Robin boundary conditions:

$$\begin{aligned}
-\nabla \cdot (\lambda \nabla u) &= f && \text{in} \ \Omega \,, \\
\lambda \, \nabla u \cdot n &= b - au && \text{on} \ \partial \Omega \,.
\end{aligned}$$

Here $\Omega$ is a bounded domain with smooth boundary and $\lambda$, $f$, $a$, $b$ are smooth functions with $\lambda \geq \lambda_0 > 0$ and $a \geq a_0 > 0$.

(a) Derive a minimization problem which is equivalent to this boundary value problem.
(b) Show that the functional $J$ to be optimized in a) is coercive in the following sense:

$$J(u) \geq c_1 \|u\|_{H^1(\Omega)}^2 - c_2 \quad \text{with} \ c_1, c_2 > 0 \,.$$

Use the following form of Poincaré's inequality

$$\int_{\Omega} |\nabla u|^2 \, dx + \int_{\partial\Omega} |u|^2 \, ds_x \geq c_P \|u\|_{H^1(\Omega)}^2 \, .$$

You do not have to prove this inequality.

(c) Is part (b) still true if one does not require $a \geq a_0 > 0$?

**Exercise 6.2** (*Existence of potentials*) Let $\Omega$ be a simply connected domain in $\mathbb{R}^3$ and let $f$ be a curl-free vector field. We want to prove that a potential $\varphi$ with $\nabla\varphi = f$ exists.

Let $a \in \Omega$ be arbitrary. We choose $\varphi(a) = 0$, as the property of being a potential does not change by adding a constant. For $b \in \Omega$ we choose a $C^{\infty}$–path

$$c : [0, 1] \to \Omega , \quad c(0) = a , \quad c(1) = b$$

from $a$ to $b$ and set

$$\varphi(b) := \int_0^1 f(c(t)) \cdot c'(t) \, dt \, .$$

Show that this definition is independent of the chosen path and it holds that $\nabla\varphi = f$.
*Hint*: Use Stokes' theorem.

**Exercise 6.3** (*Complex velocity potential I*) Outside of the ball $B_a(0)$ with radius $a > 0$ around 0 we define the complex potential

$$w(z) := U \left( z + \frac{a^2}{z} \right)$$

with a given $U \in \mathbb{R}$.

(a) Let $w = \varphi + i\psi$ with a real potential $\varphi$ and the stream function $\psi$. Determine $\varphi$ and $\psi$ in cylindrical coordinates $(r, \beta) \in [a, \infty) \times [0, 2\pi)$.
    In order to do so use the fact that $z = r \exp(i\beta) = r(\cos\beta + i \sin\beta)$.
(b) Compute the complex velocity $F$ and determine the real and imaginary part $v_1$ and $v_2$ in cylindrical coordinates.
(c) Now consider the boundary $\partial B_a(0)$. In which points is the modulus of $F$ the largest and the lowest. What property holds in these points for the pressure?
(d) Sketch the streamlines and the lines on which $\varphi$ is constant.
(e) Show that no forces act on the ball.

**Exercise 6.4** (*Complex velocity potential II*)

(a) Consider outside of $B_a(0)$ the complex potential

$$\tilde{w}(z) := \frac{\Gamma}{2\pi i} \ln z$$

with $\Gamma \in \mathbb{R}$ and solve for this $\tilde{w}$ (a) and (b) as in Exercise 6.3.

(b)  We now assume that the following potential is given

$$\widehat{w}(z) := w(z) + \widetilde{w}(z) = U\left(z + \frac{a^2}{z}\right) + \frac{\Gamma}{2\pi i}\ln z.$$

Determine all stagnation points with $\widehat{F} = 0$. Here $\widehat{F}$ is the complex velocity related to $\widehat{w}$.

*Hint*: It holds that $\widehat{v} = \nabla_x \widehat{\varphi}$. Hence it is sufficient to consider $\partial_r \widehat{\varphi} = 0$ and $\partial_\beta \widehat{\varphi} = 0$.

**Exercise 6.5**  (*Streamlines*) Let $F = v_1 - iv_2$ be a complex velocity with complex potential $w$ in a domain $\Omega \subset \mathbb{C}$. Furthermore, let $c : [0, 1] \to \Omega$ be a smooth embedded curve. Such a curve is called *segment* of a streamline $s$ if a reparameterization $\beta$ of the interval $[0, 1]$ to the domain of definition of $s$ exists, such that $s(\beta(\tau)) = c(\tau)$.

We assume that $F$ does not vanish on $c$. Prove that the following assertions are equivalent:

(i)   $v_1(c(\tau)) c_2'(\tau) = v_2(c(\tau)) c_1'(\tau)$ for $0 \leq \tau \leq 1$,
(ii)  $v(c(\tau)) = (v_1, v_2)(c(\tau))$ is parallel to $c'(\tau) = (c_1', c_2')(\tau)$ for $0 \leq \tau \leq 1$,
(iii) $c$ is a segment of a streamline,
(iv)  the stream function $\psi$ is constant on $c$.

**Exercise 6.6**  (*D'Alembert's paradox*) Show that the potential

$$\varphi(x) = \left(\frac{a^3}{2|x|^3} + 1\right) U \cdot x$$

in $\mathbb{R}^3 \setminus B_a(0)$, $a > 0$, leads to a divergence-free flow which is a solution of the stationary Euler equations. Here $U$ is the velocity for $|x| \to \infty$. Prove that the force acting on the ball

$$f = -\int_{\partial B_a(0)} \varphi n \, ds$$

is equal to zero.

**Exercise 6.7**  Let $u_-, u_+, w \in L_2(\Omega)$ with $u_- \leq u_+$ be given. We now define

$$K = \{v \in L_2(\Omega) \mid u_-(x) \leq v(x) \leq u_+(x) \quad \text{almost everywhere}\}.$$

We search for the solution $Pw$ of the minimization problem

$$\min\{\|v - w\|_{L_2(\Omega)} \mid v \in K\}.$$

Show that there exists a unique solution of the minimization problem and that $(v - Pw, w - Pw)_{L_2(\Omega)} \leq 0$ holds for all $v \in K$. The solution $Pw$ is called the orthogonal projection of $w$ onto $K$.

In addition to this formulate and prove a related result for general Hilbert spaces $H$. In doing so you should replace $L_2(\Omega)$ by $H$ and $K$ by a closed, convex set.

**Exercise 6.8** Show that the variational inequality (6.39) is equivalent to the statement that the mapping (6.40) has its minimum at $\bar{u}(x)$.

**Exercise 6.9** Let $\bar{u}, \bar{p}, u_-, u_+ : \overline{\Omega} \to \mathbb{R}$ be continuous and $u_-(x) < u_+(x)$ hold for all $x \in \overline{\Omega}$. Show that

$$\int_\Omega (\alpha \bar{u}(x) + \bar{p}(x))(u(x) - \bar{u}(x)) \, dx \geq 0$$

holds for all continuous $u : \overline{\Omega} \to \mathbb{R}$ with $u_-(x) \leq u(x) \leq u_+(x)$ if and only if for all $x \in \overline{\Omega}$ the following variational inequality holds

$$(\alpha \bar{u}(x) + \bar{p}(x))(v - \bar{u}(x)) \geq 0 \quad \text{for all} \quad v \in [u_-(x), u_+(x)].$$

Additional question: What is the corresponding statement if all functions are just square-integrable?

**Exercise 6.10** Formulate a minimum principle for the problem $(P_2)$ from Sect. 6.1.7 analogously to Theorem 6.13.

**Exercise 6.11** (a) Let $u(t, x)$ be the probability that a particle at time $t = 0, \tau, 2\tau, \ldots$ is at the lattice point $x = 0, \pm h, \pm 2h, \ldots$. We assume that at the next time step the particle moves with a probability $\alpha_1$ to the left and with a probability $\alpha_2$ to the right. The probability to remain at the original point is $(1 - \alpha_1 - \alpha_2)$. Which partial differential equation do we obtain to leading order in $\tau$ and $h$ if we choose $\alpha_1, \alpha_2$ such that

$$\frac{(\alpha_2 - \alpha_1)h}{\tau} = E \quad \text{and} \quad \frac{(\alpha_1 + \alpha_2)h^2}{2\tau} = D ?$$

*Hint*: Argue as at the end of Sect. 6.2.9.
(b) Generalize the approach at the end of Sect. 6.2.9 to the case of several space dimensions.

**Exercise 6.12** Let $V \leq -2$. Show that there exists no function $U : \mathbb{R} \to \mathbb{R}$, such that

$$U'' + V U' + U(1 - U) = 0,$$

$$\lim_{z \to -\infty} U(z) = 0, \quad \lim_{z \to \infty} U(z) = 1.$$

*Hint*: Multiply the differential equation with $U'(z)$.

**Exercise 6.13** Let $A$ be a real $(2 \times 2)$-matrix without real eigenvalues and $u + iv$, $u, v \in \mathbb{R}^2$ is assumed to be an eigenvector for the eigenvalue $\gamma + i\omega$.

(a) Show that the vectors $u$ and $v$ are linearly independent and that all solutions of $x' = Ax$ are given as

$$c_1 x_1(t) + c_2 x_2(t) , \quad c_1, c_2 \in \mathbb{R}$$

with

$$x_1(t) = e^{\gamma t} (\cos(\omega t) u - \sin(\omega t) v)$$
$$x_2(t) = e^{\gamma t} (\sin(\omega t) u + \cos(\omega t) v) .$$

(b) Sketch the phase portrait of the system of differential equations $x' = Ax$ for

$$A = \begin{pmatrix} -1 & -1 \\ 1 & -1 \end{pmatrix} .$$

**Exercise 6.14** Show the existence of a solution $(U, W)$ for (6.79), (6.80) with

$$\lim_{z \to -\infty} (U, W)(z) = (0, 0) , \quad \lim_{z \to \infty} (U, W)(z) = (1, 0) , \tag{6.137}$$

with the help of the following approach:

(a) Use continuity arguments to show that a $\overline{W} \in \mathbb{R}_+$ exists, such that the initial value problem

$$\frac{dW}{dU}(U) = -V - \frac{U(1 - U)}{W} , \quad W(\tfrac{1}{2}) = \overline{W} \tag{6.138}$$

has a solution $W : [\tfrac{1}{2}, 1] \to [0, \infty)$ with $W(1) = 0$.
(b) Show that this solution $W$ can be extended to a solution $W^* : [0, 1] \to [0, \infty)$ with $W^*(0) = 0$.
(c) Show that the solution of

$$U'(z) = W^*(U(z)) , \quad U(0) = \tfrac{1}{2}$$

now yields a solution $(U(z), W^*(U(z)))$ of (6.79), (6.80) with (6.137).

**Exercise 6.15** Let $u : [0, \infty) \times \Omega \to \mathbb{R}$ be a smooth solution of the initial value problem

$$\begin{aligned} \partial_t u &= d \, \Delta u & \text{for } x \in \Omega , \, t \geq 0 , \\ \nabla u \cdot n &= 0 & \text{for } x \in \partial\Omega , \, t \geq 0 , \\ u(0, x) &= u_0(x) & \text{for } x \in \Omega . \end{aligned}$$

Show that:

(a) $\dfrac{d}{dt} \displaystyle\int_{\Omega} u(t, x)\, dx = 0.$

(b) There exists a constant $\bar{c} > 0$, such that

$$\int_{\Omega} (u(t, x) - \bar{c})^2\, dx \to 0 \quad \text{as} \quad t \to \infty.$$

*Hint:* There exists a constant $C_0 > 0$, such that for all functions $v \in H^1(\Omega)$ with $\int_{\Omega} v(x)\, dx = 0$ the following variant of the Poincaré inequality holds

$$\int_{\Omega} v^2(x)\, dx \le C_0 \int_{\Omega} |\nabla v(x)|^2\, dx . \tag{6.139}$$

This assertion is then true in particular for functions $v \in C^1(\overline{\Omega})$ with $\int_{\Omega} v(x)\, dx = 0$. You do not need to prove the inequality (6.139).

**Exercise 6.16** Determine for the Schnakenberg system (6.93), (6.94) with periodic boundary conditions the positive stationary solutions $(U_0, V_0)$. Derive necessary conditions for a situation in which a Turing instability occurs. How large do we have to choose $\Omega = (0, a)$ in order to obtain a Turing instability in $\Omega$?

**Exercise 6.17** Prove that for periodic solutions of the Cahn–Hilliard equation (6.99) on $\Omega = [0, \ell]^d$, $\ell > 0$ the mass conservation (6.100) and the inequality for the energy (6.101) hold. Are there other boundary conditions for which (6.100) and (6.101) are true?

**Exercise 6.18** Compute the entropy solutions for the hyperbolic conservation law

$$\partial_t u + u\, \partial_x u = 0 \quad \text{on } \mathbb{R}_+ \times \mathbb{R}$$

with the initial conditions

(a) $u(0, x) = \begin{cases} 0 & \text{for } x < -1, \\ x + 1 & \text{for } -1 \le x \le 1, \\ 2 & \text{for } x > 1, \end{cases}$

(b) $u(0, x) = \begin{cases} 2 & \text{for } x < -1, \\ 1 - x & \text{for } -1 \le x \le 1, \\ 0 & \text{for } x > 1. \end{cases}$

**Exercise 6.19** Compute for the following differential equations all possible solutions $u = u(t, x)$, $t > 0$, $x \in \mathbb{R}$, which are of the type of a traveling wave, i.e., $u(t, x) = w(x - vt)$ with a suitable function $w : \mathbb{R} \to \mathbb{R}$ and a suitable propagation velocity $v \in \mathbb{R}$:

(a) $\partial_t u + \partial_x u = 0$,

(b) $\partial_t u - \partial_x^2 u = 0$,

(c) $\partial_t^2 u - \partial_x^2 u = 0$,
(d) $\partial_t^2 u + \partial_x^2 u = 0$.

For which equations does one obtain solutions for *all* choices of the function $w$?

**Exercise 6.20** Construct a sequence $u_n : [0, \infty) \times [0, 1] \to \mathbb{R}$ of solutions of the equation $\partial_t u + \partial_{xx} u = 0$ with $u(t, 0) = u(t, 1) = 0$ for all $t > 0$, such that

$$\lim_{n \to \infty} \left( \sup_{x \in [0,1]} |u_n(0, x)| \right) = 0$$

and

$$\lim_{n \to \infty} \left( \sup_{x \in [0,1]} |u_n(t, x)| \right) = \infty \quad \text{for all} \quad t > 0.$$

This shows that solutions of the problem $\partial_t u + \partial_{xx} u = 0$ with $u(t, 0) = u(t, 1) = 0$ for all $t > 0$ do not depend continuously on the initial data.
*Hint*: Make a separation ansatz $u(t, x) = v(t) w(x)$.

# Chapter 7
# Free Boundary Problems

Many applications in the sciences and in technology lead to problems, in which the geometry of the domain, on which an equation has to be solved, is a priori unknown. If a partial differential equation has to be solved in a domain, from which a part of the boundary is unknown, then one speaks of a *free boundary problem*. In addition to the usual boundary conditions, which are necessary to solve a partial differential equation, in this case further conditions at the free boundary have to be imposed. Free boundary problems appear among other cases in the following areas: melting and solidification phenomena (Stefan problems), obstacle problems for elastic membranes, contact problems with elastic deformation, growth of tumors, flows with free surfaces and the modeling of financial derivative products.

In this chapter we want to discuss some classical free boundary problems, which have initiated many of the developments in this area. We treat obstacle problems for elastic membranes as contact problems for elastic bodies and we will see that variational approaches are also very successful for free boundary problems. Additionally we discuss the Stefan problem, which among others describes melting processes and crystal growth, as an example for a time-dependent free boundary problem. Further free boundary problems, which will be discussed in this chapter, stem from the modeling of flows in porous media and of flows with free surfaces.

In Chap. 6 we have treated boundary value problems for elliptic differential equations. For a fixed domain $\Omega \subset \mathbb{R}^d$ with boundary $\Gamma = \partial\Omega$ and a given function $f : \Omega \to \mathbb{R}$ we have considered the following problem: Find a function $u : \overline{\Omega} \to \mathbb{R}$, such that

$$-\Delta u = f \quad \text{in } \Omega\,,$$
$$u = 0 \quad \text{on } \Gamma = \partial\Omega\,.$$

We obtain a free boundary problem if the domain $\Omega$ and hence the boundary $\Gamma = \partial\Omega$ is a priori unknown. In such a case for example a function $f : \mathbb{R}^d \to \mathbb{R}$ is given, and we are looking for a set $\Gamma$, which is the boundary of a domain $\Omega \subset \mathbb{R}^d$ and a function $u : \overline{\Omega} \to \mathbb{R}$, such that

© Springer International Publishing AG 2017
C. Eck et al., *Mathematical Modeling*, Springer Undergraduate
Mathematics Series, DOI 10.1007/978-3-319-55161-6_7

$$\begin{cases} -\Delta u = f & \text{in } \Omega\,, \\ \quad u = 0 & \text{on } \Gamma = \partial\Omega\,, \\ \nabla u \cdot n = 0 & \text{on } \Gamma = \partial\Omega\,. \end{cases} \tag{7.1}$$

Here $n$ denotes the exterior unit normal on $\partial\Omega$. As the boundary is a further degree of freedom in the problem formulation, we have to impose a further boundary condition to determine this degree of freedom.

In this chapter we will assume some knowledge about hypersurfaces in $\mathbb{R}^d$ at several places. In particular the familiarity with the following notions and concepts is assumed: tangent space, tangential derivative, and integration on manifolds. In this context we refer to advanced courses in analysis or to the book of Morgan [100].

## 7.1　Obstacle Problems and Contact Problems

We consider an elastic membrane, which can be described as the graph of a function $u : \Omega \to \mathbb{R}$, where $\Omega \subset \mathbb{R}^d$ is a domain, and which is firmly fixed at the boundary. The latter means that $u(x) = u_0(x)$ for $x \in \partial\Omega$ with a given function $u_0 : \partial\Omega \to \mathbb{R}$. Furthermore we assume that forces act on the membrane. Let these forces be given by a function $f : \Omega \to \mathbb{R}$, which denotes how strong the forces act from below in the vertical direction. Now we want to denote the potential energy of the membrane: to enlarge the surface of the membrane, energy must be spent such that the potential energy can be described up to an additive constant by the term

$$\int_\Omega \left[ \lambda\left(\sqrt{1 + |\nabla u|^2} - 1\right) - fu \right] dx\,.$$

Here $\lambda > 0$ is a parameter which denotes the density of the surface energy. The term $\int_\Omega fu\,dx$ describes the work done by the exterior forces. If we assume that $|\nabla u|$ is small, then after Taylor expansion and omission of terms of higher order we can instead consider the energy

$$\int_\Omega \left( \frac{\lambda}{2}|\nabla u|^2 - fu \right) dx$$

(compare Sect. 4.7 for the case $d = 1$). Now we are looking for states of minimal potential energy, as these are observed in nature. This task leads to the minimization problem

$$\min\left\{ \int_\Omega \left( \frac{\lambda}{2}|\nabla u|^2 - fu \right) dx \mid u \in V \right\} \tag{7.2}$$

with $V = \{v \in H^1(\Omega) \mid v = u_0 \text{ on } \partial\Omega\}$. We have treated this problem already in Sect. 6.1.

If the motion of the membrane is constrained from below by an obstacle which is given by a function $\psi : \Omega \to \mathbb{R}$ then the problem

$$\min \left\{ \int_\Omega \left( \frac{\lambda}{2} |\nabla u|^2 - fu \right) dx \mid u \in V \text{ and } u \geq \psi \right\} \tag{7.3}$$

is obtained. In the following we assume $\psi \in L_2(\Omega)$ and define the set of admissible functions as

$$K = \{ v \in H^1(\Omega) \mid v = u_0 \text{ on } \partial\Omega \text{ and } v \geq \psi \},$$

where $u_0 \in H^1(\Omega)$ defines the boundary values and $u_0 \geq \psi$ is also assumed.
   Furthermore let

$$E(v) := \int_\Omega \left( \frac{\lambda}{2} |\nabla v|^2 - fv \right) dx$$

and $u \in K$ is assumed to be an absolute minimum of $E$ on $K$, i.e.,

$$E(u) = \min_{v \in K} E(v),$$

see Fig. 7.1. As $K$ is convex, for every $v \in K$ also the convex combination $(1 - \varepsilon)u + \varepsilon v = u + \varepsilon(v - u)$ lies in $K$ for $\varepsilon \in [0, 1]$. Therefore it holds that $E(u + \varepsilon(v - u)) \geq E(u)$ and we obtain for all $v \in K$

$$0 \leq \frac{d}{d\varepsilon} E(u + \varepsilon(v - u))|_{\varepsilon=0} = \int_\Omega \left( \lambda \nabla u \cdot \nabla(v - u) - f(v - u) \right) dx . \tag{7.4}$$

We call a function $u \in K$ a solution of the *variational inequality* for $E$ and $K$, if the inequality (7.4) holds true for all $v \in K$. We have the following lemma (see also [4]).

**Lemma 7.1** *Let* $\lambda \geq 0$. *Then the absolute minima of* $E$ *on* $K$ *coincide with the solutions of the variational inequalities* (7.4) *of* $E$ *on* $K$.

*Proof* We have already shown that every absolute minimum is a solution of the variational inequality. Now let $u \in K$ be a solution of the variational inequality. Then we have

$$E(v) = E(u) + \int_\Omega \left( \lambda \nabla u \cdot \nabla(v - u) - f(v - u) \right) dx + \int_\Omega \frac{\lambda}{2} |\nabla(v - u)|^2 dx$$

$$\geq E(u) + \int_\Omega \frac{\lambda}{2} |\nabla(v - u)|^2 dx$$

for all $v \in K$. As $\lambda$ is nonnegative we can conclude the assertion.                    □

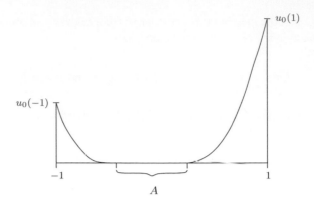

**Fig. 7.1**  Solution of an obstacle problem for a negative $f$, $\Omega = (-1, 1)$, and $\psi \equiv 0$. The active set is given by $A = \{x \in (-1, 1) \mid u(x) = 0\}$

We may also allow the coefficient $\lambda$ to depend on $x$. In the following we always assume that constants $\lambda_0$ and $\lambda_1$ exist such that $0 < \lambda_0 \leq \lambda(x) < \lambda_1 < \infty$ for all $x \in \Omega$. Under sufficient smoothness assumptions on $\lambda$, $f$, $\psi$, $u_0$, and $\partial\Omega$ it can be shown (see [42, 79]) that solutions $u$ of the above mentioned variational inequality are in $H^2(\Omega)$. This means that also second derivatives exist in a weak sense and these are square integrable. In fact it holds true that the second derivatives lie in $L_\infty(\Omega)$ and that the first derivatives exist in a classical sense and are Lipschitz continuous. Then from the variational inequality and the fact that $u$ and $v$ attain the same boundary data it follows that

$$\int_\Omega [\nabla \cdot (\lambda\nabla u) + f](v - u)\, dx \leq 0 \quad \text{for all } v \in K\,. \tag{7.5}$$

If $u$ and $\psi$ are continuous, then the set $N := \{x \in \Omega \mid u(x) > \psi(x)\}$ is open. For functions $\zeta \in C_0^\infty(N)$ and $\varepsilon$ small enough the functions $v = u \pm \varepsilon\zeta$ lie in $K$. If we plug these $v$ in (7.5), we obtain

$$\int_\Omega [\nabla \cdot (\lambda\nabla u) + f] \cdot \zeta\, dx = 0 \quad \text{for all } \zeta \in C_0^\infty(N)$$

and therefore, as $\zeta \in C_0^\infty(N)$ can be chosen arbitrarily,

$$\nabla \cdot (\lambda\nabla u) + f = 0 \quad \text{in } N\,.$$

If $\zeta \in C_0^\infty(\Omega)$ has a support which is not fully contained in $N$, then in general $v = u + \zeta \in K$ only holds true if $\zeta \geq 0$. We obtain

$$\int_\Omega [\nabla \cdot (\lambda\nabla u) + f] \cdot \zeta\, dx \leq 0 \quad \text{for all } \zeta \in C_0^\infty(\Omega) \text{ with } \zeta \geq 0$$

and therefore
$$\nabla \cdot (\lambda \nabla u) + f \leq 0 \quad \text{for almost every } x \in \Omega .$$

As solutions of the variational inequality in the case of smooth data lie in $C^1(\Omega)$, we have in this case $u = \psi$ and $\nabla u = \nabla \psi$ on $A := \Omega \setminus N$. In particular, if

$$\Gamma := \partial N \cap \partial A$$

possesses a unit normal $n$ in particular it follows that

$$u = \psi, \quad \lambda \nabla u \cdot n = \lambda \nabla \psi \cdot n \text{ on } \Gamma .$$

In summary we have obtained

$$\left.\begin{aligned} \nabla \cdot (\lambda \nabla u) + f &\leq 0 \\ u &\geq \psi \\ (\nabla \cdot (\lambda \nabla u) + f)(u - \psi) &= 0 \end{aligned}\right\} \quad \text{in} \quad \Omega ,$$

$$\begin{aligned} u &= u_0 && \text{on } \partial\Omega , \\ u &= \psi && \text{on } \Gamma , \\ \lambda \nabla u \cdot n &= \lambda \nabla \psi \cdot n && \text{on } \Gamma . \end{aligned}$$

This formulation of the obstacle problem is also called the *complementarity formulation*. As the obstacle is active in $A := \Omega \setminus N$ we call $A$ the *active set* and $N$ the *inactive* ("non-active") set. Some authors also call $A$ the coincidence set. The common boundary $\Gamma = \partial A \cap \partial N$ is denoted as the free boundary as it is unknown a priori.

If we define a so called *Lagrange multiplier* $\mu \subset L_2(\Omega)$ by

$$\mu(x) = \begin{cases} 0 & \text{if } x \in N , \\ -\nabla \cdot (\lambda(x)\nabla u(x)) - f(x) & \text{if } x \in A , \end{cases}$$

we can reformulate the first three conditions of the complementarity system as follows:

$$\left.\begin{aligned} \nabla \cdot (\lambda \nabla u) + f + \mu &= 0 \\ u &\geq \psi \\ \mu &\geq 0 \\ \mu(u - \psi) &= 0 \end{aligned}\right\} \quad \text{in} \quad \Omega .$$

In the following we want to derive a reformulation of the obstacle problem as a *free boundary problem* for the operator $\nabla \cdot (\lambda \nabla u)$.

We are looking for a domain $N \subset \Omega$, a free boundary $\Gamma = \partial N \cap \Omega$, and a function $u : N \to \mathbb{R}$ such that

$$\nabla \cdot (\lambda \nabla u) + f = 0 \qquad \text{in } N \, ,$$
$$u = u_0 \qquad \text{on } \partial \Omega \cap \overline{N} \, ,$$
$$u = \psi, \; \lambda \nabla u \cdot n = \lambda \nabla \psi \cdot n \qquad \text{on } \Gamma \, ,$$
$$u \geq \psi \qquad \text{in } N \, ,$$
$$\nabla \cdot (\lambda \nabla \psi) + f \leq 0 \qquad \text{in } \Omega \setminus N \, .$$

Therefore the obstacle problem can be written as

- *a minimization problem,*
- *a variational inequality,*
- *a complementarity problem,*
- *a free boundary problem*

or using

- *Lagrange multipliers.*

In Exercise 7.2 we will develop a further formulation which does not need inequalities.

If we set $\psi \equiv 0$, $\lambda \equiv 1$ and $\Omega = \mathbb{R}^d$, the solution of the obstacle problem is just the solution of the problem in (7.1) formulated at the beginning of the chapter. Here $\Omega$ plays the role of the inactive set and $\mathbb{R}^d \setminus \Omega$ the role of the active set. To this end the function $u$ has to be extended by $u = 0$ in $\mathbb{R}^d \setminus \Omega$. If we consider the necessary conditions for this case we see that all requirements formulated in (7.1) are fulfilled.

For the analysis of an obstacle problem the formulation (7.3) as an optimization problem and (7.4) as a variational inequality are particularly helpful. The uniqueness of a solution can be derived easily from the variational inequality (7.4): If $u_1$ and $u_2$ are two solutions, setting $v = u_2$ in the variational inequality for the solution $u_1$ and $v = u_1$ in the variational inequality for the solution $u_2$ and adding the results leads to

$$0 \leq \int_{\Omega} \left( \lambda \, \nabla u_1 \cdot \nabla (u_2 - u_1) + \lambda \, \nabla u_2 \cdot \nabla (u_1 - u_2) \right) dx$$
$$= - \int_{\Omega} \lambda |\nabla (u_1 - u_2)|^2 \, dx \, .$$

Because of $\lambda(x) \geq \lambda_0 > 0$ we conclude $\nabla u_1 = \nabla u_2$ in $\Omega$. As $u_1$ and $u_2$ attain the same values at the boundary of $\Omega$ it follows that $u_1 = u_2$. The existence of solutions can be shown by applying the direct method of the calculus of variations to (7.3), for details we refer to [42, 79].

A further problem class which leads to variational inequalities of very similar structure are *contact problems* for elastic bodies. We consider a body made from a linear elastic material, whose deformation is described by a displacement field $u$. The material properties are given by Hook's law

$$\sigma_{ij}(u) = \sum_{k,\ell=1}^{d} a_{ijk\ell} \, \varepsilon_{k\ell}(u)$$

with a linearized strain tensor $\varepsilon_{ij}(u) = \frac{1}{2}(\partial_i u_j + \partial_j u_i)$. If an exterior force is exerted on the body described by a volume force density $f$, and if the displacements at the boundary $\partial\Omega$ of the body are given by a function $u_0$, then the state of equilibrium is a solution of the minimization problem

$$\min\{J(u) \mid u - u_0 \in H_0^1(\Omega)^d\}$$

with the functional

$$J(u) = \int_\Omega \left(\tfrac{1}{2}a(u, u) - f \cdot u\right) dx,$$

where

$$a(u, v) = \sigma(u) : \varepsilon(v) = \sum_{i,j,k,\ell=1}^{d} a_{ijk\ell}\, \varepsilon_{ij}(u)\, \varepsilon_{k\ell}(v).$$

Here it is assumed that $u_0$ is extended to the whole domain $\Omega$ and that $u_0 \in H^1(\Omega)^d$ holds.

In the case of a *contact problem* at a part $\Gamma_C$ of the boundary an obstacle is given such that the component $u_\nu$ in direction of the obstacle is bounded by a given function $g$,

$$u_\nu := u \cdot \nu \le g \quad \text{on } \Gamma_C.$$

Here $\nu$ denotes a unit vector which describes the direction of the smallest distance to the obstacle. Often $\nu$ is also chosen as the normal vector on $\partial\Omega$. On the remaining part $\Gamma_U = \partial\Omega \setminus \Gamma_C$ the displacement field $u = u_0$ is prescribed. The static equilibrium of elastic bodies then is given by a solution of the minimization problem

$$\min \left\{ J(u) \mid u \in K \right\} \tag{7.6}$$

with *the same* functional $J$ but with another set of admissible functions

$$K := \left\{ u \in H^1(\Omega)^d \mid u = u_0 \text{ on } \Gamma_U,\ u_\nu \le g \text{ on } \Gamma_C \right\}.$$

This set, as it is easily seen, is convex. If $u$ is a solution of (7.6) and $v \in K$, then for every $\delta \in (0, 1)$ also $u + \delta(v - u) \in K$ and therefore we obtain with the same consideration as for the obstacle problem

$$0 \le \frac{d}{d\delta} J(u + \delta(v - u))\big|_{\delta=0} = \int_\Omega \left(\sigma(u) : \varepsilon(v - u) - f \cdot (v - u)\right) dx. \tag{7.7}$$

This is the formulation of the contact problem in form of a *variational inequality*. It can be shown with a proof analogous to the one of Lemma 7.1 that the formulations (7.6) and (7.7) are equivalent.

Now we derive an equivalent formulation in form of a complementarity problem. For $w \in C_0^\infty(\Omega)^d$ obviously we have $u \pm w \in K$, therefore it follows that

$$\int_\Omega \left( \sigma(u) : \varepsilon(w) - f \cdot w \right) dx = 0$$

and after integration by parts

$$\int_\Omega \left( -\nabla \cdot \sigma(u) - f \right) \cdot w \, dx = 0 .$$

As this holds true for arbitrary $w \in C_0^\infty(\Omega)^d$, we obtain the system of equations of static linear elasticity

$$-\nabla \cdot \sigma(u) = f \quad \text{in } \Omega . \tag{7.8}$$

Furthermore we obtain by integration by parts for an arbitrary test function $v \in K$

$$0 \leq \int_\Omega \left( -\nabla \cdot \sigma(u) - f \right) \cdot (v - u) \, dx + \int_\Gamma \sigma^n(u) \cdot (v - u) \, ds_x = \int_{\Gamma_C} \sigma^n(u) \cdot (v - u) \, ds_x$$

with normal stress $\sigma^n(u) = \sigma(u)n$ at the boundary. Here $n$ denotes the normal vector on $\partial\Omega$. The integral over $\Gamma_U$ is zero because of $u = v = u_0$ on $\Gamma_U$. If we now use a test function $v = u \pm w$ with $w_\nu = 0$ on $\Gamma_C$, then we obtain

$$\int_{\Gamma_C} \sigma_\tau^n(u) \cdot w \, ds_x = 0 ,$$

where $\sigma_\tau^n = \sigma^n - \sigma^n \cdot \nu \, \nu$ is the part of $\sigma^n(u)$ which is orthogonal to $\nu$. As $w$ such that $w_\nu = 0$ can be chosen arbitrarily, it follows that

$$\sigma_\tau^n = 0 \quad \text{on } \Gamma_C . \tag{7.9}$$

For an arbitrary test function we obtain by using $\sigma_\nu^n := \sigma^n \cdot \nu$

$$\int_{\Gamma_C} \sigma_\nu^n(u) \cdot (v_\nu - u_\nu) \, ds_x \geq 0 .$$

If we now choose $\zeta$ such that $\zeta \cdot \nu \leq 0$ on $\Gamma_C$ and set $v = u + \zeta$, then we have $v \cdot \nu \leq g$ and we obtain

$$\sigma_\nu^n(u) \cdot \zeta_\nu \geq 0 \quad \text{on } \Gamma_C$$

for all $\zeta$ such that $\zeta \cdot \nu \leq 0$. In points where $u_\nu < g$ holds, $\zeta_\nu$ can have an arbitrary sign and $\sigma_\nu^n(u) = 0$ holds. At points where $u_\nu = g$ holds, only $\sigma_\nu^n(u) \leq 0$ follows. Altogether the complementary condition

$$u_\nu \le g, \quad \sigma_\nu^n \le 0, \quad \sigma_\nu^n(u_\nu - g) = 0 \tag{7.10}$$

holds. Hence, the contact problem is given by equations (7.8), (7.9), the complementary condition (7.10), and the boundary condition $u = u_0$ on $\Gamma_U$.

The uniqueness of a solution of the contact problem can be shown with the same proof as for the obstacle problem using Korn's inequality, see Exercise 7.4. The existence of a solution follows by using the direct method of the calculus of variations.

## 7.2   Free Boundaries in Porous Media

In this section we will consider flow and transport processes in porous media. The most important example of a *porous medium* is the soil. It is a *multiphase system*, consisting of a solid skeleton, the so-called *matrix*, and of fluids which fill the voids, the so-called *pore space*. The fluid phases are present as a liquid phase (the *soil water*) and as a gaseous phase (the *soil air*). Because of the inherent microscopic heterogeneity of a porous medium at least two different levels of consideration have to be distinguished. The first one is the *microscopic level* or microscale, in which an entity consisting of several connected pores defines the domain of consideration. Here continuum mechanics can be applied by, e.g., formulating the laws of fluid mechanics. This level of consideration, however is not sufficient, as neither the pore geometry with its highly irregular boundaries between the pore space and the pore matrix can be described in a general deterministic sense, nor do the quantities appearing in such a microscopic description can be related to quantities which can be measured. The measured quantities rather are to be interpreted as averages of corresponding microscopic quantities over small volumes, which always contain a part of the pore space and of the pore matrix. To achieve a coincidence of modeled and measured quantities, the transition to a second, *macroscopic* level (macroscale) by means of this averaging process has to be performed. This can be done under various assumptions and with various rigorousness by means of *volume averaging* or by means of homogenization (see Sect. 6.1.6). The now appearing quantities have to interpreted as averages over a so-called *representative elementary volume* (REV), whose size is small in relation to the total macroscopic domain, but is large compared to the characteristic size of the microscale as the size of the grain or of the volume of a pore.

A macroscopic model for flows in porous media is *Darcy's law*

$$v = -K\nabla(p + G) \tag{7.11}$$

with the average velocity $v$, the so-called *Darcy velocity*, the pressure $p$, and the conductivity tensor $K$, which in general is a symmetric, positive-definite matrix, and a gravitational potential

$$G = \varrho g x_3 ,$$

where $x = (x_1, x_2, x_3)$ denotes the spatial coordinates. The average velocity $v$ can be interpreted as a surface-specific flow rate, more precisely $v \, \Delta A \, \Delta t$ denotes the volume of fluid which passes through a surface orthogonal to the flow direction with surface area $\Delta A$ and in a time interval of length $\Delta t$. Instead of $v$ often the notation $q$ is used. For a homogeneous, isotropic porous medium $K$ is a scalar, which in the following will be denoted by $\lambda$. Darcy's law has been discovered already in the 19th century from experimental data. It also can be derived in a rigorous mathematical fashion by means of homogenization from a system of differential equations to describe a flow at the scale of the pore structure, for example this has been worked out in [71, Chaps. 1 and 3] for the Stokes equations as a model for flow in a pore geometry.

A further heuristic justification can be deduced from the Hagen-Poiseuille law for flow in a pipe, which has been derived in Exercise 5.21. In an idealized sense a porous medium can be perceived more or less as a uniform network of pipes, where one third of the pipes is oriented in one of the three coordinate directions. According to the Hagen-Poiseuille law the flow rate through the pipes is proportional to the ratio of pressure difference and length. If one assumes a continuous differentiable distribution of pressure in a porous medium, then for the flow rate $q_j$ in coordinate direction $j$ we just obtain the relation

$$q_j = -k_j \, \partial_j (p + G)$$

with a constant $k_j$, which depends on the pore geometry and the viscosity of the fluid. In a pore geometry uniformly distributed in all directions the parameter $k_j$ is not dependent on the coordinate direction $j$, and the flow rate $q_j$ in direction $j$ is proportional to the component of the velocity in direction $j$. In this way we can conclude Darcy's law (7.11). As the characteristic length of the pore structure is very small, and the pore diameter enters the Hagen–Poiseuille law with its fourth power, $K$ and therefore also the velocity $v$ are very small. This justifies the usage of the Stokes equations as a model for flow in a pore geometry even for the flow of gases, as the Reynolds number $LV/\eta$ with characteristic length $L$, characteristic velocity $V$, and kinematic viscosity $\eta$ is a small number even for small viscosities.

We now consider the flow of a *gas* in a porous medium and neglect the gravitational potential. Neglecting variations in temperature the classical equations of state for gases are given by

$$p = k\varrho^\gamma, \quad k > 0, \quad \gamma > 0,$$

where $\varrho$ denotes the density. Using the mass conservation law

$$\partial_t \varrho + \nabla \cdot (\varrho v) = 0$$

and Darcy's law $v = -\lambda \nabla p$ we obtain the equation

$$\partial_t \varrho - \lambda k \nabla \cdot (\varrho \nabla \varrho^\gamma) = 0.$$

If we choose a normalized density $u = c\varrho$ with $c^\gamma = \lambda k \frac{\gamma}{\gamma+1}$ as the function to be determined, we obtain

$$\partial_t u - \Delta u^m = 0, \quad m = \gamma + 1 > 1. \tag{7.12}$$

The Eq. (7.12) is called the *porous medium equation* and we want to show that this equation leads to a free boundary problem. We are looking for a solution, which is invariant under simple dilation transformations. Analogously to Sect. 6.2.7 which is dealing with invariant transformations of the heat equation, we make the ansatz

$$u(t, x) = t^{-\alpha} U(t^{-\beta} x) \quad \text{for } x \in \mathbb{R}^n, \ t > 0,$$

with $\alpha, \beta > 0$.

We want to obtain solutions which conserve the total mass and therefore require

$$\int_{\mathbb{R}^d} t^{-\alpha} U(t^{-\beta} x) \, dx = \int_{\mathbb{R}^d} U(x) \, dx \quad \text{for all } t > 0.$$

From the transformation formula we conclude $t^{-\alpha+d\beta} = 1$ and therefore

$$\alpha = d\beta.$$

If we set $y = t^{-\beta} x$ and denote the gradient with respect to $y$ by $\nabla_y$, then we obtain

$$\nabla u^m(t, x) = t^{-\alpha m} t^{-\beta} \nabla_y U^m(t^{-\beta} x),$$
$$\Delta u^m(t, x) = t^{-\alpha m} t^{-2\beta} \Delta_y U^m(t^{-\beta} x),$$
$$\partial_t u(t, x) = (-\alpha) t^{-\alpha-1} U(t^{-\beta} x) + t^{-\alpha}(-\beta) t^{-\beta-1} x \cdot \nabla_y U(t^{-\beta} x).$$

This means that $u(t, x) = t^{-\alpha} U(t^{-\beta} x)$ fulfills the porous medium equation if and only if

$$0 = \alpha t^{-(\alpha+1)} U(y) + \beta t^{-(\alpha+1)} y \cdot \nabla_y U(y) + t^{-(\alpha m+2\beta)} \Delta_y U^m(y).$$

This equation can only be correct for all $t > 0$ if all $t$-powers coincide. Therefore we require

$$\alpha + 1 = \alpha m + 2\beta.$$

If we are now looking for a radial solution $U(y) = v(|y|)$, then we can conclude from the identities

$$\nabla_y U(y) = v'(|y|) \frac{y}{|y|}, \quad \Delta_y U^m(y) = (v^m)''(|y|) + \frac{d-1}{|y|} (v^m)'(|y|),$$

and $\alpha = d\beta$, using the notation $r := |y|$

$$0 = \beta d\, v(r) + \beta r\, v'(r) + (v^m)''(r) + \frac{d-1}{r}(v^m)'(r).$$

After multiplication by $r^{d-1}$ we obtain

$$0 = \beta(r^d v)' + (r^{d-1}(v^m)')'.$$

The integration of this equation leads to existence of a constant $a$ such that

$$a = \beta r^d v + r^{d-1}(v^m)'.$$

As we want to obtain solutions $U$ which are smooth at the origin, necessarily $a = 0$ must hold and therefore we obtain

$$0 = \beta r v + (v^m)'.$$

Because of $(v^m)' = \frac{m}{m-1} v(v^{m-1})'$,

$$(v^{m-1})' = -\frac{m-1}{m}\beta r$$

follows. An elementary integration leads to a constant $b > 0$, such that

$$v^{m-1}(r) = b - \frac{m-1}{2m}\beta r^2,$$

as long as $v$ is positive. We extend this solution by zero and obtain

$$v(r) = \left(b - \frac{m-1}{2m}\beta r^2\right)_+^{1/(m-1)}.$$

Here we use the notation

$$(a)_+ := \max(a, 0).$$

We take into account only the positive part of the bracket to guarantee the nonnegativity of the solution.

It is easy to verify that the resulting function

$$u(t, x) = \frac{1}{t^\alpha}\left(b - \frac{m-1}{2m}\beta\frac{|x|^2}{t^{2\beta}}\right)_+^{1/(m-1)} \tag{7.13}$$

fulfills the porous medium equation in the distributional sense (see Exercise 7.5). The function (7.13) is called the *Barenblatt solution* and it has a compact support.

The solution shows that the porous medium equation has solutions with finite speed of propagation. This is in contrast to the situation for the heat equation where we have seen that solutions for initial data with compact support are strictly positive for all positive times. The porous medium equation however allows for solutions with compact support, as the equation for $u$ and correspondingly for $\varrho$ degenerates as $u, \varrho \to 0$. This is reflected in the fact that the flux $-\nabla u^m = -m\, u^{m-1}\, \nabla u$ vanishes for $u \to 0$, as the corresponding solution dependent diffusion coefficient $m\, u^{m-1}$ converges to zero. "Exceptional" behavior of the solution $u$ is to be expected at most at $u = 0$. Because of the diffusion coefficient vanishing at $u = 0$ one also speaks of *slow diffusion* in the context of such equations.

For a general solution $u$ of (7.12) we denote by

$$\Gamma(t) = \partial\{x \in \mathbb{R}^d \mid u(t, x) > 0\}$$

the free boundary of the porous medium equation. The Barenblatt solution has the free boundary

$$\Gamma(t) = \partial B_{r(t)}(0) \quad \text{with} \quad r(t) = t^\beta \left(\frac{2m}{m-1}\frac{b}{\beta}\right)^{1/2}.$$

The importance of particular solutions with finite speed of propagation lies in the fact that by a comparison of a general solution with an appropriate particular solution the property of finite speed of propagation can also be verified for general solutions. To do so the differential equation has to fulfill a *comparison principle*, saying that a (pointwise) order of the initial and the boundary data implies a corresponding order of solutions. For the heat equation (also with the convective part) such a comparison principle follows immediately from the maximum principle (see Sect. 6.2.4). Also for (7.12) and the other differential equations to be treated in this section such comparison principles are valid, and the properties of particular solutions with respect to finite speed of propagation in principle show the general picture. For solutions of (7.12) for arbitrary initial data with compact support we have that the support spreads with finite velocity, and that the velocity of the support for large time has the same temporal asymptotics as the support of the Barenblatt solution. Additionally the decay behavior with respect to the $L_\infty$-norm corresponds to the one of the Barenblatt solution.

To avoid a discussion of the involved existence and regularity theory (which is not yet completed for the more general to be discussed differential equations for $d > 1$), in the following we only consider particular solutions. A type even simpler than (7.13) is given by traveling waves (for $d = 1$ or for planar spreading, respectively) as they have been already discussed in the form of wave fronts in Sect. 6.2.10. For (7.12) (and $d = 1$) the ansatz $u(t, x) = U(x - Vt)$ first leads to

$$-VU' - (U^m)'' = 0$$

and taking into account

$$\lim_{x \to \infty} u(t, x) = u_+ = 0 \tag{7.14}$$

after integration the equation

$$VU + (U^m)' = 0 \tag{7.15}$$

arises. Wave fronts, i.e., solutions which are also bounded for $x \to -\infty$, can at most
exist for $V = 0$ but for the mentioned usage as comparison functions also *semi-wave
fronts* are sufficient, which are monotone traveling waves which fulfill (7.14). Due
to the interpretation of $U$ as a density we restrict ourself to nonnegative solutions.
For $V \geq 0$ we obtain such a semi-wave front explicitly by

$$U(\xi) := \begin{cases} \left(-\frac{m-1}{m} V \xi\right)^{1/(m-1)} & \text{if } \xi < 0, \\ 0 & \text{if } \xi \geq 0. \end{cases} \tag{7.16}$$

Thus $U$ is a solution with a free boundary $x = Vt$, which is continuous there and
has a continuous flux because of

$$\left(U^{(m)}\right)'(0-) = 0.$$

Because of

$$U'(0-) = \frac{V}{m} \left(-\frac{m-1}{m} V \xi\right)^{(2-m)/(m-1)}, \tag{7.17}$$

$U$ is for $m \geq 2$ not a classical solution with continuous spatial derivative $u_x$. An
implicit representation of $U$ from (7.15) for $V > 0$ is given by

$$-\frac{m}{V} \int_0^{U(\xi)} s^{m-2} ds = \xi \quad \text{if } \xi < 0,$$

and the appearance of a free boundary here is caused by the integrability near 0 of
the function $f(s) = s^{m-2}$ for $m > 1$. As equations of exactly this form also appear
for other flow and transport problems in porous media we consider the *generalized
porous medium equation with convection* in one space dimension:

$$\partial_t u - \partial_{xx}(a(u)) - \partial_x(b(u)) = 0. \tag{7.18}$$

Typical examples for $a$ and $b$ are

$$a(u) = u^m, \ m > 1, \ b(u) = \lambda u^n, \ \lambda \in \mathbb{R}, \ n > 0, \tag{7.19}$$

and we assume

$a, b$ are continuous for $u \geq 0$, continuously differentiable for $u > 0$

$$a(0) = b(0) = 0, \ a'(u) > 0 \text{ for } u > 0, \tag{7.20}$$

as we are interested only in nonnegative solutions and the consequences of $a'(0) = 0$ for the solution behavior at $u = 0$. A semi-wave front for (7.18) having $u_+ = 0$ has to fulfill

$$VU + (a(U))' + b(U) = 0. \tag{7.21}$$

To obtain a monotone solution we must have

$$Vs + b(s) > 0 \text{ for } 0 < s \leq \delta \tag{7.22}$$

for a certain constant $\delta > 0$. Therefore monotone semi-front solutions can only exist for

$$V > V^* := \limsup_{s \to 0+} \left( -\frac{b(s)}{s} \right) \tag{7.23}$$

and in fact they always exist for $V > V^*$. An implicit representation of the solution is then given for $U(\xi) > 0$ and $\xi < 0$ by

$$-\int_0^{U(\xi)} \frac{a'(s)}{Vs + b(s)} \, ds = \xi, \tag{7.24}$$

such that a semi-wave front with a free boundary appears if the integrand on the left-hand side of (7.24) is integrable close to 0, which here is equivalent to (see Theorem 5.2 in [51])

$$\int_0^\delta \frac{a'(s)}{\max(s, b(s))} \, ds < \infty \text{ for } \delta > 0. \tag{7.25}$$

For the model problem (7.19) this leads to the following result:

There exists a wave front for $V > 0 \iff n > 1$ and $\lambda < 0$. \hfill (7.26)

In that case we have

$$u_- := \lim_{x \to -\infty} U(t, x) = (-V/\lambda)^{1/(n-1)}.$$

Additionally unbounded semi-wave fronts exist for $V > 0$ in the case

$$n \neq 1, \ \lambda > 0 \text{ or } n = 1, \ \lambda > -V.$$

The wave front has a free boundary $U = 0$, if and only if $m > 1$; the unbounded semi-wave front has a free boundary if and only if $m > 1$ for $n > 1$, $\lambda > 0$ or if $m > n$.

Now we consider the flow of water or of water and air in a porous medium. In hydrology usually the pressure is given in the unit of length, i.e., it is scaled by $1/(\varrho g)$, where $\varrho$ is the (constant) density of water and $g$ is the gravitational acceleration, and the scaled pressure will be denoted in the following by $\Psi$. In the case where not the whole pore space is filled with water (i.e., it is not *saturated*) but if one assumes for simplification that the gas phase is connected and has constant atmospheric pressure, then the gas phase has not to be considered explicitly furthermore. But then (7.11) has to be extended for this *unsaturated* situation, in which microscopically air and water coexist. We assume that the pressure of the gas phase is scaled to $\Psi = 0$. By means of capillary forces suction power appears leading to a negative pressure $\Psi$ in the unsaturated region, therefore we have

$$\text{saturated} \quad \Leftrightarrow \Psi \geq 0,$$
$$\text{unsaturated} \Leftrightarrow \Psi < 0.$$

During successive draining of pores for decreasing $\Psi < 0$ also the conductivity decreases, such that $K$ is now a monotone non-decreasing function of $\Psi$ for $\Psi \leq 0$. As a result Darcy's law takes the form

$$q = -Kk(\Psi)\big(\nabla(\Psi + x_3)\big) \tag{7.27}$$

for the volumetric flow rate $q$ which now has the dimension of the velocity, and with a fixed conductivity tensor $K$ and a scalar *unsaturated* conductivity function $k$. The latter fulfills

$$k \text{ is strictly monotonically increasing for } \Psi < 0, \lim_{\Psi \to -\infty} k(\Psi) = 0. \tag{7.28}$$

For $\Psi \geq 0$ the function $k$ will be extended by $k_s := k(0)$ and again we obtain (7.11). Mass conservation here takes the form of volume conservation and reads as

$$\partial_t \theta + \nabla \cdot q = 0, \tag{7.29}$$

where $\theta$ is the *water content*, i.e., the volume of water in the representative elementary volume in relation to its total volume. As the decrease of $\Psi < 0$ is connected with a successive draining of pores $\theta$ turns out to be a function of $\Psi$,

$$\theta = \Theta(\Psi),$$

which fulfills:

$$\Theta \text{ is strictly monotonically increasing for } \Psi < 0, \lim_{\Psi \to -\infty} \Theta(\Psi) := \theta_r \geq 0. \tag{7.30}$$

For $\Psi \geq 0$ the function $\Theta$ is extended by $\theta_s := \Theta(0)$, which is just the *porosity*. In the following we assume that (by scaling) $\theta_r = 0$ holds. Summarizing we obtain as a model for saturated-unsaturated flow the *Richards equation*

$$\partial_t \big(\Theta(\Psi)\big) - \nabla \cdot \big(K\, k(\Psi)\, \nabla(\Psi + x_3)\big) = 0\,. \tag{7.31}$$

Besides the Darcy velocity $q$ also the *pore velocity*

$$v := q/\theta$$

is of importance and thus (7.29) gets a form analogous to (5.3). The Eq. (7.31) is not uniformly parabolic because of two reasons: In the transition from $\Psi < 0$ to $\Psi \geq 0$ it is reduced to the elliptic equation

$$-\nabla \cdot \big(K\, k_s \nabla(\Psi + x_3)\big) = 0\,, \tag{7.32}$$

the model for groundwater flow.

Differentiating the term formally we obtain

$$\partial_t \Psi - \big(\Theta'(\Psi)\big)^{-1} \nabla \cdot \big(K k(\Psi) \nabla(\Psi + x_3)\big) = 0\,,$$

showing that for $\Psi \to 0$ in general the diffusion coefficient is unbounded. Therefore in the transition from unsaturated to saturated the term *fast diffusion* is used.

Now we consider the transition from saturated to dry, i.e., the limit $\Psi \to -\infty$, and restrict ourselves to $\Psi < 0$. Then it seems advantageous to consider $\theta$ instead of $\Psi$ as the independent variable and thus one obtains a nonlinear diffusion-convection equation

$$\partial_t \theta - \nabla \cdot \big(K\big(D(\theta)\,\nabla\theta + \widetilde{k}(\theta)\, e_3\big)\big) = 0\,, \tag{7.33}$$

where

$$\widetilde{k}(\theta) := k\big(\Theta^{-1}(\theta)\big)\,,$$
$$D(\theta) : = \widetilde{k}(\theta)\frac{d}{d\theta}\Theta^{-1}(\theta)\,.$$

Experiments indicate that $\widetilde{k}$ is not only strictly monotonically increasing (with $\widetilde{k}(0) = 0, \widetilde{k}(\theta_s) = k_s$), but also convex. For the transition from unsaturated to dry the behavior at $\theta \to 0$ has to be investigated which shall be done again for planar traveling waves, i.e., for solutions of the form

$$\theta(t, x) = \theta(x \cdot n - Vt) \tag{7.34}$$

for a given direction $n \in \mathbb{R}^d$ of length 1. This is equivalent to the investigation of traveling waves for the one-dimensional problem, which has the form (7.18), where

$$a'(u) = n \cdot (K\, D(u)\, n) = (n \cdot Kn)\, \widetilde{k}(u)\,\tfrac{d}{du}\Theta^{-1}(u) =: \widetilde{D}(u)$$
$$b(u) = \lambda\widetilde{k}(u) \quad \text{with } \lambda = n \cdot Ke_3\,. \tag{7.35}$$

In the isotropic case $K = I$ we have

$$\lambda = \sin \alpha,$$

where $\alpha$ is the angle between the direction of flow and the horizontal plane, therefore

$$\lambda > 0 \iff \text{against the direction of the gravity},$$
$$\lambda < 0 \iff \text{in direction of gravity},$$
$$\lambda = 0 \iff \text{horizontal}.$$

It is sufficient to consider wave speeds $V > 0$. An application of the above mentioned results concerning semi-wave solutions (with free boundary), here shortly denoted as *moistening fronts*, shows:

- For $\lambda > 0$: moistening fronts exist if and only if

$$\widetilde{D}(\theta)/\max\left(\theta, \widetilde{k}(\theta)\right) \text{ is integrable close to } \theta = 0. \tag{7.36}$$

- For $\lambda = 0$: moistening fronts exist if and only if

$$\widetilde{D}(\theta)/\theta \text{ is integrable close to } \theta = 0. \tag{7.37}$$

- For $\lambda < 0$: moistening fronts exist if and only if

$$\widetilde{D}(\theta)/\max(\theta, \widetilde{k}(\theta)) \text{ is integrable}$$
$$\text{and } -\lambda\widetilde{k}(\theta) < V\theta \text{ for some } V > 0 \text{ in the neighborhood of } \theta = 0.$$

The last condition is fulfilled if $\widetilde{k}'(0) = 0$, which is suggested by convexity.

Summarizing, and also in general situations, under the assumption (7.36) we can expect moistening fronts in all directions. Assumption (7.37) is sufficient for moistening fronts and we obtain the particular relation

$$\widetilde{D}(\theta)/\theta = n \cdot Kn \frac{k(\Psi)}{\Theta'(\Psi)} \text{ for } \theta = \Theta(\Psi).$$

Concerning the properties of the free boundary $\Gamma(t)$ in the general situations not very much is yet known. In one space dimension $x = \Gamma(t)$ is continuous and the free boundary condition to be expected is

$$\Gamma'(t) = \lim_{x \to \Gamma(t)+} q(t, x)/\theta(t, x),$$

(see [50]).

Fronts also appear for reactive solute transport in porous media, for which traveling waves come to existence similar to Sect. 6.2.11, and additionally missing Lipschitz continuity of the nonlinearity may lead to the finiteness of the speed of propagation.

If a dissolved substance is transported by an underlying flow regime with volumetric flux $q$ (for example described by the Richards equation (7.31)) then in the case without chemical reactions at the macroscopic level the linear *convection diffusion equation*

$$\partial_t(\theta c) - \nabla \cdot (\theta D \nabla c - qc) = 0 \qquad (7.38)$$

arises for the dissolved concentration $c$ (measured in relation to the water-filled part REV). Here $\theta$ denotes the water content and the term in the middle includes molecular diffusion and the macroscopic phenomenon of *dispersion*, by which $D$ becomes dependent on $q$ and is matrix-valued. A nonlinear variant of this type of equation in one space dimension is (6.70). If a chemical reaction only takes place in solution and can be described only in terms of the concentration $c$ of one species then the right-hand side of (6.70) is to be substituted by a (nonlinear) function $F(c)$ generalizing (6.70). In porous media reactions which involve both the fluid and the solid phase are of more importance. Such a process is of great importance for the transport of contaminants in groundwater is *sorption*, i.e., the attachment of dissolved particles to the inner surfaces of the porous skeleton. Assuming that it can be described in a quasi-stationary way, i.e., by a dynamic equilibrium then one has

$$F(c) = -\partial_t \varphi(c), \qquad (7.39)$$

where $\varphi(c)$ denotes the *sorption isotherm*, which represents the sorbed mass (in relation to the total volume of an REF). Here we assume that the mass density of the solid is constant. The function $\varphi$ is monotonically increasing and fulfills $\varphi(0) = 0$. A form of $\varphi$ describing many experimental data is given by the *Freundlich* isotherm

$$\varphi(c) = a\, c^p, \ a > 0, \ 0 < p < 1. \qquad (7.40)$$

Therefore the derived differential equation reads for $d = 1$ and constant $\theta(=1)$, $q$, and $D$ in the variable $u = c$:

$$\partial_t\left(u + \varphi(u)\right) + q\, \partial_x u - D\, \partial_{xx} u = 0. \qquad (7.41)$$

Analogously to Sect. 6.2.10 this equation has wave fronts which connect $u_- > u_+$: The wave velocity $V$ then necessarily is given by

$$V = \frac{u_+ - u_-}{u_+ - u_- + \varphi(u_+) - \varphi(u_-)}\, q,$$

from which the retarding effect of sorption is visible leading to the importance of this process. A wave front with velocity $V$ exist if and only if the condition (6.75) holds. Analogously to the above considerations we have for $u_- = 0$: The wave front has a free boundary $u = 0$ if and only if $1/\varphi$ is integrable in a neighborhood of $u = 0$. This excludes functions $\varphi$ which are Lipschitz continuous at $u = 0$, but it is fulfilled by

(7.40). Analogous results can be developed for kinetic or multi-component reaction processes (see [80]).

In general traveling waves (with or without free boundary) are a simple tool to estimate the propagation velocity resulting from the interplay of competing processes.

## 7.3   The Stefan Problem

In the chapter on continuum mechanics we derived the heat equation

$$\varrho c_V \partial_t T - \nabla \cdot (\lambda \nabla T) = 0$$

for the absolute temperature $T > 0$. The heat equation has to be modified if phase transitions like melting and solidification appear. Different phases differ in the constitutive relation between internal energy and temperature. We will now discuss this aspect for the solid-liquid phase transition with the help of a simple constitutive relation. For the internal energy we set

$$u(T) = \begin{cases} c_V T & \text{in the solid phase}, \\ c_V T + L & \text{in the liquid phase}. \end{cases} \tag{7.42}$$

This constitutive relation reflects the following experimentally verified fact. There are temperatures at which energy can be supplied to a system without an increase of the temperature. At such a temperature a phase transition occurs. The energy needed to change the phase is called *latent heat*. In the above constitutive relation the latent heat is called $L$. At the transition from solid to liquid at the melting temperature the latent heat is needed to enable the body to melt and hence to change its phase. This effect is used when you cool a liquid with ice cubes as the melting ice cubes withdraw heat from the surrounding. Let us consider a body at rest, with constant density, without any heat sources or forces. In this case the energy conservation law from Sect. 5.5 is given as

$$\frac{d}{dt} \int_\Omega \varrho u \, dx + \int_{\partial \Omega} q \cdot n \, ds_x = 0, \tag{7.43}$$

where we assume that no internal energy is concentrated on the free boundary. We will now analyze which local equations follow from the energy conservation law. In points $x$ which lie in the fluid or in the solid phase we can derive under suitable smoothness assumptions on $\varrho$, $T$, and $q$ as in Chap. 6 the following differential equation

$$\partial_t (\varrho u) + \nabla \cdot q = 0.$$

In the following we always assume that

$$q = -\lambda \nabla T . \tag{7.44}$$

In addition we assume for simplicity that $\lambda$ and the specific heat $c_V$ are constant and are in particular the same in both phases. This implies that the heat equation

$$\varrho c_V \partial_t T = \lambda \Delta T \tag{7.45}$$

is fulfilled in both phases.

Which equations should be postulated at the phase boundary? If phase transitions do not happen too rapidly the system locally has time to reach a thermodynamic equilibrium. Motivated by the considerations in Chap. 3 we require

*the temperature T is continuous at the phase boundary .* (7.46)

The internal energy and hence the integrand $\varrho u$ are discontinuous at the phase boundary, cf. (7.42), and in the term $\frac{d}{dt} \int_\Omega \varrho u \, dx$ we cannot interchange the time derivative and the integral.

We will now derive an additional condition at the free boundary from the energy conservation law. In order to do so we need a transport theorem which we will derive in the following. We set

$$Q := (0, T) \times \Omega = Q_\ell \cup \Gamma \cup Q_s ,$$

where the sets $Q_\ell$, $\Gamma$, and $Q_s$ are mutually disjoint. In addition $\Omega \subset \mathbb{R}^d$ and $Q_\ell, Q_s \subset \mathbb{R}^{d+1}$ are domains with Lipschitz boundary and the phase boundary $\Gamma$ is assumed to be a smooth evolving hypersurface in $\mathbb{R}^d$, see the following definition, see Fig. 7.2.

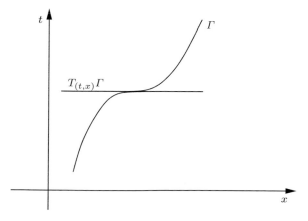

**Fig. 7.2** This situation of a tangent space is excluded in the definition of a smooth evolving hypersurface

**Definition 7.2**  $\Gamma$ is called a *smooth evolving hypersurface* in $\mathbb{R}^d$, if a $\mathcal{T} > 0$ exists such that

(i)   $\Gamma$ is a smooth hypersurface in $\mathbb{R} \times \mathbb{R}^d$,
(ii)  smooth hypersurfaces $\Gamma(t)$ in $\mathbb{R}^d$ exist, such that

$$\Gamma = \{(t, x) \mid t \in (0, \mathcal{T}), \ x \in \Gamma(t)\},$$

(iii)  the tangent spaces $T_{(t,x)}\Gamma$ of $\Gamma$ are never space-like, i.e.,

$$T_{(t,x)}\Gamma \neq \{0\} \times \mathbb{R}^d \quad \text{for all } (t, x) \in \Gamma.$$

For simplicity in the following we always assume that $\Gamma$ is a smooth evolving hypersurface and that $\Gamma(t) \subset\subset \Omega$ for all $t \in (0, \mathcal{T})$ holds. Here $\Gamma(t) \subset\subset \Omega$ means that $\Gamma(t)$ is bounded and that the closure $\overline{\Gamma(t)}$ of $\Gamma(t)$ is contained in $\Omega$. As $\Omega$ is open, this means in particular that $\overline{\Gamma(t)} \cap \partial\Omega = \emptyset$. In order to formulate the transport theorem we also need the unit normal $\nu$ at $\Gamma(t)$ which we choose such that the normal points into the domain $\Omega_\ell(t) := \{x \mid (t, x) \in Q_\ell\}$. We also need the *normal velocity* $V$. In order to define the normal velocity at a point $(t_0, x_0) \in \Gamma$ we choose a curve $x : (t_0 - \delta, t_0 + \delta) \to \mathbb{R}^d$ with $x(t) \in \Gamma(t)$ and $x(t_0) = x_0$. With this $x$ we define

$$V(t_0, x_0) = \nu(t_0, x_0) \cdot \frac{dx}{dt}(t_0). \tag{7.47}$$

Here $V$ does not depend on the choice of the curve (see Exercise 7.6).

**Theorem 7.3**  (Transport theorem) *Let* $u : Q \to \mathbb{R}$ *be such that* $u_{|Q_\ell}$ *and* $u_{|Q_s}$ *have extensions on* $\overline{Q}_\ell$ *and* $\overline{Q}_s$ *which are continuously differentiable. Under these assumptions it holds that*

$$\frac{d}{dt} \int_\Omega u(t, x)\, dx = \int_{\Omega_\ell(t)} \partial_t u(t, x)\, dx + \int_{\Omega_s(t)} \partial_t u(t, x)\, dx$$
$$- \int_{\Gamma(t)} [u]_s^\ell V\, ds_x. \tag{7.48}$$

*Here we set for* $x \in \Gamma(t)$

$$[u]_s^\ell(t, x) := \lim_{\substack{y \to x \\ y \in \Omega_\ell(t)}} u(t, y) - \lim_{\substack{y \to x \\ y \in \Omega_s(t)}} u(t, y).$$

*Proof*  The assertion (7.48) follows from the fundamental theorem of calculus if we can show

$$\int_\Omega u(t_2, x)\, dx - \int_\Omega u(t_1, x)\, dx = \int_{t_1}^{t_2} \int_{\Omega_\ell(t)} \partial_t u(t, x)\, dx\, dt$$
$$+ \int_{t_1}^{t_2} \int_{\Omega_s(t)} \partial_t u(t, x)\, dx\, dt - \int_{t_1}^{t_2} \int_{\Gamma(t)} [u]_s^\ell V\, ds_x\, dt. \tag{7.49}$$

To show (7.48) the identity (7.49) has to be differentiated with respect to $t_2$. Using the divergence theorem for the set

$$\{(t, x) \mid t \in (t_1, t_2),\ x \in \Omega_s(t)\}$$

and the vector field $F = (u, 0, \ldots, 0)$ we obtain

$$\int_{t_1}^{t_2} \int_{\Omega_s(t)} \partial_t u(t, x)\, dx\, dt = \int_{\Omega_s(t_2)} u(t_2, x)\, dx - \int_{\Omega_s(t_1)} u(t_1, x)\, dx$$
$$+ \int_{\Gamma_{t_1, t_2}} u(t, x)\, \nu_t\, ds_{(t,x)}\,.$$

Here we set

$$\Gamma_{t_1, t_2} = \{(t, x) \mid t \in (t_1, t_2),\ x \in \Gamma(t)\}\,,$$

$\nu_\Gamma = (\nu_t, \nu_x) \in \mathbb{R} \times \mathbb{R}^d$ is the space-time unit normal to the interface $\Gamma$ pointing into $Q_\ell$, and $ds_{(t,x)}$ denotes the integration with respect to the $d$-dimensional surface measure in $\mathbb{R}^{d+1}$. We will now write the integral over $\Gamma_{t_1, t_2}$ in a different form. After an orthogonal transformation we can always express $\Gamma$ in the neighborhood of a point $(t_0, x_0) \in \Gamma$ as the graph of a function

$$h : (t_1, t_2) \times D \to \mathbb{R}\,,\quad t_1 < t_0 < t_2,\ D \subset \mathbb{R}^{d-1}\ \text{open}\,.$$

This means that a point $(t, x', x_d) \in \mathbb{R} \times \mathbb{R}^{d-1} \times \mathbb{R}$ from a suitable neighborhood of $(t_0, x_0)$ lies on $\Gamma$ if and only if

$$x_d = h(t, x')\quad \text{with}\ t \in (t_1, t_2)\,,\ x' \in D\,.$$

Without loss of generality we can assume that the liquid phase lies above the graph. It hence follows that

$$\nu_\Gamma = \frac{1}{\sqrt{1 + |\partial_t h|^2 + |\nabla_{x'} h|^2}}\,(-\partial_t h, -\nabla_{x'} h, 1)^\top\,.$$

In addition, in the parametrization given by $h$ the surface element of $\Gamma$ is given by $\sqrt{1 + |\partial_t h|^2 + |\nabla_{x'} h|^2}$. We now consider $(t_0, x', h(t_0, x')) \in \Gamma$. Then we obtain in the point $(x', h(t_0, x'))$

$$\nu = \frac{1}{\sqrt{1 + |\nabla_{x'} h|^2}}\,(-\nabla_{x'} h, 1)^\top\,,\quad V = \partial_t h / \sqrt{1 + |\nabla_{x'} h|^2}\,.$$

The second identity follows if we consider the curve

$$x(t) = (x', h(t, x'))\,.$$

It follows that

$$\frac{d}{dt}x(t) = (0, \partial_t h(t, x'))$$

and with (7.47)

$$V = \partial_t h / \sqrt{1 + |\nabla_{x'} h|^2}.$$

It follows that for functions $u$ which have a support locally around $(t_0, x_0)$ we have

$$\int_{\Gamma_{t_1, t_2}} u(t, x)\, \nu_t\, ds_{(t,x)} = -\int_{t_1}^{t_2} \int_D u(t, x', h(t, x'))\, \partial_t h(t, x')\, dx'\, dt$$

$$= -\int_{t_1}^{t_2} \int_{\Gamma(t)} uV\, ds_x\, dt.$$

The last identity is obtained from the facts that $V = \partial_t h / \sqrt{1 + |\nabla_{x'} h|^2}$ and that the surface element with respect to integration on $\Gamma(t)$ is given by $\sqrt{1 + |\nabla_{x'} h|^2}$. With the help of a partition of unity we can put the local identity together to obtain a global identity. Using the same argument for $\Omega_\ell(t)$ we have to take into account that the sign of $\nu_t$ changes. This now proves (7.49) and hence the transport theorem.     □

From the identity

$$\frac{d}{dt} \int_\Omega \varrho u\, dx + \int_{\partial\Omega} q \cdot n\, ds_x = 0$$

it follows with the help of the transport theorem, the divergence theorem, and the energy conservation in the solid and liquid phase

$$0 = \int_\Omega (\varrho\, \partial_t u + \nabla \cdot q)\, dx + \int_{\Gamma(t)} \left( -\varrho[u]_s^\ell V + [q]_s^\ell \cdot \nu \right) ds_x$$

$$= \int_{\Gamma(t)} \left( -\varrho[u]_s^\ell V + [q]_s^\ell \cdot \nu \right) ds_x.$$

The above identity also holds for subsets of the given volume and hence we can choose suitable subsets of $\Gamma(t)$ and we obtain the following local form of the energy conservation law on the free boundary (Fig. 7.3)

$$\varrho[u]_s^\ell V = [q]_s^\ell \cdot \nu \quad \text{on } \Gamma(t). \tag{7.50}$$

This condition corresponds to the Rankine–Hugoniot condition for hyperbolic conservation laws and in the context of phase transitions it is called the *Stefan condition*. It now holds that

$$[u]_s^\ell = (c_V T + L - c_V T) = L.$$

Now let $q_s$ and $q_\ell$ be the fluxes in $\overline{\Omega}_s$ and $\overline{\Omega}_\ell$, respectively, and $\nu_s = \nu$, $\nu_\ell = -\nu$ the outer unit normals to $\Omega_s$ and $\Omega_\ell$, respectively. Altogether we obtain

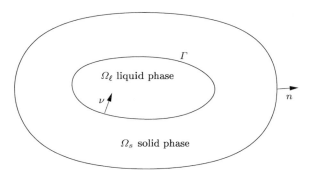

**Fig. 7.3** Illustration of the geometry in the Stefan problem

$$q_\ell \cdot \nu_\ell + q_s \cdot \nu_s = -\varrho L V .$$

The expression on the left-hand side gives the total heat supplied by the two phases at the phase boundary. The more heat enters the phase boundary the quicker the melting process will be and the velocity of the phase boundary is proportional to the total energy flux. The heat entering the phase boundary yields the latent heat that is needed for the formation of the new liquid phase region. If the total energy flux into the phase boundary is negative heat is withdrawn from the phase boundary and the liquid solidifies. This implies that the latent heat is set free which compensates the negative total energy flux.

We so far derived the two conditions (7.46) and (7.50). For the heat equation in the liquid and in the solid phase we need one condition on the free boundary each. An additional degree of freedom appears due to the fact that the phase boundary is free to move.

In the classical Stefan problem one requires that at each point the phase with the smaller free energy is attained. In the following we will consider the case in which the densities of the internal energy $u$, of the free energy $f$ and of the entropy $s$ are given for the solid ($\alpha = s$) and the liquid ($\alpha = \ell$) phase as follows:

$$u(T) = c_V T + L^\alpha ,$$
$$f(T) = -c_V T \left( \ln \frac{T}{T_M} - 1 \right) - L^\alpha \frac{T - T_M}{T_M} ,$$
$$s(T) = c_V \ln \frac{T}{T_M} + \frac{L^\alpha}{T_M} .$$

Here $T_M > 0$ and $L^\ell$ and $L^s$ are latent heats of the solid and liquid phase. This choice of $u$, $f$, and $s$ correspond to the definition of the internal energy $u$ in (7.42) and the fact that the Gibbs identities $s = -f_{,T}$ and $u = f + Ts$ hold true. As only the latent heat $L = L^\ell - L^s$ of the phase transition will play a role in the following we set $L^\ell = L$ and $L^s = 0$. As $L = L^\ell > L^s = 0$ we obtain that the solid phase has a smaller free energy for $T < T_M$ and for $T > T_M$ the liquid phase has a smaller free energy. Consequently $T_M$ is the melting temperature, and points with $T > T_M$ will be in the liquid phase while points with $T < T_M$ will be in the solid phase. In addition it holds true that

$$T = T_M \quad \text{on } \Gamma(t),$$

i.e., on the phase boundary the melting temperature is attained.

Altogether we obtain the following problem: Find a liquid phase $Q_\ell$, a solid phase $Q_s$, a free boundary $\Gamma$, and a temperature $T : Q \to \mathbb{R}$ such that

$$\varrho c_V \partial_t T - \lambda \Delta T = 0 \quad \text{in } Q_s \text{ and } Q_\ell, \tag{7.51}$$

$$\varrho L V + [\lambda \nabla T]_s^\ell \cdot \nu = 0 \quad \text{on } \Gamma(t), \tag{7.52}$$

$$T = T_M \text{ on } \Gamma(t). \tag{7.53}$$

In addition we need to specify initial conditions for $T$ and $\Gamma$, and for $T$ we also require boundary conditions on $\partial\Omega$. In the following we set the heat flux through the boundary to zero, i.e.,

$$-\lambda \nabla T \cdot n = 0 \quad \text{on } \partial\Omega \times (0, T).$$

Considering the condition (7.52) we notice that $\nabla T$ has to jump across $\Gamma$ whenever $V \neq 0$ holds. This is the case if the phase boundary is moving in time.

The free boundary problem (7.51)–(7.53) is the classical Stefan problem for melting and solidification. In this model, as stated above, one requires that in the solid phase $T < T_M$ holds and that in the liquid phase $T > T_M$ is true. In this case we can write the Stefan problem in a compact form which is called the *enthalpy formulation*

$$\partial_t \big(\varrho(c_V T + L\chi_{\{T>T_M\}})\big) = \lambda \Delta T. \tag{7.54}$$

The expression $\chi_{\{T>T_M\}}$ is the characteristic function of the set $\{(t, x) \mid T(t, x) > T_M\}$, i.e., $\chi_{\{T>T_M\}}$ is 1 in the liquid phase and 0 in the solid phase. This formulation follows from the identity $u(T) = c_V T + L\chi_{\{T>T_M\}}$. Due to the fact that $\chi_{\{T>T_M\}}$ is not differentiable it is not possible to interpret the Eq. (7.54) in a classical sense. Hence we interpret the identity (7.54) in a distributional sense, i.e., for all $\zeta \in C_0^\infty(Q)$, $Q = (0, \infty) \times \Omega$ we require

$$\int_Q \big(\varrho(c_V T + L\chi_{\{T>T_M\}})\partial_t \zeta + \lambda T \Delta\zeta\big) \, dx \, dt = 0. \tag{7.55}$$

We now seek a function $T(t, x)$, which fulfills (7.54) in a distributional sense. Having determined $T$ we obtain the liquid and the solid phase a posteriori as the sets $\Omega_\ell(t) = \{x \mid T(t, x) > T_M\}$ and $\Omega_s(t) = \{x \mid T(t, x) < T_M\}$. The phase boundary is given as $\Gamma(t) = \{x \mid T(t, x) = T_M\}$. However, there are situations in which $\Gamma(t)$ is not a hypersurface anymore and has a nonempty interior. In such a situation one says that a "mushy region" has formed. Introducing the quantity

$$e = \begin{cases} \varrho c_V T & \text{for } T \le T_M, \\ \varrho(c_V T + L) & \text{for } T > T_M \end{cases}$$

and defining

$$\beta(e) := \begin{cases} e/(\varrho c_V) & \text{for } e < \varrho c_V T_M, \\ T_M & \text{for } \varrho c_V T_M \le e \le \varrho(c_V T_M + L), \\ (e/\varrho - L)/c_V & \text{for } e > \varrho(c_V T_M + L), \end{cases}$$

we can formally rewrite the Eq. (7.54) as

$$\partial_t e = \lambda \, \Delta \beta(e). \tag{7.56}$$

Here it is important to notice that $\beta$ is not strictly monotonically increasing. As a consequence the formulation (7.56) leads to a degenerate parabolic equation. For the numerical approximation of solutions to the Stefan problem the formulation (7.56) on the other hand has many advantages. In particular, a simple explicit time discretization can be used to construct approximate solutions.

## 7.4 Entropy Inequality for the Stefan Problem

We now always assume

$$q \cdot n = 0 \quad \text{on } \partial \Omega,$$

i.e., there is no heat flux into the domain $\Omega$. This implies the following balance for the internal energy, see (7.43),

$$\frac{d}{dt} \int_\Omega \varrho u \, dx = 0.$$

We will now also derive an entropy inequality. For the total entropy $\int_\Omega \varrho s \, dx$ we obtain with the help of the transport theorem (Theorem 7.3) and integration by parts

$$\begin{aligned}
\frac{d}{dt} \int_\Omega \varrho s \, dx &= \int_{\Omega_\ell(t)} \varrho c_V \frac{\partial_t T}{T} \, dx + \int_{\Omega_s(t)} \varrho c_V \frac{\partial_t T}{T} \, dx - \int_{\Gamma(t)} [\varrho s]_s^\ell V \, ds_x \\
&= \int_{\Omega_\ell(t)} \frac{\lambda \Delta T}{T} \, dx + \int_{\Omega_s(t)} \frac{\lambda \Delta T}{T} \, dx - \int_\Gamma [\varrho s]_s^\ell V \, ds_x \\
&= - \int_{\Omega_\ell(t)} \lambda \nabla T \cdot \nabla\left(\tfrac{1}{T}\right) dx - \int_{\Omega_s(t)} \lambda \nabla T \cdot \nabla\left(\tfrac{1}{T}\right) dx
\end{aligned}$$

$$- \int_{\Gamma(t)} \lambda \frac{1}{T} [\nabla T]_s^\ell \cdot v \, ds_x - \int_{\Gamma(t)} [\varrho s]_s^\ell V \, ds_x$$

$$= \int_\Omega \lambda T^2 |\nabla (\tfrac{1}{T})|^2 \, dx + \int_{\Gamma(t)} \varrho \left[\tfrac{u}{T} - s\right]_s^\ell V \, ds_x$$

$$= \int_\Omega \lambda T^2 |\nabla (\tfrac{1}{T})|^2 \, dx \geq 0 .$$

Here, one has to use the free boundary condition (7.52) and the last identity follows from the fact that at the free boundary the free energy $f = u - Ts$ is continuous, i.e.,

$$\left[\tfrac{u}{T} - s\right]_s^\ell = \left[\tfrac{f}{T}\right]_s^\ell = \frac{c_V T_M + L}{T_M} - \frac{L}{T_M} - \frac{c_V T_M}{T_M} = 0 .$$

Altogether, we showed that the total entropy cannot decrease. This condition is necessary in order to fulfill the second law of thermodynamics. In fact, another inequality is true which has important consequences for the analysis of the Stefan problem. With the transport theorem we obtain

$$\frac{d}{dt} \int_\Omega \varrho c_V \tfrac{1}{2} (T - T_M)^2 \, dx = \int_{\Omega_\ell(t)} \varrho c_V (T - T_M) \, \partial_t T \, dx$$

$$+ \int_{\Omega_s(t)} \varrho c_V (T - T_M) \, \partial_t T \, dx - \int_{\Gamma(t)} \left[\varrho c_V \tfrac{1}{2} (T - T_M)^2\right]_s^\ell V \, ds_x$$

$$= \int_{\Omega_\ell(t)} \lambda (T - T_M) \Delta T \, dx + \int_{\Omega_s(t)} \lambda (T - T_M) \Delta T \, dx$$

$$= - \int_{\Omega_\ell(t)} \lambda |\nabla T|^2 \, dx - \int_{\Omega_s(t)} \lambda |\nabla T|^2 \, dx - \int_{\Gamma(t)} \lambda [\nabla T]_s^\ell \cdot v \, (T - T_M) \, ds_x .$$

The integrals with respect to $\Gamma(t)$ vanish due to the fact that $T = T_M$ on $\Gamma(t)$. Altogether we obtain

$$\frac{d}{dt} \int_\Omega \varrho c_V \tfrac{1}{2} (T - T_M)^2 \, dx + \int_\Omega \lambda |\nabla T|^2 \, dx = 0 .$$

This equality is the basis for an existence proof using techniques of functional analysis.

## 7.5 Undercooled Liquids

So far we assumed that the liquid phase is characterized by $T > T_M$ and that the solid phase is characterized by $T < T_M$. In fact, it is possible that liquids are undercooled and that solids are superheated. It may happen, for instance, that liquids remain in the liquid phase even when the temperature in the liquid is below the melting temperature.

We now want to study this situation for an idealized model system. Now we set $u = T - T_M$. Notice that $u$ here and in the Sects. 7.5–7.7 will *not* denote the density of the internal energy. We consider the growth of a crystal nucleus which at time $t = 0$ only consists of one point. In addition, we assume the following:

- Spatially the temperature only varies in one direction and we have $u = u(t, x)$ with $x \in [0, 1]$.
- The phase boundary is orthogonal to the $x$-axis.
- The location of the phase boundary is denoted by $y(t) \in [0, 1]$ and it holds $y(0) = 0$. We assume that points $x < y(t)$ are in the solid phase and points with $x > y(t)$ are in the liquid phase.
- The diffusion constant is large, or more precisely, $\frac{\varrho c_V}{\lambda}$ is small. In this case we can consider the quasi-stationary equation

$$u_{xx}(t, x) = 0, \ x \in (0, 1), \ x \neq y(t), \ t > 0$$

instead of the heat equation.
- The left boundary is isolated and at the right boundary we prescribe the constant temperature $U$, i.e., for a $U \in \mathbb{R}$ we require

$$u_x(t, 0) = 0, \ u(t, 1) = U, \ t > 0.$$

Denoting by $u(t, x-)$ and $u(t, x+)$ the limit with respect to $x$ from the left and the right, respectively, we obtain the following problem:

$$u_{xx}(t, x) = 0 \ \text{for} \ x \in (0, 1), \ x \neq y(t), \ t > 0,$$
$$u(t, y(t)) = 0 \ \text{for} \ t > 0,$$
$$\varrho L V = \varrho L y'(t) = \lambda(u_x(t, y(t)-) - u_x(t, y(t)+)) \ \text{for} \ t > 0,$$
$$y(0) = 0,$$
$$u_x(t, 0) = 0, \ u(t, 1) = U.$$

The solution is

$$u(t, x) = 0 \ \text{for} \ 0 \leq x \leq y(t),$$
$$u(t, x) = U \frac{x - y(t)}{1 - y(t)} \ \text{for} \ y(t) < x \leq 1.$$

In addition, it holds

$$\varrho L y'(t) = -\lambda \frac{U}{1 - y(t)},$$
$$y(0) = 0.$$

Hence, for $U = 0$ we have no movement of the phase boundary. For $U > 0$ we obtain $y'(0) \leq 0$ and the phase boundary cannot move into the liquid region. On the other hand if $U < 0$ we obtain

$$y(t) = 1 - \left(1 + 2\frac{\lambda}{\varrho L} U t\right)^{1/2}.$$

We hence observe that the crystal nucleus can only grow if the liquid is undercooled.

For solidification the form of the phase boundary plays an important role. If the phase boundary is curved then the temperature at the phase boundary will deviate from the melting temperature $T_M$ due to capillary effects. For a round solid nucleus the temperature at which solidification happens is below the melting temperature. The temperature will in fact depend on the curvature of the phase boundary (Gibbs–Thomson effect). We will see later that the Gibbs–Thomson effect has a stabilizing effect on solidification. In order to formulate the Gibbs–Thomson condition at the phase boundary we need the notion of mean curvature which is introduced in Appendix B.

## 7.6  Gibbs–Thomson Effect

We will now consider undercooling and superheating and in addition take into account that the melting temperature at the phase boundary can depend on the curvature. More precisely, we consider the *Mullins–Sekerka problem* for the scaled temperature $u = \frac{\lambda}{\varrho L}(T - T_M)$

$$\begin{cases} \Delta u = 0 & \text{in } \Omega_s(t) \cup \Omega_\ell(t), \\ V = -[\nabla u]_s^\ell \cdot \nu & \text{on } \Gamma(t), \\ u = \gamma \kappa & \text{on } \Gamma(t). \end{cases} \tag{7.57}$$

Here $\gamma$ is a suitable scaled surface tension and $\kappa$ is the sum of the principle curvatures which we will call the mean curvature in what follows, cf. Appendix B. The sign of $\kappa$ is chosen such that $\kappa < 0$ for a strictly convex solid. The Laplace equation is obtained as a quasi-static variant of the Stefan problem which we derived in the previous sections. This means we assume that $\frac{\varrho c_V}{\lambda}$ is small, which implies that the time derivative in the heat equation can be neglected. The last equation in (7.57) is the *Gibbs–Thomson condition* and this condition allows for an undercooling due to the curvature. This means for example that the temperature at the phase boundary for a convex nucleus can be below the melting temperature.

We will now analyze under which conditions a crystal nucleus can grow. In order to do so we now consider the Mullins–Sekerka problem (7.57) with

$$\Omega = B_R(0) \subset \mathbb{R}^3, \ \Omega_s(0) = B_{r_0}(0), \ \Gamma(0) = \partial B_{r_0}(0), \ \Omega_\ell(0) = B_R(0) \setminus \overline{B_{r_0}(0)},$$

where $0 < r_0 < R < \infty$. On the exterior boundary we prescribe a fixed temperature underneath the melting temperature:

$$u(t, x) = u_0 \equiv \text{const} < 0 \quad \text{for} \quad x \in \partial\Omega, \, t > 0. \tag{7.58}$$

We are looking for a radial solution $u(t, x) = v(t, |x|)$ with $\Gamma(t) = \partial B_{r(t)}(0)$. As $\kappa(t) = -\frac{2}{r(t)}$ on $\Gamma(t)$ we obtain in the solid phase

$$v(t, r) = -\frac{2\gamma}{r(t)} \quad \text{for all} \quad r \in [0, r(t)].$$

For $d = 3$ we obtain $\Delta_x v(|x|) = v''(|x|) + \frac{2}{|x|} v'(|x|)$, cf. (6.12). From (7.57) we deduce

$$0 = v'' + \tfrac{2}{r} v' \quad \text{or} \quad 0 = (r^2 v')'.$$

We hence obtain

$$v(t, r) = c_1(t)\tfrac{1}{r} + c_2(t), \quad c_1(t), c_2(t) \in \mathbb{R}.$$

The Gibbs–Thomson condition and the boundary condition (7.58) yield the following boundary conditions in the liquid phase

$$c_1(t)\tfrac{1}{R} + c_2(t) = u_0 \quad \text{and} \quad c_1(t)\tfrac{1}{r(t)} + c_2(t) = -\tfrac{2\gamma}{r(t)}.$$

From this we compute

$$c_1(t) = \left(u_0 + \tfrac{2\gamma}{r(t)}\right) / \left(\tfrac{1}{R} - \tfrac{1}{r(t)}\right) = (\gamma \kappa(t) - u_0) / \left(\tfrac{1}{r(t)} - \tfrac{1}{R}\right),$$
$$c_2(t) = -\tfrac{1}{r(t)}(2\gamma + c_1).$$

As $V = \dot{r}(t)$ the Stefan condition in (7.57) implies

$$\dot{r}(t) = c_1(t)\tfrac{1}{r(t)^2} = (\gamma \kappa(t) - u_0) / \left(\tfrac{1}{r(t)^3} - \tfrac{1}{r(t)^2 R}\right). \tag{7.59}$$

The right-hand side is zero if and only if

$$\gamma\kappa = u_0 \quad \text{and hence} \quad r = -\frac{2\gamma}{u_0}.$$

The radius $r_{\text{crit}} = -\frac{2\gamma}{u_0}$ is called *critical radius*. For $r_0 < r_{\text{crit}}$ The right-hand side in (7.59) is negative at initial time and a nucleus which is too small will vanish in finite time. For $r_0 > r_{\text{crit}}$ we obtain that the nucleus will grow. In addition, it holds for large undercooling $r_{\mathrm{crit}}$ is small,

for small undercooling $r_{\mathrm{crit}}$ is large.

A nucleus has to have a certain size to be able to grow. For large undercooling also small nuclei can grow.

## 7.7   Mullins–Sekerka Instability

In the previous section we saw that a nucleus can only grow if the surrounding is undercooled. The analysis in this section will show that a phase boundary which grows into an undercooled fluid is unstable.

We consider a solidification process in $\mathbb{R}^2$ where the coordinates are denoted by $(x, y)^\top$. We consider a solidification front which is parallel to the $x$-axis and which grows with velocity $V_0$ in the direction of the $y$-axis. For $y \to +\infty$ and $y \to -\infty$ we prescribe the heat fluxes

$$-\nabla u = -g_\ell \begin{pmatrix} 0 \\ 1 \end{pmatrix} \quad \text{and} \quad -\nabla u = -g_s \begin{pmatrix} 0 \\ 1 \end{pmatrix}.$$

We consider a coordinate system which moves with the solidification front, such that

$$\Omega_\ell(t) = \{(x, y)^\top \mid y > 0\},$$
$$\Omega_s(t) = \{(x, y)^\top \mid y < 0\}.$$

In this geometry we consider the following solution of the Mullins–Sekerka problem (7.57)

$$u_s(x, y) = g_s\, y \text{ for } y < 0,$$
$$u_\ell(x, y) = g_\ell\, y \text{ for } y > 0.$$

With $\nabla u_s = (0, g_s)^\top$ for $y < 0$, $\nabla u_\ell = (0, g_\ell)^\top$ for $y > 0$ and $\nu = (0, 1)^\top$ we obtain

$$[\nabla u]_s^\ell \cdot \nu = g_\ell - g_s,$$

and hence

$$V_0 = g_s - g_\ell.$$

In addition, the condition $u = 0$ holds on the phase boundary $\Gamma = \{(x, y) \mid y = 0\}$.

We now analyze whether the phase boundary is stable with respect to small perturbations. We now perturb the flat phase boundary at initial time with a sinusoidal profile with a small amplitude $\delta(0)$ and a period $\frac{2\pi}{\omega}$. In the moving coordinate system such a perturbed phase boundary is given by

$$y = h(t, x) = \delta(t) \sin(\omega x) \quad \text{with} \quad \omega > 0.$$

Since we perturbed the phase boundary with a certain wavelength we also make an ansatz for the solution with the same wavelength. This ansatz would also be obtained from an ansatz with separation of variables. Hence we make the ansatz

$$u_s(t, x, y) = g_s\, y + c_s(t)\, e^{\omega y} \sin(\omega x),$$
$$u_\ell(t, x, y) = g_\ell\, y + c_\ell(t)\, e^{-\omega y} \sin(\omega x).$$

These functions solve the Laplace equation $\Delta u = 0$ and we ensure that for large $|y|$ the perturbation decays. Far away from the phase boundary we prescribe the temperature gradient and hence we prescribe the heat flux.

We want to guarantee that $u_\ell$ and $u_s$ fulfill the conditions at the free boundary at least approximately. In order to solve the equation $u = \gamma \kappa$ on $\Gamma$ approximately we compute $u_s$ and $u_\ell$ on the phase boundary as

$$u_s(t, x, h(t, x)) = g_s\, h(t, x) + c_s(t)\, e^{\omega h(t,x)} \sin(\omega x)$$
$$= g_s\, \delta(t) \sin(\omega x) + c_s(t)\, e^{\omega h(t,x)} \sin(\omega x),$$
$$u_\ell(t, x, h(t, x)) = g_\ell\, h(t, x) + c_\ell(t)\, e^{-\omega h(t,x)} \sin(\omega x)$$
$$= g_\ell\, \delta(t) \sin(\omega x) + c_\ell(t)\, e^{-\omega h(t,x)} \sin(\omega x).$$

Denoting by $\kappa(t, x)$ the curvature at $(x, h(t, x))$ we obtain (cf. Appendix B)

$$\kappa(t, x) = \frac{\partial_{xx} h}{(1 + (\partial_x h)^2)^{3/2}} = \partial_{xx} h + \mathcal{O}(\delta^2) \approx -\delta(t)\, \omega^2 \sin(\omega x).$$

In order to fulfill $u = \gamma \kappa$ up to an error of the order $\delta^2$, we require

$$c_s(t) = -(\gamma \omega^2 + g_s)\delta(t), \qquad (7.60)$$
$$c_\ell(t) = -(\gamma \omega^2 + g_\ell)\delta(t). \qquad (7.61)$$

We now evaluate the jump condition $V = -[\nabla u]_s^\ell \cdot \nu$. With the normal vector

$$\nu = \left(1/\left(1 + (\partial_x h)^2\right)^{1/2}\right)(-\partial_x h, 1)^\top$$

it follows that

$$\nabla u_s = \left(\omega\, c_s(t)\, e^{\omega y} \cos(\omega x),\ g_s + \omega\, c_s(t)\, e^{\omega y} \sin(\omega x)\right)^\top$$

and then with (7.60),

$$\nabla u_s \cdot \nu_{|y=h} = g_s - \delta(t)(\gamma \omega^2 + g_s)\omega \sin(\omega x) + \mathcal{O}(\delta^2).$$

Analogously we compute with (7.61)

$$\nabla u_\ell \cdot \nu_{|y=h} = g_\ell + \delta(t)(\gamma\omega^2 + g_\ell)\omega \sin(\omega x) + \mathcal{O}(\delta^2) \,.$$

It follows that

$$-[\nabla u]_s^\ell \cdot \nu = g_s - g_\ell - \delta(2\gamma\omega^2 + g_s + g_\ell)\omega \sin(\omega x) \,. \tag{7.62}$$

Besides we obtain as in Sect. 7.3 for the normal velocity $V(t, x)$ in the point $(x, h(t, x))$

$$
\begin{aligned}
V(t, x) &= \begin{pmatrix} 0 \\ V_0 + \partial_t h \end{pmatrix} \cdot \begin{pmatrix} -\partial_x h \\ 1 \end{pmatrix} \frac{1}{(1 + (\partial_x h)^2)^{1/2}} \\
&= \frac{V_0 + \partial_t h}{(1 + (\partial_x h)^2)^{1/2}} = V_0 + \delta'(t)\sin(\omega x) + \mathcal{O}(\delta^2) \,.
\end{aligned}
$$

Together with (7.62) we obtain

$$\delta'(t) = -\lambda(\omega)\,\delta(t)$$

with

$$\lambda(\omega) = \omega\{2\gamma\omega^2 + g_s + g_\ell\} \,.$$

Hence the perturbations will grow exponentially in time if $\lambda(\omega) < 0$ and will be damped if $\lambda(\omega) > 0$. Now let $g_s \geq 0$, i.e., the temperature in the solid phase is negative. If $g_\ell \geq 0$ then the fluid is not undercooled and we obtain $\lambda(\omega) > 0$ and hence this is a stable situation.

In the case $g_\ell < 0$ the liquid phase is supercooled. There are two cases:

(i)  The case $\gamma = 0$ without capillary term. In case of undercooling

$$g_\ell + g_s < 0$$

all wavelengths are amplified and we obtain a strongly unstable situation. For

$$g_\ell + g_s > 0 \,,$$

we obtain a stable situation.

(ii)  The case $\gamma > 0$. In this case all small wavelengths, which are perturbations with a large $\omega$, are damped. In case of strong undercooling, i.e., in a case where $g_\ell$ is strongly negative, certain large wavelength are unstable. The most unstable wavelengths are obtained for $\omega$ near the positive local minimum $\omega_{si}$ of $\lambda$. We obtain

$$\omega_{si} = \sqrt{\frac{-(g_s + g_\ell)}{6\gamma}}$$

and with our experience from Sects. 6.2.12 and 6.2.13 we expect that the length scale defined by $\omega_{si}^{-1}$ plays an important role in real solidification scenarios with planar fronts.

Summarizing we observe in the case without capillary term $\gamma\kappa$ and strong undercooling lead to a very unstable phase boundary. As the formation of new surface costs energy one obtains that the capillary term has a stabilizing effect such that small wavelengths are damped.

The fact that strongly undercooled fluids have very unstable phase boundaries yield very bifurcated phase boundaries. Many solidification fronts lead to dendritic (tree-like) structures. Variants of the Mullins–Sekerka stability analysis from above are used to explain the diverse patterns observed in snow crystal growth (see, e.g., Libbrecht [88], Barrett et al. [11]).

## 7.8  A Priori Estimates for the Stefan Problem with Gibbs–Thomson Condition

We introduced the Gibbs–Thomson condition in order to allow for a curvature undercooling. In addition, one often considers a kinetic undercooling which takes into account that the system needs time to evolve into the local equilibrium conditions $T = T_M$ or $T - T_M = \gamma\kappa$. Hence, we now consider the relaxation dynamics

$$\beta V = \gamma\kappa - \varrho L(T - T_M) \quad \text{with} \quad \beta \geq 0. \tag{7.63}$$

Here $\gamma$ is a suitable scaled surface energy density. We note that the scaling here is different to the scaling in Sect. 7.6. The parameter $\beta$ reflects the strength of the kinetic undercooling. We can easily show that the energy conservation law (7.43) also holds with the condition (7.63). We now want to understand the regularizing effect of the terms $\beta V$ and $\gamma\kappa$ a little better. In order to do so, we show that the law (7.63) allows for an estimate of the surface area of the phase boundary. This implies that the phase boundary cannot become arbitrarily irregular. It is a general observation in nature that capillarity has a smoothing effect. To proceed we need the following result.

**Theorem 7.4** *For a smooth, compact evolving hypersurface without boundary it holds*

$$\frac{d}{dt}\int_{\Gamma(t)} 1\, ds_x = -\int_{\Gamma(t)} \kappa V\, ds_x.$$

*Proof* We first give a proof in the similar case that the phase boundary is given by a closed curve in the plane. In this case the proof can be done quite elementary. We consider a smooth function

$$y : [t_1, t_2] \times [a, b] \to \mathbb{R}^2, \quad (t, p) \mapsto y(t, p)$$

with $t_1 < t_2$ and $a < b$ such that $y(t, \cdot)$ is a regular parametrization of $\Gamma(t)$. In this case the length element is given by $|\partial_p y|$ and $\frac{1}{|\partial_p y|}\partial_p\left(\frac{\partial_p y}{|\partial_p y|}\right) = \kappa\nu$, cf. Appendix B. We can now compute

$$\frac{d}{dt}\int_{\Gamma(t)} 1\, ds_x = \frac{d}{dt}\int_a^b |\partial_p y(t, p)|\, dp = \int_a^b \left(\frac{\partial_p y}{|\partial_p y|}\cdot\partial_t\partial_p y\right)(t, p)\, dp$$

$$= -\int_a^b \left(\partial_p \frac{\partial_p y}{|\partial_p y|}\cdot\partial_t y\right)(t, p)\, dp = -\int_a^b \left(\kappa\nu\cdot\partial_t y\,|\partial_p y|\right)(t, p)\, dp$$

$$= -\int_{\Gamma(t)} \kappa V\, ds_x.$$

In the general case we choose a partition of unity $\{\zeta_i\}_{i=1}^M$ of the $\mathbb{R}^{1+d}$, such that the surface patches $\Gamma \cap \operatorname{supp}\zeta_i$ can be written as a graph as in the proof of Theorem 7.3. We then obtain

$$\frac{d}{dt}\int_{\Gamma(t)} 1\, ds_x = \sum_{i=1}^M \frac{d}{dt}\int_{\Gamma(t)} \zeta_i(t, x)\, ds_x.$$

As $\operatorname{supp}\zeta_i \cap \Gamma$ can be written as a graph of

$$h : (t_1, t_2) \times D \to \mathbb{R},\ t_1 < t_2,\ D \subset \mathbb{R}^{d-1}\ \text{open},$$

we can compute

$$\frac{d}{dt}\int_{\Gamma(t)} \zeta_i(t, x)\, ds_x = \frac{d}{dt}\int_D \zeta_i(t, x', h(t, x'))\sqrt{1 + |\nabla_{x'}h(t, x')|^2}\, dx'$$

$$= \int_D (\partial_t\zeta_i + \partial_{x_d}\zeta_i\,\partial_t h)\sqrt{1 + |\nabla_{x'}h|^2}\, dx'$$

$$+ \int_D \frac{\zeta_i}{\sqrt{1 + |\nabla_{x'}h|^2}}\nabla_{x'}h\cdot\nabla_{x'}\partial_t h\, dx'$$

$$= \int_D (\partial_t\zeta_i + \partial_{x_d}\zeta_i\,\partial_t h)\sqrt{1 + |\nabla_{x'}h|^2}\, dx'$$

$$- \int_D \partial_t h\,\zeta_i\,\nabla_{x'}\cdot\left(\frac{\nabla_{x'}h}{\sqrt{1 + |\nabla_{x'}h|^2}}\right) dx'$$

$$- \int_D (\nabla_{x'}\zeta_i + \partial_{x_d}\zeta_i\,\nabla_{x'}h)\cdot\frac{\nabla_{x'}h}{\sqrt{1 + |\nabla_{x'}h|^2}}\partial_t h\, dx'.$$

Using $\nu = \frac{1}{\sqrt{1+|\nabla_{x'}h|^2}}(-\nabla_{x'}h, 1)^\top$, $\kappa = \nabla_{x'}\cdot\left(\frac{\nabla_{x'}h}{\sqrt{1+|\nabla_{x'}h|^2}}\right)$ and $V = \frac{\partial_t h}{\sqrt{1+|\nabla_{x'}h|^2}}$ we obtain

$$\frac{d}{dt} \int_{\Gamma(t)} \zeta_i(t, x)\, ds_x = \int_{\Gamma(t)} (\nabla \zeta_i \cdot \nu\, V + \partial_t \zeta_i - \zeta_i V \kappa)\, ds_x \, .$$

By summation with respect to $i$ we obtain the claim by using $\sum_{i=1}^{M} \zeta_i = 1$.   □

To obtain an estimate for the Stefan problem (7.51), (7.52), with Gibbs–Thomson condition and kinetic undercooling (7.63) we first of all compute

$$\frac{d}{dt} \int_{\Gamma(t)} \gamma\, ds_x = -\int_{\Gamma(t)} \gamma \kappa V\, ds_x$$

$$= -\int_{\Gamma(t)} \beta V^2\, ds_x - \int_{\Gamma(t)} \varrho L V (T - T_M)\, ds_x \, .$$

Here and in the following we always assume that the phase boundary does not intersect the outer boundary $\Gamma$. As in Sect. 7.4 one can deduce, using that $T$ is continuous across $\Gamma(t)$,

$$\frac{d}{dt} \int_{\Omega} \varrho c_V \tfrac{1}{2}(T - T_M)^2\, dx = -\int_{\Omega} \lambda |\nabla T|^2\, dx - \int_{\Gamma(t)} (T - T_M)\lambda [\nabla T]_s^\ell \cdot \nu\, ds_x \, .$$

Combining the last two identities we obtain with (7.52)

$$\frac{d}{dt} \left( \int_{\Omega} \varrho c_V \tfrac{1}{2}(T - T_M)^2\, dx + \int_{\Gamma(t)} \gamma\, ds_x \right)$$

$$= -\int_{\Gamma(t)} \beta V^2\, ds_r - \int_{\Omega} \lambda |\nabla T|^2\, dx \, . \tag{7.64}$$

The capillary term in the Gibbs–Thomson equation hence yields the surface area of the phase boundary as additional term in the a priori estimate. As additional dissipative term we obtain besides the Dirichlet integral of the temperature the squared $L_2$-norm of the velocity on the boundary.

We here only considered a simple variant of the Stefan problem with undercooling including capillary and kinetic terms which both lead to a regularization. For a more systematic derivation also of more general models we refer to the book of Gurtin [60].

The variant of the Stefan problem with Gibbs–Thomson condition studied so far is the one which is typically used in practice. However, the version studied above is only an approximate theory and in particular does not fulfill the second law of thermodynamics. A simple modification, which then fulfills the second law, is given as follows. The relaxation dynamics (7.63) is replaced by

$$\widehat{\beta} V = \widehat{\gamma} \kappa + \varrho L \left( \frac{1}{T} - \frac{1}{T_M} \right) \tag{7.65}$$

where $\widehat{\beta}, \widehat{\gamma}$ are constants which now have a different physical dimension as the quantities $\beta$ and $\gamma$ in (7.63). It can be shown that (Exercise 7.13)

$$\frac{d}{dt}\left(\int_\Omega \varrho s\, dx - \int_{\Gamma(t)} \widehat{\gamma}\, ds_x\right) \geq 0\,, \tag{7.66}$$

where $s$ is the entropy density from Sect. 7.3. The constant $-\widehat{\gamma}$ in this case is an entropy density per area. The above inequality shows that with the modified condition at the free boundary the total entropy can only increase.

## 7.9   Phase Field Equations

In the Stefan problem the phase boundary is described in the form of a hypersurface. If the free boundary changes its topology, which happens for example in cases where two regions occupied by the solid phase join or in case where a connected region of a solid phase splits into two disconnected regions then this approach has some disadvantages. In all possible descriptions of the hypersurface one will obtain a singularity in the representation of the hypersurface if the topology changes. This is the reason why in the last decades phase-field approaches have been used to describe solidification phenomena. In a phase field model a smooth phase field variable $\varphi$ is introduced which in the solid phase attains values close to $-1$ and in the liquid phase attains values close to 1. In regions where the phase changes interior layers appear in which the values of $\varphi$ change rapidly. Solutions of the phase field equation remain smooth even in cases where the topological structure of the phase boundary changes.

The internal energy in the phase field model is given as $u(T, \varphi) = c_V T + L\frac{\varphi+1}{2}$, i.e., we interpolate the latent heat in a linear fashion. We then obtain the energy balance

$$\partial_t\left(\varrho c_V T + \varrho L\frac{\varphi+1}{2}\right) = \lambda\, \Delta T\,. \tag{7.67}$$

The Eq. (7.63) for the free boundary is replaced by the equation

$$\varepsilon\beta\,\partial_t\varphi = \varepsilon\gamma\,\Delta\varphi - \frac{\gamma}{\varepsilon}\psi'(\varphi) + \frac{\varrho L}{2}(T - T_M)\,. \tag{7.68}$$

Here, $\psi$ is given as $\psi(\varphi) = \frac{9}{32}(\varphi^2 - 1)^2$, see Fig. 7.4, and $\varepsilon > 0$ is a parameter which later will turn out to be proportional to the thickness of the interfacial layer. We will see in addition that in an asymptotic expansion for $\varepsilon \to 0$ the term $\partial_t\varphi$ will lead to the negative normal velocity and the term $\varepsilon\Delta\varphi - \frac{1}{\varepsilon}\psi'(\varphi)$ will lead to the negative mean curvature.

Assuming homogeneous Neumann boundary conditions $\nabla T \cdot n = \nabla\varphi \cdot n = 0$ we obtain for smooth solutions of (7.67), (7.68)

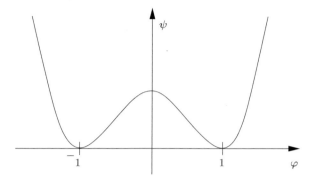

**Fig. 7.4** The energy contribution $\psi(\varphi)$ in (7.70) penalizes values of $\varphi$ which differ strongly from $\pm 1$

$$\frac{d}{dt}\left(\int_{\Omega} \varrho c_V \tfrac{1}{2}(T - T_M)^2\, dx + \int_{\Omega} \gamma\big(\tfrac{\varepsilon}{2}|\nabla\varphi|^2 + \tfrac{1}{\varepsilon}\psi(\varphi)\big)\, dx\right)$$

$$= \int_{\Omega} \varrho c_V (T - T_M)\, \partial_t T\, dx + \int_{\Omega} \gamma\big(\varepsilon\nabla\varphi \cdot \partial_t\nabla\varphi + \tfrac{1}{\varepsilon}\psi'(\varphi)\partial_t\varphi\big)\, dx$$

$$= -\int_{\Omega} \varrho(T - T_M)\frac{L}{2}\partial_t\varphi\, dx + \int_{\Omega} \lambda\Delta T\, (T - T_M)\, dx \qquad (7.69)$$

$$+ \int_{\Omega} \gamma\big(-\varepsilon\Delta\varphi + \tfrac{1}{\varepsilon}\psi'(\varphi)\big)\partial_t\varphi\, dx$$

$$= -\int_{\Omega} \varepsilon\beta|\partial_t\varphi|^2\, dx - \int_{\Omega} \lambda|\nabla T|^2\, dx .$$

Comparing this identity with (7.64) we can conjecture that the term

$$\int_{\Omega} \gamma\left(\frac{\varepsilon}{2}|\nabla\varphi|^2 + \frac{1}{\varepsilon}\psi(\varphi)\right)\, dx \qquad (7.70)$$

corresponds to the capillary term $\int_{\Gamma} \gamma\, ds_x$. This can be shown with the help of a suitable limiting process, see [99]. In this context we remark that the gradient term $\frac{\varepsilon}{2}|\nabla\varphi|^2$ in (7.70) penalizes strong spatial changes in the phase field variables. The term $\frac{1}{\varepsilon}\psi(\varphi)$ in (7.70) will become large for small $\varepsilon$ if $\varphi$ attains values which differ strongly from $\pm 1$, compare Fig. 7.4. Solutions $(T, \varphi)$ of the phase field system (7.67), (7.68) hence typically lead to "diffuse" phase boundaries, see Figs. 7.5 and 7.6.

We now want to show with the help of formal asymptotic analysis that solutions of (7.67)–(7.68) converge to solutions of the Stefan problem with Gibbs–Thomson condition and kinetic undercooling. We now consider solutions $(T^\varepsilon, \varphi^\varepsilon)$ of (7.67), (7.68). For simplicity we consider an $\Omega \subset \mathbb{R}^2$ and we assume that the sets

$$\Gamma^\varepsilon = \{(t, x) \in [0, T] \times \Omega \mid \varphi^\varepsilon(t, x) = 0\} \text{ for } \varepsilon > 0$$

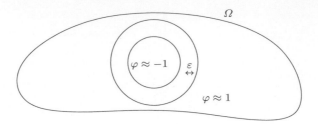

**Fig. 7.5**  Typical behavior of the phase field variable $\varphi$. Regions in which $\varphi \approx \pm 1$ are separated by a "diffuse" phase boundary with a thickness which is proportional to $\varepsilon$

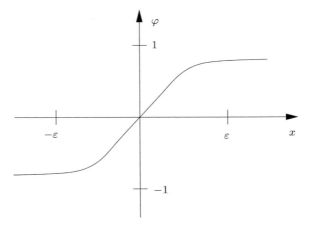

**Fig. 7.6**  The phase field variable forms a phase boundary whose thickness is proportional to $\varepsilon$

are evolving hypersurfaces which for all times do not intersect the boundary of $\Omega$. In addition we assume that the $\Gamma^\varepsilon$ converge towards an evolving hypersurface $\Gamma^0$ in a sense which will be made precise later. Now let $y(t, s)$ be a smooth function which for all $t$ is an arc-length parametrization of $\Gamma^0(t)$. In addition, let $\nu(t, s)$ be the unit normal at $\Gamma^0(t)$ in the point $y(t, s)$ which points into the liquid phase. The parametrization is assumed to be oriented such that $(\nu, \tau)$ with the tangent

$$\tau(t, s) = \partial_s y(t, s)$$

is positively oriented.

In addition, we assume that we can parametrize $\Gamma^\varepsilon(t)$ with the help of $\Gamma^0(t)$ as follows

$$y^\varepsilon(t, s) = y(t, s) + d^\varepsilon(t, s)\, \nu(t, s).$$

The hypersurfaces $\Gamma^\varepsilon(t)$ converge to $\Gamma^0(t)$ if $d^\varepsilon \to 0$ as $\varepsilon \to 0$. We make the ansatz

$$d^\varepsilon(t, s) = d_0(t, s) + \varepsilon\, d_1(t, s) + \cdots.$$

The convergence of $\Gamma^\varepsilon$ to $\Gamma^0$ as $\varepsilon \to 0$ means that

$$d_0 \equiv 0 \,.$$

In the set $\Omega \setminus \Gamma^0(t)$ we will make an outer expansion and in the neighborhood of $\Gamma^0$ we make an ansatz with an inner expansion in a similar way as in Sect. 6.6. For the outer expansion we make the ansatz

$$T^\varepsilon(t, x) = T_0(t, x) + \varepsilon\, T_1(t, x) + \cdots ,$$
$$\varphi^\varepsilon(t, x) = \varphi_0(t, x) + \varepsilon\, \varphi_1(t, x) + \cdots .$$

Plugging this ansatz in the Eq. (7.68) we obtain to leading order after a Taylor expansion

$$\psi'(\varphi_0) = 0 \,.$$

This equation has the solutions $\varphi_0 = -1$, $\varphi_0 = 0$ and $\varphi_0 = 1$. The solutions $\varphi_0 = 1$ and $\varphi_0 = -1$ correspond to the liquid and the solid phase. Since $\psi''(0) < 0$ one obtains that $\varphi_0 = 0$ is an unstable solution of $\varepsilon \beta\, \partial_t \varphi = -\frac{\gamma}{\varepsilon}\psi'(\varphi)$, and as we expect that unstable solutions are not observed in nature we will not consider this solution any longer. The liquid phase at time $t$ is given by the set

$$\Omega_\ell(t) := \{x \in \Omega \mid \varphi_0(t, x) = 1\}$$

and the solid phase at time $t$ is the set

$$\Omega_s(t) := \{x \in \Omega \mid \varphi_0(t, x) = -1\} \,.$$

In both phases the Eq. (7.67) gives to leading order

$$\varrho c_V\, \partial_t T_0 = \lambda \Delta T_0 \,.$$

### Inner Expansion

In the neighborhood of the phase boundary $\Gamma^0$ we introduce a new coordinate system. We consider the parameter transformation

$$F^\varepsilon(t, s, z) := (t,\, y(t, s) + \varepsilon z\, \nu(t, s)) \,.$$

The variable $z$ describes the $\varepsilon$-scaled signed distance to $\Gamma^0$ where we choose a negative sign for points in the solid phase (cf. Exercise 7.14). We now define

$$V = \partial_t y \cdot \nu \quad \text{and} \quad V_{\tan} = \partial_t y \cdot \tau$$

as the normal velocity and the tangential velocity of the parameterization of the evolving curve $\Gamma^0$ and in addition $\kappa$ is defined to be its curvature. Then (cf. Exercise 7.15)

$$\partial_s \tau = \kappa \nu \quad \text{and} \quad \partial_s \nu = -\kappa \tau \quad \text{(Frenet formula)} \tag{7.71}$$

holds. With this we obtain

$$D_{(t,s,z)} F^\varepsilon(t, s, z) = \begin{pmatrix} 1 & 0 & 0 \\ \partial_t y + \varepsilon z \, \partial_t \nu & (1 - \varepsilon z \kappa) \tau & \varepsilon \nu \end{pmatrix}. \tag{7.72}$$

From the identity $\nu \cdot \partial_t \nu = \frac{1}{2} \partial_t |\nu|^2 = \frac{1}{2} \partial_t 1 = 0$ we derive with $(t, x) = F^\varepsilon(t, s, z)$ and $(F^\varepsilon)^{-1}(t, x) =: (t, s(t, x), z(t, x))$

$$\begin{pmatrix} \partial_t t(t, x) & D_x t(t, x) \\ \partial_t s(t, x) & D_x s(t, x) \\ \partial_t z(t, x) & D_x z(t, x) \end{pmatrix} = (DF_\varepsilon)^{-1}(t, s(t, x), z(t, x))$$

$$= \begin{pmatrix} 1 & (0, 0) \\ -\frac{1}{1-\varepsilon\kappa z}(V_{\tan} + \varepsilon z \, \tau \cdot \partial_t \nu) & \frac{1}{1-\varepsilon\kappa z}\tau^{\mathsf{T}} \\ -\frac{1}{\varepsilon}V & \frac{1}{\varepsilon}\nu^{\mathsf{T}} \end{pmatrix}. \tag{7.73}$$

Using the chain rule we obtain for a scalar function $b(t, s, z)$ and a vector field $j(t, s, z)$

$$\frac{d}{dt} b(t, s(t, x), z(t, x)) = -\frac{1}{\varepsilon} V \partial_z b + \mathcal{O}(1),$$

$$\nabla_x b(t, s(t, x), z(t, x)) = \frac{1}{\varepsilon} \partial_z b \, \nu + \left(1 + \varepsilon \kappa z + \mathcal{O}(\varepsilon^2)\right) \partial_s b \, \tau,$$

$$\nabla_x \cdot j(t, s(t, x), z(t, x)) = \frac{1}{\varepsilon} \partial_z j \cdot \nu + (1 + \varepsilon \kappa z) \partial_s j \cdot \tau + \mathcal{O}(\varepsilon^2),$$

$$\Delta_x b(t, s(t, x), z(t, x)) = \nabla_x \cdot \left(\nabla_x b(t, s(t, x), z(t, x))\right)$$

$$= \frac{1}{\varepsilon^2} \partial_{zz} b - \frac{1}{\varepsilon} \kappa \, \partial_z b - \kappa^2 z \, \partial_z b + \partial_{ss} b + \mathcal{O}(\varepsilon).$$

In the neighborhood of the phase boundary we introduce the new coordinates and define functions $\Theta^\varepsilon$ and $\Phi^\varepsilon$ such that

$$T^\varepsilon(t, x) = \Theta^\varepsilon(t, s(t, x), z(t, x)),$$

$$\varphi^\varepsilon(t, x) = \Phi^\varepsilon(t, s(t, x), z(t, x)).$$

We expand $\Theta^\varepsilon$ and $\Phi^\varepsilon$ in these new variables

$$\Theta^\varepsilon(t, s, z) = \Theta_0(t, s, z) + \varepsilon \, \Theta_1(t, s, z) + \cdots,$$

$$\Phi^\varepsilon(t, s, z) = \Phi_0(t, s, z) + \varepsilon \, \Phi_1(t, s, z) + \cdots.$$

The phase field Eq. (7.68) yields to leading order

$$0 = \partial_{zz} \Phi_0 - \psi'(\Phi_0). \tag{7.74}$$

Now and also later we need matching conditions relating inner and outer solutions to obtain boundary conditions for $z \to \pm\infty$. We will now derive these conditions for all orders that we will need later.

### Matching Conditions

Let us introduce the variable $r = z\varepsilon$ and write the function $\varphi_k$ in the variables $(t, s, r)$ as follows

$$\varphi_k(t, x) = \widehat{\varphi}_k(t, s, r), \quad k = 0, 1, 2, \dots .$$

In addition, we assume that the functions $\varphi_k$ from both sides can be smoothly extended onto $\Gamma^0$. Taylor expansion close to $r = 0$ yields

$$\widehat{\varphi}_k(t, s, r) = \widehat{\varphi}_k(t, s, 0+) + \partial_r \widehat{\varphi}_k(t, s, 0+)r + \cdots \quad \text{for} \quad r > 0 \quad \text{small},$$
$$\widehat{\varphi}_k(t, s, r) = \widehat{\varphi}_k(t, s, 0-) + \partial_r \widehat{\varphi}_k(t, s, 0-)r + \cdots \quad \text{for} \quad r < 0 \quad \text{small}.$$

We expect that the functions $\Phi_k(t, s, z)$ for large $z$ can be approximated by a polynomial in $z$, i.e., for all $k \in \mathbb{N}_0$ there exists an $n_k \in \mathbb{N}$ such that

$$\Phi_k(t, s, z) \approx \Phi^\pm_{k,0}(t, s) + \Phi^\pm_{k,1}(t, s) z + \cdots + \Phi^\pm_{k,n_k}(t, s) z^{n_k} \quad \text{for} \quad z \to \pm\infty .$$

Here we use the notation $f \approx g$ for $z \to \pm\infty$ if $f - g$ together with all its derivatives with respect to $z$ converge for $z \to \pm\infty$ to zero. To obtain the matching condition we consider as in Sect. 6.6 an "intermediate variable" $r_\eta = \frac{r}{\varepsilon^\alpha} = \varepsilon^{1-\alpha}z$ with $0 < \alpha < 1$ and analyze for fixed $r_\eta$ the limit $\varepsilon \to 0$. To simplify the notation we suppress the dependence on $t$ and $s$ in what follows. On the side of the liquid phase we obtain

$$\widehat{\varphi}^\varepsilon(\varepsilon^\alpha r_\eta) = \widehat{\varphi}_0(0+) + \varepsilon^\alpha \partial_r \widehat{\varphi}_0(0+) r_\eta + \mathcal{O}(\varepsilon^{2\alpha}) + \varepsilon \widehat{\varphi}_1(0+) + \mathcal{O}(\varepsilon^{1+\alpha}) + \cdots$$

and

$$\widehat{\Phi}^\varepsilon(\varepsilon^{\alpha-1} r_\eta) = \Phi^+_{0,0} + \varepsilon^{\alpha-1} \Phi^+_{0,1} r_\eta + \cdots + \varepsilon^{n_0(\alpha-1)} \Phi^+_{0,n_0} (r_\eta)^{n_0}$$
$$+ \varepsilon \Phi^+_{1,0} + \varepsilon^\alpha \Phi^+_{1,1} r_\eta + \cdots + \varepsilon^{1+n_1(\alpha-1)} \Phi^+_{1,n_1} (r_\eta)^{n_1} + \cdots .$$

We say that the expansions match if $\widehat{\varphi}^\varepsilon$ and $\widehat{\Phi}^\varepsilon$ coincide to all orders in $\varepsilon$ and $r_\eta$. Comparing the coefficients yields

$$\Phi^+_{0,0} = \widehat{\varphi}_0(0+), \quad \Phi^+_{0,i} = 0 \text{ for } 1 \le i \le n_0,$$
$$\widehat{\Phi}^+_{1,0} = \widehat{\varphi}_1(0+), \quad \Phi^+_{1,1} = \partial_r \widehat{\varphi}_0(0+), \quad \Phi^+_{1,i} = 0 \text{ for } 2 \le i \le n_1.$$

Corresponding equations hold for the solid phase.

For $x \in \Gamma_t$ we denote by $\varphi_0(x\pm)$, $\nabla\varphi_0(x\pm)$ the limits of the quantities $\varphi_0$, $\nabla\varphi_0$ from the liquid and the solid phase, respectively, where we suppressed the $t$-dependence. As $\partial_r \widehat{\varphi}_0(0\pm) = (\nabla\varphi_0(x\pm) \cdot \nu)$ and since

$$\Phi_1(t, s, z) \approx \Phi_{1,0}^{\pm}(t, s) + \Phi_{1,1}^{\pm}(t, s)\, z \quad \text{for } z \to \pm\infty$$

we obtain the matching conditions

$$\Phi_0(z) \approx \varphi_0(x\pm)\,,\ \partial_z\Phi_0(z) \approx 0\,,$$
$$\Phi_1(z) \approx \varphi_1(x\pm) + (\nabla\varphi_0(x\pm) \cdot \nu)z\,,$$
$$\partial_z\Phi_1(z) \approx \nabla\varphi_0(x\pm) \cdot \nu$$

which have to hold for $z \to \pm\infty$. Analogous conditions can be shown for the temperature.

### Conditions at the Phase Boundary

For $\Phi_0$ we obtain the conditions

$$\Phi_0(t, s, z) \to \pm 1 \quad \text{for } z \to \pm\infty\,.$$

Multiplying (7.74) by $\partial_z\Phi_0$ yields

$$0 = \frac{d}{dz}\left(\tfrac{1}{2}(\partial_z\Phi_0)^2 - \psi(\Phi_0)\right)\,.$$

Using the matching condition it follows that

$$\tfrac{1}{2}(\partial_z\Phi_0)^2 = \psi(\Phi_0)$$

and as $\Phi_0$ is monotonically increasing we obtain

$$\partial_z\Phi_0 = \sqrt{2\psi(\Phi_0)}\,.$$

Choosing, as stated above, $\psi(\varphi) = \tfrac{9}{32}(\varphi^2 - 1)^2$ we obtain

$$\Phi_0(t, s, z) = \tanh\left(\tfrac{3}{4}z\right)$$

as a solution. In principle any translation with respect to $z$ is also a solution. However we will see that the final result will be independent of this translation. Considering (7.67) to leading order yields

$$0 = \partial_{zz}\Theta_0\,.$$

The matching conditions yield that $\Theta_0(z)$ has to be bounded for $z \to \pm\infty$. We hence obtain that $\Theta_0(z)$ has to be constant and the matching condition yields

$$T_0 \text{ is continuous across } \Gamma_0\,.$$

The Eq. (7.68) yields to leading order $\mathcal{O}(1)$ the condition

$$\partial_{zz}\Phi_1 - \psi''(\Phi_0)\Phi_1 = \left(\kappa - \frac{\beta}{\gamma}V\right)\partial_z\Phi_0 - \frac{\varrho L}{2\gamma}(T_0 - T_M).\qquad(7.75)$$

The left-hand side is the linearization of the operator on the right-hand side in (7.74). The Eq. (7.74) is translation invariant and this fact implies that the linearization has a nontrivial kernel: If $\Phi_0(z)$ is a solution of (7.74) then we obtain that $\Phi_0(z+y)$ is for all $y \in \mathbb{R}$ a solution of (7.74). This implies

$$0 = \frac{d}{dy}\left[\partial_{zz}\Phi_0(z+y) - \psi'(\Phi_0(z+y))\right]$$

$$= \partial_{zz}(\partial_z\Phi_0(z+y)) - \psi''(\Phi_0(z+y))\,\partial_z\Phi_0(z+y).$$

Setting $y = 0$ we observe: $\partial_z\Phi_0$ lies in the kernel of the differential operator $\partial_{zz}\Phi - \psi''(\Phi_0)\Phi$. We obtain for all solutions $\Phi_1$ of (7.75) with $\Phi_1(\pm\infty) = 0$ with the help of integration by parts that

$$\int_{\mathbb{R}}\left(\partial_{zz}\Phi_1 - \psi''(\Phi_0)\Phi_1\right)\partial_z\Phi_0\,dz = \int_{\mathbb{R}}\Phi_1\left(\partial_{zz}(\partial_z\Phi_0) - \psi''(\Phi_0)\partial_z\Phi_0\right)dz = 0$$

holds. This implies that the right-hand side in (7.75) has to fulfill a solvability condition. This condition reads as

$$\left(\kappa - \frac{\beta}{\gamma}V\right)\int_{\mathbb{R}}|\partial_z\Phi_0|^2\,dz - \frac{\varrho L}{2\gamma}(T_0 - T_M)\int_{\mathbb{R}}\partial_z\Phi_0\,dz = 0.$$

Using

$$\int_{\mathbb{R}}|\partial_z\Phi_0|^2\,dz = \int_{\mathbb{R}}\partial_z\Phi_0\sqrt{2\psi(\Phi_0)}\,dz$$

$$= \int_{-1}^{1}\sqrt{2\psi(y)}\,dy = \int_{-1}^{1}\tfrac{3}{4}(1 - y^2)\,dy = 1,$$

it follows that

$$\gamma\kappa - \beta V - \varrho L(T_0 - T_M) = 0.$$

We now want to derive the Stefan condition at the phase boundary. The energy conservation (7.67) yields to the order $\mathcal{O}(\frac{1}{\varepsilon})$

$$-V\varrho\frac{L}{2}\partial_z\Phi_0 = \lambda\,\partial_{zz}\Theta_1.$$

It follows that

$$-V \frac{\varrho L}{2} \Phi_0 = \lambda \partial_z \Theta_1 + \alpha(t, s) .$$

The conditions $\Phi_0(z) \to \pm 1$ and $\partial_z \Theta_1(z) \to \nabla T_0(x\pm) \cdot \nu$ for $z \to \pm\infty$ yield

$$-\varrho L V = [\lambda \nabla T_0 \cdot \nu]_s^\ell .$$

By now we derived all conditions which have to hold for the Stefan problem.

Analogous to (7.64) for the Stefan problem, the estimate (7.69) does not correspond to the entropy inequality. We now want to derive an entropy inequality. In order to achieve this we need to replace the phase field equation by

$$\varepsilon \widehat{\beta} \, \partial_t \varphi = \varepsilon \widehat{\gamma} \, \Delta \varphi - \frac{\widehat{\gamma}}{\varepsilon} \psi'(\varphi) - \frac{\varrho L}{2} \left( \frac{1}{T} - \frac{1}{T_M} \right) \qquad (7.76)$$

(cf. Exercise 7.18). The system (7.67), (7.76) has been introduced by Penrose and Fife, see [15, 107].

The phase field method has found many applications and we refer to Chen [19], Eck, Garcke and Stinner [31], Garcke, Nestler and Stinner [47], Garcke [45] and [46] for more details.

## 7.10   Free Surfaces in Fluid Mechanics

In this section we want to study the motion of a fluid with a free surface. To simplify the situation we consider the following geometry: At time $t$ the fluid fills the domain

$$\Omega(t) = \left\{ (x', x_3) = (x_1, x_2, x_3) \in \mathbb{R}^3 \mid x' \in \Omega', \ 0 < x_3 < h(t, x') \right\} \qquad (7.77)$$

where $\Omega' \subset \mathbb{R}^2$ is a bounded domain and $h : [0, T] \times \Omega' \to \mathbb{R}$ is a smooth function which only attains positive values. The fluid hence fills a part of a cylindrical volume with base $\Omega'$. We assume that the Navier–Stokes equations for incompressible fluids

$$\varrho(\partial_t v + (v \cdot \nabla)v) = \mu \Delta v - \nabla p + \varrho f , \qquad (7.78)$$
$$\nabla \cdot v = 0 \qquad (7.79)$$

are fulfilled in

$$Q = \{(t, x) \mid x \in \Omega(t)\} .$$

We only have to specify the conditions on $\partial\Omega(t)$. On the fixed boundary, i.e., for points $x = (x', x_3)$ with $x' \in \partial\Omega'$ or $x_3 = 0$ we specify the no-slip condition

$$v(t, x', x_3) = 0 \ \text{ if } \ x' \in \partial\Omega' \ \text{ or } \ x_3 = 0 . \qquad (7.80)$$

The fluid hence adheres to the container boundary. This condition is not always meaningful and in the literature also slip-conditions are discussed. These conditions allow for a velocity different from zero at the boundary, see Exercise 7.24.

On the free surface

$$\Gamma(t) = \left\{ (x', x_3) \in \mathbb{R}^3 \mid x' \in \Omega', \ x_3 = h(t, x') \right\}$$

we demand a kinetic condition. This condition makes sure that the free boundary moves with the velocity of the fluid. A point on the surface which moves with the velocity of the fluid can be described as $(x'(t), h(t, x'(t)))$. This gives

$$\partial_t x'(t) = (v_1, v_2)(t, x'(t), h(t, x'(t))),$$

$$\frac{d}{dt} h(t, x'(t)) = v_3(t, x'(t), h(t, x'(t))).$$

We hence obtain the *kinetic boundary condition*

$$v_3 = \partial_t h + v_1 \, \partial_1 h + v_2 \, \partial_2 h \quad \text{on} \quad \Gamma(t). \tag{7.81}$$

The forces which act from the fluid on the surface $\Gamma(t)$ are given, as discussed in the Sects. 5.5 and 5.6, by $\sigma(-\nu)$ with the stress tensor

$$\sigma = \mu(Dv + (Dv)^\top) - pI \tag{7.82}$$

and the normal vector $\nu = \frac{1}{\sqrt{1+|\nabla_{x'}h|^2}} \left( -\nabla_{x'}h, 1 \right)^\top$. On the free surface these forces have to be balanced by capillary forces. The capillary forces try to reduce the total surface area. They act in the direction of the normal $\nu$ and they are proportional to the mean curvature. We hence obtain the force balance

$$\sigma\nu = \gamma\kappa\nu. \tag{7.83}$$

The tangential parts of $\sigma\nu$ hence have to be zero and the normal part is proportional to the mean curvature. We now show that for solutions of the Navier–Stokes equations in $\Omega$ with boundary conditions (7.80)–(7.83) an energy inequality holds if for the angle between the free surface and the boundary of the container the angle condition

$$\sphericalangle(\Gamma(t), \partial\Omega) = 90° \tag{7.84}$$

holds. The condition (7.84) can be interpreted as boundary condition for the curvature operator $\kappa$. From Appendix B it follows that

$$\kappa = \nabla_{x'} \cdot \left( \frac{\nabla_{x'}h}{\sqrt{1 + |\nabla_{x'}h|^2}} \right).$$

The condition (7.84) yields

$$\nabla_{x'} h \cdot n_{\partial \Omega'} = 0 \tag{7.85}$$

and this can be interpreted as a boundary condition for the operator

$$\nabla_{x'} \cdot \left( \frac{\nabla_{x'} h}{\sqrt{1 + |\nabla_{x'} h|^2}} \right).$$

In Exercise 7.21 it is discussed how models with more general angle conditions can be treated.

**Theorem 7.5** *A smooth solution of* (7.78)–(7.84) *fulfills in the case* $f = 0$ *the identity*

$$\frac{d}{dt} \left[ \int_{\Omega(t)} \frac{\varrho}{2} |v(t,x)|^2 \, dx + \int_{\Gamma(t)} \gamma \, ds_x \right] = - \int_{\Omega(t)} \sigma : Dv \, dx$$

$$= - \int_{\Omega(t)} \frac{\mu}{2} \left( Dv + Dv^\top \right) : \left( Dv + Dv^\top \right) dx$$

$$\leq 0 \, .$$

*Proof* We only give a proof for the case that $\Gamma(t)$ can be written as a graph for all times $t$, cf. (7.77). With Theorem 7.3 and the identity $\nabla \cdot \sigma = \mu \Delta v - \nabla p$ it follows that

$$\frac{d}{dt} \int_{\Omega(t)} \frac{\varrho}{2} |v|^2 \, dx = \int_{\Omega(t)} \varrho v \cdot \partial_t v \, dx + \int_{\Gamma(t)} \frac{\varrho}{2} |v|^2 V \, ds_x$$

$$= \int_{\Omega(t)} \left( - \varrho v \cdot (v \cdot \nabla) v + (\nabla \cdot \sigma) \cdot v \right) dx + \int_{\Gamma(t)} \frac{\varrho}{2} |v|^2 V \, ds_x \, .$$

Here, $\Omega$ plays the role of $\Omega_s$ in Theorem 7.3 and $\mathbb{R}^3 \setminus \Omega$ takes the role of $\Omega_\ell$. In addition, $\frac{\varrho}{2} |v|^2$ is extended by zero to $\mathbb{R}^3 \setminus \Omega$. It holds that $v \cdot (v \cdot \nabla) v = \frac{1}{2} v \cdot \nabla |v|^2$ and as $v$ is divergence free we obtain after integration by parts and by using the boundary conditions

$$\frac{d}{dt} \int_{\Omega(t)} \frac{\varrho}{2} |v|^2 \, dx = - \int_{\Omega(t)} \sigma : Dv \, dx + \int_{\Gamma(t)} \frac{\varrho}{2} |v|^2 (V - v \cdot \nu) \, ds_x$$

$$+ \int_{\Gamma(t)} v \cdot \sigma \nu \, ds_x$$

$$= - \int_{\Omega(t)} \sigma : Dv \, dx + \int_{\Gamma(t)} (v \cdot \nu) \gamma \kappa \, ds_x \, .$$

Here we used in the last step that $\nu = \frac{1}{\sqrt{1+|\nabla_{x'}h|^2}}(-\nabla_{x'}h, 1)$ implies

$$v \cdot \nu = \frac{1}{\sqrt{1+|\nabla_{x'}h|^2}}(-v_1\partial_1 h - v_2\partial_2 h + v_3) = \frac{\partial_t h}{\sqrt{1+|\nabla_{x'}h|^2}} = V.$$

Using the angle condition (7.84) or (7.85) it follows that

$$\frac{d}{dt}\int_{\Gamma(t)} \gamma \, ds_x = \frac{d}{dt}\int_{\Omega'} \gamma \sqrt{1+|\nabla_{x'}h|^2}\, dx'$$

$$= \int_{\Omega'} \frac{\gamma}{\sqrt{1+|\nabla_{x'}h|^2}} \nabla_{x'}h \cdot \nabla \partial_t h \, dx' = -\int_{\Omega'} \gamma \nabla_{x'} \cdot \left(\frac{\nabla_{x'}h}{\sqrt{1+|\nabla_{x'}h|^2}}\right) \partial_t h \, dx'$$

$$= -\int_{\Omega'} \gamma \kappa \frac{\partial_t h}{\sqrt{1+|\nabla_{x'}h|^2}} \sqrt{1+|\nabla_{x'}h|^2}\, dx' = -\int_{\Gamma(t)} \gamma \kappa V \, ds_x.$$

Altogether we obtain with (7.79), (7.82)

$$\frac{d}{dt}\left[\int_{\Omega(t)} \frac{\varrho}{2}|v|^2 \, dx + \int_{\Gamma(t)} \gamma \, ds_x\right] = -\int_{\Omega(t)} \sigma : Dv \, dx$$

$$= \int_{\Omega(t)} p\, \nabla \cdot v \, dx - \int_{\Omega(t)} \mu(Dv + Dv^\top) : Dv \, dx$$

$$= -\int_{\Omega(t)} \frac{\mu}{2}(Dv + Dv^\top) : (Dv + Dv^\top)\, dx,$$

where in the last identity we used that $Dv^\top : Dv^\top = Dv : Dv$. ⊔

Taking into account capillary terms for free surfaces in fluid mechanics hence yields a similar situation as in the Stefan problem with a surface energy term.

For free surfaces with the Euler equations we set $\mu = 0$ in (7.78) and the boundary condition (7.83) simplifies to

$$p = -\gamma \kappa. \tag{7.86}$$

In addition, the condition (7.80) changes to

$$v(t, x', x_3) \cdot n = 0 \quad \text{if} \quad x' \in \partial\Omega' \quad \text{or} \quad x_3 = 0, \tag{7.87}$$

where $n$ is the exterior unit normal. In this case we obtain for the free surface problem for the Euler equations (see Exercise 7.22)

$$\frac{d}{dt}\left[\int_{\Omega(t)} \frac{\varrho}{2}|v|^2 \, dx + \int_{\Gamma(t)} \gamma \, ds_x\right] = 0. \tag{7.88}$$

## 7.11 Thin Films and Lubrication Approximation

In many applications the height of a fluid domain is small when compared to the horizontal extension. This is the case for example for wetting films or paint coats on a wall. In this case it is often possible to simplify the problem with free surface studied in the previous section considerably.

Let $H$ be a typical height of the film, $L$ be a typical horizontal length scale and $\overline{V}$ a typical size of the vertical velocity components. In addition, we assume

$$\varepsilon := \tfrac{H}{L} \ll 1 \,.$$

We now denote the nondimensional variables with a $\widehat{\phantom{x}}$ and obtain

$$\begin{aligned}
(x_1, x_2) &= L(\widehat{x}_1, \widehat{x}_2)\,, & x_3 &= H\widehat{x}_3\,, & t &= (L/\overline{V})\widehat{t}\,,\\
(v_1, v_2)(t, x) &= \overline{V}(\widehat{v}_1, \widehat{v}_2)(\widehat{t}, \widehat{x})\,, & v_3(t, x) &= \varepsilon\overline{V}\widehat{v}_3(\widehat{t}, \widehat{x})\,,\\
h(t, x_1, x_2) &= H\widehat{h}(\widehat{t}, \widehat{x}_1, \widehat{x}_2)\,, & p(t, x) &= \overline{p}\,\widehat{p}(\widehat{t}, \widehat{x})\,.
\end{aligned}$$

We will specify a typical value $\overline{p}$ for the pressure later.

In the following we also denote differential operators with a $\widehat{\phantom{x}}$ if they correspond to the variables $\widehat{x}$ and $\widehat{t}$. We hence denote

$$\begin{aligned}
\partial_t v_i &= \overline{V}\tfrac{\overline{V}}{L}\partial_{\widehat{t}}\widehat{v}_i\,, & i &= 1, 2\,,\\[4pt]
\partial_t v_3 &= \varepsilon\overline{V}\tfrac{\overline{V}}{L}\partial_{\widehat{t}}\widehat{v}_3\,,\\[4pt]
\partial_{x_i} v_j &= \tfrac{\overline{V}}{L}\partial_{\widehat{x}_i}\widehat{v}_j\,, & i, j &= 1, 2\,,\\[4pt]
\partial_{x_3} v_j &= \tfrac{\overline{V}}{L}\varepsilon^{-1}\partial_{\widehat{x}_3}\widehat{v}_j\,, & j &= 1, 2\,,\\[4pt]
\partial_{x_i} v_3 &= \tfrac{\overline{V}}{L}\varepsilon\,\partial_{\widehat{x}_i}\widehat{v}_3\,, & i &= 1, 2\,,\\[4pt]
\partial_{x_3} v_3 &= \varepsilon\,\overline{V}\tfrac{1}{H}\,\partial_{\widehat{x}_3}\widehat{v}_3 = \tfrac{\overline{V}}{L}\partial_{\widehat{x}_3}\widehat{v}_3\,,\\[4pt]
\partial_{x_i x_i} v_j &= \tfrac{\overline{V}}{L^2}\partial_{\widehat{x}_i\widehat{x}_i}\widehat{v}_j\,, & i, j &= 1, 2\,,\\[4pt]
\partial_{x_3 x_3} v_j &= \tfrac{\overline{V}}{L^2}\varepsilon^{-2}\partial_{\widehat{x}_3\widehat{x}_3}\widehat{v}_j\,, & j &= 1, 2\,,\\[4pt]
\partial_{x_i x_i} v_3 &= \tfrac{\overline{V}}{L^2}\varepsilon\,\partial_{\widehat{x}_i\widehat{x}_i}\widehat{v}_3\,, & i &= 1, 2\,,\\[4pt]
\partial_{x_3 x_3} v_3 &= \tfrac{\overline{V}}{L^2}\varepsilon^{-1}\partial_{\widehat{x}_3\widehat{x}_3}\widehat{v}_3\,,\\[4pt]
\partial_{x_i} p &= \tfrac{\overline{p}}{L}\partial_{\widehat{x}_i}\widehat{p}\,, & i &= 1, 2\,,\\[4pt]
\partial_{x_3} p &= \tfrac{\overline{p}}{L}\varepsilon^{-1}\partial_{\widehat{x}_3}\widehat{p}\,.
\end{aligned}$$

In addition, it holds that

$$\nabla h = \frac{H}{L} \widehat{\nabla} \, \widehat{h} = \varepsilon \, \widehat{\nabla} \, \widehat{h} \,,$$

$$\nu = \frac{1}{\sqrt{1 + |\nabla h|^2}} \begin{pmatrix} -\nabla h \\ 1 \end{pmatrix} = \begin{pmatrix} 0 \\ 0 \\ 1 \end{pmatrix} + \mathcal{O}(\varepsilon) \,,$$

$$\kappa = \nabla \cdot \left( \frac{\nabla h}{\sqrt{1 + |\nabla h|^2}} \right) = \frac{1}{L} \varepsilon \, \widehat{\Delta} \, \widehat{h} + \mathcal{O}(\varepsilon^2) \,.$$

Let us neglect terms of higher order in $\varepsilon$. Here $\overline{V}$ and $\overline{p}$ need to be scaled appropriately. In the Navier–Stokes equations the viscous part of the stress tensor and the pressure have to be scaled in the same way, i.e.,

$$\mu \, \Delta v \approx \nabla p \,.$$

From the first two components we can deduce that the scaling

$$\mu \frac{\overline{V}}{L^2} \varepsilon^{-2} = \frac{\overline{p}}{L}$$

is meaningful, that is,

$$\overline{V} = \frac{\varepsilon^2 L}{\mu} \overline{p} \,.$$

The tangential component of the force balance (7.83) on the upper boundary yields to leading order

$$\partial_{\widehat{x}_3} \widehat{v}_j = 0 \quad \text{for} \quad j = 1, 2 \,.$$

In (7.83) the terms of lowest order have the factors $\gamma \varepsilon / L$ and $\overline{p}$. Hence we choose

$$\overline{p} = \frac{\varepsilon \gamma}{L} \,.$$

From the normal component of (7.83) it then follows that

$$\widehat{p} = -\widehat{\Delta} \, \widehat{h} + \mathcal{O}(\varepsilon) \quad \text{for} \quad \widehat{x}_3 = \widehat{h}(\widehat{t}, \widehat{x}_1, \widehat{x}_2) \,.$$

The Navier–Stokes equations yield after multiplication by $\frac{L^2 \varepsilon^2}{\overline{V} \mu}$

$$\varepsilon^2 \mathrm{Re} \big( \partial_{\widehat{t}} \widehat{v}_1 + \widehat{v}_1 \, \partial_{\widehat{x}_1} \widehat{v}_1 + \widehat{v}_2 \, \partial_{\widehat{x}_2} \widehat{v}_1 + \widehat{v}_3 \, \partial_{\widehat{x}_3} \widehat{v}_1 \big)$$
$$= -\partial_{\widehat{x}_1} \widehat{p} + \varepsilon^2 \partial_{\widehat{x}_1 \widehat{x}_1} \widehat{v}_1 + \varepsilon^2 \partial_{\widehat{x}_2 \widehat{x}_2} \widehat{v}_1 + \partial_{\widehat{x}_3 \widehat{x}_3} \widehat{v}_1$$

with the Reynolds number $\mathrm{Re} = \varrho L \overline{V}/\mu$. In addition, we get an analogous equation for $\widehat{v}_2$. For $\widehat{v}_3$ it holds that

$$\varepsilon^3 \mathrm{Re}\left(\partial_{\widehat{t}}\widehat{v}_3 + \widehat{v}_1\,\partial_{\widehat{x}_1}\widehat{v}_3 + \widehat{v}_2\,\partial_{\widehat{x}_2}\widehat{v}_3 + \widehat{v}_3\,\partial_{\widehat{x}_3}\widehat{v}_3\right)$$
$$= -\varepsilon^{-1}\partial_{\widehat{x}_3}\widehat{p} + \varepsilon^3\partial_{\widehat{x}_1\widehat{x}_1}\widehat{v}_3 + \varepsilon^3\partial_{\widehat{x}_2\widehat{x}_2}\widehat{v}_3 + \varepsilon\,\partial_{\widehat{x}_3\widehat{x}_3}\widehat{v}_3\,.$$

The equation $\nabla \cdot v = 0$ yields after multiplication by $L/\overline{V}$

$$\widehat{\nabla} \cdot \widehat{v} = 0\,.$$

Under the assumptions

$$\varepsilon^2 \mathrm{Re} \ll 1 \ \text{ and } \ \varepsilon \ll 1$$

we obtain as equations to leading order in the fluid region

$$\partial_{\widehat{x}_3\widehat{x}_3}\widehat{v}_i = \partial_{\widehat{x}_i}\widehat{p} \ \text{ for } i = 1, 2\,,$$
$$0 = \partial_{\widehat{x}_3}\widehat{p}\,,$$

and on the free boundary $\widehat{x}_3 = \widehat{h}(\widehat{t}, \widehat{x}_1, \widehat{x}_2)$ we obtain

$$\partial_{\widehat{x}_3}\widehat{v}_i = 0 \ \text{ for } i = 1, 2\,,$$
$$\widehat{p} = -\widehat{\Delta}\,\widehat{h}\,, \tag{7.89}$$
$$\widehat{v}_3 = \partial_{\widehat{t}}\widehat{h} + (\partial_{\widehat{x}_1}\widehat{h})\widehat{v}_1 + (\partial_{\widehat{x}_2}\widehat{h})\widehat{v}_2\,,$$

where the last condition is just the kinematic boundary condition (7.81) which under the given scaling does not change its form. In the following we will drop the $\widehat{\phantom{x}}$. As $p$ does not depend on $x_3$ we obtain

$$p(t, x_1, x_2, x_3) = -\Delta h(t, x_1, x_2)\,.$$

Integration of $\nabla \cdot v = 0$ with respect to $x_3$ yields

$$0 = \int_0^{h(t,x_1,x_2)} (\partial_{x_1} v_1 + \partial_{x_2} v_2)\,dx_3 + v_3(t, x_1, x_2, h(t, x_1, x_2)) - v_3(t, x_1, x_2, 0)\,.$$

As $v_3 = 0$ for $x_3 = 0$ we deduce from the kinematic boundary condition (7.89)

$$\partial_t h = -\int_0^{h(t,x_1,x_2)} (\partial_{x_1} v_1 + \partial_{x_2} v_2)\,dx_3 - (\partial_{x_1} h)v_1 - (\partial_{x_2} h)v_2$$
$$= -\nabla \cdot \left(\int_0^{h(t,x_1,x_2)} \begin{pmatrix} v_1 \\ v_2 \end{pmatrix} dx_3\right)\,.$$

The equations $\partial_{x_3 x_3} v_i = \partial_{x_i} p$, $i = 1, 2$, can be solved using $v_i = 0$ for $x_3 = 0$ and $\partial_{x_3} v_i = 0$ for $x_3 = h(t, x_1, x_2)$ as follows

$$v_i(t, x_1, x_2, x_3) = \partial_{x_i} p(t, x_1, x_2) \left[ \frac{x_3^2}{2} - h(t, x_1, x_2) x_3 \right].$$

We then obtain

$$\int_0^{h(t, x_1, x_2)} v_i \, dx_3 = \partial_{x_i} p(t, x_1, x_2) \left( -\frac{h^3}{3} \right).$$

Altogether this leads to

$$\partial_t h = \nabla \cdot \left( \frac{h^3}{3} \nabla p \right)$$

or

$$\partial_t h = -\nabla \cdot \left( \frac{h^3}{3} \nabla \Delta h \right). \tag{7.90}$$

We have hence deduced from the free surface problem an equation which is formulated solely for the height. The Eq. (7.90) is called the *thin film equation*. Equations of the form

$$\partial_t h = -\nabla \cdot (ch^n \nabla \Delta h) \text{ with } c > 0, \, n > 0,$$

allow for solutions with a compact support which is similar to the porous medium equation (cf. Exercise 7.25). Such solutions describe fluid films on a horizontal surface which do not wet the whole surface. The boundary of the support can as in the case of the porous medium equation be interpreted as a *free boundary*.

## 7.12   Literature

Classical books on problems with free boundaries and variational inequalities are the books of Duvaut and Lions [30], Friedman [42], Kinderlehrer and Stampacchia [79], and Elliott and Ockendon [33]. A good reference for the analysis of convex optimization problems with side constraints is the book of Ekeland and Temam [32]. More information on the Stefan problem and on phase field equations can be found in the books of Brokate and Sprekels [15], Davis [26] and Visintin [136]. A nice introduction into the theory of phase transitions is given by Gurtin [60]. For information on the geometry of curves and surfaces we refer to books on differential geometry [9, 18, 35]. More information on free surfaces in fluid dynamics can be found in the book of Segel [118].

## 7.13   Exercises

**Exercise 7.1** Determine the solution of the minimization problem

$$\min\left\{\int_{-1}^{1}\left((u')^2 + 2u\right)dx \mid u \in H^1(-1, 1), \ u \geq 0, \ u(-1) = a, \ u(1) = b\right\}.$$

State the active and inactive sets and the free boundary in dependence on $a$ and $b$. Compute the corresponding Lagrange multiplier $\mu$.

**Exercise 7.2** We consider the minimization problem

$$\min\left\{\int_{\Omega}\left(\frac{\lambda}{2}|\nabla u|^2 - fu\right)dx \mid u \in V \ \text{and} \ u \geq \psi\right\} \tag{7.91}$$

with $V = \left\{v \in H^1(\Omega) \mid v = u_0 \ \text{on} \ \partial\Omega\right\}$ under the assumptions that $f$, $\lambda$, $\psi$, $u_0$, $\partial\Omega$ are smooth and $\lambda(x) \geq \lambda_0 > 0$ for all $x \in \Omega$.

Show that $u \in H^2(\Omega) \cap C^0(\overline{\Omega}) \cap K$ with $K = \left\{v \in V \mid v \geq \psi\right\}$ is a solution of (7.91), if and only if a $\mu \in L_2(\Omega)$ exists such that

$$\int_{\Omega}\lambda\nabla u \cdot \nabla v \, dx + \int_{\Omega}fv \, dx + \int_{\Omega}\mu v \, dx = 0 \ \text{for all} \ v \in H_0^1(\Omega),$$

$$\mu = \max(0, \mu - c(u - \psi)) \ \text{almost everywhere in} \ \Omega,$$

where we assume that $c > 0$.
Remark: This formulation does not require any inequality.

**Exercise 7.3** We consider the minimization problem

$$\min\left\{\int_{\Omega}\left(\frac{\lambda}{2}|\nabla u|^2 - fu\right)dx \mid u \in V \ \text{and} \ u \geq \psi\right\}$$

with $V = \left\{v \in H^1(\Omega) \mid v = u_0 \ \text{on} \ \partial\Omega\right\}$ under the assumption that $f$, $\lambda$, $\psi$, $u_0$ are smooth and $\Omega$ has a smooth boundary. With the notation from Sect. 7.1 we assume that $\Gamma = \partial A \cap \partial N$ is a smooth hypersurface which is contained in $\Omega$. In addition, it is assumed that $u_{|\overline{N}} \in C^2(\overline{N})$. Show that under these conditions

$$u = \psi, \quad \lambda\nabla u \cdot n = \lambda\nabla\psi \cdot n \ \text{on} \ \Gamma.$$

Remark: $u \in C^1(\Omega)$ is not an assumption in this exercise.

**Exercise 7.4** Let $\Omega \subset \mathbb{R}^d$ be a domain with a smooth boundary which is the disjoint union of two surfaces $\Gamma_U$ and $\Gamma_C$ both with positive surface area. In addition, let $f$, $u_0$ : $\Omega \to \mathbb{R}^d$, $g : \Gamma_C \to \mathbb{R}$, $\nu : \Gamma_C \to \mathbb{R}^d$ be smooth enough. We consider the following contact problem: Find $u \in K = \{v \in H^1(\Omega)^d \mid v = u_0 \text{ on } \Gamma_U, \, v \cdot \nu \le g \text{ on } \Gamma_C\}$, such that for all $v \in K$

$$\int_\Omega \sigma(u) : \varepsilon(v - u) \, dx \ge \int_\Omega f \cdot (v - u) \, dx \,.$$

Here we assume that Hooke's law (5.50) is true with coefficients which are bounded, symmetric and positive definite in the sense of (5.51), (5.52).

Show that the above contact problem has at most one solution.

*Hint*: Show that the intersection of all rigid deformations $\mathcal{R} := \{u \in H^1(\Omega)^d \mid \varepsilon(u) = 0\}$ with the set $\{u \in H^1(\Omega)^d \mid u = 0 \text{ on } \Gamma_U\}$ only contains the function $v = 0$. This follows from the fact that $\Gamma_U$ has positive surface area and that $\Omega$ is connected.

**Exercise 7.5** Show that the Barenblatt solution (7.13) solves the porous medium equation $\partial_t u = \Delta u^m$ in a distributional sense, i.e.,

$$\int_0^\infty \int_{\mathbb{R}^d} (u \, \partial_t \zeta + u^m \Delta \zeta) \, dx \, dt = 0$$

holds for all $\zeta \in C_0^\infty\big((0, \infty) \times \mathbb{R}^d\big)$.

**Exercise 7.6** Let $\Gamma$ be an evolving hypersurface in $\mathbb{R}^d$. Show that the definition (7.47) of the normal velocity in Sect. 7.3 does not depend on the curve chosen in the definition.

*Hint*: Describe $\Gamma$ locally around $(t_0, x_0)$ as a level surface of a function $v : \mathbb{R} \times \mathbb{R}^d \to \mathbb{R}$.

**Exercise 7.7** Show that the smallest and the largest principal curvatures $\kappa_1$ and $\kappa_{d-1}$ of a hypersurface in $\mathbb{R}^d$ minimizes or maximizes, respectively, the second fundamental form.

**Exercise 7.8** Let $\Omega \subset \mathbb{R}^d$ be a domain and $T : (0, T) \times \Omega \to \mathbb{R}$ a continuous distributional solution of

$$\partial_t \big(\varrho(c_V T + L\chi_{\{T > T_M\}})\big) = \lambda \Delta T \,,$$

cf. (7.55). Let us assume that the set $\Gamma = \{(t, x) \mid T(t, x) = T_M\}$ is a smooth evolving hypersurface with $\Gamma(t) \subset\subset \Omega$ for all $t \in (0, T)$. We now set $\Omega_s(t) = \{x \in \Omega \mid T < T_M\}$ and $\Omega_\ell(t) = \{x \in \Omega \mid T > T_M\}$. We assume that $T_{|\Omega_s}$ and $T_{|\Omega_\ell}$ individually have twice continuously differentiable extensions onto $\Gamma(t)$.

Show that under these assumptions

$$\varrho L V = [-\lambda \nabla T]_s^\ell \cdot \nu \quad \text{on} \quad \Gamma(t)$$

holds.

**Exercise 7.9**  A liquid in a half space $\Omega = (0, \infty) \times \mathbb{R}$ with melting temperature $T_M$ and initial temperature $T_M$ is cooled at the boundary of $\Omega$ to a temperature $T_0 < T_M$ below the melting temperature. Determine a solution of the corresponding Stefan problem

$$
\begin{aligned}
\varrho c_V \, \partial_t T - \lambda \partial_x^2 T &= 0 && \text{for } 0 < x < a(t)\,, \\
T &= T_M && \text{for } x = a(t)\,, \\
\lambda \partial_x T &= \varrho L \dot{a}(t) && \text{for } x = a(t)\,,
\end{aligned}
$$

where $a(t)$ describes the position of the phase boundary. To compute the solution the ansatz

$$
T(t, x) = T_0 + (T_M - T_0)\, u(s)
$$

with rescaled variable $s = \sqrt{\frac{c_V \varrho}{4 \lambda t}}\, x$ should be used.

**Exercise 7.10**  Let $u$ be a solution of the following one-dimensional one-sided Stefan problem

$$
\begin{aligned}
\partial_t u &= \partial_{xx} u && \text{for } 0 < x < s(t)\,, \\
u &= 0 && \text{for } x \geq s(t)\,, \\
\partial_x u &= -s'(t) && \text{for } x = s(t)\,, \\
\partial_x u &= 0 && \text{for } x = 0\,, \\
s(0) = s_0\,, \ u(0, x) &= u_0(x) < 0 \text{ for } 0 < x < s_0\,.
\end{aligned}
$$

Here we have for $0 < x < s(t)$ an undercooled liquid and for $x > s(t)$ a solid. Show the following assertions.

(a)  It holds that

$$
\frac{d}{dt}\left( \int_0^{s(t)} u(t, x)\, dx + s(t) \right) = 0\,.
$$

   Remark: $s(t)$ is the length of the set occupied by the fluid phase at time $t$.

(b)  It holds that

$$
\frac{d}{dt} \int_0^{s(t)} u^2\, dx + \int_0^{s(t)} 2|\partial_x u|^2\, dx = 0\,.
$$

**Exercise 7.11**  Let $u$ be a solution of the one-dimensional Stefan problem in Exercise 7.10. Show that

(a)  It holds

$$
u(t, x) \leq 0 \ \text{ for all } t > 0\,, \ 0 < x < s(t)\,.
$$

*Hint*: Compute $\frac{d}{dt} \int_0^{s(t)} u_+^2\, dx$ with $u_+ = \max(u, 0)$.

(b) Show that $s'(t) \leq 0$.
(c) Show that

$$\frac{d}{dt} \left( \frac{1}{2}(s(t))^2 + \int_0^{s(t)} x \, u(t, x) \, dx \right) = u(t, 0) \leq 0$$

and deduce from this: If

$$\frac{1}{2} s_0^2 + \int_0^{s_0} x \, u_0(x) \, dx < 0 \tag{7.92}$$

holds there exists a time $t_0$ for which either $u(t_0, x) \equiv 0$ or $s(t_0) = 0$ holds.
(d) Assume (7.92) and show that the convergences

$$s(t) \to s_\infty \quad \text{for } t \to \infty,$$
$$u(t, \cdot) \to u_\infty \quad \text{for } t \to \infty \text{ in the } H^1 - \text{norm}$$

leads to a contradiction.
*Hint*: Use the time integrated assertion of Exercise 7.10, part (b).

**Exercise 7.12** Consider the Mullins–Sekerka problem (7.57) on the domain $\Omega = B_R(0)$. On $\partial\Omega$ we assume Neumann boundary conditions $\nabla u \cdot n = 0$ and at time $t = 0$ the fluid phase is assumed to be given by $\Omega_\ell(0) = B_{r_2}(0) \setminus \overline{B_{r_1}(0)}$ with $0 < r_1 < r_2 < R$. The fluid phase for positive time will have the form $\Omega_\ell(t) = B_{r_2(t)}(0) \setminus \overline{B_{r_1(t)}(0)}$.
  Which ordinary differential equations are fulfilled by $r_1(t)$ and $r_2(t)$?
*Hint*: Assume that $u$ is radially symmetric.

**Exercise 7.13** Consider the Stefan problem with a relaxation dynamic (7.65) instead of (7.63) and derive the entropy inequality (7.66).

**Exercise 7.14** Let $y : [a, b] \to \mathbb{R}^2$ be an arc-length parametrization of a closed smooth curve $\Gamma$ and let $\nu : [a, b] \to \mathbb{R}^2$ be a unit normal to $\Gamma$.
Show that:

(a) The mapping

$$F : [a, b] \times (-\varepsilon, \varepsilon) \to \mathbb{R}^2$$
$$(s, d) \mapsto y(s) + d \, \nu(s)$$

is continuously differentiable for $\varepsilon$ small enough.
(b) $F$ is invertible,
(c) If $x = F(s, d)$ holds, $|d|$ is the distance of $x$ to $\Gamma$.

**Exercise 7.15** (a) Prove the Frenet formulas (7.71).
(b) Show the equalities (7.72) and (7.73).

**Exercise 7.16**   Consider the mean curvature flow in the plane $\mathbb{R}^2$

$$V = \kappa \quad \text{on } \Gamma(t),$$

where $\Gamma = \cup_{t \geq 0} \Gamma(t)$ is an evolving curve in $\mathbb{R}^2$. We would like to find a solution in the form of a "traveling wave". To find such a solution make the ansatz

$$\Gamma(t) = \{(x, u(t, x)) \mid x \in I\}$$

with a function $u$ of the form

$$u(t, x) = h(x) + ct, \quad c \in \mathbb{R}.$$

Derive a differential equation for $h$ and solve this equation.

**Exercise 7.17**   (a)   We want to determine a radially symmetric solution of the mean curvature flow

$$V = \kappa \quad \text{on } \Gamma(t),$$

where $\Gamma = \cup_{t \geq 0} \Gamma(t)$ is an evolving surface in $\mathbb{R}^d$. Make the ansatz $\Gamma(t) = \partial B_{r(t)}(0)$ and derive a differential equation for $r(t)$ and solve this equation. What happens for increasing $t$?

(b)   Consider the ansatz in (a) for the evolution law:

$$V = \kappa + c \quad \text{on } \Gamma(t),$$

where $c \in \mathbb{R}$ is constant. Make qualitative statements on the behavior of the solution for growing $t$ in dependence on the parameter $c$. Distinguish between the cases: There exists a solution for $t \to \infty$ and there exists no solution for $t \to \infty$.

**Exercise 7.18**   Let $s(T, \varphi) := c_V \ln \frac{T}{T_M} + \frac{L}{T_M} \frac{1+\varphi}{2}$ and let $T, \varphi$ be be a solution of (7.67), (7.76) with Neumann boundary condition for $T$ and $\varphi$. Show the estimate

$$\frac{d}{dt} \left( \int_\Omega \left[ \varrho s - \widehat{\gamma} \left( \frac{\varepsilon}{2} |\nabla \varphi|^2 + \frac{1}{\varepsilon} \psi(\varphi) \right) \right] dx \right) \geq 0.$$

**Exercise 7.19**   Perform a formal asymptotic expansion for (7.67), (7.76) analogous to the discussion in Sect. 7.9.

**Exercise 7.20**   Perform a formal asymptotic expansion for the Cahn–Hilliard equation

$$\partial_t c = L \Delta w,$$
$$w = -\gamma \varepsilon \Delta c + \frac{\gamma}{\varepsilon} f'(c)$$

with $f(c) = \alpha(c^2 - a^2)^2$ in a similar way as for the phase field model.

**Exercise 7.21**  Let $\Omega \subset \mathbb{R}^d$ be a bounded domain with smooth boundary. In addition let $\gamma_1, \gamma_2, \gamma_3 > 0$ with $|\gamma_1 - \gamma_2| < \gamma_3$, $m \in \mathbb{R}$ and

$$K = \left\{ h \in C^2(\overline{\Omega}) \mid \int_\Omega h\, dx = m \right\}.$$

Consider the functional

$$E(h) = \gamma_3 \int_\Omega \sqrt{1 + |\nabla h|^2}\, dx + (\gamma_2 - \gamma_1) \int_{\partial\Omega} h\, ds_x \quad \text{for } h \in K.$$

Show that: If $h \in K$ is a minimizer of $E$, it follows that the graph of $h$, i.e., the surface

$$\Gamma = \{(x, h(x)) \mid x \in \Omega\} \subset \mathbb{R}^{d+1}$$

has constant mean curvature. In addition, the graph intersects the cylinder

$$Z = \{(x, z) \mid x \in \partial\Omega,\ z \in \mathbb{R}\} = \partial\Omega \times \mathbb{R}$$

with an angle arccos $\frac{\gamma_1 - \gamma_2}{\gamma_3}$.

*Hint*: Rewrite the boundary condition in the Euler–Lagrange equation into a condition on the normals to $Z$ and $\Gamma$.

**Exercise 7.22**  For the Euler equations with free surface derive the identity (7.88). Here we assume the boundary conditions (7.81), (7.86), (7.87) and the angle condition (7.85) which were specified in Sect. 7.10.

**Exercise 7.23**  Consider a flow with free surface as discussed in Sect. 7.10 where the angle condition (7.84) is replaced by the condition

$$\sphericalangle(\Gamma(t), \partial\Omega) = \varphi$$

with $\varphi = \arccos\left(\frac{\gamma_1 - \gamma_2}{\gamma_3}\right)$ and the force balance on the free surface is replaced by $\sigma\nu = \gamma_3 \kappa \nu$. Here we assume that $\gamma_1, \gamma_2, \gamma_3 > 0$ fulfill $|\gamma_1 - \gamma_2| < \gamma_3$.

Under the assumption that for all $t$ the set $\Gamma(t)$ can be written as a graph show the following inequality

$$\frac{d}{dt}\left( \int_{\Omega(t)} \frac{\varrho}{2} |v(t, x)|^2\, dx + \int_{\Gamma(t)} \gamma_3\, ds_x + (\gamma_2 - \gamma_1) \int_{\partial\Omega'} h(t, x')\, ds_{x'} \right) \leq 0$$

holds. How can the result be shown if $\Gamma$ is an evolving hypersurface which cannot be written globally as a graph? Discuss in addition the case of more general container geometries.

Remark: The quantities $\gamma_1$ and $\gamma_2$ are the surface energy densities of the interfaces between solid and gas and solid and fluid phase, respectively. The term $(\gamma_2 - \gamma_1) \int_{\partial \Omega'} h(t, x') \, ds_{x'}$ hence describes a contribution to the energy which consists of contributions from the solid-gas and solid-liquid interfaces.

**Exercise 7.24** Consider the fluid-flow problem from Sect. 7.10 with a slip condition, i.e., replace in (7.78)–(7.85) the no-slip condition (7.80) by the condition

$$(\sigma n)_\tau (t, x', x_3) = -\beta \, v(t, x', x_3) \quad \text{for } x' \in \partial \Omega' \text{ or } x_3 = 0.$$

Here $\beta > 0$ is a constant and for a vector field $j$ on $\partial \Omega(t)$ we define $(j)_\tau = j - (j \cdot n)n$ where $n$ is the outer unit normal on $\partial \Omega(t)$. Show in the case $f = 0$ the energy inequality

$$\frac{d}{dt} \left( \int_{\Omega(t)} \frac{\varrho}{2} |v(t, x)|^2 \, dx + \int_{\Gamma(t)} \gamma \, ds_x \right) \le 0.$$

**Exercise 7.25** Show that the function

$$h(t, x) = t^{-d\beta} f(r) \quad \text{with } r = \frac{|x|}{t^\beta}, \quad \beta = \frac{1}{4 + d},$$

and

$$f(r) = \begin{cases} \frac{1}{8(2+d)(4+d)} (a^2 - r^2)^2, & \text{if } 0 \le r < a, \\ 0, & \text{elsewhere} \end{cases}$$

solves the equation $\partial_t h + \nabla \cdot (h \nabla \Delta h) = 0$ on $\mathbb{R}_+ \times \mathbb{R}^d$ in a weak sense, i.e.,

$$\int_0^\infty \int_{\mathbb{R}^d} (h \, \partial_t \varphi + h \, \nabla \Delta h \cdot \nabla \varphi) \, dx \, dt = 0$$

for all $\varphi \in C_0^\infty ((0, \infty) \times \mathbb{R}^d)$. Here $\nabla \Delta h$ has to be defined piecewise. Show that there exists an $\alpha > 0$ such that $h(t, \cdot) \to \alpha \delta_0$ as $t \searrow 0$ in the sense of distributions, i.e., for all $\zeta \in C_0^\infty (\mathbb{R}^d)$ it holds that

$$\int_{\mathbb{R}^d} h(t, x) \zeta(x) \, dx \to \alpha \zeta(0) \quad \text{as } t \searrow 0.$$

Compute an $a > 0$, such that $\int_{\mathbb{R}^d} h(t, x) \, dx = 1$.

**Exercise 7.26** (*Thin films on a vertical or inclined plane*)
Perform a lubrication approximation as in Sect. 7.11 for the equations of fluid dynamics in Sect. 7.10 under the following assumptions: In the Navier–Stokes equations in the fixed direction $(\ell_1, \ell_2, \ell_3) \in \mathbb{R}^3$ gravitation will be taken into account, i.e., for $g > 0$ constant it holds that

$$\varrho(\partial_t v + (v \cdot \nabla)v) = \mu \Delta v - \nabla p - \varrho g (\ell_1, \ell_2, \ell_3)^\top ,$$
$$\nabla \cdot v = 0 .$$

The force balance on the free boundary is assumed to be

$$\sigma \nu = 0 ,$$

i.e., we choose $\gamma = 0$ in (7.83). The no-slip condition (7.80) and the kinematic condition (7.81) are not changed.

With the notation in Sect. 7.11 now set $\overline{p} = \mu \overline{V}/(\varepsilon^2 L)$ and multiply the Navier–Stokes equations by $L^2 \varepsilon^2/(\overline{V}\mu)$ and the force balance by $L\varepsilon^2/(\overline{V}\mu)$. Derive in the following two cases the equations to leading order where the fact that the velocity has zero divergence, the no-slip condition and the kinematic condition are preserved (the $\widehat{\phantom{x}}$ in Sect. 7.11 is dropped):

(i) For $\overline{V} = \frac{\varepsilon^3 L^2 \varrho g}{\mu}$:
   In the fluid:

$$\partial_{x_3 x_3} v_i = \partial_{x_i} p , \quad i = 1, 2 ,$$
$$-\partial_{x_3} p = \ell_3 .$$

On the free boundary ($x_3 = h(t, x_1, x_2)$):

$$\partial_{x_3} v_i = 0 , \quad i = 1, 2 ,$$
$$p = 0 .$$

(ii) For $\overline{V} = \frac{\varepsilon^2 L^2 \varrho g}{\mu}$:
   In the fluid:

$$\partial_{x_3 x_3} v_i = \partial_{x_i} p + \ell_i , \quad i = 1, 2 ,$$
$$-\partial_{x_3} p = 0 .$$

On the free boundary ($x_3 = h(t, x_1, x_2)$):

$$\partial_{x_3} v_i = 0 , \quad i = 1, 2 ,$$
$$p = 0 .$$

**Exercise 7.27** Derive for the two scalings considered in Exercise 7.26 a differential equation for the height $h$.

# Appendix A
# Function Spaces

We denote the vector space of all real-valued continuous functions on an open set $\Omega \subset \mathbb{R}^d$ with $C^0(\Omega)$. Continuous functions with a domain of definition $\Omega$ and values in $\mathbb{R}^m$, $\mathbb{C}$ or $\mathbb{C}^m$ are denoted by $C^0(\Omega, \mathbb{R}^m)$, or by $C^0(\Omega, \mathbb{C})$, or $C^0(\Omega, \mathbb{C}^n)$, respectively, and we will use for the other function spaces which we will consider later a corresponding notation. For each number $k \in \mathbb{N}$ we define $C^k(\Omega)$ as the space of all $k$–times continuously differentiable functions on $\Omega$. The set of all arbitrarily often differentiable functions is denoted by $C^\infty(\Omega)$. For a function $\varphi : \Omega \to \mathbb{R}$ we define its support to be

$$\operatorname{supp} \varphi := \overline{\{x \in \Omega \mid \varphi(x) \neq 0\}} \,.$$

The set $C_0^\infty(\Omega)$ is the set of all arbitrarily often differentiable functions whose support is a compact subset of $\Omega$.

In addition, we will use the function spaces

$$C^k(\overline{\Omega}) := \big\{u : \overline{\Omega} \to \mathbb{R} \mid u \text{ is continuous, } u_{|\Omega} \in C^k(\Omega)$$
$$\text{and all partial derivatives up to}$$
$$\text{the order } k \text{ can be continuously extended to } \overline{\Omega}\big\} \,.$$

If $\overline{\Omega}$ is bounded one obtains that the spaces $C^k(\overline{\Omega})$ are Banach spaces, i.e., they are complete normed vector spaces. A *normed space* is a vector space $X$ with a mapping that maps each $x \in X$ to a real number $\|x\|$ such that

(1)     $\|x\| = 0$ if and only if $x = 0$,
(2)     $\|\alpha x\| = |\alpha|\, \|x\|$ for all scalars $\alpha$ and all $x \in X$,
(3)     $\|x + y\| \le \|x\| + \|y\|$ for all $x, y \in X$ (*triangle inequality*).

A sequence $(x_k)_{k \in \mathbb{N}}$ in a normed space $X$ converges if and only if an $x \in X$ exists such that the following holds: For all $\varepsilon > 0$ there exists an $M \in \mathbb{N}$ such that

$$\|x_k - x\| < \varepsilon \ \text{ for all } \ k > M \,.$$

© Springer International Publishing AG 2017
C. Eck et al., *Mathematical Modeling*, Springer Undergraduate
Mathematics Series, DOI 10.1007/978-3-319-55161-6

A sequence $(x_k)_{k\in\mathbb{N}}$ in a normed space $X$ is a Cauchy sequence, if and only if for all $\varepsilon > 0$ an $M \in \mathbb{N}$ exists such that

$$\|x_k - x_\ell\| < \varepsilon \quad \text{for all } k, \ell > M.$$

A normed space $X$ is complete and hence a Banach space if each Cauchy sequence in $X$ has a limit $x \in X$. If $\overline{\Omega}$ is bounded it can be shown that $C^k(\overline{\Omega})$ with the norm

$$\|u\|_{C^k(\overline{\Omega})} := \max_{\substack{\alpha \in \mathbb{N}_0^d \\ |\alpha| \le k}} \sup_{x \in \Omega} |\partial^\alpha u(x)|$$

is a Banach space. Here $\alpha = (\alpha_1, \ldots, \alpha_d) \in \mathbb{N}_0^d$ is a multi-index and $|\alpha| = \alpha_1 + \cdots + \alpha_d$. We remark that $\|u\|_{C^0(\overline{\Omega})}$ is the supremum of $|u|$.

By $L_p(\Omega), 1 \le p \le \infty$ we denote the set of all real-valued measurable functions such that $|u|^p$ is Lebesgue–integrable. $L_p(\Omega)$ with the norm

$$\|u\|_{L_{p(\Omega)}} := \left( \int_\Omega |u(x)|^p \, dx \right)^{1/p} \quad \text{for } 1 \le p < \infty,$$

$$\|u\|_{L_{\infty(\Omega)}} := \inf_{\substack{N \subset \Omega \\ N \text{ null set}}} \sup_{x \in \Omega \setminus N} |u(x)| \quad \text{for } p = \infty$$

is a Banach space. The condition (1) in the definition of the norm is fulfilled if we identify functions which differ on a set of measure zero with each other. We define equivalence classes as follows. We say that a function $v$ is in the equivalence class $[u]$ of the function $u$, i.e., $v \sim u$, if

$$u = v \quad \text{almost everywhere in } \Omega.$$

The set of all locally integrable functions is

$$L_{1,\text{loc}}(\Omega) := \{u : \Omega \to \mathbb{R} \mid \text{for all } D \subset\subset \Omega \text{ it holds } u_{|D} \in L_1(D)\}.$$

Here $D \subset\subset \Omega$ means that $D$ is bounded and $\overline{D} \subset \Omega$. For functions $u \in L_p(\Omega)$, $v \in L_q(\Omega)$ with $\frac{1}{p} + \frac{1}{q} = 1$ *Hölder's inequality*

$$\int_\Omega |u(x)\, v(x)| \, dx \le \left( \int_\Omega |u(x)|^p \, dx \right)^{1/p} \left( \int_\Omega |v(x)|^q \, dx \right)^{1/q}$$

holds. For $p = q = 2$ this inequality is called the *Cauchy–Schwarz inequality*.

On the space $L_2(\Omega)$ one can introduce the inner product

$$\langle u, v \rangle_{L_2(\Omega)} := \int_\Omega u(x)\, v(x) \, dx.$$

An *inner product* on a real vector space $H$ is a real-valued function $(x, y) \mapsto \langle x, y \rangle$ such that

(i)    $\langle x, y \rangle = \langle y, x \rangle$ for all $x, y \in H$,
(ii)   $\langle \alpha x + \beta y, z \rangle = \alpha \langle x, z \rangle + \beta \langle y, z \rangle$ for all $\alpha, \beta \in \mathbb{R}$ and $x, y, z \in H$,
(iii)  $\langle x, x \rangle \geq 0$ for all $x \in X$,
(iv)   $\langle x, x \rangle = 0$ if and only if $x = 0$.

In a real vector space $H$ with inner product the quantity

$$\|x\| = \sqrt{\langle x, x \rangle}$$

defines a norm. If $H$ is complete with respect to this norm we call $H$ a *Hilbert space*. The space $L_2(\Omega)$ with the above inner product is a Hilbert space.

Often it makes sense to consider a weaker form of the classical notion of derivative. If for an integrable function $u : \Omega \to \mathbb{R}$ and a multi-index $\alpha$ an integrable function $v_\alpha : \Omega \to \mathbb{R}$ exists such that

$$\int_\Omega u(x) \, \partial^\alpha \varphi(x) \, dx = (-1)^{|\alpha|} \int_\Omega v_\alpha(x) \, \varphi(x) \, dx$$

for all $\varphi \in C_0^\infty(\Omega)$, we call $\partial^\alpha u := v_\alpha$ the $\alpha^{\text{th}}$ weak partial derivative of $u$.

The *Sobolev space* $H^1(\Omega)$ consists of all functions $u \in L_2(\Omega)$ whose first partial derivative exists in a weak sense and for which these partial derivatives are square integrable. The space $H^1(\Omega)$ with the inner product

$$\langle u, v \rangle_{H^1(\Omega)} := \int_\Omega (u(x) \, v(x) + \nabla u(x) \cdot \nabla v(x)) \, dx$$

is a Hilbert space.

Similarly one can define Sobolev spaces $H^m(\Omega)$, $m \geq 2$ by demanding that also higher weak partial derivatives are square integrable.

In $L_2(\Omega)$ and $H^1(\Omega)$ functions belonging to an equivalent class differ on a set of measure zero. This means that it does not make sense to define *boundary values* as the boundary typically is a set of measure zero. However under suitable assumptions on the boundary $\partial \Omega$ one can introduce boundary values for functions in $H^1(\Omega)$. A simple possibility to define zero boundary values is to consider the completion of $C_0^\infty(\Omega)$ in $H^1(\Omega)$:

$$H_0^1(\Omega) := \{u \in H^1(\Omega) \mid \text{there exists a sequence } (u_k)_{k \in \mathbb{N}} \text{ in } C_0^\infty(\Omega),$$
$$\text{such that } \|u_k - u\|_{H^1(\Omega)} \to 0\}.$$

One can interpret this space as the set of all functions in $H^1(\Omega)$ which have zero boundary values. The fact that functions in $H_0^1(\Omega)$ have zero boundary values in a certain sense can also be justified more rigorously.

If $\Omega$ is bounded one can show the *Poincaré inequality* for functions in $H_0^1(\Omega)$, i.e., there exists a constant $c_P > 0$ which only depends on $\Omega$ such that

$$\int_\Omega |v|^2 \, dx \le c_P \int_\Omega |\nabla v|^2 \, dx$$

for all $v \in H_0^1(\Omega)$. If $\Omega$ is connected we also obtain such an inequality for functions $u \in H^1(\Omega)$ with $\int_\Omega u(x) \, dx = 0$ or for functions $u \in H^1(\Omega)$ which are zero only on a part of $\partial\Omega$ having positive surface measure.

Finally, we remark that for functions $u \in H^m(\Omega)$ in the case $m - \frac{d}{2} > 0$ a continuous function $v \in C^0(\Omega)$ exists such that $u = v$ holds almost everywhere in $\Omega$.

For further details and proofs of the stated results we refer to the books of Adams [2] and Alt [4].

# Appendix B
# Curvature of Hypersurfaces

The curvature of a curve in the plane is the change of direction per length. Let $I \subset \mathbb{R}$ be an interval and $y : I \to \mathbb{R}^2$ be a differentiable curve which is parameterized by arc-length, i.e., $|y'(s)| = 1$ for all $s \in I$. Then $\tau := y'(s)$ is a unit tangent of the curve and $y''(s)$ measures the local change of the tangent per length. If $|y''(s)|$ is larger the tangent changes more quickly. The quantity $y''(s)$ hence describes how much the curve is curved.

For a curve $\Gamma \subset \mathbb{R}^2$ with unit normal $\nu$ we choose an arc-length parametrization $y$ such that $(\nu, y'(s))$ is positively oriented, i.e., $\det(\nu, y'(s)) > 0$. Differentiating $|y'(s)|^2 = 1$ leads to

$$0 = \frac{d}{ds}|y'(s)|^2 = 2\,y''(s) \cdot y'(s)\,.$$

This implies that $y''(s)$ is a multiple of $\nu$ and we define

$$\kappa = y''(s) \cdot \nu$$

as the *curvature of the curve* $\Gamma$. Choosing instead of $\nu$ the normal $-\nu$ we change the sign of the curvature $\kappa$.

We now consider a smooth surface $\Gamma$ in $\mathbb{R}^3$ with unit normal $\nu$. In this case we can define a normal curvature for each tangential direction $\tau$. This normal curvature is defined as the curvature of the plane curve which is obtained by intersecting the plane spanned by $\nu$ and $\tau$ with the surface $\Gamma$. Now let $\kappa_1$ and $\kappa_2$ be the minimal and maximal values of the *normal curvature*. The values $\kappa_1$ and $\kappa_2$ are called the *principal curvatures* and the *mean curvature* $\kappa$ is defined as

$$\kappa = \kappa_1 + \kappa_2\,. \tag{B.1}$$

Strictly speaking $\frac{1}{2}(\kappa_1 + \kappa_2)$ is the mean curvature. However, nowadays most mathematicians use (B.1) as the definition of mean curvature as this leads to simpler expressions in most cases.

© Springer International Publishing AG 2017
C. Eck et al., *Mathematical Modeling*, Springer Undergraduate
Mathematics Series, DOI 10.1007/978-3-319-55161-6

We now want to define curvature generally for hypersurfaces $\Upsilon$ in $\mathbb{R}^d$, i.e., for smooth $(d-1)$–dimensional surfaces of $\mathbb{R}^d$. In order to do so we first of all consider *tangential derivatives* of functions which are defined on $\Upsilon$. For $x \in \Upsilon$ we define by $T_x\Upsilon$ the space of all tangential vectors at a point $x \in \Upsilon$. Let $f : \Upsilon \to \mathbb{R}^m$, $m \in \mathbb{N}$ be a smooth function, $x \in \Upsilon$ and $\tau \in T_x\Upsilon$. Let us choose a curve $y : (-\varepsilon, \varepsilon) \to \Upsilon$ with $y(0) = x$ and $y'(0) = \tau$ and define

$$\partial_\tau f(x) = \tfrac{d}{ds} f(y(s))\big|_{s=0}.$$

One can show that this definition does not depend on the specific choice of the curve $y$ as long as $y'(0) = \tau$ holds. In addition, it follows that the mapping $\tau \mapsto \partial_\tau f$ is *linear* with respect to $\tau$, i.e., for $\tau_1, \tau_2, \tau \in T_x\Upsilon$ with $\tau = \alpha_1\tau_1 + \alpha_2\tau_2$ it holds that

$$\partial_\tau f(x) = \alpha_1 \, \partial_{\tau_1} f(x) + \alpha_2 \, \partial_{\tau_2} f(x) \,.$$

We can now consider tangential derivatives of the *normal vector* $\nu$. The mapping

$$W_x : T_x\Upsilon \to T_x\Upsilon \,,$$
$$\tau \mapsto -\partial_\tau \nu$$

is called *Weingarten map* of $\Upsilon$ in the point $x$. As $\partial_\tau$ is linear with respect to $\tau$ we obtain that the Weingarten map $W_x$ is linear. The identity

$$(\partial_\tau \nu) \cdot \nu = \partial_\tau \tfrac{1}{2}|\nu|^2 = \partial_\tau \tfrac{1}{2} = 0$$

implies that $\partial_\tau \nu$ lies in the tangent space $T_x\Upsilon$. This implies that $-\partial_\tau \nu$ is a mapping from the tangent space $T_x\Upsilon$ into itself. For two smooth vector fields $\tau_1, \tau_2 : \Upsilon \to \mathbb{R}^d$ with $\tau_1(x), \tau_2(x) \in T_x\Upsilon$ for all $x$ it follows with the help of $\tau_2 \cdot \nu = 0$

$$(\partial_{\tau_1}\tau_2) \cdot \nu = (\partial_{\tau_1}\tau_2) \cdot \nu - \partial_{\tau_1}(\tau_2 \cdot \nu) = -\tau_2 \cdot \partial_{\tau_1}\nu \,. \tag{B.2}$$

We now consider a local parametrization $F : D \to \Upsilon$. The symmetry of the second derivative of $F$ implies the symmetry of the Weingarten map with respect to the Euclidean inner product in $\mathbb{R}^d$ which we will demonstrate in what follows. The vectors

$$t_1 = \partial_1 F, \dots, t_{d-1} = \partial_{d-1} F$$

form a basis of the tangent space. For arbitrary tangential vectors we find a representation

$$\tau_1 = \sum_{i=1}^{d-1} \alpha_i \, \partial_i F \,, \quad \tau_2 = \sum_{i=1}^{d-1} \beta_i \, \partial_i F$$

and we compute with the help of $\nu \cdot \partial_i F = 0$, $\partial_{t_i} F = \partial_i F$ for $i = 1, \ldots, d-1$ and the linearity of the tangential derivative

$$W_x(\tau_1) \cdot \tau_2 = -\left(\partial_{\tau_1} \nu\right) \cdot \tau_2 = \nu \cdot \partial_{\tau_1} \tau_2 = \nu \cdot \sum_{i,j=1}^{d-1} \alpha_i \beta_j \, \partial_{t_i} \left(\partial_j F\right)$$

$$= \nu \cdot \sum_{i,j=1}^{d-1} \alpha_i \beta_j \, \partial_i \partial_j F = \nu \cdot \sum_{i,j=1}^{d-1} \alpha_i \beta_j \, \partial_j \partial_i F$$

$$= \nu \cdot \partial_{\tau_2} \tau_1 = -\tau_1 \cdot \partial_{\tau_2} \nu = \tau_1 \cdot W_x(\tau_2) .$$

As the Weingarten map is symmetric real eigenvalues $\kappa_1 \leq \cdots \leq \kappa_{d-1}$ exist with corresponding orthonormal eigenvectors $v_1, \ldots, v_{d-1} \in T_x \Upsilon$. It holds that

$$\kappa_i = \kappa_i |v_i|^2 = -\left(\partial_{v_i} \nu\right) \cdot v_i = \left(\partial_{v_i} v_i\right) \cdot \nu .$$

This implies that $\kappa_i$ is the normal curvature in the direction $v_i$. The quantities $\kappa_1, \ldots, \kappa_{d-1}$ are called *principal curvatures*. The curvatures $\kappa_1$ and $\kappa_{d-1}$, respectively, minimize and maximize the following quadratic form which corresponds to the Weingarten map

$$(\tau, \tau) \mapsto (-\partial_\tau \nu) \cdot \tau = (\partial_\tau \tau) \cdot \nu$$

under all $\tau \in T_x \Upsilon$ with $|\tau| = 1$, see Exercise 7.7. The bilinear form $(\tau_1, \tau_2) \mapsto (-\partial_{\tau_1} \nu) \cdot \tau_2$ is called the *second fundamental form* of $\Upsilon$. One obtains that

$$\kappa = \text{trace } W_x = \sum_{i=1}^{d-1} \kappa_i$$

and for $d = 3$ this is just the definition (B.1). In the following, the quantity $\kappa$ is called *mean curvature* of $\Upsilon$ at $x$.

For an orthonormal basis $\{\tau_1, \ldots, \tau_{d-1}\}$ of the tangent space $T_x \Upsilon$ we can compute the mean curvature as follows

$$\kappa = \text{trace } W_x = -\sum_{i=1}^{d-1} \left(\partial_{\tau_i} \nu\right) \cdot \tau_i = \sum_{i=1}^{d-1} \left(\partial_{\tau_i} \tau_i\right) \cdot \nu .$$

Defining the tangential divergence $\nabla_\Upsilon \cdot$ of a vector field $f$ on $\Upsilon$ by

$$\nabla_\Upsilon \cdot f = \sum_{i=1}^{d-1} \tau_i \cdot \partial_{\tau_i} f$$

one obtains the compact form

$$\kappa = -\nabla_\Upsilon \cdot \nu \, .$$

Hence, the mean curvature is just the negative tangential divergence of the normal.

Often the mean curvature has to be computed in a situation in which a natural but not orthonormal basis $\{t_1, \dots, t_{d-1}\}$ of the tangent space is present. We will shortly discuss the linear algebra needed in this case. As a special case we will compute the mean curvature of a graph. Let $\{\tau_1, \dots, \tau_{d-1}\}$ be an orthonormal basis of the tangent space. Then there exists a matrix $A = (a_{ij})_{i,j=1}^{d-1}$ such that

$$\tau_k = \sum_{i=1}^{d-1} a_{ki} t_i \, .$$

As $\{\tau_1, \dots, \tau_{d-1}\}$ is orthonormal it follows

$$\delta_{k\ell} = \tau_k \cdot \tau_\ell = \sum_{i,j=1}^{d-1} a_{ki} \, a_{\ell j} \, t_i \cdot t_j = (AGA^\top)_{k\ell}$$

with

$$G = (g_{ij})_{i,j=1}^{d-1} := (t_i \cdot t_j)_{i,j=1}^{d-1} \, .$$

It hence holds that $AGA^\top = \text{Id}$ and hence

$$G = A^{-1}(A^{-1})^\top \, , \quad G^{-1} =: (g^{ij})_{i,j=1}^{d-1} = A^\top A \, .$$

For a vector field $f : \Upsilon \to \mathbb{R}^d$ we now obtain from the fact that $\partial_\tau f$ is linear with respect to $\tau$,

$$\nabla_\Upsilon \cdot f = \sum_{k=1}^{d-1} \tau_k \cdot \partial_{\tau_k} f = \sum_{i,j,k=1}^{d-1} a_{ki} \, a_{kj} \, t_j \cdot \partial_{t_i} f = \sum_{i,j=1}^{d-1} g^{ij} \, t_j \cdot \partial_{t_i} f \, .$$

We now assume that $\Upsilon$ is given as a graph

$$\Upsilon = \left\{ x = (x', x_d) \in \mathbb{R}^{d-1} \times \mathbb{R} \mid x_d = h(x') \, , \ x' \in D \right\},$$

where

$$h : D \to \mathbb{R} \, , \quad D \subset \mathbb{R}^{d-1} \ \text{open} \, ,$$

is twice continuously differentiable. A basis of the tangent space is given as

$$t_1 = (e_1, \partial_1 h), \dots, t_{d-1} = (e_{d-1}, \partial_{d-1} h) \, ,$$

where $e_1, \ldots, e_{d-1} \in \mathbb{R}^{d-1}$ are the standard basis vectors of $\mathbb{R}^{d-1}$. In this case we have $g_{ij} = \delta_{ij} + \partial_i h\, \partial_j h$ and a simple computation yields

$$g^{ij} = \delta_{ij} - \frac{\partial_i h\, \partial_j h}{1 + |\nabla_{x'} h|^2} \, .$$

As unit normal we obtain

$$\nu = \frac{1}{\sqrt{1 + |\nabla_{x'} h|^2}} (-\nabla_{x'} h, 1)^\top \, .$$

For a quantity $f : \Upsilon \to \mathbb{R}$ with $f(x) = f(x', h(x')) = \widehat{f}(x')$ one obtains $\partial_{t_i} f = \partial_i \widehat{f}$ and we compute

$$
\begin{aligned}
\kappa &= -\sum_{i,j=1}^{d-1} g^{ij}\, t_j \cdot \partial_{t_i} \nu = \frac{1}{\sqrt{1 + |\nabla_{x'} h|^2}} \sum_{i,j=1}^{d-1} g^{ij}\, t_j \cdot \partial_i (\nabla_{x'} h, -1)^\top \\
&= \frac{1}{\sqrt{1 + |\nabla_{x'} h|^2}} \sum_{i,j=1}^{d-1} g^{ij} \partial_{ij} h \\
&= \frac{\Delta_{x'} h}{\sqrt{1 + |\nabla_{x'} h|^2}} - \sum_{i,j=1}^{d-1} \frac{\partial_i h}{(1 + |\nabla_{x'} h|^2)^{3/2}} \partial_j h\, \partial_{ij} h \\
&= \nabla_{x'} \cdot \left( \frac{\nabla_{x'} h}{\sqrt{1 + |\nabla_{x'} h|^2}} \right) .
\end{aligned}
$$

# References

1. I. Aavatsmark, *Mathematische Einführung in die Thermodynamik der Gemische* (Akademie Verlag, Berlin, 1995)
2. R.A. Adams, *Sobolev Spaces* (Academic Press, Orlando, 1975)
3. M. Alonso, E.J. Finn, *Physik* (Addison–Wesley, Reading, 1977)
4. H.W. Alt, *Linear Functional Analysis, an Application Oriented Introduction* (Springer, London, 2016)
5. H.W. Alt, I. Pawlow, *On the entropy principle of phase transition models with a conserved order parameter.* Adv. Math. Sci. Appl. **6**(1), 291–376 (1996)
6. H. Amann, *Ordinary Differential Equations: An Introduction to Nonlinear Analysis* (Walter de Gruyter & Co., Berlin, 1995), pp. xiv+458
7. S.S. Antman, *Nonlinear Problems of Elasticity* (Springer, New York, 1995)
8. H. Babovsky, *Die Boltzmann-Gleichung* (Teubner, Stuttgart, 1998)
9. C. Bär, *Elementare Differentialgeometrie* (De Gruyter, Berlin, 2001)
10. H.T. Banks, K. Kunisch, *Estimation Techniques for Distributed Parameter Systems.* Systems and Control: Foundations and Applications, 1. Birkhäuser (Boston, MA, 1989)
11. J.W. Barrett, H. Garcke, R. Nürnberg, *Numerical computations of facetted pattern formation in snow crystal growth.* Phys. Rev. E **86**(1), 011604 (2012)
12. E. Becker, W. Bürger, *Kontinuumsmechanik* (Teubner, Stuttgart, 1997)
13. R.B. Bird, W.E. Steward, E.N. Lightfoot, *Transport Phenomena* (Wiley, New York, 1960)
14. W. Bolton, *Linear Equations and Matrices. Mathematics for Engineers* (Longman Scientific & Technical, Harlow, Essex, 1995)
15. M. Brokate, J. Sprekels, *Hysteresis and Phase Transitions* (Springer, New York, 1996)
16. G. Buttazzo, M. Giaquinta, S. Hildebrandt, *One-dimensional Variational Problems* (Clarendon Press, Oxford, 1998)
17. E. Casas, *Control of an elliptic problem with pointwise state constraints.* SIAM J. Cont. Optim. **24**, 1309–1322 (1986)
18. M.P. do Carmo, *Differential Geometry of Curves and Surfaces*, 2nd edn. (Dover Publications, New York, 2016)
19. L.-Q. Chen, *Phase-field models for microstructure evolution.* Annu. Rev. Mater. Res. **32**, 113–1140 (2002)
20. A.J. Chorin, J.E. Marsden, *A Mathematical Introduction to Fluid Mechanics* (Springer, New York, 1979)
21. S.N. Chow, J.K. Hale, *Methods of Bifurcation Theory* (Springer, New York, 1982)

© Springer International Publishing AG 2017
C. Eck et al., *Mathematical Modeling*, Springer Undergraduate
Mathematics Series, DOI 10.1007/978-3-319-55161-6

22. P.G. Ciarlet, *Mathematical Elasticity*. Vol. 1: Three Dimensional Elasticity, North–Holland, Amsterdam, 1988; Vol. 2: Theory of Plates, Elsevier, Amsterdam, 1997; Vol. 3: Theory of Shells, North–Holland, Amsterdam, 2000

23. D. Cioranescu, P. Donato, *An Introduction to Homogenization* (Oxford University Press, Oxford, 1999)

24. D. Cioranescu, J. Saint Jean Paulin, *Homogenization of Reticulated Structures* (Springer, New York, 1999)

25. R. Courant, D. Hilbert, *Methods of Mathematical Physics*, vol. 1 (Wiley-VCH, Germany, 1989)

26. S.H. Davis, *Theory of Solidification* (Cambridge University Press, Cambridge, 2001)

27. K. Deckelnick, G. Dziuk, C.M. Elliott, *Computation of geometric partial differential equations and mean curvature flow*. Acta Numer. **14**, 139–232 (2005)

28. P. Deuflhard, F. Bornemann, *Scientific Computing with Ordinary Differential Equations* (Springer, Berlin, 2010)

29. A. Deutsch (Hrsg.), *Muster des Lebendigen* (Vieweg, 1994)

30. G. Duvaut, J.-L. Lions, *Inequalities in Mechanics and Physics* (Springer, Berlin, 1976)

31. C. Eck, H. Garcke, B. Stinner, *Multiscale problems in solidification processes*, in Analysis, Modeling and Simulation of Multiscale Problems, ed. by By A. Mielke (Springer, Berlin, 2006), pp. 21–64

32. I. Ekeland, R. Temam, *Convex Analysis and Variational Problems* (Classics in Applied Mathematics (SIAM, Philadelphia, 1990)

33. C.M. Elliott, J.R. Ockendon, *Weak and Variational Methods for Moving Boundary Problems* (Pitman, Boston, 1982)

34. H.W. Engl, M. Hanke, A. Neubauer, *Regularization of Inverse Problems* (Kluwer, Dordrecht, 2000)

35. J.-H. Eschenburg, J. Jost, *Differentialgeometrie und Minimalflächen* (Springer, Berlin, 2007)

36. L.C. Evans, *An Introduction to Mathematical Optimal Control Theory*. Lecture notes, http://math.berkeley.edu/~evans/control.course.pdf

37. L.C. Evans, *Partial Differential Equations* (AMS, Providence, 1998)

38. B. Fiedler, *Romeo and Juliet, spontaneous pattern formation, and Turing's instability*, in Mathematics Everywhere, (ed.) by M. Aigner et al. (AMS, Providence, 2010) pp. 57–75

39. G. Fischer, *Analytische Geometrie* (Vieweg, Braunschweig, 1978)

40. G. Fischer, *Lineare Algebra* (Vieweg, Braunschweig, 1979)

41. N.D. Fowkes, J.J. Mahony, *An Introduction to Mathematical Modelling* (Wiley, Chichester, 1994)

42. A. Friedman, *Variational Principles and Free Boundary Problems* (Robert E. Krieger Publishing Company, Malabar, 1988)

43. A. Friedman, *Free boundary problems in science and technology*. Not. AMS **47**(8), 854–861 (2000)

44. G.P. Galdi, *An Introduction to the Mathematical Theory of the Navier–Stokes Equations. vol. 1: Linearized Steady Problems, vol. 2: Nonlinear Steady Problems* (Springer, New York, 1994)

45. H. Garcke, *Mechanical effects in the Cahn–Hilliard model: a review on mathematical results*, in Mathematical Methods and Models in Phase Transitions, (ed.) by A. Miranville (Nova Science Publishers, New York, 2005), pp. 43–77

46. H. Garcke *Curvature Driven Interface Evolution*. Jahresbericht der Deutschen Mathematiker-Vereinigung **115**(2), 63–100 (2013). doi:10.1365/s13291-013-0066-2

47. H. Garcke, B. Nestler, B. Stinner, *A diffuse interface model for alloys with multiple components and phases*. SIAM J. Appl. Math. **64**(3), 775–799 (2004)

48. A. Gierer, H. Meinhardt, *A theory of biological pattern formation*. Kybernetik **12**, 30–39 (1972)

49. D. Gilbarg, N. Trudinger, *Elliptic Partial Differential Equations of Second Order*, 2nd edn. (Springer, Berlin, 1983)

50. B.H. Gilding, *The occurrence of interfaces in nonlinear diffusion-advection processes*. Arch. Ration. Mech. Anal. **100**(3), 243–263 (1988)

51. B.H. Gilding, R. Kersner, *Travelling Waves in Nonlinear Diffusion-Convection-Reaction* (Birkhäuser, Basel, 2004)

52. M. Griebel, T. Dornseifer, T. Neunhoeffer, *Numerische Simulation in der Strömungslehre Eine praxisorientierte Einführung* (Vieweg, Wiesbaden, 1995)

53. D.J. Griffith, *Introduction to Electrodynamics* (Prentice Hall, New Jersey, 1981)

54. E. van Groesen, J. Molenaar, *Continuum Modeling in the Physical Sciences* (SIAM, 2007)

55. D. Gross, W. Hauger, J. Schröder, W.A. Wall, N. Rajapakse, *Engineering Mechanics*, vol. 1 (Springer Dordrecht, 2009; 2013)

56. C. Großmann, H.-J. Roos., M. Stynes, *Numerical Treatment of Partial Differential Equations* (Springer, 2007; Teubner, Stuttgart, 1992)

57. M. Günther, A. Jüngel, *Finanzderivate mit MATLAB* (Vieweg, Wiesbaden, 2003)

58. J. Guckenheimer, P. Holmes, *Nonlinear Oscillations, Dynamical Systems, and Bifurcations of Vector Fields*. Applied Mathematical Sciences, vol. 42 (Springer, New York, 1990), pp. xvi+459

59. M.E. Gurtin, *An Introduction to Continuum Mechanics* (Academic Press, New York, 1981)

60. M.E. Gurtin, *Thermomechanics of Evolving Phase Boundaries in the Plane* (Oxford University Press, New York, 1993)

61. R. Haberman, *Mathematical Models* (Prentice-Hall, New Jersey, 1977)

62. W. Hackbusch, *Elliptic Differential Equations: Theory and Numerical Treatment* (Springer, Berlin, 1992)

63. W. Hackbusch, *Iterative Solution of Large Sparse Systems of Equations*, 2nd edn. (Springer, Berlin, 2016)

64. G. Hagmann, *Grundlagen der Elektrotechnik*, 8th edn. (AULA-Verlag, Wiebelsheim, 2001)

65. M. Hanke-Bourgeois, *Grundlagen der Numerischen Mathematik und des Wissenschaftlichen Rechnens* (B.G. Teubner, Wiesbaden, 2006)

66. S. Hildebrandt, *Analysis 2* (Springer, Berlin, 2003)

67. M. Hinze, R. Pinnau, M. Ulbrich, S. Ulbrich, *Optimization with PDE Constraints*. Mathematical Modelling: Theory and Applications, vol. 23 (Springer, New York, 2009)

68. M.H. Holmes, *Introduction to Perturbation Methods* (Springer, New York, 1995)

69. M.H. Holmes, *Introduction to the Foundations of Applied Mathematics*. Texts in Applied Mathematics, vol. 56 (Springer, New York, 2009), pp. xiv+467

70. J. Honerkamp, H. Römer, *Grundlagen der Klassischen Theoretischen Physik* (Springer, Berlin, 1986)

71. U. Hornung (ed.), *Homogenization and Porous Media* (Springer, New York, 1997)

72. K. Hutter, K. Jöhnk, *Continuum Methods of Physical Modeling. Continuum Mechanics, Dimensional Analysis, Turbulence* (Springer, Berlin, 2004)

73. J.D. Jackson, *Classical Electrodynamics* (Wiley, New York, 1962)

74. C. Johnson, *Numerical Solution of Partial Differential Equations by the Finite Element Method* (Cambridge University Press, Cambridge, 1987)

75. J. Jost, *Partial Differential Equations* (Springer, New York, 2013)

76. J. Kevorkian, *Perturbation Methods in Applied Mathematics* (Springer, Heidelberg, 1981)

77. J. Kevorkian, J.D. Cole, *Multiple Scale and Singular Perturbation Methods* (Springer, New York, 1996)

78. H. Kielhöfer, *Bifurcation Theory. An Introduction with Applications to PDEs* (Springer, New York, 2004)

79. D. Kinderlehrer, G. Stampacchia, *An Introduction to Variational Inequalities and Their Applications* (Academic Press, New York, 1980)

80. P. Knabner, *Mathematische Modelle für Transport und Sorption gelöster Stoffe in porösen Medien* (Peter Lang, Frankfurt, 1988)

81. P. Knabner, L. Angermann, *Numerical Methods for Elliptic and Parabolic Partial Differential Equations* (Springer, New York, 2003)

82. K. Königsberger, *Analysis II* (Springer, Berlin, 2004)

83. D. Kondepudi, I. Prigogine, *Modern Thermodynamics. From Heat Engines to Dissipative Structures* (Wiley, Chichester, 1998)
84. D. Kröner, *Numerical Schemes for Conservation Laws* (Wiley, Chichester, 1997)
85. H. Kuchling, *Taschenbuch der Physik* (Harri Deutsch, Thun & Frankfurt, 1988)
86. L. Landau, E. Lifšic, *Course of Theoretical Physics*, vols. 1–10 (Pergamon, 1959)
87. R.J. Leveque, *Finite Volume Methods for Hyperbolic Problems* (Cambridge University Press, Cambridge, 2002)
88. K.G. Libbrecht, *Morphogenesis on Ice: The Physics of Snow Crystals*. Engineering and Science, vol. 1 (2001)
89. C.-C. Lin, L.A. Segel, *Mathematics Applied to Deterministic Problems in the Natural Sciences* (Macmillan, New York, 1974)
90. J.L. Lions, *Optimal Control of Systems Governed by Partial Differential Equations* (Springer, Berlin, 1971)
91. I-Shih Liu, *Method of Lagrange multipliers for exploitation of the entropy principle*. Arch. Ration. Mech. Anal. **46**, 131–148 (1972)
92. J. Macki, A. Strauss, *Introduction to Optimal Control Theory* (Springer, New York, 1982)
93. P.A. Markowich, *Applied Partial Differential Equations* (Springer, Berlin, 2007)
94. J.E. Marsden, T.J.R. Hughes, *Mathematical Foundations of Elasticity* (Prentice-Hall, New Jersey, 1983)
95. H. Meinhardt, *The Algorithmic Beauty of Sea Shells* (Springer, Berlin, 2003)
96. A.M. Meirmanov, *The Stefan Problem* (De Gruyter, Berlin, 1992)
97. T. Meis, U. Marcowitz, *Numerische Behandlung partieller Differentialgleichungen* (Springer, Berlin, 1978)
98. A. Meister, *Numerik linearer Gleichungssysteme* (Vieweg, Wiesbaden, 1999)
99. L. Modica, *The gradient theory of phase transitions and the minimal interface criterion*. Arch. Ration. Mech. Anal. **98**(2), 123–142 (1987)
100. F. Morgan, *Riemannian Geometry: A beginner's Guide*, 2nd edn. (A K Peters, Ltd., Wellesley, 1998), pp. x+156
101. I. Müller, *Grundzüge der Thermodynamik* (Springer, Berlin, 2001)
102. I. Müller, *Thermodynamics* (Pitman Advanced Publishing Program, London, 1985)
103. I. Müller, W.H. Müller, *Fundamentals of Thermodynamics and Applications* (Springer, Berlin, 2009)
104. I. Müller, W. Weiss, *Entropy and Energy* (Springer, Berlin, 2005)
105. J.D. Murray, *Mathematical Biology I+II* (Springer, Berlin, 1989)
106. D.R. Owen, *A First Course in the Mathematical Foundations of Thermodynamics* (Springer, New York, 1984)
107. O. Penrose, P. Fife, Thermodynamically consistent models of phase-field type for the kinetics of phase transitions. Phys. D **43**(1), 44–62 (1990)
108. E. Pestel, J. Wittenburg, *Technische Mechanik Band 2: Festigkeitslehre* (B.I.-Wissenschaftsverlag, Mannheim, 1992)
109. A. Quarteroni, A. Valli, *Numerical Approximation of Partial Differential Equations* (Springer, Berlin, 1994)
110. M. Renardy, R.C. Rogers, *An Introduction to Partial Differential Equations*. Texts in Applied Mathematics, vol. 13 (Springer, New York, 1993)
111. B.N. Roy, *Fundamentals of Classical and Statistical Thermodynamics* (Wiley, Chichester, 2002)
112. Y. Saad, *Iterative Methods for Sparse Linear Systems* (PWS Publishing Company, Boston, 1996)
113. F. Saaf, *A Study of Reactive Transport Phenomena in Porous Media*, Ph.D. thesis, Rice University, Houston, 1996
114. A.A. Samarskii, A.P. Mikhailov, *Principles of Mathematical Modeling* (Taylor & Francis, London, 2002)
115. M.B. Sayir, J. Dual, S. Kaufmann, *Ingenieurmechanik 1 Grundlagen und Statik* (Teubner, Stuttgart, 2004)

116. Ch. Schmeiser, *Angewandte Mathematik*. Lecture notes, Institut für Angewandte und Numerische Mathematik, TU Wien
117. H.R. Schwarz, *Numerical Analysis* (Wiley, New York, 1989)
118. L.A. Segel, *Mathematics Applied to Continuum Mechanics* (Macmillan Publ. Co., New York, 1977)
119. J. Smoller, *Shock Waves and Reaction-Diffusion Equations* (Springer, New York, 1994)
120. H. Sohr, *The Navier-Stokes Equations An Elementary Functional Analytic Approach* (Birkhäuser, Basel, 2001)
121. Th Sonar, *Angewandte Mathematik, Modellierung und Informatik* (Vieweg, Braunschweig, 2001)
122. O. Steinbach, *Lösungsverfahren für lineare Gleichungssysteme* (Teubner, Stuttgart, 2005)
123. J. Stoer, R. Bulirsch, *Introduction to Numerical Analysis* (Springer, New York, 2002)
124. G. Strang, *Introduction to Applied Mathematics* (Wellesley-Cambridge, Wellesley, 1986)
125. G. Strang, *Introduction to Linear Algebra* (Wellesley-Cambridge Press, Wellesley, 2016)
126. N. Straumann, *Thermodynamik* (Springer, Berlin, 1986)
127. E. Süli, D.F. Mayers, *An Introduction to Numerical Analysis* (Cambridge University Press, Cambridge, 2007)
128. R. Szilard, *Finite Berechnungsmethoden der Strukturmechanik. Band 1: Stabwerke* (Verlag von Wilhelm Ernst & Sohn, Berlin, 1982)
129. R. Temam, *Navier–Stokes Equations. Theory and Numerical Analysis*. Repr. with corr (AMS, Providence, RI, 2001; North Holland, Amsterdam, 1984)
130. R. Temam, A.M. Miranville, *Mathematical Modelling in Continuum Mechanics* (Cambridge University Press, Cambridge, 2001)
131. F. Tröltzsch, *Optimal Control of Partial Differential Equations*. Graduate Studies in Mathematics, vol. 112 (American Mathematical Society, Providence, RI, 2010), pp. xvi+399
132. C. Truesdell, *A First Course in Rational Continuum Mechanics*, 2nd edn., vol. 1 (Academic Press, Boston, 1991)
133. C. Truesdell, *The Mechanical Foundations of Elasticity and Fluid Dynamics* (Gordon & Breach, New York, 1966)
134. C. Truesdell, *The Elements of Continuum Mechanics* (Springer, Berlin, 1966)
135. A. Turing, *The chemical basis of morphogenesis*. Phil. Trans. R. Soc. Lond. B **237**, 37–72 (1952)
136. A. Visintin, *Models of Phase Transitions. Progress in Nonlinear Differential Equations and their Applications*, vol. 28 (Birkhäuser, Boston, 1996)
137. T. Witelski, M. Bowen, *Methods of Mathematical Modelling Continuous Systems and Differential Equations*. Undergraduate Mathematics Series. (Springer, Cham, 2015) pp. xviii+305
138. K. Yosida, *Functional Analysis, Fundamental Principles of Mathematical Sciences*, vol. 123 (Springer, Berlin, 1980), pp. xii+501
139. E. Zeidler, *Nonlinear Functional Analysis and its Applications*, vol. III (Springer, New York, 1988)
140. E. Zeidler, *Nonlinear Functional Analysis and its Applications*, vol. IV (Springer, New York, 1988)
141. V.V. Zhikov, S.M. Kozlov, O.A. Oleinik, *Homogenization of Differential Operators and Integral Functionals* (Springer, Berlin, 1994)

# Index

© Springer International Publishing AG 2017
C. Eck et al., *Mathematical Modeling*, Springer Undergraduate
Mathematics Series, DOI 10.1007/978-3-319-55161-6

Printed in the United States
By Bookmasters